Spon's External Works and Landscape Price Book

2023

Spon's External Works and Landscape Price Book

Edited by

AECOM

2023

Forty-second edition

CRC Press
Taylor & Francis Group

First edition 1978
Forty-second edition published 2023
by CRC Press
4 Park Square, Milton Park, Abingdon, Oxon, OX14 4RN

and by CRC Press
Taylor & Francis, 6000 Broken Sound Parkway, NW, Suite 300, Boca Raton, FL 33487

CRC Press is an imprint of the Taylor & Francis Group, an informa business

British Library Cataloguing in Publication Data
A catalogue record for this book is available from the British Library

ISBN13: 978-1-032-33176-8
Ebook: 978-1-003-31891-0
ISSN: 0267-4181

Typeset in Arial by Taylor & Francis Books

Printed and bound by CPI Group (UK) Ltd, Croydon, CR0 4YY

Contents

Preface to the Forty-Second Edition

In the past 12 months pressure on construction costs has been exceptional. First came COVID lockdowns – which triggered abrupt shifts in demand, along with disruption to supply and distribution. Then came Brexit and a fundamental shift in trade and international recruitment policy. And then Russia invaded Ukraine which shocked energy markets and withdrew valuable supply of key construction materials and components from the market overnight.

Each of these exceptional events has implications for the cost of delivering construction projects in the UK market and presently the outlook for pricing is unusually uncertain.

Prices presented in this book have a base date of Q1 2022 – meaning indirect impacts of the Ukraine war have largely been excluded. Immediate, direct impacts, such as initial disruption to steel and aluminium billet on structural steel and aluminium cladding panels, have generally been captured but the indirect impacts, from higher energy costs on manufactured goods for example, are likely to be significant and will take time to feed through the supply chain.

The Market Update, included in this edition, provides commentary on recent drivers of construction cost inflation, pricing volatility, the inflationary outlook and risks.

Readers are recommended to register for the supplementary Updates which will highlight changes to market rates for materials which will inevitably follow.

Labour Rates

There is pressure on landscape labour in the sector with low availability of skilled staff. We have increased the overall rate of labour in this year's book by 5% to £24.31 per man-hour on average which includes the overhead of the contracting organization. Please see the calculation for this in the Wages section of this book.

Going forward, labour markets may be affected by further pressures and shortages which may affect prices up to the time of the next publication in September 2024.

Supplier Prices

At the time of compiling this year's book it should be highlighted that many material suppliers are extending delivery periods and fixing prices with noticeably short acceptance periods. Plastics, timber, steel and packaging materials have had an impact on the rates shown.

Prices for Suppliers and Services

We acknowledge the support afforded to us by the suppliers of products and services who issue us with the base information used. Their contact details are published in the acknowledgements section at the front of this book. We advise that readers wishing to evaluate the cost of a product or service should approach these suppliers directly should they wish to confirm prices prior to submission of tenders or quotations.

Profits and Overheads

Rates in this publication assume that all company overhead is included in the labour rate. Additionally, profit is added to the rates shown at 15% for material, labour and plant and 10% for subcontractors. All works are assumed to be executed by a main landscape contractor who might employ specialist subcontractors and mark up their quotations accordingly (10%)

Whilst every effort is made to ensure the accuracy of the information given in this publication, neither the editors nor the publishers in any way accept liability for loss of any kind resulting from the use made by any person of such information. We specifically note that this year's price rises given to us by suppliers might be temporary but they reflect the prices at the time of compiling this publication – Q1 2022.

We remind readers that we would be grateful to hear from them during the year with suggestions or comments on any aspect of the contents of this year's book and suggestions for improvements. Our contact details are shown below.

AECOM Ltd
Aldgate Tower
2 Leman Street
London
E1 8FA

Tel: 020 7645 2000
e-mail: spons.europe@aecom.com

Acknowledgements

This list has been compiled from the latest information available but as firms frequently change their names, addresses and telephone numbers due to reorganization, users are advised to check this information before placing orders.

ABG Ltd
Tel: 01484 852096
Website: www.abg-geosynthetics.com
Email: enquiries@abgltd.com
Reinforced turf

ACO Technologies Plc
Tel: 01462 816666
Website: www.aco.co.uk
Email: technologies@aco.co.uk
Linear drainage

Addagrip Terraco Ltd
Tel: 01825 761333
Website: www.addagrip.co.uk
Email: sales@addagrip.co.uk
Resin bound & bonded surfacing

Adezz BV
Netherlands
Tel: 0031 413 745 423
Website: www.adezz.com
Email: info@adezz.uk
Corten edgings

Agripower Ltd
Tel: 01494 866776
Website: www.agripower.co.uk
Email: info@agripower.co.uk
Land drainage and sports grounds

AHS Ltd
Tel: 01797 252728
Website: www.ahs-ltd.co.uk
Email: via website
Safety surfaces

Albion Stone Plc
Tel: 01737 771772
Website: www.albionstone.com
Email: enquiries@albionstone.com
Portland stone

ALS Contract Services
Tel: 01952 259281
Website: www.alscontracts.co.uk
Email: contracts@amenity.co.uk
Horticultural products supplier

Anderton Concrete Products Ltd
Tel: 01606 535300
Website: www.andertonconcrete.co.uk
Email: civils@andertonconcrete.co.uk
Concrete retaining wall systems

Arbour Landscape Solutions
Tel: 0208 953 6177
Website: www.arbourlandscapesolutions.co.uk
Solidblock

Architectural Heritage Ltd
Tel: 01386 584414
Website: www.architectural-heritage.co.uk
Email: office@architectural-heritage.co.uk
Ornamental stone buildings and water features

Bituchem Group
Tel: 01594 826768
Website: www.bituchem.com
Email: tech@bituchem.com
Coloured hard landscaping materials

Breedon Special Aggregates
(A division of Breedon Aggregates England Ltd)
Tel: 01332 694001
Website: www.breedon-special-aggregates.co.uk
Email: specialaggs@breedongroup.com
Specialist gravels

British Sugar Topsoil
Tel: 0870 240 2314
Website: www.bstopsoil.co.uk
Email: topsoil@britishsugar.com
Topsoil

Broxap Street Furniture
Tel: 01782 564411
Website: www.broxap.com
Email: web.sales@broxap.com
Street furniture

Buzon UK
Tel:020 8614 0874
Website: www.buzonuk.com
Email: sales@buzonuk.com
Pedestals

Cedar Nursery
Tel: 01932 862473
Website: www.landscaping.co.uk
Email: sales@landscaping.co.uk
Aggregate containment grids

Chelmer Valley
Tel: 01277 568840
Website: www.chelmervelley.co.uk
Email: info@chelmervalley.co.uk
Paving products

Coveya Ltd
Tel: 0800 278 5389
Website: www.coveya.co.uk
Email: support@coveya.co.uk
Conveyors

DLF Seeds Ltd
Tel: 01386 793135
Website: www.dlf.co.uk
Email: amenity@dlf.co.uk
Grass and wildflower seed

Earth Anchors Ltd
Tel: 020 8684 9601
Website: www.earth-anchors.com
Email: info@earth-anchors.com
Anchors for site furniture

Easigrass
Tel: 020 8843 4180
Website: www.easigrass.com
Email: reception@easigrass.com
Artificial grass

Elliott Workspace Ltd
Tel: 01733 298700
Website: www.elliottuk.com
Email: hirediv@elliottuk.com
Site offices

Elite Balustrade Systems
Tel: 01254 825594
Website: www.elitebalustrade.com
Email: neil@elitebalustrade.com
Glass balustrades

English Woodlands
Tel: 01435 862992
Website: www.ewburrownursery.co.uk
Email: info@ewburrownursery.co.uk
Plant protection

En Tout Cas Tennis Court Ltd
Tel: 01832 274199
Website: www.tenniscourtsuk.co.uk
Email: sales@etc.courts.com
Tennis courts

Exterior Solutions Ltd
Tel: 01494 711800
Website: www.exterior.supplies.co.uk
Email: via website
Decking

Fairwater Ltd
Tel: 01903 892228
Website: www.fairwater.co.uk
Email: info@fairwater.co.uk
Water garden specialists

Five Rivers
Tel: 01722 783 041
Website: www.five-rivers.com
Email: office@five-rivers.com
Environmental specialists

Flagpole Express Ltd
Tel: 01159 44 22 55
Website: www.flagpoleexpress.co.uk
Email: sales@flagpoleexpress.co.uk
Flagpoles

FP McCann Ltd
Tel: 01530 240013
Website: www.fpmccann.co.uk
Email: sales@fpmccann.co.uk
Precast concrete pipes

George Lines Ltd
Tel: 01753 685354
Website: www.georgelines.co.uk
Email: sales@georgelines.co.uk
Building materials

Germinal GB
Tel: 01522 868714
Website: www.germinalamenity.com
Email: expert@germinalamenity.com
Grass and wildflower seed

Grass Concrete Ltd
Tel: 01924 379443
Website: www.grasscrete.com
Email: info@grasscrete.com
Grass block paving

Greenfix Soil Stabilization and Erosion Control Ltd
Tel: 01386 881493
Website: www.greenfix.co.uk
Email: info@greenfix.co.uk
Seeded erosion control mats

Greenblue Urban Ltd
Tel: 0800 018 7797
Website: www.greenblue.com
Email: enquiries@greenblueurban.com
Root directors

Greentech Ltd
Tel: 01423 332100
Website: www.green-tech.co.uk
Email: sales@green-tech.co.uk
Landscaping suppliers

Grundon Waste Management Ltd
Tel: 01494 834311
Website www.grundon.com
Email: info@grundon.com
Gravel

Guncast Swimming Pools Ltd
Tel: 01798 343725
Website: www.guncast.com
Email: info@guncast.com
Swimming pools

Hags -SMP Ltd
Tel: 01784 489100
Website: www.hags.co.uk
Email: sales@hags.co.uk
Playground equipment

Haddonstone Ltd
Tel: 01604 770711
Website: www.haddonstone.com
Email: info@haddonstone.co.uk
Landscape ornaments and architectural cast stone

Hahn Plastics Ltd
Tel: 0161 850 1965
Website: www.hahnplastics.com
Email: info@hahnplastics.co.uk
Recycled products

Headland Amenity Ltd
Tel: 01223 491090
Website: www.headlandamenity.com
Email: info@headlandamenity.com
Chemicals

Hepworth Wavin Plc
Tel: 01709 856200
Website: www.wavin.com
Email: info@wavin.com
Drainage

Hillier Nurseries
Tel: 01794 368733
Website: www.hillier.co.uk
Email: trees@hillier.co.uk
Trees

HSS Hire
Tel: 020 8936 7307
Website: www.hss.com
Email: hire@hss.com
Tool and plant hire

ICL (formerly Everris Limited)
Tel: 01473 237100
Website: www.icl-sf.com
Email: prof.sales@icl-group.com
Landscape chemicals

Inturf
Tel: 01759 321000
Website: www.inturf.com
Email: info@inturf.co.uk
Turf

J Toms Ltd
Tel: 01233 771923
Website: www.jtoms.co.uk
Email: jtoms@btopenworld.com
Plant protection

Jacksons Fencing
Tel: 01233 750393
Website: www.jacksons-fencing.co.uk
Email: sales@jacksons-fencing.co.uk
Fencing

James Coles & Sons (Nurseries) Ltd
Tel: 01162 412115
Website: www.colesnurseries.co.uk
Email: sales@colesnurseries.co.uk
Tree nurseries

Johnsons Wellfield Quarries Ltd
Tel: 01484 652311
Website: www.johnsons-wellfield.co.uk
Email: sales@johnsons-wellfield.co.uk
Natural Yorkstone pavings

Kent Stainless
Tel: 0800 376 8377
Website: www.kentstainless.com
Email: info@kentstainless.com
Stainless steel products

Kinley
Tel: 01580 830688
Website: www.kinley.co.uk
Email: sales@kinley.co.uk
Edging

Kompan Ltd
Tel: 01908 201002
Website: www.kompan.co.uk
Email: kompan.uk@kompan.com
Play equipment

Landmark Timber Products
Tel: 0808 129 3773
Website: www.landmarktimber.co.uk
Email: enquiries@madebylandmark.com
Fingerpost signs

Landscape Plus
Tel: 01666 577577
Website: www.landscapeplus.com
Email: help@landscapeplus.com
Lighting

Lang and Fulton
Tel: 0131 441 1255
Website: www.langandfulton.co.uk
Email: sales@langandfulton.co.uk
Steel gratings and louvres

Lappset UK Ltd
Tel: 01536 412612
Website: www.lappset.com
Email: lappset@lappset.com
Play equipment

Leaky Pipe Systems Ltd
Tel: 01622 746495
Website: www.leakypipe.co.uk
Email: sales@leakypipe.co.uk
Irrigation systems

Leca UK
Tel: 08443 351770
Website: www.leca.co.uk
Email: enquiries@leca.co.uk
Light aggregates

Lindum Turf
Tel: 01904 448675
Website: www.turf.co.uk
Email: lindum@turf.co.uk
Meadow and specialized turf

LiveTrakway
Tel: 08700 767676
Website: www.livetrakway.com
Email: marketing@livetrakway.com
Portable roads

London Stone
Tel: 01753 212950
Website: www.londonstone.co.uk
Email: info@londonstone.co.uk
Natural stone paving

Lorenz von Ehren
Tel: 01865 910261
Website: www.lve.de
Email: sales@lve.de
Topiary

Marshalls Plc
Tel: 01422 312000
Website: www.marshalls.co.uk
Email: info@marshalls.co.uk
Concrete products, hard landscape materials, linear drainage, street furniture and street lighting

Melcourt Industries Ltd
Tel: 01666 502711
Website: www.melcourt.co.uk
Email: mail@melcourt.co.uk
Mulch and compost

Meon UK
Tel: 023 9220 0606
Website: www.meonuk.com
Email: mail@meonuk.com
Resin surfaces and safety surfaces

Milton Precast
Tel: 01795 425191
Website: www.miltonprecast.co.uk
Email: sales@miltonprecast.com
Soakaway rings

Mobilane UK
Tel: 0203 741 8049
Website: www.mobilane.com
Email: sales@mobilane.co.uk
Green screens

Neptune Outdoor Furniture Ltd
Tel: 01962 777799
Website: www.neptunestreetfurniture.co.uk
Email: enquiries@neptunestreetfurniture.co.uk
Street furniture

Nomix Enviro
A Division of Frontier Agriculture Ltd
Tel: 0800 032 4352
Website: www.nomixenviro.co.uk
Email: sales@nomixenviro.co.uk
Herbicides

Oakover Nurseries Ltd
Tel: 01233 713016
Website: www.oakovernurseries.co.uk
Email: enquiries@oakovernurseries.co.uk
Native plant nurseries

Outdoor Design
Tel: 01903 716960
Website: www.outdoordesign.co.uk
Email: info@outdoordesign.co.uk
Metal fabricator

PBA Solutions
Tel: 01202 816134
Website: www.pba-solutions.com
Email: info@pba-solutions.com
Japanese Knotweed consultants

Pennyhill Timber Ltd
Tel: 01483 486 730
Website: www.pennyhilltimber.co.uk
Email: david@pennyhilltimber.co.uk
Timber products

Phi Group
Tel: 01242 707600
Website: www.phigroup.co.uk
Email: southern@phigroup.co.uk
Retaining wall specialists

Platipus Anchors Ltd
Tel: 01737 762300
Website: www.platipus-anchors.com
Email: info@platipus-anchors.com
Tree anchors

Pop up Power Supplies Ltd
Tel: 020 8551 8363
Website: www.popuppower.co.uk
Email: info@popuppower.co.uk
Public area power supplies

Readyhedge Ltd
Tel: 01386 750585
Website: www.readyhedgeltd.com
Email: info@readyhedgeltd.com
Pre-grown hedging

Rigby Taylor Ltd
Tel: 0800 424 919
Website: www.rigbytaylor.com
Email: sales@rigbytaylor.com
Horticultural supply

RIW Ltd
Tel: 01344 397777
Website: www.riw.co.uk
Email: enquiries@riw.co.uk
Waterproofing products

Rolawn Ltd
Tel: 01904 757372
Website: www.rolawn.co.uk
Email: info@rolawn.co.uk
Industrial turf

Saint-Gobain PAM UK Ltd
Tel: 0115 930 5000
Website: www.saint-gobain-pam.co.uk
Email: sales.uk.pam@saint-gobain.com
Cast iron drainage

Scotscape Ltd
Tel: 02082 545000
Website: www.scotscape.net
Email: sales@scotscape.net
Living walls

Screwfix Ltd
Tel: 03330 112 112
Website: www.screwfix.com
Email: online@screwfix.com
Hardware

Steelway Brickhouse
Tel: 01215 214500
Website: www.steelway.co.uk
Email: sales@steelwaybrickhouse.co.uk
Access covers

Steelway Fensecure Ltd
Tel: 01902 490919
Website: www.steelway.co.uk
Email: sales@fensecure.co.uk
Fencing

SteinTec
Tel: 01708 860049
Website: www.steintec.co.uk
Email: webenquiries@tuffbau.com
Specialized mortars

SureSet UK Limited
Tel: 0800 612 6501
Website: www.sureset.co.uk
Email: website@sureset.co.uk
Solid steel products

Tensar International
Tel: 01254 262431
Website: www.tensar.co.uk
Email: info@tensar.co.uk
Erosion control, soil stabilization

Terranova Lifting Ltd
Tel: 0118 931 2345
Website: www.terranovagroup.co.uk
Email: cranes@terranovagroup.co.uk
Crane hire

The Garden Trellis Company
Tel: 01255 688361
Website: www.gardentrellis.co.uk
Email: info@gardentrellis.co.uk
Garden joinery experts

TK Solutions AG Ltd
Tel: 020 3853 1285
Email: tksukltdnew@gmail.com
Skype Name: tomer.kniznik
Irrigation supplies

TLC Electrical Supplies
Tel: 01293 565630
Website: www.tlc-direct.co.uk
Email: sales@tlc-direct.co.uk
Electrical goods

Townscape Products Ltd
Tel: 01623 513355
Website: www.townscape-products.co.uk
Email: sales@townscape-products.co.uk
Hard landscaping

Valley Provincial Ltd
Tel: 01322 279799
Website: www.valleyprovincial.com
Email: sales@valleyprovincial.com
Playground installer

Wade International Ltd
Tel: 01787 475151
Website: www.wadedrainage.co.uk
Email: sales@wadeint.co.uk
Stainless steel drainage

Wavin Plastics Limited
Tel: 0844 856 9400
Website: www.wavin.com
Email: info@wavin.com
Drainage products

Weavo Fencing Products Ltd
Tel: 01823 480 571
Website: www.weavo.co.uk
Email: sales@weavo.com
Fencing

Wicksteed Leisure Ltd
Tel: 01536 517028
Website: www.wicksteed.co.uk
Email: sales@wicksteed.co.uk
Play equipment

Woodscape Ltd
Tel: 01254 685185
Website: www.woodscape.co.uk
Email: sales@woodscape.co.uk
Street furniture

Wybone Ltd
Tel: 01226 352333
Website: www.wybone.co.uk
Email: hello@wybone.co.uk
Street furniture

Yeoman Aggregates Ltd
Tel: 020 8896 6820
Website: www.aggregate.com
Email: via website
Aggregates

Zinco Green Roof Systems Ltd
Tel: 01223 229700
Website: www.zinco-greenroof.com
Email: info@zinco-greenroof.com
Green roof systems

How this Book is Compiled

INTRODUCTION

First-time users of *Spon's External Works and Landscape Price Book* and others who may not be familiar with the way in which prices are compiled may find it helpful to read this section before starting to calculate the costs of landscape works.

Please also refer to the 'How to Use this Book' notes on page 33.

The cost of an item of landscape construction or planting is made up of many components:

- the cost of the product
- the labour and additional materials needed to carry out the job
- the cost of running the contractor's business

These are described more fully below.

IMPORTANT NOTES ON THE PROFIT ELEMENT OF RATES IN THIS BOOK

The rates shown in the Approximate Estimates and Measured Works sections of this book include a profit element. Prices are shown at marked up rates. This is the assumed installed price including the costs of company overhead and profit required to perform this task or project. For reference please see the tables in the 'Labour Rates Used in this Edition' section on pages 7–13.

Analysed Rates Versus Subcontractor Rates

As a general rule if a rate is shown as an analysed rate in the Measured Works section, i.e. it has figures shown in the columns other than the 'Unit' and 'Total Rate' column, it can be assumed that this has the subcontractor's profit and overhead element as well as the landscape subcontractor's profit on top of this. This calculation is shown as a direct labour/material supply/plant task being performed by a contractor.

On the other hand, if a rate is shown as a Total Rate only, this would normally be a subcontractor's rate and would contain the profit and overhead for the subcontractor and the landscape contractor's profit element for management of the subcontractor.

The foregoing applies for the most part to the Approximate Estimates section, however in some items there may be an element of subcontractor rates within direct works build-ups.

As an example of this, to excavate, lay a base and place a macadam surface, a general landscape contractor would normally perform the earthworks and base installation. The macadam surfacing would be performed by a specialist subcontractor to the landscape contractor.

The Approximate Estimate for this item uses rates from the Measured Works section to combine the excavation, disposal, base material supply and installation with all associated plant. There is profit included on these elements. The cost of the surfacing, however, as supplied to us in the course of our annual price enquiries from the subcontractors, does include the subcontractor's profit element plus a portion for the landscape contractor of 10%.

The landscape contractor would add this to their section of the work but would normally apply a mark-up at a lower percentage to the subcontract element of the composite task. Users of this book should therefore allow for the profit element at the prevailing rates. Please see the worked examples below and the notes on overheads for further clarification.

INTRODUCTORY NOTES ON PRICING CALCULATIONS USED IN THIS BOOK

There are two pricing sections to this book:

Approximate Estimating Rates – Average contract value £100,000.00 – £500,000.00

Prices for Measured Works – Average contract value £100,000.00 – £500,000.00

Typical Project Profile Used in this Book

Contract value	£100,000.00 – £3,000,000.00
Labour rate (see page 7)	£24.31 per hour
Labour rate for maintenance contracts	£25.20 per hour
Number of site staff	5–35
Project area	6000 m^2
Project location	Outer London
Project components	50% hard landscape 50% soft landscape and planting
Access to works areas	Very good
Contract	Main contract
Delivery of materials	Full loads

As explained in more detail later the prices are generally based on wage rates and material costs current at June 2021. They do not allow for preliminary items, which are dealt with in the Preliminaries section of this book, or for any Value Added Tax (VAT) which may be payable.

Adjustments should be made to standard rates for time, location, local conditions, site constraints and any other factors likely to affect the costs of a specific scheme.

Term contracts for general maintenance of large areas should be executed at rates somewhat lower than those given in this section.

There is now a facility available to readers that enables a comparison to be made between the level of prices given and those for projects carried out in regions other than outer London; this is dealt with on page 31.

Units of Measurement

The units of measurement have been varied to suit the type of work and care should be taken when using any prices to ascertain the basis of measurement adopted.

The prices per unit of area for executing various mechanical operations are for work in the following areas and under the following conditions:

Prices per m^2	Relate to all areas where the rate is not too low to display a figure
Prices per $100\,m^2$	Relate to areas where the rate per unit is too low to show a meaningful figure
Prices per ha	Generally used on clear areas suitable for the use of tractors and tractor-operated equipment

The prices per unit area for executing various operations by hand generally vary in direct proportion to the change in unit area.

Approximate Estimating

These are combined Measured Work prices which give an approximate cost for a complete section of landscape work. For example, the construction of a car park comprises excavation, levelling, road-base, surfacing and marking parking bays. Each of these jobs is priced separately in the Measured Works section, but a comprehensive price is given in the Approximate Estimates section, which is intended to provide a quick guide to the cost of the job. It will be seen that the more items that go to make up an approximate estimate, the more possibilities there are for variations in the PC prices and the user should ensure that any PC price included in the estimate corresponds to their specification.

The figures given in *Spon's External Works and Landscape Price Book* are intended for general guidance, and if a significant variation appears in any one of the cost groups, the Price for Measured Work should be recalculated.

Measured Works

Prime Cost: Commonly known as the 'PC'. Prime Cost is the actual price of the material item being addressed such as paving, shrubs, bollards or turves, as sold by the supplier. Prime Cost is given 'per square metre', 'per 100 bags' or 'each' according to the way the supplier sells their product. In researching the material prices for the book we requested that the suppliers price for full loads of their product delivered to a site close to the M25 in London. Spon's rates do not include VAT. Some companies may be able to obtain greater discounts on the list prices than those shown in the book. Prime Cost prices for those products and plants which have a wide cost range will be found under the heading of Market Prices in the main sections of this book, so that the user may select the product most closely related to their specification.

Materials: The PC material plus the additional materials required to fix the PC material. Every job needs materials for its completion besides the product bought from the supplier. Paving needs sand for bedding, expansion joint strips and cement pointing; fencing needs concrete for post setting and nails or bolts; tree planting needs manure or fertilizer in the pit, tree stakes, guards and ties. If these items were to be priced out separately, *Spon's External Works and Landscape Price Book* (and the Bill of Quantities) would be impossibly unwieldy, so they are put together under the heading of Materials.

Labour: This figure covers the cost of planting shrubs or trees, laying paving, erecting fencing etc. and is calculated on the wage rate (skilled or unskilled) and the time needed for the job. Extras such as highly skilled craft work, difficult access, intermittent working and the need for labourers to back up the craftsman all add to the cost. Large regular areas of planting or paving are cheaper to install than smaller intricate areas, since less labour time is wasted moving from one area to another.

Plant, consumable stores and services: This rather impressive heading covers all the work required to carry out the job which cannot be attributed exactly to any one item. It covers the use of machinery ranging from JCBs to shovels and compactors, fuel, static plant, water supply (which is metered on a construction site), electricity and rubbish disposal. The cost of transport to site is deemed to be included elsewhere and should be allowed for elsewhere or as a preliminary item. Hired plant is calculated on an average of 36 hours working time per week.

Subcontract rates: Where there is no analysis against an item, this is deemed to be a rate supplied by a subcontractor. In most cases these are specialist items where most external works contractors would not have the expertise or the equipment to carry out the task described. It should be assumed that subcontractor rates include for the subcontractor's profit. An example of this may be found for the tennis court rates in Part 3, section 35.

Labour Rates Used in this Edition

The rates for labour used in this edition have been based on surveys carried out on a cross section of external works contractors. These rates include for company overheads such as employee administration, transport insurance, and oncosts such as National Insurance. Please see the calculation of these rates shown on pages 7–13. The rates do now include for profit.

Overheads: An allowance for this is included in the labour rates which are described above. The general overheads of the contract such as insurance, site huts, security, temporary roads and the statutory health and welfare of the labour force are not directly assignable to each item, so they are distributed as a percentage on each, or as a separate preliminary cost item. The contractor's and subcontractor's profits are now included in this group of costs. Site overheads, which will vary from contract to contract according to the difficulties of the site, labour shortages, inclement weather or involvement with other contractors, have not been taken into account in the build-up of these rates, while overhead (or profit) may have to take into account losses on other jobs and the cost to the contractor of unsuccessful tendering.

In many instances a composite team rate is used in the calculations.

ADJUSTMENT AND VARIATION OF THE RATES IN THIS BOOK

It will be appreciated that a variation in any one item in any group will affect the final Measured Works price. Any cost variation must be weighed against the total cost of the contract. A small variation in Prime Cost where the items are ordered in thousands may have more effect on the total cost than a large variation on a few items. A change in design that necessitates the use of earth moving equipment which must be brought to the site for that one job will cause a dramatic rise in the contract cost. Similarly, a small saving on multiple items will provide a useful reserve to cover unforeseen extras.

Worked examples

These are shown on page 35.

HOW THIS BOOK IS UPDATED EACH YEAR

The basis for this book is a database of Material, Labour, Plant and Subcontractor resources, each with its own area of the database.

Material, Plant and Subcontractor Resources

Each year the suppliers of each material, plant or subcontract item are approached and asked to update their prices to reflect the second quarter of that year.

These resource prices are individually updated in the database. Each resource is then linked to one or many tasks. The tasks in the task library section of the database are automatically updated by changes in the resource library. A quantity of the resource is calculated against the task. The calculation is generally performed once and the links remain in the database.

On occasions where new information or method or technology are discovered or suggested, these calculations would be revisited. A further source of information is simple time and production observations made during the course of the last year.

Labour Resource Update

Most tasks, except those shown as subcontractor rates (see above), employ an element of labour. The Data Department at AECOM conducts ongoing research into the costs of labour in various parts of the country.

Tasks or entire sections are then re-examined and recalculated. Comments on the rates published in this book are welcomed and may be submitted to the contact addresses shown in the Preface.

Copr Bay, Swansea – Phase One
Developer: Swansea Council
Image courtesy of: AFL Architects

PART 1

General

This part of the book contains the following sections:

Market Update

Introduction

Brexit, the coronavirus pandemic and the Ukraine war have, together, generated exceptional inflationary pressure in the UK construction sector. These factors have proved to be the ultimate collective test of globalized supply chains and just in time delivery models – and vulnerabilities have been exposed. In the past 12 months we have seen overall construction cost inflation hit historic highs and uncertainty around the near-term inflationary outlook is heightened.

Prices presented in this book have a base date of Q1 2022 – meaning indirect impacts of the Ukraine war have largely been excluded. Immediate, direct impacts of the Ukraine war have been factored into the rates but indirect impacts are likely to be significant and will take time to feed through the supply chain.

It's an exceptional, rapidly evolving time, and pricing decisions require a greater degree of judgement.

This article provides a summary of current inflationary drivers and identifies the most prominent risk factors as at July 2022. Updates, reporting on market developments, will be issued throughout the lifespan of this book to provide guidance on adjustments it would be sensible to make, and any factors that should be taken into consideration when interpreting the material presented.

Inflationary trends and drivers

Construction materials prices are under significant pressure as energy price inflation feeds through. Supply of some core materials and components is still restricted due to manufacturing capacity not yet returning to pre-pandemic levels and the impact of the war on Ukrainian and Russian supply.

Recent data tracking factory gate price movements suggests inflationary pressure on manufactured product prices is extremely strong, and accelerating, as lagged impacts from supply chain disruption and high energy prices feed through as stocks are replenished and supply agreements renegotiated.

In May 2022, construction materials price inflation reported by BEIS, rose at an annual rate of 27%. The list of materials with factory gate inflation exceeding 20% in May is long, with energy intensive steel and concrete reinforcing bars increasing at the highest rate.

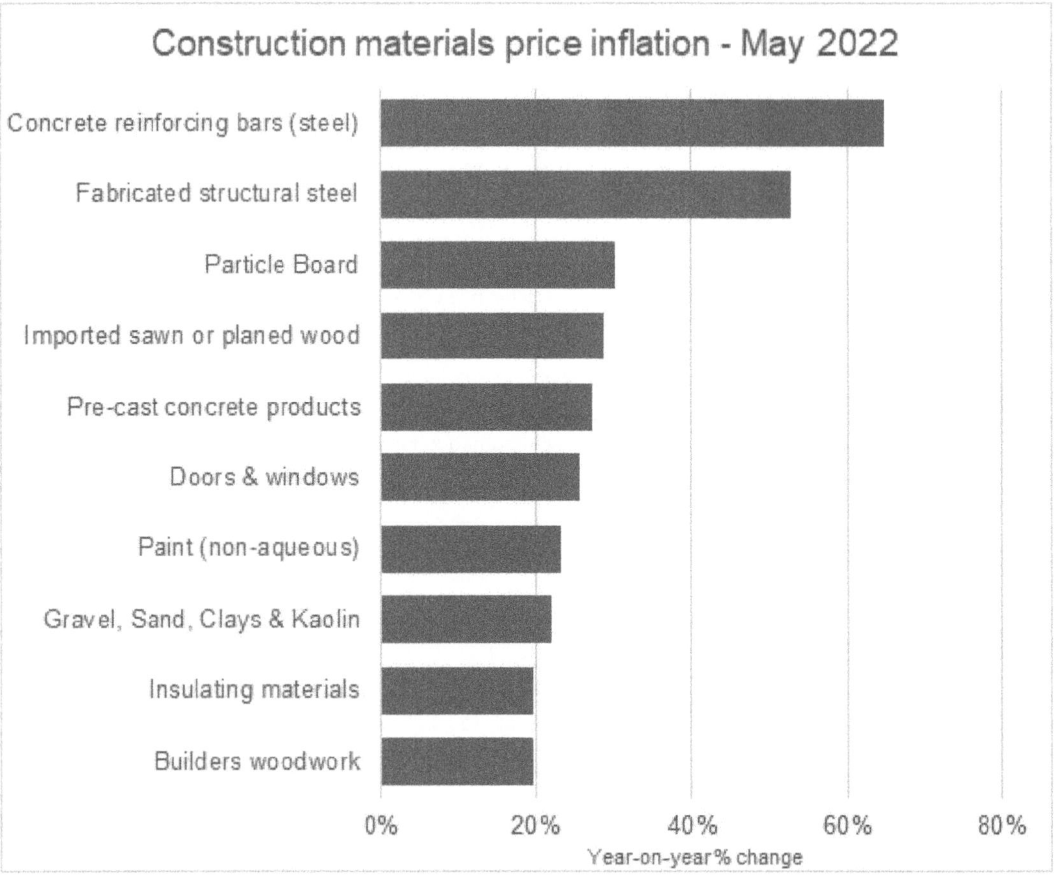

Source: BEIS

Price pressures on raw materials have eased slightly, although absolute levels remain very high relative to 2019. Oil prices have exceeded $100 per barrel for the past three months, around 65% up on levels a year ago. European gas prices have fallen back from their March peak but are still around 230% higher year-on-year. Lagged impacts from this exceptional inflationary price shock will impact prices of energy-intensive construction products for several quarters as manufacturers' supply agreements come up for renegotiation. Furthermore, wholesale gas prices may have softened slightly but the situation remains precarious and significant upside risks still exist.

Pressure on construction wage rates is adding to inflationary woes as workers face the biggest squeeze on real income in decades. With vacancies for construction jobs at record highs, an aging workforce, a COVID-induced increase in the early retirement rate and restrictions on economic migration, pressure on wages will persist. Currently annual construction sector wage inflation falls in the 5% to 7% range.

The inflationary outlook

AECOM's latest forecast suggests building cost inflation, incorporating anticipated trends in labour, plant and materials, will average 11% during 2022 and 10% in 2023. Tender price inflation is forecast to average 8.7% during 2022, reducing to 6.1% in 2023.

This forecast assumes the UK construction sector will avoid recession and that industry demand will continue to grow at a modest pace. Against this backdrop, demand will be such that pipeline visibility is sufficiently stable to enable cost increases to be passed through the supply chain.

A chart presenting AECOM's latest building and tender price forecasts is presented in Building Cost Indices; Tender Price Indices and Location Factors.

Risks

Risks to the outlook are numerous and broad. From uncertainty around the domestic policy agenda, and heightened geopolitical tension, to the potential impact of the generational reduction in real household disposable income, rising interest rates on economic activity and, ultimately, UK construction demand, the outlook is exceptionally uncertain.

Energy prices continue to pose a risk to the inflationary outlook. Proposed strike action by Norwegian gas workers recently threatened to stop supply to the UK. Action was averted but it highlights the extreme vulnerability of supply stability. With prices still at extremely high levels by historical standards, a further supply-side shock would have serious ramifications.

Insolvencies in the construction sector increased sharply in the first quarter of 2022. Construction contractors and specialists will continue to face financial headwinds over the next 12 months as inflation remains stubbornly high. Delivering on fixed price contracts agreed before the Ukraine war is challenging and maintaining cash flow is essential.

Plans to replace EU CE testing certification for construction products sold in the UK with a domestic testing regime have been relaxed slightly in response to industry warnings. New testing requirements will still take effect in January 2023, but CE marked products imported into the UK by the end of 2022 will no longer need to be retested. While this is welcomed by manufacturers, fundamental concerns remain about limited testing capacity, along with the ability to hamper product supply.

From a demand perspective, risks are also exceptionally high. Orders remain healthy and projects are progressing but household and business confidence has plummeted. With the Help to Buy scheme – which supported close to 40% of all newly built homes in the private market between 2013 and 2021 – ending just as mortgage interest rates are rising and household disposable income is reducing, headwinds in the house building sector are strong.

Labour Rates Used in this Edition

Introduction

Based on surveys carried out on a cross section of external works contractors, the following rates for labour have been used in this edition.

These rates include for company overheads such as employee administration, transport insurance, and oncosts such as National Insurance.

The rates used include for overhead & profit. Please see the calculations on pages 7–13.

The rates used may include a subcontractor's profit where subcontractor's rates are used to build up the costs of an item in the book.

Please use the Regional Variations table published on page 31 to adjust the costs of the Item being selected from this book for the geographical area of interest. The factor should be applied to the whole project cost and not just the labour rate. The rates in this book are based on a project on the London outer periphery.

This year's **'at-cost'** rates are calculated on the rounded value of a 30-person working team rounded to £31.00.

Labour Type	Unit	Cost incl. Overhead	Mark-up %	Rate incl.. Overhead & Profit
General contracting	/man-hour	£31.00	15%	£35.65
Brickwork 1+1 man gang	/team-hour	£71.00	15%	£81.65
Brickwork 1+2 man gang	/team-hour	£102.11	15%	£117.43
Maintenance contracting	/man-hour	£25.20	15%	£28.98

Other resource categories

Materials generally	15%
Plant	15%
Subcontractors generally	10%

How the Labour Rate has been calculated

NOTES:

1 Absence due to sickness has been assumed to be for periods not exceeding 3 days for which no payment is due.

2 EasyBuild Stakeholder Pension effective from 1 July 2002. Death and accident benefit cover is provided free of charge. The minimum employer contribution shall be £3.00 per week. Where the operative contributes between £3.00 and £13.50 per week the employer shall increase the minimum contribution to match that of the operative up to a maximum of £13.50 per week.

3 All N.I. Payments are at not-contracted out rates applicable from April 2021. National Insurance is paid for 52 weeks.

4 Public holiday pay includes for 8 normal public holidays.

NOTES: HOURS WORKED CALCULATIONS

Actual hours worked			
Normal hours worked/week	39.0 hrs		
Normal weeks worked/year	52.0 weeks		**2028 hrs**
Less Holidays			
Annual holidays	−4.4 weeks	−171.6 hrs	
Public holidays	−1.6 weeks	−62.4 hrs	
		−234.0 hrs	**1794 hrs**
Less Sick			
1 week	−1.0 weeks	−39.0 hrs	
		−39.0 hrs	
		Balance	**1755.0 hrs**
Less time lost for inclement weather, say	−2%		−35.1 hrs
			1719.9 hrs
Actual hours worked per annum		**1719.9 hrs**	
Actual weeks worked per annum		**44.1 weeks**	

A survey of typical landscape/external works companies indicates that they are currently paying above average wages for multi-skilled operatives regardless of specialist or supervisory capability. In our current overhaul of labour constants and costs we have departed from our previous policy of labour rates based on national awards and used instead a gross rate per hour for all rate calculations. Estimators can readily adjust the rate if they feel it is inappropriate for their work.

The productive labour of any organization must return their salary plus the cost of the administration which supports the labour force.

The labour rate used in this edition is calculated as follows:

Team size

A three-man team with annual salaries as shown and allowances for National Insurance, uniforms, site tools and a company vehicle, returns a basic labour rate of £24.31 per hour to which company overhead is added as per the following tables.

Basic labour rate calculation for Spon's 2023 – See table on next page.

Standard Working Hours per year (2022/2023)
= 1794

Less allowance for sick leave and down time for inclement weather
= −74

Actual Working Hours per year (2022/2023)
=1720 (1719.9) hrs

Labour rate calculation

Working hours/year 2022–2023 = 1720 *Allows sick and down time of 39 hours*						
	Number of	**Basic**	**Net Cost incl. NI & Pension**	**Clothing**	**Site Tools**	**TOTAL 3 Man Team**
Foreman	1	38300.06	47580.29	200.00	150.00	109046.67
Craftsman	1	30747.94	35190.37	200.00	150.00	
Labourer	1	22116.94	25226.02	200.00	150.00	
	3	91164.94	107996.67	600.00	450.00	
Overtime hours based on working nominal hours /annum at 1.5 times normal standard rate						
	40	33.40	1505.33			
	40	26.82	1172.70			
	40	19.29	840.26			
Totals	3		111763.99	600.00	450.00	112813.99
Vehicle costs inclusive of fuel insurances etc.						
	Working Days	£ /Day				
	252	50				12600.00
						£ 125,413.99
Net average cost per man hour			24.31			

The basic average labour rate excluding overhead costs (below) is

$$\frac{£\ 125,413.99}{\text{Total staff in team (3) x Working Hours per year (1720)}} = £24.31 \text{ per hour}$$

Add to the above the overhead costs as per the table on the next page

Overhead costs

COST CENTRE	Number of	Cost	Total
MD: Salary only: excludes profits + NI + Vehicle	1	80000.00	80000.00
Senior contracts managers + NI + Vehicle £51375 K basic	1	88936.12	88936.12
Other contracts managers + NI + Vehicle £35975 K basic	1	49816.42	49816.42
Business administrator/ secretary incl. NI & pension	1	32218.42	32218.42
Bookkeeper incl.. NI & pension	1	32218.42	32218.42
Rental	12	1300.00	15600.00
Insurances	1	5250.00	5250.00
Telephone and Mobiles	12	250.00	3000.00
Office Equipment	1	1500.00	1500.00
Stationery	12	50.00	600.00
Advertising	12	400.00	4800.00
Other vehicles not allocated to contract teams	1	9500.00	9500.00
Other consultants	1	6000.00	6000.00
Accountancy	1	2500.00	2500.00
Lights heating water	12	200.00	2400.00
Other expenses	1	9000.00	9000.00
TOTAL OFFICE OVERHEAD			£348,139.37

Add to the above basic rate the company overhead costs for a small company as per the table below. These costs are absorbed by the number of working men multiplied by the number of working hours supplied in the table below. This then generates an hourly overhead rate which is added to the net cost rate above.

Illustrative overhead for company employing 12–50 landscape operatives

The final labour rate used is then generated as follows:

$$\text{Net labour rate} = \frac{\text{Total Office Overhead (£348139.37)}}{\text{Working hours per year (1720)} \times \text{Labour resources employed}}$$

Total Nr of Site Staff	Admin Cost per hour	Total Rate per Man Hour
12	16.87	41.17
15	13.49	37.80
18	11.24	35.55
20	10.12	34.43
25	8.10	32.40
30	6.75	31.05
35	5.78	30.09
40	5.06	29.37
50	4.05	28.35

Hourly labour rates used for *Spon's External Works and Landscape 2023.*

This year's **'at-cost'** rates are calculated on the rounded value of a 30-person working team rounded to £31.00.

Labour Type	Unit	Cost incl. Overhead	Mark-up %	Rate incl. Overhead & Profit
General contracting	/man-hour	£31.00	15%	£35.65
Brickwork 1+1 man gang	/team-hour	£71.00	15%	£81.65
Brickwork 1+2 man gang	/team-hour	£102.11	15%	£117.43
Maintenance contracting	/man-hour	£25.20	15%	£28.98
Other resource categories				
Materials generally			15%	
Plant			15%	
Subcontractors generally			10%	
Some high cost material or subcontract items may be at lower rates of 5%				

Smaller organizations

Smaller organizations should adjust the rates shown in this book to suit a labour rate which is generally slightly higher due to the balance of company overhead and number of employees. The costs for smaller organizations are no longer shown in this edition but users may adjust the rates using the resultant labour rates shown below.

The calculation of this labour rate is as follows.

- A typical smaller works labour calculation is based on the above, but the administration cost of the organization is £181,251.26 per annum.
- The owner of the business does not carry out site works and is assumed to be paid a salary of £100,711.26 including National insurance, pension contributions and their vehicle costs. The balance of their income is taken in dividend and profit share figure constitutes part of the £181,251.26 shown.
- There are between 6 and 15 site operatives.

Illustrative overhead for small company employing 6–15 landscape operatives.

Cost Centre	Number of	Cost	Total
MD: Salary only: excludes profits + NI + Vehicle	1.00	100711.26	100711.26
Secretary	1.00	38000.00	38000.00
Bookkeeper (part time)	1.00	10500.00	10500.00
Rental	12.00	325.00	3900.00
Insurances	1.00	4000.00	4000.00
Telephone and Mobiles	12.00	200.00	2400.00
Office Equipment	1.00	1000.00	1000.00
Stationary	12.00	20.00	240.00
Advertising	12.00	200.00	2400.00
Other vehicles not allocated to contract teams	1.00	4000.00	4000.00
Other consultants	1.00	1000.00	1000.00
Accountancy	1.00	1500.00	1500.00
Lights heating water	12.00	300.00	3600.00
Other expenses	1.00	8000.00	8000.00
TOTAL OFFICE OVERHEAD			£181,251.26

The final labour rate used is then generated as follows:

$$\text{Net labour rate} = \frac{\text{Total Office Overhead (£181,251.26)}}{\text{Working hours per year (1720)} \times \text{Labour resources employed}}$$

Total Nr of Site Staff	Admin Cost per hour	Total Rate per Man Hour
3.00	35.13	59.43
6.00	17.56	41.87
9.00	11.71	36.01
12.00	8.78	33.09
15.00	7.03	31.33

Hourly labour rates for a typical small company working in London for 2021 could be calculated on the rounded value of a 9-man working team (£29.26).

General and Maintenance Contracting (Smaller organizations)	£32.00

Calculation of Annual Hours Worked see full table on page 8.

Normal hours worked		
Normal hours worked per week		39 hours
Normal hours worked per year	39 hrs × 52 wks.	2028 hours
Less holidays		
Annual holidays	39 hrs × 4.2 wks.	−163.8 hours
Public holidays	8 days @7.8 hours	−62.4 hours
Less sick and downtime	39 hrs × 1 wk.	−74 hours
Actual hours worked per annum		1720 hours

Building Industry Wages

BUILDING CRAFT AND GENERAL OPERATIVES

England, Wales and Scotland

The following basic rates of pay are assumed effective from June 2022.

	Rate per 39-hour week (£)	Rate per hour (£)
Craft Rate	531.94	13.64
Skill Rate 1	506.14	12.98
Skill Rate 2	487.71	12.51
Skill Rate 3	456.18	11.70
Skill Rate 4	431.20	11.06
General Operative	400.08	10.26

Notes on the New Rules of Measurement (NRM) Used in this Edition

New Rules of Measurement (NRM)

The NRM suite covers the life cycle of cost management and means that, at any point in a building's life, quantity surveyors will have a set of rules for measuring and capturing cost data. In addition, the BCIS Standard Form of Cost Analysis (SFCA) 4th edition has been updated so that it is aligned with the NRM suite.

The three volumes of the NRM are as follows.

- **NRM1** – Order of cost estimating and cost planning for capital building works.
- **NRM2** – Detailed measurement for building works.
- **NRM3** – Order of cost estimating and cost planning for building maintenance works.
  The latest editions of the NRM suite were effective from 1 December 2021. The NRM suite is published by the RICS and readers are strongly advised to read the documents.

NRM1

First launched in May 2009, the New Rules of Measurement (NRM1) Order of cost estimating and elemental cost planning were introduced by the RICS. The NRM have been written to provide a standard set of measurement rules that are understandable by all those involved in a construction project, including the employer; thereby aiding communication between the project/design team and the employer. In addition, the NRM should assist the quantity surveyor/cost manager in providing effective and accurate cost advice to the employer and the rest of the project team.

NRM2

The NRM2 has been developed to provide a detailed set of measurement rules to support the procurement of construction works. These rules are a direct replacement for the *Standard Method of Measurement for Building Works (SMM7).*

NRM3

The NRM3 has been developed to provide a consistent approach for measuring building maintenance costs throughout the life cycle of a construction project. NRM 3 is linked to the International Property Measurement Standards (IPMS) for measuring floor areas, and can also be mapped to the international standard ICMS: Global Consistency in Presenting Construction Life Cycle Costs and Carbon Emissions. NRM 3 also takes account of both the RIBA Plan of Work 2020 and the RIBA Digital Plan of Work (DPoW).

Computation of the Cost of Materials

Percentages of default waste are placed against material resources within the supplier database. These range from 2.5% (for bricks) to 20% (for topsoil). An allowance for the cost of unloading, stacking etc. should be added to the cost of materials.

The following are typical hours of labour for unloading and stacking some of the more common building materials.

Material	Unit	Labourer hours
Cement	tonne	0.67
Lime	tonne	0.67
Common bricks	1000	1.70
Light engineering bricks	1000	2.00
Heavy engineering bricks	1000	2.40

The Landfill Tax

The Tax

The Landfill Tax came into operation on 1 October 1996. It is levied on operators of licensed landfill sites in England, Wales and Northern Ireland at the following rates with effect from 1 April 2022:

Inactive or inert wastes [Lower Rate]	£3.15 / tonne	Included are soil, stones, brick, plain and reinforced concrete, plaster and glass – lower rate
All other taxable wastes [Standard Rate]	£98.60 / tonne	Included are timber, paint and other organic wastes generally found in demolition work and builders skips – standard rate

The 2022 increase in the standard and lower rates of Landfill Tax are in line with the RPI, rounded to the nearest 5 pence and compare with previous rates as follows:

Lower Rate	From 1 April 2019	£2.90 / tonne
Lower Rate	From 1 April 2020	£3.00 / tonne
Lower Rate	From 1 April 2021	£3.10 / tonne
Standard Rate	From 1 April 2019	£91.35 / tonne
Standard Rate	From 1 April 2020	£94.15 / tonne
Standard Rate	From 1 April 2021	£96.70 / tonne

There has been no change in the loss on ignition threshold for fines which is a major eligibility factor for lower rate Landfill Tax since its introduction in April 2015. The threshold remains at 10%. However, Excise Notice LFT1: a general guide to Landfill Tax was updated on 9 April 2020 in response to the current COVID-19 pandemic to include the following temporary change to the loss of ignition re-testing condition requiring a re-test to be carried out within 21 days of the disposal of the material (Section 6.9): However from January 2022 this no longer applies.

> 'If you're unable to get the retest processed within this time frame due to disruptions caused by coronavirus, you can carry this out as soon as is reasonably possible and you should keep evidence to demonstrate this.'

Please note that this change is temporary and Excise Notice LFT1 will be re-issued when the change ends. Nothing has been issued to date.

The Landfill Tax (Qualifying Material) Order 2011 came into force on 1 April 2011 and set out the qualifying criteria that are eligible for the lower rate of Landfill Tax. A waste will be lower rated for Landfill Tax only if it is listed as a qualifying material in the Landfill Tax (Qualifying Material) Order 2011 and is disposed of at an authorized waste site.

The principal qualifying materials are:

- Naturally occurring rocks and subsoils
- Ceramic or concrete materials
- Processed or prepared minerals
- Furnace slags
- Ash arising from wood or waste combustion and from the burning of coal and petroleum coke (including when burnt with biomass)
- Low activity inorganic compounds
- Calcium sulphate
- Calcium hydroxide and brine

From 1 April 2018, Landfill Tax is due on disposals of material at unauthorized sites in England and Northern Ireland. This also applies to disposals made prior to 1 April 2018 which are still on the site on 1 April 2018. Currently disposals on a site operated as a 'relevant regulated facility' which (as defined under the Environmental Permitting (England and Wales) Regulations 2016) includes sites which are 'a waste operation, such as a transfer station or treatment facility' are not considered to be taxable disposals. The Landfill Tax (Disposals of Material) Order 2018 provides more clarity on what constitutes a 'taxable disposal'. In particular it clarifies which elements of a landfill cell are not taxable under the requirements of Part 3 of the Finance Act 1996, listing them as:

1. material that forms a layer which performs the function of drainage at the base of a landfill cell;
2. material that forms the impermeable layer that delimits the landfill cell;
3. a pipe, pump or associated infrastructure inserted into a landfill cell for the purposes of the extraction or control of surplus liquid or gas from or within that cell;
4. material used for restoration of a landfill cell that only contains inert material; or
5. material placed in an information area within the meaning given by regulation 16 A(1) of the Landfill Tax Regulations 1996.

Calculating the Weight of Waste

There are two options:

1. If licensed sites have a weighbridge, tax will be levied on the actual weight of waste.
2. If licensed sites do not have a weighbridge, tax will be calculated by one of the three Specified Methods described in Excise Notice LFT1: a general guide to Landfill Tax (updated on 9 April 2020 and available via the HMRC website).

Water

If water greater than or equal to 25% of the material by weight has been added to the material to facilitate disposal, for the extraction of minerals or in the course of an industrial process, a landfill operator may apply to discount the water content of the material before tax is levied.

Effect on Prices

The tax is paid by landfill site operators only. Tipping charges reflect this additional cost.

As an example, Spon's A & B rates for mechanical disposal will be affected as follows:

Inactive waste	Spon's A & B 2022 net rate	£28.24	per m^3
	Tax, 1.9 tonne per m^3 (unbulked) @ £3.15	£5.89	per m^3
	Spon's gross rate including tax	£34.13	per m^3
Active waste	Active waste will normally be disposed of by skip and will probably be mixed with inactive waste. The tax levied will depend on the weight of materials in the skip which can vary significantly.		

Exemptions

The following disposals are exempt from Landfill Tax subject to meeting certain conditions:

- dredgings which arise from the maintenance of inland waterways and harbours
- naturally occurring materials arising from mining or quarrying operations
- pet cemeteries
- inert waste used to restore landfill sites and to fill working and old quarries where a planning condition or obligation is in existence
- waste from visiting NATO forces

Devolution of Landfill Tax to Scotland from 1 April 2015

The Scotland Act 2012 provided for Landfill Tax to be devolved to Scotland. From 1 April 2015, operators of landfill sites in Scotland are no longer liable to pay UK Landfill Tax for waste disposed at their Scottish sites. Instead, they are liable to register and account for Scottish Landfill Tax (SLfT).

Operators of landfill sites only in Scotland were deregistered from UK Landfill Tax with effect from 31 March 2015.

The Scottish Government announced rates of Landfill Tax from 1 April 2022 to be as follows:

- Lower Rate – £3.15 / tonne
- Standard Rate – £98.60 / tonne

Both rates mirror those applicable in England and Northern Ireland.

Devolution of Landfill Tax to Wales from April 2018

The Wales Act 2014 provides for Landfill Tax to be devolved to Wales. This took effect from 1 April 2018. Operators of landfill sites in Wales are liable to register and account for Landfill Disposals Tax (LDT); information is available on the Welsh Revenue Authority website, but in summary there are 3 rates of LDT:

- a lower rate of £3.15 / tonne
- a standard rate of £98.60 / tonne
- an unauthorized disposals rate of £145.05 / tonne (set at 150% of the standard rate)

For further information contact the HMRC VAT and Excise Helpline, telephone 0300 200 3700.

The Devolution of Landfill Tax (Wales) (Consequential Transitional and Savings Provisions) Order 2018 provides for the closure of the Landfill Communities Fund (LCF) in Wales and the arrangements for a two-year transitional period during which environmental bodies may continue to spend funds on projects in Wales.

The LCF was replaced by the Landfill Disposals Tax (LDT) Communities Scheme which is distributed under contract for the period 1 April 2018 to 31 March 2022 by the Wales Council for Voluntary Action (WCVA). More information is available on the WCVA website.

The Aggregates Levy

Introduction

The Aggregates Levy came into operation on 1 April 2002 in the UK, except for Northern Ireland where it was phased in over five years from 2003.

It was introduced to ensure that the external costs associated with the exploitation of aggregates are reflected in the price of aggregate, and to encourage the use of recycled aggregate. There continues to be strong evidence that the levy is achieving its environmental objectives, with sales of primary aggregate down and production of recycled aggregate up. The Government expects that the rates of the levy will at least keep pace with inflation over time, although it accepts that the levy is still bedding in.

The rate of the levy remains at £2.00 per tonne from 1 April 2022 and is levied on anyone considered to be responsible for commercially exploiting 'virgin' aggregates in the UK and should naturally be passed by price increase to the ultimate user.

All materials falling within the definition of 'Aggregates' are subject to the levy unless specifically exempted.

It does not apply to clay, soil, vegetable or other organic matter.

The intention is that it will:

- Encourage the use of alternative materials that would otherwise be disposed of to landfill sites
- Promote development of new recycling processes, such as using waste tyres and glass
- Promote greater efficiency in the use of virgin aggregates
- Reduce noise and vibration, dust and other emissions to air, visual intrusion, loss of amenity and damage to wildlife habitats

Definitions

'Aggregates' means any rock, gravel or sand which is extracted or dredged in the UK for aggregates use. It includes whatever substances are for the time being incorporated in it or naturally occur mixed with it.

'Exploitation' is defined as involving any one or a combination of any of the following:

- Being removed from its original site, a connected site which is registered under the same name as the originating site or a site where it had been intended to apply an exempt process to it, but this process was not applied
- Becoming subject to a contract or other agreement to supply to any person
- Being used for construction purposes
- Being mixed with any material or substance other than water, except in permitted circumstances

The definition of 'aggregate being used for construction purposes' is when it is:

- Used as material or support in the construction or improvement of any structure
- Mixed with anything as part of a process of producing mortar, concrete, tarmacadam, coated roadstone or any similar construction material

Incidence

It is a tax on primary aggregates production – i.e. 'virgin' aggregates won from a source and used in a location within the UK territorial boundaries (land or sea). The tax is not levied on aggregates which are exported or on aggregates imported from outside the UK territorial boundaries.

It is levied at the point of sale.

Exemption from tax

An 'aggregate' is exempt from the levy if it is:

- Material which has previously been used for construction purposes
- Aggregate that has already been subject to a charge to the Aggregates Levy
- Aggregate which was previously removed from its originating site before the start date of the levy
- Aggregate which is moved between sites under the same Aggregates Levy Registration
- Aggregate which is removed to a registered site to have an exempt process applied to it
- Aggregate which is removed to any premises where china clay or ball clay will be extracted from the aggregate
- Aggregate which is being returned to the land from which it was won, provided that it is not mixed with any material other than water
- Aggregate won from a farm land or forest where used on that farm or forest
- Rock which has not been subjected to an industrial crushing process
- Aggregate won by being removed from the ground on the site of any building or proposed building in the course of excavations carried out in connection with the modification or erection of the building and exclusively for the purpose of laying foundations or of laying any pipe or cable
- Aggregate won by being removed from the bed of any river, canal or watercourse or channel in or approach to any port or harbour (natural or artificial), in the course of carrying out any dredging exclusively for the purpose of creating, restoring, improving or maintaining that body of water
- Aggregate won by being removed from the ground along the line of any highway or proposed highway in the course of excavations for improving, maintaining or constructing the highway otherwise than purely to extract the aggregate
- Drill cuttings from petroleum operations on land and on the seabed
- Aggregate resulting from works carried out in exercise of powers under the New Road and Street Works Act 1991, the Roads (Northern Ireland) Order 1993 or the Street Works (Northern Ireland) Order 1995
- Aggregate removed for the purpose of cutting of rock to produce dimension stone, or the production of lime or cement from limestone.
- Aggregate arising as a waste material during the processing of the following industrial minerals:
    - anhydrite
    - ball clay
    - barites
    - calcite
    - china clay
    - clay, coal, lignite and slate
    - feldspar
    - flint
    - fluorspar
    - fuller's earth
    - gems and semi-precious stones
    - gypsum
    - any metal or the ore of any metal
    - muscovite
    - perlite
    - potash
    - pumice
    - rock phosphates
    - sodium chloride
    - talc
    - vermiculite
- Spoil from the separation of the above industrial minerals from other rock after extraction
- Material that is mainly but not wholly the spoil, waste or other by-product of any industrial combustion process or the smelting or refining of metal.

Anything that consists 'wholly or mainly' of the following is exempt from the levy (note that 'wholly' is defined as 100% but 'mainly' as more than 50%, thus exempting any contained aggregates amounting to less than 50% of the original volumes:

- clay, soil, vegetable or other organic matter
- drill cuttings from oil exploration in UK waters
- material arising from utility works, if carried out under the New Roads and Street Works Act 1991

However, when ground that is more than half clay is mixed with any substance (for example, cement or lime) for the purpose of creating a firm base for construction, the clay becomes liable to Aggregates Levy because it has been mixed with another substance for the purpose of construction.

Anything that consists completely of the following substances is exempt from the levy:

- Spoil, waste or other by-products from any industrial combustion process or the smelting or refining of metal – for example, industrial slag, pulverised fuel ash and used foundry sand. If the material consists completely of these substances at the time it is produced it is exempt from the levy, regardless of any subsequent mixing
- Aggregate necessarily arising from the footprint of any building for the purpose of laying its foundations, pipes or cables. It must be lawfully extracted within the terms of any planning consent
- Aggregate necessarily arising from navigation dredging
- Aggregate necessarily arising from the ground in the course of excavations to improve, maintain or construct a highway or a proposed highway
- Aggregate necessarily arising from the ground in the course of excavations to improve, maintain or construct a railway, monorail or tramway

Relief from the levy either in the form of credit or repayment is obtainable where:

- it is subsequently exported from the UK in the form of aggregate
- it is used in an exempt process
- it is used in a prescribed industrial or agricultural process
- it is waste aggregate disposed of by dumping or otherwise, e.g. sent to landfill or returned to the originating site

An exemption for aggregate obtained as a by-product of railway, tramway and monorail improvement, maintenance and construction was introduced in 2007.

Discounts

Water which is added to the aggregate after the aggregate has been won (washing, dust dampening etc.) may be discounted from the tax calculations. There are two accepted options by which the added water content can be calculated.

The first is to use HMRC's standard added water percentage discounts listed below:

- washed sand 7%
- washed gravel 3.5%
- washed rock/aggregate 4%

Alternatively a more exact percentage can be agreed for dust dampening of aggregates.

Whichever option is adopted, it must be agreed in writing in advance with HMRC.

Impact

The British Aggregates Association suggested that the additional cost imposed by quarries is more likely to be in the order of £3.40 per tonne on mainstream products, applying an above average rate on these in order that by-products and low grade waste products can be held at competitive rates, as well as making some allowance for administration and increased finance charges.

With many gravel aggregates costing in the region of £20.00 per tonne, there is a significant impact on constructioncosts.

Avoidance

An alternative to using new aggregates in filling operations is to crush and screen rubble which may become available during the process of demolition and site clearance as well as removal of obstacles during the excavation processes.

More detailed information can be found on the HMRC website (www.hmrc.gov.uk) in Notice AGL1 Aggregates Levy published 1 April 2014 (updated 9 February 2017).

Contact

For further information or assistance, please contact one of the Fiscal Incentives team:

help.fiscalincentives@aecom.com

Cost Indices

Introduction

The purpose of this section is to show changes in the cost of carrying out landscape work (hard surfacing and planting) since 2015. It is important to distinguish between costs and tender prices: the following table reflects the change in cost to contractors but does not necessarily reflect changes in tender prices. In addition to changes in labour and material costs, which are reflected in the indices given below, tender prices are also affected by factors such as the degree of competition at the time of tender and in the particular area where the work is to be carried out, the availability of labour and materials, and the general economic situation. This can mean that in a period when work is scarce, tender prices may fall despite the fact that costs are rising, and when there is plenty of work available, tender prices may increase at a faster rate than costs.

The Constructed Cost Index

A Constructed Cost Index has been developed based on Price Adjustment Formula Indices (PAFI) (Building Series 3). Cost indices for the various building trades are published monthly by BCIS and are available via a subscription service.

Although the PAFI indices are prepared for use with price-adjustment formulae for calculating reimbursement of increased costs during the course of a contract, they also present a time series of cost indices for the main components of landscaping projects. They can therefore be used as the basis of an index for landscaping costs.

The method used here is to construct a composite index by allocating weightings to the indices representing the usual work categories found in a landscaping contract, the weightings being established from an analysis of actual projects.

Constructed Landscaping Cost Index REBASE 2015 = 100

Year	Quarter 1	Quarter 2	Quarter 3	Quarter 4	Annual average	Annual change
2015	99.2	100.0	100.6	100.2	100.0	4.0%
2016	99.9	100.9	102.3	103.3	101.6	1.6%
2017	103.9	104.0	105.7	106.3	105.0	3.2%
2018	107.2	108	110.5	110.8	109.1	3.8%
2019	112.2	113.6	114.9	115.1	114.0	4.2%
2020	115.6	115.7	116.4	118.1	116.5	2.1%
2021	120.9	123.1	129.1	132.1	126.3	7.0%
2022	135.4 (p)	136.1 (f)	136.8	137.6	136.5	7.5%
2023	138.3	139.0	139.7	140.4	139.4	2.1%

(p) = provisional; f() = forecast. This index is updated every quarter in Spon's Price Book Update. The updating service is available, free of charge, to all purchasers of Spon's Price Books (complete the reply-paid card onclocod).

Regional Variations

Prices in *Spon's External Works and Landscape Price Book* are based upon conditions prevailing for a competitive tender in the Outer London area. For the benefit of readers, this edition includes regional variation adjustment factors which can be used for an assessment of price levels in other regions.

Special further adjustment may be necessary when considering city centre or very isolated locations.

The regional variations shown in the following table are based on our forecast of price differentials in each of the economic planning regions at the time the book was published. Historically the adjustments do not vary much year-on-year; however, there can be wider variations within any region, particularly between urban and rural locations. The following table shows suggested adjustments required to cost plans or estimates prepared using the *Spon's External and Landscape Price Book.*

Please note: these adjustments need to be applied to the total cost plan value, not to individual material or project resources or items.

Region	Adjustment Factor
Outer London (Spon's 2021 Q4)	1.00
Inner London	1.05
South East	0.96
South West	0.90
East Midlands	0.88
West Midlands	0.87
East of England	0.93
Yorkshire & Humberside	0.86
North East	0.83
North West	0.88
Scotland	0.88
Wales	0.88
Northern Ireland	0.76

The following example illustrates the adjustment of prices for regions other than Outer London, by use of regional variation adjustment factors.

A Value of items priced using *Spon's External Works and Landscape Price Book* for Outer London £100,000

B. Adjustment to value of A. to reflect North West region price level 100,000 × 0.88 £ 88,000

HOW TO USE THIS BOOK

Item	Unit	Range £
How this book is presented		
Prices in this book generally include overhead and profit. Market prices where shown exclude overhead and profit		
The price information sections of this book are split into two subsections as follows:		
Approximate Estimates – This is presented in a 3 column format. This section combines prices from the Measured Works items to form a composite rate for the work described.		
Measured Works – This is presented in an 8 column format and is a detailed analysis of single tasks commonly employed throughout the external works and landscape sector.		
Column formats		
Description of the column formats used in this book.		
Approximate Estimates		
Only the 3 column format is used; users of the book cannot see the components which make up the prices in the book format. An electronic version is supplied free with this book. Users of the electronic system can see more detail in the electronic format.		
Cost and Marked up values shown in this edition		
Prices shown include overhead and profit in the 2021 edition at the following rates:		
Materials	15%	–
Labour	15%	–
Plant	15%	–
Subcontractors (generally)	10%	–
Note some high value subcontractors such as swimming pools and sportsfields and some street furniture and playground items which have a high unit cost may be marked up at lower rates (5%)		
Measured Works		
Where an 8 column analysis is shown these are tasks carried out directly by the external works or landscape contractor and are shown at cost.		
Where prices are shown in the 8 column section without analysis, prices have been supplied by specialist subcontractors and are assumed to have the specialist subcontractor's selling price of the item included.		
Explanation of the columns used in the Measured Works analysis		
Prime Cost		
Commonly known as the 'PC'. Prime Cost is the actual price of the material item being addressed, such as paving, shrubs, bollards or turves, as sold by the supplier. Prime Cost is given 'per square metre', 'per 100 bags' or 'each' according to the way the supplier sells his product. In researching the material prices for the book we requested that the suppliers price for full loads of their product delivered to a site close to the M25 in London. Spon's rates do not include VAT. Some companies may be able to obtain greater discounts on the list prices than those shown in the book. Prime Cost prices for those products and plants which have a wide cost range will be found under the heading of Market Prices in the main sections of this book, so that the user may select the product most closely related to their specification.		

Item	Unit	Range £
Column formats – cont		
Explanation of the columns used in the Measured Works analysis – cont		
Labour Hours		
This is the total amount of man hours required to carry out the quantity and unit of the measured works item which is being analysed.		
Labour £ – Shown inclusive of Overhead & Profit – Generally 15%		
This is the value of the labour used. Please refer to the labour rates used in this edition which are detailed preceding the Approximate Estimates section		
Plant – Shown inclusive of Overhead & Profit – Generally 15%		
This covers the cost of machinery used to work out the cost of the item addressed. Plant can range from heavy excavators to light electrical breakers. All plant is assumed as hired in except for concrete mixers and normal site tools which a contractor would be expected to carry in his or her company vehicle (hand drills, grinders etc.). Plant includes for fuel and down time. The rates used are shown in the section of this book 'Hired in Plant' section on page 154		
Delivery and removal to site costs are not included in any rates.		
Materials – Shown inclusive of Overhead & Profit – Generally 15%		
The PC material plus the additional materials required to fix the PC material. Every job needs materials for its completion besides the product bought from the supplier. Paving needs sand for bedding, expansion joint strips and cement pointing; fencing needs concrete for post setting and nails or bolts; tree planting needs manure or fertilizer in the pit, tree stakes, guards and ties. If these items were to be priced out separately, *Spon's External Works and Landscape Price Book* (and the Bill of Quantities) would be impossibly unwieldy, so they are put together under the heading of Materials.		
Subcontract rates or subcontractors used within an item build-up – Shown inclusive of Overhead & Profit – Generally 10%		
Where only a figure is shown in the columns		
This is a subcontract rate which has been received from a specialist subcontractor.		
This rate would generally include the subcontractor's overhead and profit but not the main contractors oncosts.		
Delivery Prices		
These are not generally included. Exceptions would be aggregates or loose materials which are delivered by the load. Our specification to suppliers of information for this book is that all materials are delivered in full loads to a site close to the M25 in North London.		
Unit		
The unit being used to measure the item.		
Total Price £		
The total price inclusive of Overhead & Profit		
Where a figure is shown in the Material column only		
This is generally the market price of a single material only.		

HOW TO USE THIS BOOK

Item	Unit	Range £
Worked examples using data in this edition		
The following operations describe the trench excavation, pipe laying and backfilling operations as contained in section NRM Section 34 of this edition.		
Example 1		
Pipe laying 100 m		
Strip topsoil 150 × 300 mm wide	100 m	50.00
Excavate for drain 150 × 450 mm; excavated material to spoil heaps on site by machine	100 m	1425.00
Lay flexible plastic pipe 100 mm dia. wide	100 m	210.00
Backfilling with gravel rejects, blinding with sand and topping with 150 mm topsoil	100 m	465.00
TOTAL FOR ALL OF THE ABOVE COMBINED	100 m	2150.00
This total above item is a composite of all of the above 'Measured Works' items combined and is typical of an entry in the Approximate Estimates section of this book.		
Example 2		
The following individual items are shown as components of a slab paving specification for one m². These individual items are all extracted from the paving section of this book and include a profit margin of 15%		
The PC of 900 × 600 × 50 mm precast concrete flags is shown in the PC (Prime cost) column as:	m^2	15.00
Costs of bedding and pointing materials are added to give a total Materials cost of:	m^2	7.00
Further costs for labour and mechanical plant and profit give a resultant price of:	m^2	40.00
If a quotation of PC £9.00/m^2 was received from a supplier of the flags the resultant price would be calculated as:		
Original price of the flags:	m^2	15.00
Gives a reduced cost including mark-up of	m^2	9.00
Less the original cost PC £10.03 plus the revised cost £9.00 + 15% profit in both cases to give	m^2	49.00
Example 3		
The following samples are extracted from the tree planting section of this edition		
The prime cost of an Acer platanoides 10–12 cm bare root tree is given as:	nr	52.00
To which the cost of handling and planting only to treepits excavated separately is added:	nr	21.00
Giving an overall cost of:	nr	67.00
The following items are also shown as component prices in the Measured Works section		
labour and mechanical plant for mechanical excavation of a 600 × 600 × 600 mm pit	nr	6.00
a single tree stake	nr	9.00
importing Topgrow compost in 75 litre bags	nr	0.00
backfilling with imported topsoil	nr	9.00
transporting the excavated material to spoil heaps on site 25 m distant	nr	1.00
and disposing of the excavated material	nr	8.00

Item	Unit	Range £
Worked examples using data in this edition – cont		
The following samples are extracted from the tree planting section of this edition – cont		
The following price which is a composite of all the above might be shown in the Approximate Estimates section:		
Tree planting; supply and plant Acer platanoides 10–12 cm in 600 × 600 mm tree pit excavated mechanically; allow for a single tree stake and Topgrow compost and backfilling with imported topsoil; all excavated material removed from site	nr	105.00
ADJUSTMENT AND VARIATION OF THE RATES IN THIS BOOK		
Example 4		
It will be appreciated that a variation in any one item in any group will affect the final Measured Works price. Any cost variation must be weighed against the total cost of the contract. A small variation in Prime Cost where the items are ordered in thousands may have more effect on the total cost than a large variation on a few items. A change in design that necessitates the use of earth moving equipment which must be brought to the site for that one job will cause a dramatic rise in the contract cost. Similarly, a small saving on multiple items will provide a useful reserve to cover unforeseen extras.		
Using the tree planting, 'Example 3', above, if the tree size was to be increased from a 10–12 cm tree to a 12–14 cm tree of the same variety, this might necessitate an increase in tree pit size, more compost, a larger tree stake and more excavated material being disposed of off site.		
The resultant variation would again be referenced from the tree planting section as follows:		
The revised PC (prime cost) of an Acer platanoides 12 –14 cm bare root tree is given as:	nr	85.00
to which the cost of handling and planting only to treepits excavated separately is added:	nr	34.00
Giving an overall cost of:	nr	120.00
Add to this the following variations as described:		
Tree pit increased in size from 600 × 600 × 600 mm to 900 × 900 × 600 mm	nr	14.00
Additional compost	nr	26.00
Increased staking requirements to two stakes	nr	13.00
Increased volume of imported topsoil	nr	20.00
Increased disposal volume	nr	21.00
Giving a resultant price for the increased tree size as	nr	195.00
The variation therefore from the original specification for the 10–12 cm tree to the revised 12–14 cm tree is calculated as an additional sum per tree of	nr	68.00
Example 5 – Composite labour rates		
The worked example below shows the build-up for Keyblok paviours as used elsewhere in this book. Readers will note that there appears to be an anomaly between the labour time and the labour rate. This however is due to the useage of a composite labour rate. In this case a 3 man team which is 3 × the hourly rate shown in the labour rates section 2022 = £29.45 / hour		

HOW TO USE THIS BOOK

Item	Unit	Range £
Precast concrete paving blocks; Keyblok Marshalls Plc; on prepared base (not included); on 50 mm compacted sharp sand bed; blocks laid in 7 mm loose sand and vibrated; joints filled with sharp sand and vibrated; level and to falls only		
Herringbone bond		
200 × 100 × 60 mm; natural grey	m^2	58.00
The labour value is shown as £28.46 but the labour hours are 0.28 hrs. The labour rate is (£29.45 x0.28 hrs) × 3 man team × 15% profit. = £28.46. Please ignore minor rounding calculations. Readers should extrapolate in these instances.		
Labour Rates Used in this Edition		
General landscape construction; The labour rates shown below are shown at cost and are deemed to include company overhead as per the analysis of labour rates in the front of this book. Profit margins are added to these rates as per the rates shown below to produce the final labour rate		
General labour rate for all works	hr	31.00
Specialist labour rate for brickwork	hr	39.00
Landscape maintenance		
General labour rate for all works	hr	25.00
Cost and Marked up values shown in this edition		
Materials	15%	–
Labour	15%	–
Plant	15%	–
Subcontractors (generally)	10%	–

PART 2

Approximate Estimating Rates

Prices in this section are based upon the Prices for Measured Works but allow for incidentals which would normally be measured separately in a Bill of Quantities. They do not include for Preliminaries which are priced elsewhere in this book.

Items shown as subcontract or specialist rates would normally include the specialist's overhead and profit. All other items which could fall within the scope of works of general landscape and external works contractors do now include profit in this edition.

Based on current commercial rates, profits of 10% to 15% have been added to these rates to indicate the likely 'with profit' values of the tasks below. The variation quoted above is dependent on the sector in which the works are taking place – domestic, public or commercial. Users should adjust these rates for their sector or location.

Labour Type	Unit	Cost incl. Overhead	Mark-up %	Rate incl. Overhead & Profit
General contracting	/man-hour	£31.00	15%	£35.65
Brickwork 1+1 man gang	/team-hour	£71.00	15%	£81.65
Brickwork 1+2 man gang	/team-hour	£102.11	15%	£117.43
Maintenance contracting	/man-hour	£25.20	15%	£28.98

Other resource categories

Materials generally	15%
Plant	15%
Subcontractors generally	10%

Some high cost material or subcontract items may be at lower rates of 5%.

COST MODELS

Item	Unit	Range £
Housing and house builder		
Cost Models for areas as described below. These models below exclude items such as demolitions, bulk groundworks, drainage unless specifically mentioned as included; Site or project preliminaries of 7.5% – 12% may also be added to the rates; These items below are based on areas of 1000 m^2		
Housing cost models meeting minimum planning requirements or to maximize sales potential at economic cost		
Soft landscaped space only space; planting: turf ratio's as shown; planting sizes and densities as indicated; 1 tree of 18–20 cm girth at 1 tree per 100 m^2 overall external area; Mulch and watering after planting		
Landscaped area; turf and trees only; imported topsoil to turf areas at 50 thick;	100 m^2	2150.00
Landscaped area 70:30 turf to planted area with imported topsoil to turf areas at 50 thick;	100 m^2	3125.00
Landscaped area 50:50 turf to planted area; imported topsoil to turf areas at 50 thick;	100 m^2	3425.00
Landscaped area; Planting and trees only; Planting of shrubs 100% at 400 mm centres	100 m^2	4200.00
Landscaped area; Planting and trees only; Planting of shrubs and groundcovers 100% Shrubs: Groundcovers 70:30 at 400:250 mm centres	100 m^2	4800.00
Landscaped area; Planting and trees only; Planting of shrubs and groundcovers 100% Shrubs: Groundcovers 30:70 at 400:250 mm centres	100 m^2	5500.00
Mixed Hard and Soft Landscaped Space with parking; with 1 tree of 18–20 cm every 100 m^2; imported topsoil to turf areas at 50 thick		
Landscaped area where 50% of the external area is hard surface of macadam parking area; with surround of PC kerbs and 50% soft landscape of 70% turf and 30% planted area at 50 thick	100 m^2	6200.00
Landscaped area where 50% of the external area is hard surface of block paved parking area; surround of PC kerbs 50% soft landscape of 70% turf and 30% planted area	100 m^2	7700.00
External area where the area is 50% parking and 50% soft landscape. The soft landscape is 50% turf and 50% planting; planting at 400 mm ccs – 2–3 litre plants per m^2; Trees at 1 nr 16–18 cm girth per 100 m^2 throughout the parking and landscape area		
macadam parking area	100 m^2	6500.00
concrete block paved parking area	100 m^2	8400.00
External area where the area is 50% parking and 50% soft landscape. The soft landscape comprises 100% planting; planting at 400 mm ccs – 2–3 litre plants per m^2; Trees at 1 nr 16–18 cm girth per 100 m^2 throughout the parking and landscape area		
macadam parking area	100 m^2	4100.00
concrete block paved parking area	100 m^2	5800.00
External area where the area is 50% parking and 50% soft landscape. The soft landscape comprises 100% planting; planting at 400 mm ccs – 2–3 litre plants per m^2; Trees at 1 nr 16–18 cm girth per 100 m^2 throughout the parking and landscape area; imported topsoil to planting areas 400 mm thick inclusive of removal of existing material and disposal offsite		
macadam parking area	100 m^2	6000.00
concrete block paved parking area	100 m^2	7800.00

COST MODELS

Item	Unit	Range £
Housing and house builder – cont		
Green infrastructure to new housing development; Comprising public greenspace, play and amenity areas, mitigation of new internal spine roads; Landscaping to individual properties not included; Prices shown relate to the green infrastructure space area indicated.		
Rural locations or not in major metropolitan areas; average property size 350 m^2; Landscape impact to suit development for first time buyers; amenity areas catering for families with young children; Tree planting of 1 nr 18–20 tree per 100 m^2; all other planting shown separately		
Rural location with 100 properties; public space, green buffers and green mitigation comprises 25% of the total area; Total site area 60000 m^2; amenity space buffers and accesses 30000 m^2; Green infrastructure space – 15000 m^2; (50%) 800 m^2 of fenced equipped play area	m^2	41.00
Economic housing – first time buyers; amenity catering for young children; Total site area 13500 m^2; public space and accesses; 5100 m^2 ; Green space including footways – 4500 m^2; fenced and gated green area with benches and litterbins with LEAP compliant play equipment	m^2	47.00
Economic housing – first time buyers; amenity catering for young children; Total site area 13500 m^2; public space and accesses; 5100 m^2 ; Green space including footways – 4500 m^2; fenced and gated green area with benches and litterbins with no play equipment; single mini football goal	m^2	41.00
Soft landscape boundary planting to housing developments; Perimeter planting only. For more soft landscape permutations please refer to the External Planting section at 8.3.2 of this book		
Native species hedging 800 mm wide	m	24.00
Native species hedging 1.20 m wide	m	36.00
Native species hedging 1.60 m wide	m	48.00
Green infrastructure and access; fenced isolated site of 25000 m^2; comprising 35 properties; prices shown include for creation of roadways 5.5 m wide with 500 mm roadway buildup inclusive of kerbs and soft verge 1 side; Allowance for development is 16250 m^2; Roadways 2300 m^2; footways 2 m wide 415 m (430 m^2); All costs to sales related development areas excluded; services excluded		
No allowance for community shared space; Seeded footway 2.00 m wide one side only; Trees 14–16 at 1 tree per 15 lm		
greenfield site	1000 m^2	10500.00
brownfield site	1000 m^2	13000.00
Public open space internally comprising 5% of the site; with graded and levelled bow fenced and gated play area for informal ball games; grass cutting for 3 years at 22 times per year		
greenfield site	1000 m^2	14000.00
brownfield site	1000 m^2	13000.00

COST MODELS

Item	Unit	Range £
Green infrastructure for housing development; Existing greenfield site; Enhancement of existing woodland boundary; Structural enhancement of existing native species boundary to one side; New native tree and shrub boundary enhancement based on tree consultant report; Native species hedge regeneration and gapping; New accent tree planting along main vehicle access path; New occasional tree planting in public open spaces; creation of enhanced recreation open spaces with 5 year development size planting; Planting and turf/Seeded/wildflower areas to open space and arrival areas; Unit frontages comprising ornamental hedges and ornamental planting; Wildflower open areas; Seeded open space areas; Play areas NEAPs and LEAPS. Private internal gardens excluded. Plan area assumed as square		
Arboriculture and clearances		
to one side of development 400 m	400 m	3475.00
to entire perimeter of development	1600 m	12000.00
Existing boundary tree works and planting enhancements		
Tree felling and tree removal; works as recommend by arboricultural consultant	400 m	5500.00
New native planting along boundary		
Structural whips and shrubs to reinforce existing wooded boundary at 1 whip / 100 m^2 and 25% planting reinfocement of native shrubs; all planting in protective tubes; total area shown – only 25% planted	4000 m^2	14500.00
Turfing Seeding and Wildflower to public openspace		
Open greenspace areas 95% of green infrastructure; cultivate and prepare for seeding and surfing to public open space areas in proportions of Seeded: Turf: Wildflower –70:10:20: (7.6 ha)	76000 m^2	560000.00
Ornamental trees within development; allowances for varying sizes of structural or ornamental trees based on placement along drive routes; and within proximities of dwelling units		
Trees of 14–16 cm girth inclusive of timber stakes and protection with watering tubes and backfilling with composts and fertilizers	75 trees	26000.00
Trees of 16–18 cm girth inclusive of timber stakes and protection with watering tubes and backfilling with composts and fertilizers	75 trees	35500.00
Ornamental planting and hedges at unit frontages, parking areas development entrances, and amenity areas		

COST MODELS

Item	Unit	Range £
Science and Office parks		
Business or Science park models		
Generally areas of more dense landscape incorporating relaxation areas and larger planting schemes; planting is generally more mature at implementation with larger trees; Planting schemes often incorporate larger swathes of groundcovers to give graded views over landscaped areas; planting density and sizes are generally of higher standard than housing		
Soft landscaped space only space; planting: turf ratio's as shown; planting sizes and densities as indicated; 1 tree of 25–30 cm girth at 1 tree per 100 m^2 overall external area; Mulch and watering after planting		
Landscaped area; Turf and trees only; imported topsoil to turf areas at 50 thick;	100 m^2	2500.00
Landscaped area 70: 30 turf to planted area with imported topsoil to turf areas at 50 thick;	100 m^2	3375.00
Landscaped area 50: 50 turf to planted area; imported topsoil to turf areas at 50 thick;	100 m^2	3675.00
Landscaped area; Planting and trees only; Planting of shrubs 100% at 400 mm centres	100 m^2	4300.00
Landscaped area; Planting and trees only; Planting of shrubs and groundcovers 100% Shrubs: Groundcovers 70:30 at 400:250 mm centres	100 m^2	4825.00
Landscaped area; Planting and trees only; Planting of shrubs and groundcovers 100% Shrubs: Groundcovers 30:70 at 400:250 mm centres	100 m^2	5500.00
Mixed hard and soft landscaped space with parking; 15% of the external space are pathways or building surrounds paved with good quality textured concrete pavings. The remaining area of 85% of the external area is divided into 50: 50 Soft landscape: Parking; 1 tree 25–30 cm per 100 m^2 of external area; Parking area treated as described		
External area where 50% of the remaining external area is parking area; with surround of PC kerbs and 50% soft landscape of 70% turf and 30% planted area; imported topsoil to turf areas at 50 thick;		
Macadam parking area	100 m^2	9800.00
Concrete block paved parking area	100 m^2	11500.00
The soft landscape is 50% turf and 50% planting; planting at 400 mm ccs – 2–3 litre plants per m^2;		
macadam parking area	100 m^2	10500.00
concrete block paved parking area	100 m^2	12000.00
The soft landscape comprises 100% planting; planting at 400 mm ccs – 2–3 litre plants per m^2;		
macadam parking area	100 m^2	9800.00
concrete block paved parking area	100 m^2	11500.00
External area where the area is 50% parking and 50% soft landscape. The soft landscape comprises 100% planting; planting at 400 mm ccs – 2–3 litre plants per m^2; imported topsoil to planting areas 400 mm thick inclusive of removal of existing material and disposal offsite		
macadam parking area	100 m^2	11000.00
concrete block paved parking area	100 m^2	13000.00

COST MODELS

Item	Unit	Range £
Civic environments and Streetscape		
Civic and Public areas in the current climate encompass higher quality surfaces for public access. Planting is generally larger. There is a need for furniture and lighting;		
Civic town centre area of paved open space; model area 500 m²		
General open area; One 30–35 cm tree with root cell planting system every 400 m²; Metal tree grid; One bench per 200 m² and one bin per 500 m²; Granite kerb surround		
Conservation concrete slab paved PC £48.41/m²	100 m²	18000.00
Granite paved	100 m²	32000.00
Granite paved with stepped terraces. 3 steps of 10 m long each in an area of 500 m²	100 m²	37000.00
Diamond sawn Yorkstone paved	100 m²	34000.00
Granite paved with stepped terraces. 3 steps of 10 m long each in an area of 500 m²	100 m²	39000.00
General open area; one 30–35 cm tree with root cell planting system every 200 m²; Metal tree grid; One bench per 200 m² and one bin per 500 m²; Granite kerb surround		
Granite paved	100 m²	35000.00
Granite paved with stepped terraces. 3 steps of 10 m long each in an area of 500 m²	100 m²	40500.00
Diamond sawn Yorkstone paved	100 m²	37000.00
Granite paved with stepped terraces. 3 steps of 10 m long each in an area of 500 m²	100 m²	42000.00
Hard landscaped civic streetscape space of 1000 m² plan area with paved amenity areas comprising kerb surrounded paved area with street furniture. Raised beds filled with planting; trees of 30–35 cm at 100 m² centres.		
Raised seating walls surrounding planted area with bollard lights at 100 m² centres; planting of low box hedges and shrubs comprising 10% of the total area		
Reconstituted stone slab paved	100 m²	23000.00
Granite paved	100 m²	37000.00
Yorkstone paved	100 m²	39000.00
Sunken area with long bullnosed steps with bollard and in-ground lighting; Planting area comprising 15% of the available area with box hedging and annual bedding plants and associated maintenance for 1 year.		
Reconstituted stone slab paved	100 m²	31000.00
Yorkstone paved	100 m²	40500.00
Hard landscaped civic streetscape space of 1000 m² plan area with paved amenity areas comprising kerb surrounded paved area with street furniture. Raised water feature of 25 m²; lighting bollards; Trees of 30–35 cm at 10 m centres to the perimeter		
Sunken area with long bullnosed steps with bollard lighting; Planting area comprising 15% of the available area with box hedging and annual bedding plants and associated maintenance for 1 year.		
Reconstituted stone slab paved	100 m²	33000.00
Yorkstone paved	100 m²	42000.00

COST MODELS

Item	Unit	Range £
Playgrounds		
New Lappset Play area 1200 m^2; Sub to one of above; Play area comprising play equipment in landscaped and fenced park environment; Picket fencing with 2 × swing gates, macadam pathway, and planting to 10% of the park area on the perimeter; safety surfaces as required by play equipment supplier; Gates (2x) single Tree 20–25 cm; bins and benches. Climbing and play tower Q10945, ABC play tower Q10159, Tractor rocker 010515; Pig rocker O10516; Octopus seasaw; Q10259; Froggy spring equipment; Swing with cradle seats		
Total area 1200 m^2	nr	105000.00
equivalent budget cost	m^2	84.00
Urban Play area Lappset Ltd; comprising fenced play equipment for children to 12 years; Cushioned fall play equipment as specified by equipment manufacturer. Bow fenced all sides with benches and litter bins; appropriate signage; Traditional equipment swings slides and spring equipment; 6 activities; Safety surfacing below and surrounding equipment; 3 benches and litter bin; turfed where not surfaced		
Local parish traditional equipment	1200 m^2	56000.00
City and urban medium sized play equipment; Modern design climbing, zipwire net climbing and medium sized theme multiplay units Lapset	1200 m^2	105000.00
City and urban but with larger multiplay themed play equipment	1500 m^2	185000.00
Mixed themed and traditional play area with fitness elements and adventure climbing equipment	3000 m^2	205000.00
LEAP Play areas for compliance with housing development; 35–100 units; Lappset Ltd; Brownfiled sites allow for additional stripping and replacement of existing assumed contaminated surface to 100 mm; Local Equipped Area for Play (LEAP) is a play area equipped for children of early school age (mainly 4–8 year olds). The activity zone should have a minimum area of 400 m^2, with grass playing space and at least five types of play equipment with appropriate safety surfacing. There should also be seating for accompanying adults. Contains at least 5 types of play equipment; Contains seating for parents and/or carers; Contains a litter bin signage fencing;		
minimum requirement	nr	49000.00
minimum requirement but with green seeded kickabout area	nr	55000.00
minimum requirement brownfield site	nr	59000.00
NEAP Play area for compliance for housing developers; Lappset Ltd; Brownfiled sites allow for additional stripping and replacement of existing assumed contaminated surface to 100 mm; A Neighbourhood Equipped Area for Play (NEAP) is a play area for 8–14 year olds which should include a grassed kickabout area, a hard surfaced area for ball games or wheeled activities, 8 types of play equipment appropriate to children in this age group and seating, including a youth shelter. This requires an activity zone of at least 1000 m^2.		
minimum requirement	nr	81000.00
minimum requirement brownfield site	nr	91000.00

COST MODELS

Item	Unit	Range £
Domestic Cost models		
Suburban new or refurbished landscape comprising; Driveways and parking; Recreational pavings; Access pavings; Perimeter fencing; Trees; Planting beds and turf; Measured at per 1000 m² of total site area		
Total property size; 3300 m² of which building footprint comprises 280 m² (8.5%) Access and parking 330 m² (10%); Closeboard fencing to perimeter; Balance of landscape (2700 m² +/- 80%) area as follows		
Hard recreation and access areas/soft landscaped areas; standard size planting	1000 m²	140000.00
Hard recreation and access areas/soft landscaped areas; with enhanced planting and larger trees	1000 m²	175000.00
Hard recreation and access areas/soft landscaped areas; with enhanced planting and larger trees but with brick entrance walls gates and piers	1000 m²	185000.00
Hard recreation and access areas/soft landscaped areas; with enhanced planting and larger trees but with brick entrance walls gates and piers and with a lighting scheme	1000 m²	195000.00
Hard recreation and access areas/soft landscaped areas; with enhanced planting and larger trees and with brick entrance walls gates and piers and with a lighting scheme; but with stone paved steps to the patio 5 steps high × 2.00 m wide	1000 m²	200000.00
Hard recreation and access areas/soft landscaped areas; with enhanced planting and larger trees and with brick entrance walls gates and piers and with a lighting scheme and with stone paved steps to the patio 5 steps high × 2.00 m wide but with timber pergolas and water feature	1000 m²	205000.00
Parking areas		
Parking area to office block; Surface of carpark stated below inclusive of all excavations bases and surface		
Macadam 625 m²; 200 Type 1 base		
Open space with kerb surround 25 × 25 m	nr	44500.00
With trees 18–20 cm girth at 10.00 m centres;	nr	50000.00
With trees 18–20 cm girth at 10.00 m centres and planting to 1.00 m boundary perimeter (external to the parking area on 3 sides – 75 m²)	nr	53000.00
With trees 18–20 cm at 10 m centres; planting to 1.00 m boundary and Marshalls Bega asymmetrical pole lighting 6.00 high	nr	66000.00
Concrete block surface of Marshalls Priora on 200 mm Type 3 base; 625 m²		
Open space with kerb surround 25 × 25 m	nr	66000.00
With trees 18–20 cm girth at 10.00 m centres;	nr	71000.00
With trees 18–20 cm girth at 10.00 m centres and planting to 1.00 m boundary perimeter; (external to the parking area on 3 sides – 75 m²)	nr	75000.00
With trees 18–20 cm at 10 m centres; planting to 1.00 m boundary and Marshalls Bega asymmetrical pole lighting 6.00 high	nr	87000.00
Clay brick; Vande Moortel; butt jointed on 200 mm Type 1 base 625 m²		
SeptimA; Vanilla; 215 × 52 × 70 mm	nr	110000.00
SeptimA; Vanilla 215 × 52 × 72 mm including trees 18–20 cm girth at 10.00 m centres;	nr	115000.00
SeptimA; Vanilla 215 × 52 × 72 mm including trees 18–20 cm girth at 10.00 m centres and planting to 1.00 m boundary perimeter; (external to the parking area on 3 sides – 75 m²)	nr	120000.00

COST MODELS

Item	Unit	Range £
Parking areas – cont		
Parking area to office block – cont		
Clay brick – cont		
SeptimA; Vanilla 215 × 52 × 72 mm including trees 18–20 cm at 10 m centres; planting to 1.00 m boundary and Marshalls Bega asymmetrical pole lighting 6.00 high	nr	130000.00
Recycled Suds; SudsTech; on 200 mm Type 3 base 625 m^2		
Open space with kerb surround 25 × 25 m	nr	88000.00
With trees 18–20 cm girth at 10.00 m centres;	nr	93000.00
With trees 18–20 cm girth at 10.00 m centres and planting to 1.00 m boundary perimeter; (external to the parking area on 3 sides – 75 m^2)	nr	96000.00
With trees 18–20 cm at 10 m centres; planting to 1.00 m boundary and Marshalls Bega asymmetrical pole lighting 6.00 high	nr	110000.00
Landscape maintenance		
Annual landscape maintenace to public spaces or landscapes; inner city housing with shared amenity space; excludes private gardens		
Frequency of visits;		
The following are the most common frequencies of vists on long termmaintence regimes allowing as follows		
Fortnightly in summer and monthly in winter		
18 visits		
Weekly in summer and monthly in winter		
26 visits		
Weekly in summer and fortnightly in winter		
32 visits		
Weekly throughout the year; maintenance of hard surfaces and litter only in December to February		
52 visits		
Grass cutting; maintenance of sight lines hedge cutting; shrub beds care and weeding; path sweeping; amenity area; paved areas including street furniture; Leaf clearance and litter picking on visits only; includes gritting – 6 occassions; and snow clearance as necessary; Annual mulch top-up;		
Hard surface 70%; Planting 20%; Hedges 10%		
32 visits	1000 m^2	15000.00
26 visits	1000 m^2	12500.00
18 visits	1000 m^2	10500.00
Hard surface 70%; Planting 20%; Hedges 10%; but excluding risk items for snow gritting and snow clearance		
32 visits	1000 m^2	11500.00
26 visits	1000 m^2	9600.00
18 visits	1000 m^2	7200.00

COST MODELS

Item	Unit	Range £
Hard surface 50%; Planting 20%; Hedges 10%; Turf 20%;		
32 visits	1000 m^2	13000.00
26 visits	1000 m^2	11000.00
18 visits	1000 m^2	8800.00
Hard surface 50%; Planting 20%; Hedges 10%; Turf 20%; but excluding risk items for snow gritting and snow clearance		
32 visits	1000 m^2	11000.00
26 visits	1000 m^2	9100.00
18 visits	1000 m^2	7000.00
Green roofs and Podiums		
INTENSIVE GREEN ROOF		
Install green roof components to podium at ground level or where whichnery has been craned for material movement; Substructure podium base waterproofing and excess water drainage not included. Containment system to podium assumed as installed to perimeter by others with 1.00 m wide service area. Allows for Irrigation system of £80.000 including pump room commissioning.		
Podiums with Turf 75% – 50% of total area		
Growing medium 400 mm deep with 75% Turf and 25% planting area ground cover plants: small shrubs at 50: 50 each of the available planting space at ground level with metal edge containment	1000 m^2	210000.00
Inclusive of street furniture; benches at 200 m^2 centres and bins at 500 m^2 centres Growing medium 400 mm deep with 75% Turf and 25% planting area ground cover plants: small shrubs at 50: 50 each of the available planting space at ground level with metal edge containment	1000 m^2	210000.00
Podiums inclusive of Hard landscaping		
Granite paving flags on pedestals to 25% of the area Turf to 50% and planting to 25%; Growing medium 400 mm deep with planting area ground cover plants: small shrubs at 50: 50 each of the available planting space at ground level with metal edge containment.	1000 m^2	260000.00
Marshalls concrete paving flags on pedestals to 25% of the area Turf to 50% and planting to 25%; Growing medium 400 mm deep with planting area ground cover plants: small shrubs at 50: 50 each of the available planting space at ground level with metal edge containment.	1000 m^2	250000.00
Yorkstone paving flags on pedestals to 25% of the area Turf to 50% and planting to 25%; Growing medium 400 mm deep with planting area ground cover plants: small shrubs at 50: 50 each of the available planting space at ground level with metal edge containment.	1000 m^2	265000.00
Podiums with paved areas and enhanced planting; Paving to 50–75 of total area; Tree planting in raised granite clad planters of Stepoc blocks; 10 nr semi mature trees		
Marshalls flag paving 75% of total area; Growing medium 400 deep with 25% planting area ground cover plants: small shrubs at 50: 50 each of the available planting space at ground level with metal edge containment; 10 nr trees 25–30 cm in raised blockwork clad containers, 1.20 × 1.20 × 1.00 Seating and bins included	1000 m^2	520000.00

COST MODELS

Item	Unit	Range £
Green roofs and Podiums – cont		
Install green roof components to podium at ground level or where whichnery has been craned for material movement – cont		
Podiums inclusive of Bega lighting standards		
Marshalls flag paving 75% of total area; ducting cabling and lighting standards at 15 m (4.44 units /1000 m^2) centres; Growing medium 400 mm deep with 25% planting area ground cover plants: small shrubs at 50: 50 each of the available planting space at ground level with metal edge containment; 10 nr trees 25–30 cm in raised blockwork clad containers; 1.20 × 1.20 × 1.00 Seating and bins included	1000 m^2	540000.00

PRELIMINARIES

Item	Unit	Range £
PRELIMINARY COST MODELS		
Preliminaries for a project value £100 000.00: Project length 6–8 weeks; Contract let as main contract; If the works are let as a subcontract some of the items listed below will be provided by the main contractor		
Management preliminaries		
Health and safety		
Health and safety file	nr	760.00
H&S Consultant visit – monthly	nr	250.00
Accommodations		
Office; welfare; Drying room; storage and toilet including collection and delivery costs	nr	1250.00
Site meetings		
Attendance by Contracts manager; monthly	nr	400.00
Services		
Temporary water supply and electrical connections	nr	770.00
Other sundry site costs		
Signage	nr	330.00
Site setup costs and demobilization costs	nr	850.00
Drawing duplication and printing costs	nr	250.00
Plant preliminaries		
Hired in plant		
Deliveries and collections	nr	550.00
Safety fencing		
Heras fencing inclusive of delivery and collection; Site dimensions 30 m × 30 m	nr	2375.00
Fuel bowser	nr	540.00
Site skips	nr	900.00
Allowances for site plant standing time	nr	2525.00
Labour preliminaries		
Labour costs not task associated		
Unloading time	nr	1125.00
Site cleaning costs	nr	640.00
Non productive time	nr	1225.00
Health and safety inductions	nr	425.00
Material preliminaries		
Costs of materials that are not task associated		
Surface protection; boards or mats for longer term contracts	nr	1175.00
General delivery allowances on materials not allowed elsewhere	nr	440.00
Performance bond; 10%; Based on Job value of £100,000.00	nr	830.00
Tree root zone protection	nr	1675.00
Samples	nr	385.00
Testing; Soil tests CBR and the like	nr	440.00
TOTAL FOR ALL THE ABOVE	nr	20000.00

PRELIMINARIES

Item	Unit	Range £
PRELIMINARY COST MODELS – cont		
Preliminaries for a project value £300 000.00: Project length 26–30 weeks; Contract let as main contract; If the works are let as a subcontract some of the items listed below will be provided by the main contractor		
Management preliminaries		
Health and safety		
Health and safety file	nr	760.00
H&S Consultant visit – monthly	nr	3000.00
Accommodations		
Office; welfare; Drying room; storage and toilet including collection and delivery costs	nr	3575.00
Site meetings		
Attendance by Contracts manager; monthly	nr	1425.00
Services		
Temporary water supply and electrical connections	nr	770.00
Internet connection	nr	660.00
Other sundry site costs		
Signage	nr	330.00
Site setup costs and demobilization costs	nr	850.00
Security guard; nights and weekends	nr	81000.00
Road cleaning machine hire fortnightly; during soft landscape portion of the works	nr	1450.00
Drawing duplication and printing costs	nr	345.00
Plant preliminaries		
Hired in plant		
Deliveries and collections	nr	2200.00
Heras fencing inclusive of delivery and collection; Site dimensions 30 m × 30 m	nr	12500.00
Fuel bowser	nr	2350.00
Site skips	nr	2975.00
Allowances for site plant standing time	nr	4225.00
Labour preliminaries		
Labour costs not task associated		
Unloading time	nr	2850.00
Site cleaning costs	nr	4275.00
Non productive time	nr	4275.00
Health and safety inductions	nr	850.00
Material preliminaries		
Costs of materials that are not task associated		
Surface protection; boards or mats for longer term contracts	nr	1975.00
General delivery allowances on materials not allowed elsewhere	nr	1750.00
Performance bond; 10%; Based on Job value of £300,000.00	nr	3300.00
Tree root zone protection	nr	1675.00
Samples	nr	880.00
Testing; Soil tests CBR and the like	nr	1750.00
TOTAL FOR ALL THE ABOVE	nr	60000.00

PRELIMINARIES

Item	Unit	Range £
Preliminaries for a project value £300 000.00: Project length 26–30 weeks; Contract let as subcontract;		
Management preliminaries		
Health and safety		
Health and safety file	nr	760.00
H&S Consultant visit – monthly	nr	3000.00
Site meetings		
Attendance by Contracts manager; monthly	nr	1425.00
Other sundry site costs		
Road cleaning machine hire fortnightly; during soft landscape portion of the works	nr	1450.00
Drawing duplication and printing costs	nr	345.00
Plant preliminaries		
Hired in plant		
Deliveries and collections	nr	2200.00
Heras fencing inclusive of delivery and collection; Site dimensions 30 m × 30 m	nr	12500.00
Fuel bowser	nr	2350.00
Site skips	nr	2975.00
Allowances for site plant standing time	nr	4225.00
Labour preliminaries		
Labour costs not task associated		
Unloading time	nr	2850.00
Site cleaning costs	nr	4275.00
Non productive time	nr	4275.00
Health and safety inductions	nr	850.00
Material preliminaries		
Costs of materials that are not task associated		
Surface protection; boards or mats for longer term contracts	nr	1975.00
General delivery allowances on materials not allowed elsewhere	nr	1750.00
Performance bond; 10%; Based on Job value of £300,000.00	nr	3300.00
Tree root zone protection	nr	1675.00
Samples	nr	880.00
Testing; Soil tests CBR and the like	nr	1750.00
TOTAL FOR ALL THE ABOVE	nr	53000.00

PRELIMINARIES

Item	Unit	Range £
CONTRACT ADMINISTRATION AND MANAGEMENT		
Prepare tender bid for external works or landscape project; measured works contract; inclusive of bid preparation, provisional program and method statements		
Measured works bid; measured bills provided by the employer; project value		
£30,000.00	nr	770.00
£50,000.00	nr	1275.00
£100,000.00	nr	2850.00
£200,000.00	nr	4525.00
£500,000.00	nr	5800.00
£1,000,000.00	nr	8400.00
Lump sum bid; scope document, drawings and specification provided by the employer; project value		
£30,000.00	nr	1700.00
£50,000.00	nr	2175.00
£100,000.00	nr	4325.00
£200,000.00	nr	7500.00
£500,000.00	nr	10500.00
£1,000,000.00	nr	15500.00
Prepare and maintain Health and Safety file; prepare risk and Coshh assessments, method statements, works programmes before the start of works on site; project value		
£35,000.00	nr	345.00
£75,000.00	nr	530.00
£100,000.00	nr	700.00
£200,000.00 – £500,000.00	nr	1750.00
Contract management		
Site management by site surveyor carrying out supervision and administration duties only; inclusive of oncosts		
full time management	week	1200.00
managing one other contract	week	600.00
Parking		
Parking expenses where vehicles do not park on the site area; per vehicle		
metropolitan area; city centre	week	200.00
metropolitan area; outer areas	week	160.00
suburban restricted parking areas	week	40.00
Congestion charging		
London only	week	58.00

PRELIMINARIES

Item	Unit	Range £
SITE SETUP AND SITE ESTABLISHMENT		
Site fencing; supply and erect temporary protective fencing and remove at completion of works		
Cleft chestnut paling		
1.20 m high 75 mm larch posts at 3 m centres	100 m	2250.00
Heras fencing		
2 week hire period	100 m	1200.00
4 week hire period	100 m	1575.00
8 week hire period	100 m	2350.00
12 week hire period	100 m	3125.00
16 week hire period	100 m	3950.00
24 week hire period	100 m	5500.00
Site compounds		
Establish site administration area on reduced compacted Type 1 base; 150 mm thick; allow for initial setup of site office and welfare facilities for site manager and maximum 8 site operatives		
temporary base or hard standing; 20 × 10 m	200 m²	4425.00
To hardstanding area above; erect office and welfare accommodation; allow for furniture, telephone connection, and power; restrict access using temporary safety fencing to the perimeter of the site administration area		
Establishment; delivery and collectioncosts only		
site office and chemical toilet; 20 m of site fencing	nr	1125.00
site office and chemical toilet; 40 m of site fencing	nr	1225.00
site office and chemical toilet; 60 m of site fencing	nr	1350.00
Weekly hire rates of site compound equipment		
site office 2.4 × 3.6 m; chemical toilet; 20 m of site fencing	week	130.00
site office 4.8 × 2.4 m; chemical toilet; 20 m of site fencing; armoured store 2.4 × 3.6 m	week	145.00
site office 2.4 × 3.6 m; chemical toilet; 40 m of site fencing	week	170.00
site office 2.4 × 3.6 m; chemical toilet; 60 m of site fencing	week	205.00
Removal of site compound		
Remove base; fill with topsoil and reinstate to turf on completion of site works	200 m²	2400.00
Surveying and setting out		
Note: In all instances a quotation should be obtained from a surveyor. The following cost projections are based on the costs of a surveyor using EDM (electronic distance measuring equipment) where an average of 400 different points on site are recorded in a day. The surveyed points are drawn in a CAD application and the assumption used that there is one day of drawing up for each day of on-site surveying.		

PRELIMINARIES

Item	Unit	Range £
SITE SETUP AND SITE ESTABLISHMENT – cont		
Surveying and setting out – cont		
Survey existing site using laser survey equipment and produce plans of existing site conditions		
site 1,000 to 10,000 m^2 with sparse detail and features; boundary and level information only; average five surveying stations	nr	750.00
site up to 1,000 m^2 with dense detail such as buildings, paths, walls and existing vegetation; average five survey stations	nr	1500.00
site up to 4000 m^2 with dense detail such as buildings, paths, walls and existing vegetation; average 10 survey stations	nr	2250.00

PRELIMINARIES

Item	Unit	Range £
PROJECT PRELIMINARY COST MODELS		
Preliminaries for a project value £100 000.00: Project length 6–8 weeks; Contract let as main contract; If the works are let as a subcontract some of the items listed below will be provided by the main contractor		
Management preliminaries		
Health and safety		
Health and safety file	nr	760.00
H&S Consultant visit – monthly	nr	225.00
Accommodations		
Office; welfare; drying room; storage and toilet including collection and delivery costs	nr	1075.00
Site meetings		
Attendance by Contracts manager; monthly	nr	345.00
Services		
Temporary water supply and electrical connections	nr	700.00
Internet connection	nr	–
Telephone	nr	–
Other sundry site costs		
Signage	nr	300.00
Site setup costs and demobilization costs	nr	750.00
Security guard	nr	–
Road cleaning machine hire fortnightly; during soft landscape portion of the works	nr	–
Drawing duplication and printing costs	nr	215.00
Plant preliminaries		
Hired in plant		
Deliveries and collections	nr	500.00
Safety fencing		
Heras fencing inclusive of delivery and collection; Site dimensions 30 m × 30 m	nr	2075.00
Fuel bowser	nr	475.00
Site skips	nr	780.00
Allowances for site plant standing time	nr	2200.00
Site logistics	nr	–
Temporary lighting	nr	–
Labour preliminaries		
Labour costs not task associated		
Unloading time	nr	990.00
Site cleaning costs	nr	560.00
Non productive time	nr	1075.00
Health and safety inductions	nr	370.00

PRELIMINARIES

Item	Unit	Range £
PROJECT PRELIMINARY COST MODELS – cont		
Material preliminaries		
Costs of materials that are not task associated		
Surface protection; boards or mats for longer term contracts	nr	1025.00
General delivery allowances on materials not allowed elsewhere	nr	400.00
Performance bond; 10%; Based on Job value of £100,0000.00	nr	750.00
Tree root zone protection	nr	1425.00
Protection of existing building features; boards	nr	–
Samples	nr	350.00
Testing; Soil tests CBR and the like	nr	400.00
TOTAL FOR ALL THE ABOVE	nr	17500.00
Preliminaries for a project value £300 000.00: Project length 26–30 weeks; Contract let as main contract; If the works are let as a subcontract some of the items listed below will be provided by the main contractor		
Management preliminaries		
Health and safety		
Health and safety file	nr	760.00
H&S Consultant visit – monthly	nr	2700.00
Accommodations		
Office; welfare; drying room; storage and toilet including collection and delivery costs	nr	3150.00
Site meetings		
Attendance by Contracts manager; monthly	nr	1225.00
Services		
Temporary water supply and electrical connections	nr	700.00
Internet connection	nr	600.00
Telephone	nr	–
Other sundry site costs		
Signage	nr	300.00
Site setup costs and demobilization costs	nr	750.00
Security guard; nights and weekends	nr	81000.00
Road cleaning machine hire fortnightly; during soft landscape portion of the works	nr	1200.00
Drawing duplication and printing costs	nr	305.00
Plant preliminaries		
Hired in plant		
Deliveries and collections	nr	2000.00
Heras fencing inclusive of delivery and collection; Site dimensions 30 m × 30 m	nr	11000.00
Fuel bowser	nr	2025.00
Site skips	nr	2600.00
Allowances for site plant standing time	nr	3675.00
Site logistics	nr	–
Temporary lighting	nr	–

PRELIMINARIES

Item	Unit	Range £
Labour preliminaries		
Labour costs not task associated		
Unloading time	nr	2475.00
Site cleaning costs	nr	3675.00
Non productive time	nr	3675.00
Health and safety inductions	nr	750.00
Material preliminaries		
Costs of materials that are not task associated		
Surface protection; boards or mats for longer term contracts	nr	1700.00
General delivery allowances on materials not allowed elsewhere	nr	1600.00
Performance bond; 10%; Based on Job value of £300,0000.00	nr	3000.00
Tree root zone protection	nr	1425.00
Protection of existing building features; boards	nr	–
Samples	nr	800.00
Testing; Soil tests CBR and the like	nr	1600.00
TOTAL FOR ALL THE ABOVE	nr	52000.00
Preliminaries for a project value £300 000.00: Project length 26–30 weeks; Contract let as subcontract;		
Management preliminaries		
Health and safety		
Health and safety file	nr	760.00
H&S Consultant visit – monthly	nr	2700.00
Site meetings		
Attendance by Contracts manager; monthly	nr	1225.00
Services		
Internet connection	nr	600.00
Telephone	nr	–
Other sundry site costs		
Road cleaning machine hire fortnightly; during soft landscape portion of the works	nr	1200.00
Drawing duplication and printing costs	nr	305.00
Plant preliminaries		
Hired in plant		
Deliveries and collections	nr	2000.00
Heras fencing inclusive of delivery and collection; Site dimensions 30 m × 30 m	nr	11000.00
Fuel bowser	nr	2025.00
Site skips	nr	2600.00
Allowances for site plant standing time	nr	3675.00
Site logistics	nr	–
Temporary lighting	nr	–
Labour preliminaries		
Labour costs not task associated		
Unloading time	nr	2475.00
Site cleaning costs	nr	3675.00
Non productive time	nr	3675.00
Health and safety inductions	nr	750.00

PRELIMINARIES

Item	Unit	Range £
PROJECT PRELIMINARY COST MODELS – cont		
Material preliminaries		
Costs of materials that are not task associated		
Surface protection; boards or mats for longer term contracts	nr	1700.00
General delivery allowances on materials not allowed elsewhere	nr	1600.00
Performance bond; 10%; Based on Job value of £300,0000.00	nr	3000.00
Tree root zone protection	nr	1425.00
Protection of existing building features; boards	nr	–
Samples	nr	800.00
Testing; Soil tests CBR and the like	nr	1600.00
TOTAL FOR ALL THE ABOVE	nr	47000.00

2.3 GREEN ROOFS

Item	Unit	Range £
GREEN ROOF NOTES		
All items below do not include for time to deliver the materials to the point of installation		
EXTENSIVE PITCH ROOF		
Zinco Green Roof Systems Ltd; suitable for planting with Sedum plugs or overlaying with pre-grown cultivated Sedum Carpet. Waterproofing layer must contain a root barrier as a separate root barrier cannot be used on pitched roofs		
System saturated weight – 115 kg/m^2; system water storage capacity – 38 L/m^2; Slopes 10–25°		
system depth 130 mm	m^2	110.00
Pitched roofs with slopes from 25° up to 35°; System height – 120 mm; system saturated weight – 155 kg/m^2; water storage capacity – 64 L/m^2 (depending on slope)		
system depth 120 mm	m^2	110.00
EXTENSIVE FLAT ROOF		
Extensive Sedum type Green Roof System; Zinco Green Roof Systems Ltd. System for planting with Sedum plugs or overlaying with pre-grown cultivated Sedum Carpet. (ETA 13/0668)		
Flat roofs with no standing water up to 10° or for Inverted roofs and as an alternative to single layer systems. Drainage, water retention and protection mat with attached filter sheet. 20 mm thick.		
system depth 120 mm	m^2	285.00
Extensive Rockery type Green Roof System; For planting with Sedum plugs or rockery type plants; flat roofs with no standing water up to 10°. (Not suitable for inverted roofs); Suitable system depth 90 mm. System saturated weight 110 kg/m^2; system water storage capacity: 36 L/M2; (ETA 13/0668)		
Suitable system depth 90 mm. System saturated weight 110 kg/m^2; system water storage capacity: 36 L/M2;		
system depth 90 mm; planted with sedum mat	m^2	300.00
system depth 90 mm; planted with wildflower	m^2	265.00
Zero° roof		
system depth 150 mm; planted with sedum mat	m^2	295.00
system depth 150 mm; planted with wildflower	m^2	260.00

2.3 GREEN ROOFS

Item	Unit	Range £
INTENSIVE GREEN ROOF		
Flat roofs up to 10°; Intensive Simple and Wildflower Green Roof System; Zinco Green Roof Systems Ltd; (not suitable for Inverted roofs); for planting with small shrubs, blooming perennials and wildflowers; can be combined with patios and walkways if required; saturated weight – 145 kg/m^2–245 Kg/m^2; water storage capacity – 62 L/m^2 to 80 L/m^2. (ETA 13/0668)		
Heather and lavender planting system		
system depth – 145 mm–200 mm;	m^2	250.00
Small shrubs		
system depth – 500 mm;	m^2	200.00
Turf		
system depth 300 mm	m^2	115.00
'LAWN' Green Roof System; Zinco Green Roof Systems Ltd. Lawns on roofs where build up depth is restricted; Flat roofs – zero degrees up to 5°; Lawn surface treatment; System depth 150 mm system saturated weight min 165 Kg/m^2. System water storage capacity; from 65 L/m^2		
Drip irrigated lawn system (excludes controller)		
system depth 150 mm	m^2	100.00
BLUE ROOFS		
Blue roof systems; ZinCo Green roof systems Ltd; (Stormwater management roof); For managment of stormwater under a green roof system; buildup of Blue roof system over roof surface with water proofing (not included) and root resistant protection		
Extensive Blue roof		
200 mm thick	m^2	64.00
300 mm thick	m^2	140.00
Intensive Blue roof		
200 mm thick	m^2	82.00
200 mm thick	m^2	160.00
Flow control unit and inspection chamber to the systems above		
RDS 28	m^2	480.00

8.1.1 SITE CLEARANCE

Item	Unit	Range £
CLEARING VEGETATION		
Clear existing scrub vegetation including shrubs and hedges and stack on site		
By machine		
light scrub and grasses	100 m^2	29.00
light scrub and grasses but remove vegetation to licensed tip	100 m^2	73.00
undergrowth brambles and heavy weed growth	100 m^2	70.00
small shrubs	100 m^2	70.00
By hand		
light scrub and grasses to spoil heaps 25 m distance	100 m^2	115.00
light scrub and grasses to spoil heap; subsequently disposed off site	100 m^2	125.00
light scrub and grasses to tip	100 m^2	60.00
Fell trees on site; grub up roots; all by machine; remove debris to licensed tip		
trees 600 mm girth	each	530.00
trees 1.5–3.00 m girth	each	1475.00
trees over 3.00 m girth	each	2275.00
Chip existing vegetation on site; remove to stockpile 100 m		
light scrub and bramble; average height 500 mm	100 m^2	41.00
dense scrub and bramble; average height 1.00 m	100 m^2	82.00
mixed scrub and shrubs average 1.50 m high	100 m^2	140.00
dense overgrown shrubs and bramble; average 2.00 m high	100 m^2	190.00
Turf stripping for preservation		
Cut and strip existing turf area by machine; turves 50 mm thick; turf to be lifted for preservation		
load to dumper or palette by hand; transport on site not exceeding 100 m travel to new location	100 m^2	130.00
all by hand; including barrowing to stacking locally 10 m distance	100 m^2	830.00
Turf stripping for disposal; spray turf with glyphosate and return to site after 3 weeks; strip dead grass; load to stockpile on site; rotavate to prepare area for new surface treatments to within 50 mm of final levels		
By knapsack sprayer; excavator and dumper; pedestrian rotavator		
5 tonne excavator and 6 tonne dumper; stock pile distance 25 m	100 m^2	84.00
5 tonne excavator and 3 tonne dumper; stock pile distance 50 m	100 m^2	87.00
5 tonne excavator and 3 tonne dumper; stock pile distance 100 m	100 m^2	92.00
By tractor drawn equipment		
spraying with glyphosate; 2 passes	ha	640.00

8.1.1 SITE CLEARANCE

Item	Unit	Range £
CLEARING FENCES GATES AND LIGHT STRUCTURES		
Clear site; demolish existing closeboard fence; clear shrubs 50% in shrub beds and 50% free-standing; spray turf and return to clear ground of sprayed vegetation; chip vegetation on site and transport to spoil heaps 50 m; works by mechanical excavator		
Rectangular site with 70% turf and 30% shrub area		
1000 m^2	nr	2775.00
2000 m^2	nr	5100.00
4000 m^2	nr	9600.00
Rectangular site with 60% turf grass/scrub area and 40% shrubs and overgrowth		
site area; 1000 m^2	nr	3225.00
site area; 2000 m^2	nr	6000.00
site area; 4000 m^2	nr	11000.00
Demolish fencing and gates		
Demolish fencing; break out posts, remove fencing stack and load mechanically for removal to tip		
Chain link fencing; vegetation not present on angle iron stakes driven in	100 m	300.00
Chain link fencing; overgrown with vegetation on angle iron stakes driven in	100 m	1050.00
DEMOLITIONS OF EXTERNAL STRUCTURES AND SURFACES		
Demolish existing free-standing walls; grub out foundations; remove arisings to tip		
Brick wall; 112 mm thick		
300 mm high	m	23.00
500 mm high	m	34.00
Brick wall; 225 mm thick		
300 mm high	m	28.00
500 mm high	m	37.00
1.00 m high	m	58.00
1.20 m high	m	61.00
1.50 m high	m	77.00
1.80 m high	m	86.00
Demolish existing structures		
Timber sheds on concrete base 150 mm thick inclusive of disposal and backfilling with excavated material; plan area of building		
10 m^2	nr	365.00
15 m^2	nr	510.00
Building of cavity wall construction; insulated with tiled roof; inclusive of foundations; allow for disconnection of electrical water and other services; grub out and remove 10 m of electrical cable, water and waste pipes		
10 m^2	nr	2550.00
20 m^2	nr	3425.00

8.1.1 SITE CLEARANCE

Item	Unit	Range £
CLEARING SURFACES		
Demolish existing surfaces		
Break up plain concrete slab; remove to licensed tip		
150 mm thick	m^2	18.00
200 mm thick	m^2	24.00
300 mm thick	m^2	37.00
Break up plain concrete slab and preserve arisings for hardcore; break down to maximum size of 200 × 200 mm and transport to location; maximum distance 50 m		
150 mm thick	m^2	12.00
200 mm thick	m^2	15.00
300 mm thick	m^2	23.00
Break up reinforced concrete slab and remove to licensed tip		
150 mm thick	m^2	26.00
200 mm thick	m^2	34.00
300 mm thick	m^2	51.00
Break out existing surface and associated 150 mm thick granular base load to 20 tonne muck away vehicle for removal off site		
macadam surface; 70 mm thick	m^2	17.00
macadam surface; 150 mm thick	m^2	24.00
block paving; 50 mm thick	m^2	16.00
block paving; 80 mm thick	m^2	17.00

8.1.2 PREPARATORY GROUNDWORKS

Item	Unit	Range £
TREE ROOT PROTECTION		
Protect tree roots from compaction in Root Protection Zone (RPZ)		
Reduce levels by hand; load to machines external to the RPZ and remove off site;		
Lay cellular protection on non woven geofabric to depth;		
75 mm	m²	28.00
100 mm	m²	33.00
150 mm	m²	42.00
Extra over to remove excavated material off site	m³	35.00
EXCAVATING GENERALLY		
Prices for excavating and reducing to levels are for work in light or medium soils; multiplying factors for other soils are as follows		
Clay	1.5	–
Compact gravel	1.2	–
Soft chalk	2.0	–
Hard rock	3.0	–
Excavating topsoil for preservation		
Remove topsoil average depth 300 mm; deposit in spoil heaps not exceeding 100 m travel to heap; turn heap once during storage		
by machine	m³	15.00
by hand	m³	215.00
Excavate to reduce levels; remove spoil to stockpile not exceeding 100 m travel; all by machine		
Average 0.25 m deep	m³	7.00
Average 0.30 m deep	m²	4.00
Average 1.00 m deep	m³	11.00
Average 1.0 m deep; using 21 tonne 360 tracked excavator	m³	6.00
Excavate to reduce levels; remove spoil to stockpile not exceeding 100 m travel; all by hand		
Average 0.10 m deep	m²	19.00
Average 0.20 m deep	m²	38.00
Average 0.30 m deep	m²	58.00
Average 100–300 mm deep	m³	190.00
Average 1.0 m deep	m³	215.00
Extra for carting spoil to licensed tip off site (20 tonnes)	m³	48.00
Extra for carting spoil to licensed tip off site (by skip; machine loaded)	m³	67.00
Extra for carting spoil to licensed tip off site (by skip; hand loaded)	m³	170.00
Spread excavated material to levels in layers not exceeding 150 mm; using scraper blade or bucket		
Average thickness 100 mm	m²	1.00
Average thickness 100 mm but with imported topsoil	m²	5.00
Average thickness 200 mm	m²	1.00
Average thickness 200 mm but with imported topsoil	m²	10.00
Average thickness 250 mm	m²	2.00
Average thickness 200–250 mm	m³	7.00
Average thickness 250 mm but with imported topsoil	m²	13.00

8.1.2 PREPARATORY GROUNDWORKS

Item	Unit	Range £
Average thickness 250 mm but with imported topsoil	m^3	51.00
Extra for work to banks exceeding 30° slope	30%	–
Rip subsoil using approved subsoiling machine to a depth of 600 mm below topsoil at 600 mm centres; all tree roots and debris over 150 × 150 mm to be removed; cultivate ground where shown on drawings to depths as shown		
100 mm	100 m^2	20.00
200 mm	100 m^2	20.00
300 mm	100 m^2	21.00
400 mm	100 m^2	26.00
Form landform from imported material; mounds to falls and grades to receive turf or seeding treatments; material dumped by 20 tonne vehicle to location; volume 100 m^3 spread over area; thickness varies from 150–500 mm; prices to not include for surface preparations		
Imported standard grade topsoil		
by large excavator	m^3	41.00
by 5 tonne excavator	m^3	42.00
by 3 tonne excavator	m^3	46.00
Form landform from excavated material; mounds to falls and grades to receive turf or seeding treatments; material moved by dumper to location maximum 50 m; volume 100 m^3 spread over a 200 m^2 area; thickness varies from 150–500 mm; prices to not include for surface preparations		
Imported recycled topsoil		
by large excavator	m^3	5.00
by 5 tonne excavator	m^3	6.00
by 3 tonne excavator	m^3	9.00
EXCAVATION OF TRENCHES		
Excavate trenches for foundations; trenches 225 mm deeper than specified foundation thickness; pour plain concrete foundations GEN 1 10 N/mm and to thickness as described; disposal of excavated material off site; dimensions of concrete foundations		
250 mm wide × 250 mm deep	m	17.00
300 mm wide × 300 mm deep	m	23.00
400 mm wide × 600 mm deep	m	54.00
600 mm wide × 400 mm deep	m	57.00
Excavate trenches for foundations; trenches 225 mm deeper than specified foundation thickness; pour plain concrete foundations GEN 1 10 N/mm and to thickness as described; disposal of excavated material off site; dimensions of concrete foundations		
250 mm wide × 250 mm deep	m^3	335.00
300 mm wide × 300 mm deep	m^3	255.00
600 mm wide × 400 mm deep	m^3	235.00
400 mm wide × 600 mm deep	m^3	205.00

8.1.2 PREPARATORY GROUNDWORKS

Item	Unit	Range £
EXCAVATING GENERALLY – cont		
Excavate trenches for services; grade bottoms of excavations to required falls; remove 20% of excavated material off site		
Trenches 300 mm wide		
600 mm deep	m	12.00
750 mm deep	m	15.00
900 mm deep	m	18.00
1.00 m deep	m	20.00
Ha-ha		
Excavate ditch 1200 mm deep × 900 mm wide at bottom battered one side 45° slope; excavate for foundation to wall and place GEN 1 concrete foundation 150 × 500 mm; construct one and a half brick wall (brick PC £541.00/1000) battered 10° from vertical; laid in cement: lime: sand (1:1:6) mortar in English garden wall bond; precast concrete coping weathered and throated set 150 mm above ground on high side; rake bottom and sides of ditch and seed with low maintenance grass at 35 g/m²		
wall 600 mm high	m	470.00
wall 900 mm high	m	600.00
wall 1200 mm high	m	760.00
Excavate ditch 1200 mm deep × 900 mm wide at bottom battered one side 45° slope; place deer or cattle fence 1.20 m high at centre of excavated trench; rake bottom and sides of ditch and seed with low maintenance grass at 35 g/m²		
fence 1200 mm high	m	37.00
EXCAVATION OF BANKS		
Excavate existing slope to vertical to form flat area; grade bank behind to 45°; no allowance for removal or backfilling of excavated materials;		
Existing slope angle 10°; excavation by 8 tonne machine; excavated material to location on-site 50 m distant; linear measurement of flat area at the base of the slope;		
2.00 m wide; height 350 mm	m	4.00
4.00 m wide; height 860 mm	m	17.00
6.00 m wide; height 1.28 m	m	39.00
8.00 m wide; height 1.71 m	m	70.00
10.00 m wide; height 2.14 m	m	110.00
Existing slope angle 20°; excavation by 8 tonne machine; excavated material to loctation on-site 50 m distant; linear measurement of flat area at the base of the slope;		
2.00 m wide; height 1.14 m	m	12.00
4.00 m wide; height 2.29 m	m	46.00
6.00 m wide; height 3.43 m	m	105.00
8.00 m wide; height 4.58 m	m	185.00
10.00 m wide; height 5.72 m	m	285.00
Excavate and grade banks to grade to receive stabilization treatments below; removal of excavated material not included		
By 21 tonne excavator	m^3	1.00
By 8 tonne excavator	m^3	2.00
By 5 tonne excavator	m^3	2.00

8.1.2 PREPARATORY GROUNDWORKS

Item	Unit	Range £
By 3 tonne excavator	m^3	5.00
Excavate sloped bank to vertical to receive retaining or stabilization treatment (priced separately); allow 1.00 m working space at top of bank; backfill working space and remove balance of arisings off site; measured as length along the top of the bank		
Bank height 1.00 m; excavation by 5 tonne excavator		
10 degree	m	120.00
30 degree	m	38.00
45 degree	m	22.00
60 degree	m	13.00
Bank height 1.00 m; excavation by 5 tonne excavator; arisings moved to stockpile on site		
10 degree	m	12.00
30 degree	m	5.00
45 degree	m	3.00
60 degree	m	2.00
Bank height 2.00 m; excavation by 5 tonne excavator		
10 degree	m	490.00
30 degree	m	145.00
45 degree	m	88.00
60 degree	m	50.00
Bank height 2.00 m; excavation by 5 tonne excavator but arisings moved to stockpile on site		
10 degree	m	70.00
30 degree	m	21.00
45 degree	m	12.00
60 degree	m	7.00
Bank height 3.00 m; excavation by 21 tonne excavator		
20 degree	m	530.00
30 degree	m	335.00
45 degree	m	185.00
60 degree	m	100.00
Excavate bank 3.00 m but arisings moved to stockpile on site		
20 degree	m	68.00
30 degree	m	43.00
45 degree	m	18.00
60 degree	m	10.00

8.1.2 PREPARATORY GROUNDWORKS

Item	Unit	Range £
SOIL STABILIZATION		
Please refer to slope excavation above for preparation of banks		
REVETMENTS		
Gabion mattress revetments		
Construct revetment of Reno mattress gabions laid on firm level ground; tightly packed with broken stone or concrete and securely wired; all in accordance with manufacturer's instructions		
gabions 6 × 2 × 0.17 m; one course	m^2	64.00
gabions 6 × 2 × 0.17 m; two courses	m^2	110.00
Grass concrete; bank revetments; excluding bulk earthworks		
Bring bank to final grade by machine; lay regulating layer of Type 1; lay filter membrane; lay 100 mm drainage layer of broken stone or approved hardcore 28–10 mm size; blind with 25 mm sharp sand; lay grass concrete surface (price does not include edgings or toe beams); fill with approved topsoil and fertilizer at 35 g/m^2; seed with dwarf rye based grass seed mix		
Grasscrete in situ reinforced concrete surfacing GC 1; 100 mm thick	m^2	80.00
Grasscrete in situ reinforced concrete surfacing GC 2; 150 mm thick	m^2	100.00
Grassblock 103 mm open matrix blocks; 406 × 406 × 103 mm	m^2	89.00
Grade slope to even grade; stabilize river bank edge with woven willow walling; lay bank stabilization mat pinned to graded embankment 5 m long (measured per m of river bank edge		
Slope edge to river bank 5 m long seeded with Greenfix stabilization matting		
Brushwood faggott and pre-planted coir revetment edge	m	230.00
Brushwood faggott revetment edge	m	205.00
StableGRASS grass reinforcement: cellular slope retention; clear existing bank of light scrub, shrubs and grasses; spray with glyphosate; grade surface only to even grade by machine and final hand grade; lay soil retaining system		
Backfilled with excavated material		
38 mm deep	m^2	34.00
Backfilled with imported topsoil and seeded		
100 mm deep	m^2	3.00

8.1.2 PREPARATORY GROUNDWORKS

Item	Unit	Range £
SOIL REINFORCEMENT		
Geogrid soil reinforcement (Note: measured per metre run at the top of the bank)		
Backfill excavated bank; lay Tensar geogrid; cover with excavated material as work proceeds		
2.00 m high bank; geogrid at 1.00 m vertical lifts		
10° slope	m	1100.00
20° slope	m	610.00
30° slope	m	445.00
2.00 m high bank; geogrid at 0.50 m vertical lifts		
45° slope	m	125.00
60° slope	m	98.00
3.00 m high bank		
10° slope	m	1100.00
20° slope	m	610.00
30° slope	m	445.00
45° slope	m	940.00
EROSION CONTROL		
Bank stabilization; erosion control mats		
Excavate fixing trench for erosion control mat 300 × 300 mm; backfill after placing mat selected below	m	11.00
Place mat to slope to be retained in anchor trench;allow extra over to the slope for the area of matting required to the anchor trench		
By machine; final grade only of excavated bank by excavator; lay erosion control solution; sow with low maintenance grass at 35 g/m^2; spread imported topsoil 25 mm thick incorporating medium grade sedge peat at 3 kg/m^2 and fertilizer at 30 g/m^2; water lightly		
seeded Covamat Standard	m^2	5.00
lay Greenfix biodegradable pre-seeded erosion control mat; fix with 6 × 300 mm steel pegs at 1.0 m centres	m^2	5.00
Tensar open textured erosion mat	m^2	8.00
By hand; final grade only of excavated bank by hand; lay erosion control solution; sow with low maintenance grass at 35 g/m^2; spread imported topsoil 25 mm thick incorporating medium grade sedge peat at 3 kg/m^2 and fertilizer at 30 g/m^2; water lightly		
seeded Covamat Standard	m^2	6.00
lay Greenfix biodegradable pre-seeded erosion control mat, fix with 6 × 300 mm steel pegs at 1.0 m centres	m^2	6.00
Tensar open textured erosion mat	m^2	9.00

8.1.2 PREPARATORY GROUNDWORKS

Item	Unit	Range £
SOIL REINFORCEMENT – cont		
Cellular retaining systems; Greenfix Ltd; final grade to excavated bank by machine and by hand; lay bank retaining system		
Geoweb; filling with angular aggregate	m	35.00
Haul road for site enablement		
Reduce existing soil to stockpile on site; Lay Tensar Biaxial geogrid on 200 mm compacted hardcore base; backfill with 200 mm Type 1 granular fill and compact	m^2	24.00
Reduce existing soil to stockpile on site; Lay 200 mm compacted hardcore base blinded with 50 mm sharp sand; Lay Cellweb load support	m^2	47.00

8.2.1 ROADS, PATHS AND PAVINGS

Item	Unit	Range £
KERBS AND EDGINGS TO PAVED SURFACES		
Excavate for pathways or roadbed to receive surface 100 mm thick (priced separately); grade; 100 mm hardcore; kerbs PC 125 × 255 mm on both sides; including foundations haunched		
Excavation 350 deep; Type 1 base 150 thick		
1.50 m wide	m	160.00
2.00 m wide	m	185.00
3.00 m wide	m	215.00
4.00 m wide	m	250.00
5.00 m wide	m	280.00
6.00 m wide	m	315.00
7.00 m wide	m	355.00
Excavation 450 mm deep; Type 1 base 250 mm thick		
1.50 m wide	m	300.00
2.00 m wide	m	360.00
3.00 m wide	m	485.00
4.00 m wide	m	610.00
5.00 m wide	m	730.00
6.00 m wide	m	870.00
7.00 m wide	m	990.00
EDGINGS TO PATHS AND PAVINGS		
PRECAST CONCRETE KERBS		
Note: excavation is by machine unless otherwise mentioned		
Excavate trench and construct concrete foundation 300 mm wide × 300 mm deep; lay precast concrete kerb units bedded in semi-dry concrete; slump 35 mm maximum; haunching one side; disposal of arisings off site		
Kerbs laid straight; site mixed concrete bases		
125 mm high × 125 mm thick; bullnosed; type BN	m	53.00
125 × 255 mm; ref HB2; SP	m	57.00
150 × 305 mm; ref HB1	m	68.00
Kerbs laid straight; ready mixed concrete bases		
125 mm high × 125 mm thick; bullnosed; type BN	m	46.00
125 × 255 mm; ref HB2; SP	m	51.00
150 × 305 mm; ref HB1	m	60.00
Straight kerbs 125 mm × 255 mm laid to radius ready mixed concrete bases		
4.5 m radius	m	64.00
6.1 m radius	m	62.00
9.15 m radius	m	61.00
12.20 m radius	m	60.00
Dropper kerbs		
125 mm × 255 –150 mm; left or right handed	m	58.00
Quadrants		
305 mm radius	nr	66.00
455 mm radius	nr	71.00

8.2.1 ROADS, PATHS AND PAVINGS

Item	Unit	Range £
KERBS AND EDGINGS TO PAVED SURFACES – cont		
Excavate trench and construct concrete foundation 300 mm wide × 150 mm deep; lay precast concrete kerb units bedded in semi-dry concrete; slump 35 mm maximum; haunching one side; disposal of arisings off site		
Conservation kerbs laid straight		
225 mm wide × 150 mm high;	m	95.00
155 mm wide × 255 mm high	m	94.00
Conservation kerbs laid to radius; 145 mm × 255 mm		
radius 3.25 m	m	120.00
radius 6.50 m	m	110.00
radius 9.80 m	m	110.00
solid quadrant; 305 × 305 mm	nr	110.00
STONE KERBS		
Excavate and construct concrete foundation 450 mm wide × 150 mm deep; lay kerb units bedded in semi-dry concrete; slump 35 mm maximum; haunching one side;		
Granite kerbs		
straight; 125 × 250 mm	m	125.00
curved; 125 × 250 mm	m	150.00
Precast concrete edging on concrete foundation 100 × 150 mm deep and haunching one side 1:2:4 including all necessary excavation disposal and formwork		
Rectangular, chamfered or bullnosed; to one side of straight path		
50 × 150 mm	m	47.00
50 × 200 mm	m	59.00
50 × 250 mm	m	64.00
Rectangular, chamfered or bullnosed to both sides of straight paths		
50 × 150 mm	m	100.00
50 × 200 mm	m	120.00
50 × 250 mm	m	125.00
Brick or concrete block edge restraint; excavate for foundation; lay concrete 1:2:4 250 mm wide × 150 mm deep; on 50 mm thick sharp sand bed; lay blocks or bricks; Haunched one side		
Blocks 200 × 100 × 60 mm; butt jointed		
header course	m	59.00
stretcher course	m	53.00
Bricks Wieinerberger Blue Baggeridge 200 × 100 × 62 mm; with flowable mortar joints		
header course; brick on flat	m	82.00
header course; brick on edge	m	83.00
stretcher course; flat or brick on edge	m	82.00
stretcher course; 2 rows brick on edge	m	64.00

8.2.1 ROADS, PATHS AND PAVINGS

Item	Unit	Range £
STONE EDGINGS		
Sawn Yorkstone edgings; excavate for foundations; lay concrete 1:2:4 150 mm deep × 33.3% wider than the edging; on 35 mm thick mortar bed		
Yorkstone 50 mm thick		
100 mm wide × random lengths	m	79.00
100 × 100 mm	m	77.00
100 mm wide × 200 mm long	m	91.00
250 mm wide × random lengths	m	92.00
500 mm wide × random lengths	m	185.00
Granite edgings; excavate for footings; lay concrete 1:2:4 150 mm deep × 33.3% wider than the edging; on 35 mm thick mortar bed		
Granite 50 mm thick		
100 mm wide × 200 mm long	m	75.00
100 mm wide × random lengths	m	83.00
250 mm wide × random lengths	m	89.00
setts; 100 × 100 mm	m	100.00
300 mm wide × random lengths	m	97.00
TIMBER EDGINGS		
Softwood treated boards; 150 mm × 22 mm; with timber pegs at 1.00 m centres; excavated and backfilled by machine		
Straight		
150 × 22 mm	m	14.00
150 × 50 mm	m	21.00
Curved; over 5 m radius		
150 × 22 mm	m	23.00
Curved; 4–5 m radius		
150 × 22 mm	m	27.00
Curved; 3–4 m radius		
150 × 22 mm	m	32.00
Curved; 1–3 m radius		
150 × 22 mm	m	40.00
Softwood treated boards; 150 mm × 22 mm; with timber pegs at 1.00 m centres; excavated and backfilled by hand		
Straight		
150 × 22 mm	m	17.00
Curved; over 5 m radius		
150 × 22 mm	m	30.00
Curved; 4–5 m radius		
150 × 22 mm	m	30.00
Curved; 3–4 m radius		
150 × 22 mm	m	35.00
Curved; 1–3 m radius		
150 × 22 mm	m	44.00

8.2.1 ROADS, PATHS AND PAVINGS

Item	Unit	Range £
KERBS AND EDGINGS TO PAVED SURFACES – cont		
METAL EDGINGS		
Everedge – Steel Edgings		
Steel edgings for macadam		
50 mm	m	24.00
65 mm	m	36.00
75 mm	m	25.00
100 mm	m	31.00
Edgings for paths; L shaped Edgings with 3.2 thick;		
100 mm; 4.75 mm exposed upper lip	m	28.00
50 mm with 3.18 exposed upper lip	m	24.00
75 mm; 5.2 mm exposed upper lip	m	25.00
Steel Edgings – Outdoor space design		
Galvanized steel; 6 mm thick		
100 mm	m	35.00
150 mm	m	47.00
For Corten Edgings 200 –600 mm – Please see section 8.4.3 – Retaining Walls		
BASES FOR PAVING		
Excavate ground and reduce levels to receive 38 mm thick slab and 25 mm mortar bed; dispose of excavated material off site		
Lay granular fill Type 1; limestone aggregate 150 mm thick laid to falls and compacted		
all by machine	m^2	25.00
all by hand except disposal by grab	m^2	86.00
Lay granular fill Type 1; recycled material; 150 mm thick laid to falls and compacted		
all by machine	m^2	19.00
all by hand except disposal by grab	m^2	80.00
Lay granular fill Type 1; limestone aggregate 200 mm thick laid to falls and compacted		
all by machine	m^2	34.00
all by hand except disposal by grab	m^2	105.00
Lay granular fill Type 1; recycled material; 200 mm thick laid to falls and compacted		
all by machine	m^2	26.00
all by hand except disposal by grab	m^2	105.00
Lay 1:2:4 concrete base 150 mm thick laid to falls		
all by machine	m^2	59.00
all by hand except disposal by grab	m^2	125.00
150 mm hardcore base with concrete base 150 mm deep		
concrete 1:2:4 site mixed	m^2	79.00
concrete PAV 1 35 ready mixed	m^2	51.00
concrete PAV 1 35 ready mixed; but for 50 mm thick slab	m^2	53.00
Hardcore base 150 mm deep; concrete reinforced with A142 mesh		
site mixed concrete 1:2:4; 150 mm deep	m^2	95.00
site mixed concrete 1:2:4; 250 mm deep	m^2	135.00
concrete PAV 1 35 N/mm^2 ready mixed; 150 mm deep	m^2	62.00
concrete PAV 1 35 N/mm^2 ready mixed; 250 mm deep	m^2	91.00

8.2.1 ROADS, PATHS AND PAVINGS

Item	Unit	Range £
CHANNELS		
Excavate and construct concrete foundation 450 mm wide × 150 mm deep; lay precast concrete kerb units bedded in semi-dry concrete; slump 35 mm maximum; haunching one side; lay channel units bedded in 1:3 sand mortar; jointed in cement: sand mortar 1:3		
Kerbs; 125 mm high × 125 mm thick; bullnosed; type BN		
dished channel; 125 × 225 mm; Ref CS	m	98.00
square channel; 125 × 150 mm; Ref CS2	m	92.00
bullnosed channel; 305 × 150 mm; Ref CBN	m	120.00
Brick channel; class B engineering bricks		
Excavate and construct concrete foundation 600 mm wide × 200 mm deep; lay channel to depths and falls bricks to be laid as headers along the channel; bedded in 1:3 sand mortar bricks close jointed in cement: sand mortar 1:3		
3 courses wide	m	130.00
3 coursed granite setts; dished		
Excavate and construct concrete foundation 400 mm wide × 200 mm deep; lay channel to depths and falls ; bedded in 1:3: sand mortar		
3 courses wide	m	125.00
CONCRETE ROADS		
Excavate weak points of excavation by hand and fill with well rammed Type 1 granular material		
100 mm thick	m^2	19.00
200 mm thick	m^2	24.00
Reinforced concrete roadbed; excavation and disposal off site; reinforcing mesh A142; expansion joints at 50 m centres; on 150 mm hardcore; sand blinded; kerbs 155 × 255 mm to both sides; haunched one side.		
Concrete 150 mm thick		
4.00 m wide	m	295.00
5.00 m wide	m	355.00
6.00 m wide	m	415.00
7.00 m wide	m	475.00
Concrete 250 mm thick		
4.00 m wide	m	375.00
5.00 m wide	m	460.00
6.00 m wide	m	540.00
7.00 m wide	m	630.00

8.2.1 ROADS, PATHS AND PAVINGS

Item	Unit	Range £
FOOTPATHS		
Note: For pathway surfaces; select appropriate finish from the Measured Works section		
Excavate 250 mm; disposal off site; grade to levels and falls; lay 150 mm Type 1 granular material; all disposal off site; Lay or fix edge as described		
Lay Permaloc Asphalt edge 100 mm metal edgings; path width		
1.50 m wide	m	86.00
3.00 m wide	m	115.00
Lay timber edging 150 × 38 mm; path width		
1.50 m wide	m	54.00
3.00 m wide	m	88.00
Lay precast edgings kerbs 50 × 150 mm on both sides; including foundations haunched one side in 11.5 N/mm² concrete with all necessary formwork; path width		
1.50 m wide	m	130.00
3.00 m wide	m	165.00
Lay concrete edging Keyblock header edging; path width		
1.50 m wide	m	81.00
3.00 m wide	m	115.00
Lay concrete edging Keyblock stretcher edging; path width		
1.50 m wide	m	85.00
3.00 m wide	m	140.00
Lay brick on flat header edging; brick £300.00/1000; path width		
1.50 m wide	m	125.00
3.00 m wide	m	160.00
Lay brick on flat header edging; brick £600.00/1000; path width		
1.50 m wide	m	140.00
3.00 m wide	m	170.00
Lay brick on edge header edging; brick £300.00/1000; path width		
1.50 m wide	m	155.00
3.00 m wide	m	185.00
Lay brick on edge header edging; brick £600.00/1000; path width		
1.50 m wide	m	160.00
3.00 m wide	m	195.00
Lay brick on flat or brick on edge stretcher edging; brick £300.00/1000; path width		
1.50 m wide	m	105.00
3.00 m wide	m	140.00
Lay brick on edge strecher edging; brick £600.00/1000; path width		
1.50 m wide	m	110.00
3.00 m wide	m	145.00
Lay new granite setts 100 × 100 × 100 mm single row; path width		
1.50 m wide	m	190.00
3.00 m wide	m	230.00
Lay new granite setts 200 × 100 × 100 mm single row; path width		
1.50 m wide	m	155.00
3.00 m wide	m	185.00

8.2.1 ROADS, PATHS AND PAVINGS

Item	Unit	Range £
BRICK PAVING		
WORKS BY MACHINE		
Excavate and lay base Type 1 150 mm thick remove arisings; all by machine; lay clay brick paving		
Wienerberger; butt jointed on 50 mm sharp sand bed		
Blue Baggeridge range; 200 × 50 × 62 mm laid flat	m^2	150.00
Vande Moortel; Tumbled and sanded pavers; laid on 50 mm sharp sand bed.		
Ancienne Belgique; Oyster; 183 × 43 × 88 mm	m^2	275.00
Elegantia; Taupe; 240 × 36 × 72 mm	m^2	215.00
SeptimA; Vanilla; 215 × 52 × 70 mm	m^2	160.00
SeptimA; Amarant; 215 × 52 × 70 mm	m^2	155.00
Excavate and lay base Type 1 250 mm thick all by machine; remove arisings; lay clay brick paving		
Vande Moortel; Tumbled and sanded pavers; laid on 50 mm sharp sand bed.		
Ancienne Belgique; Oyster; 183 × 43 × 88 mm	m^2	290.00
Elegantia; Taupe; 240 × 36 × 72 mm	m^2	230.00
SeptimA; Vanilla; 215 × 52 × 70 mm	m^2	180.00
Blue Baggeridge range; 200 × 50 × 62 mm laid flat	m^2	185.00
SeptimA; Amarant; 215 × 52 × 70 mm	m^2	170.00
PATHS		
Brick paved paths; brick paviors with butt joints; prices are inclusive of excavation, 150 mm Type 1 and 50 mm sharp sand; jointing in kiln dried sand brushed in; inclusive of edge restraints of brick headers		
Path 1.50 mm wide; edging course of headers		
Vande Moortel handmade pavers 243 × 1833 × 88 mm; laid on edge	m^2	395.00
Vande Moortel Elegantia Taupe; 215 × 51 × 70 mm laid on edge	m^2	380.00
Blue Baggeridge range; 200 × 50 × 62 mm laid flat	m^2	325.00
WORKS BY HAND		
Excavate and lay base Type 1 150 mm thick by hand; arisings barrowed to spoil heap maximum distance 25 m and removal off site by grab; lay clay brick paving; edgings not included		
Vande Moortel; butt jointed on 50 mm sharp sand bed		
Ancienne Belgique; Oyster; 183 × 43 × 88 mm	m^2	335.00
Elegentia; Taupe: 240 × 36 × 72 mm	m^2	280.00
SeptimA; Vanilla; 215 × 52 × 70 mm	m^2	225.00
SeptimA; Amarant: 215 × 52 × 70 mm	m^2	215.00
Blue Baggeridge range; 200 × 50 × 62 mm laid flat	m^2	210.00
Excavate and lay 150 mm concrete base; 1:3:6: site mixed concrete reinforced with A393 mesh; remove arisings to stockpile and then off site by grab; lay clay brick paving		
Vande Moortel; butt jointed on 50 mm sharp sand bed		
Ancienne Belgique; Oyster; 183 × 43 × 88 mm	m^2	360.00
Elegantia; Taupe; 240 × 36 × 72 mm	m^2	300.00
SeptimA; Vanilla; 215 × 52 × 70 mm	m^2	250.00
SeptimA; Amarant; 215 × 52 × 70 mm	m^2	240.00

8.2.1 ROADS, PATHS AND PAVINGS

Item	Unit	Range £
BLOCK PAVING		
CONCRETE SETT PAVING		
Excavate ground and bring to levels; supply and lay 150 mm granular fill Type 1 laid to falls and compacted; supply and lay setts bedded in 50 mm sand and vibrated; joints filled with dry sand and vibrated; all excluding edgings or kerbs measured separately		
Marshalls Mono; Tegula precast concrete setts		
random sizes 60 mm thick	m^2	88.00
single size 60 mm thick	m^2	84.00
random size 80 mm thick	m^2	95.00
single size 80 mm thick	m^2	120.00
Excavate 450 mm and prepare base of 100 hardcore and 250 mm Type 1; work to falls, crossfalls and cambers not exceeding 15° mark out car parking bays 5.0 × 2.4 m with thermoplastic road paint; surfaces all mechanically laid; edgings not included		
Marshalls Keyblock concrete blocks 200 × 100 × 80 mm		
grey	m^2	115.00
colours	m^2	120.00
Marshalls Keyblock concrete blocks 200 × 100 × 60 mm		
grey	m^2	115.00
colours	m^2	115.00
Marshalls Tegula concrete blocks 200 × 100 × 80 mm		
random sizes	m^2	130.00
single size	m^2	120.00
BLOCK PAVING EDGINGS		
Edge restraint to block paving; excavate for groundbeam; lay concrete 1:2:4 200 mm wide × 150 mm deep; on 50 mm thick sharp sand bed; inclusive of haunching one side		
Blocks 200 × 100 × 60 mm; PC £14.01 /m^2; butt jointed		
header course	m	33.00
stretcher course	m	27.00
Excavate ground; supply and lay granular fill Type 1 150 mm thick laid to falls and compacted; supply and lay block pavers; laid on 50 mm compacted sharp sand; vibrated; joints filled with loose sand excluding edgings or kerbs measured separately		
Concrete blocks		
200 × 100 × 60 mm	m^2	79.00
200 × 100 × 80 mm	m^2	83.00

8.2.1 ROADS, PATHS AND PAVINGS

Item	Unit	Range £
Reduce levels; lay 150 mm granular material Type 1; lay vehicular block paving to 90° herringbone pattern; on 50 mm compacted sand bed; vibrated; jointed in sand and vibrated; excavate and lay precast concrete edging 50 × 150 mm; on concrete foundation 1:2:4		
Blocks 200 × 100 × 60 mm		
1.0 m wide clear width between edgings	m	175.00
1.5 m wide clear width between edgings	m	210.00
2.0 m wide clear width between edgings	m	250.00
Blocks 200 × 100 × 60 mm but blocks laid 45° herringbone pattern including cutting edging blocks		
1.0 m wide clear width between edgings	m	185.00
1.5 m wide clear width between edgings	m	220.00
2.0 m wide clear width between edgings	m	260.00
Blocks 200 × 100 × 60 mm but all excavation by hand disposal off site by grab		
1.0 m wide clear width between edgings	m	385.00
1.5 m wide clear width between edgings	m	445.00
2.0 m wide clear width between edgings	m	500.00
PERMEABLE BLOCK PAVING		
Excavate 450 mm and prepare base of 150 mm Type 3 open; work to falls, crossfalls and cambers not exceeding 15° mark out car parking bays 5.0 × 2.4 m with thermoplastic road paint; surfaces all mechanically laid; edgings not included		
Tegula Priora		
160 × 120 × 80 mm	m^2	79.00
240 × 160 × 80 mm; traditional herringbone	m^2	85.00
Conservation Priora – Silver Grey		
200 × 400 × 65 mm	m^2	120.00
400 × 400 × 65 mm	m^2	105.00
PORCELAIN PAVING		
Porcelain paving; London Stone; including excavation and blinded reinforced concrete base 150 mm;		
Porcelain pavings; exluding any cutting		
596 × 596 mm	m^2	175.00
748 × 748 mm	m^2	170.00
1194 × 596 mm	m^2	170.00
MACADAM SURFACES		
Macadam roadway over 1000 m^2; Centar Surfacing		
Excavate 350 mm for pathways or roadbed; bring to grade; lay 100 mm well-rolled hardcore; lay 150 mm Type 1 granular material; lay precast edgings kerbs 50 × 150 mm on both sides; including foundations haunched one side in 11.5 N/ mm^2 concrete with all necessary formwork; machine lay surface of 80 mm macadam 50 mm base course and 30 mm wearing course; all disposal off site		
1.50 m wide	m	195.00
2.00 m wide	m	200.00
3.00 m wide	m	280.00

8.2.1 ROADS, PATHS AND PAVINGS

Item	Unit	Range £
MACADAM SURFACES – cont		
Macadam roadway over 1000 m² – cont		
Excavate 350 mm for pathways or roadbed – cont		
4.00 m wide	m	340.00
5.00 m wide	m	400.00
6.00 m wide	m	465.00
7.00 m wide	m	530.00
Macadam roadway 400–1000 m²; Centar Surfacing		
Excavate 350 mm for pathways or roadbed; bring to grade; lay 100 mm well-rolled hardcore; lay 150 mm Type 1 granular material; lay precast edgings kerbs 50 × 150 mm on both sides; including foundations haunched one side in 11.5 N/mm² concrete with all necessary formwork; machine lay surface of 90 mm macadam 60 mm base course and 30 mm wearing course; all disposal off site		
1.50 m wide	m	205.00
2.00 m wide	m	240.00
3.00 m wide	m	305.00
4.00 m wide	m	380.00
5.00 m wide	m	445.00
6.00 m wide	m	520.00
7.00 m wide	m	590.00
MACADAM FOOTPATH; Centar Surfacing		
Excavate footpath to reduce level; remove arisings to tip on site maximum distance 25 m; lay Type 1 granular fill 100 mm thick; lay base course of 28 mm size dense bitumen macadam 50 mm thick; wearing course of 10 mm size dense bitumen macadam 20 mm thick; timber edge 150 × 22 mm		
Areas over 1000 m²		
1.0 m wide	m	62.00
1.5 m wide	m	83.00
2.0 m wide	m	105.00
Areas 400–1000 m²		
1.0 m wide	m	72.00
1.5 m wide	m	98.00
2.0 m wide	m	145.00

8.2.1 ROADS, PATHS AND PAVINGS

Item	Unit	Range £
CAR PARKS Centar Surfacing		
Excavate 450 mm and prepare base of 100 mm hardcore and 250 mm Type 1; Work to falls, crossfalls and cambers not exceeding 15° mark out car parking bays 5.0 × 2.4 m with thermoplastic road paint; surfaces all mechanically laid; edgings not included		
Macadam base course 60 mm thick and wearing course 40 mm thick		
per bay; 5.0 × 2.4 m	each	1125.00
Gangway	m^2	89.00
Bay for disabled people; Macadam base course 60 mm thick and wearing course 40 mm thick		
per bay; 5.80 × 3.25 ambulant	each	1725.00
per bay; 6.55 × 3.80 wheelchair	each	2250.00
Excavate 280 mm and prepare base of 150 mm Type 1;		
Surface treatments as shown		
Macadam 60:40	m^2	64.00
Tegula setts 80 mm; random sizes	m^2	105.00
Tegula setts 80 mm; single size	m^2	99.00
Keyblock 80 mm; Colours	m^2	92.00
Keyblock 80 mm; Grey	m^2	91.00
IN SITU CONCRETE PAVINGS AND SURFACES		
CONCRETE PAVING		
Pedestrian areas		
Excavate to reduce levels; lay 100 mm Type 1; lay PAV 1 air entrained concrete; joints at maximum width of 6.0 m cut out and sealed with sealant; inclusive of all formwork and stripping		
100 mm thick	m^2	83.00
Trafficked areas		
Excavate to reduce levels; lay 150 mm Type 1 lay PAV 2 40 N/mm² air entrained concrete; reinforced with steel mesh 200 × 200 mm square at 2.22 kg/m²; joints at max width of 6.0 m cut out and sealed with sealant		
150 mm thick	m^2	115.00
Reinforced concrete paths; excavate path and dispose of excavated material off site; bring to grade; lay reinforcing mesh A142 lapped and joined; lay in situ reinforced concrete PAV1 35 N/mm²; with 15 impregnated fibreboard expansion joints sealant at 50 m centres; on 100 mm hardcore blinded sand; edgings 150 × 50 mm to both sides; including foundations haunched one side in 11.5 N/mm² concrete with all necessary formwork		
Concrete 150 mm thick		
1.00 m wide	m	105.00
2.00 m wide	m	155.00
3.00 m wide	m	206.00

8.2.1 ROADS, PATHS AND PAVINGS

Item	Unit	Range £
IN SITU CONCRETE PAVINGS AND SURFACES – cont		
WORKS BY HAND; Ready mixed concrete – mixed on site		
Reinforced concrete paths; excavate path by hand barrow to spoil heap 25 m and dispose by grab offsite; bring to grade; lay reinforcing mesh A142 lapped and joined; lay in situ reinforced ready mixed concrete mixed on site (1:2:4:); with 15 impregnated fibreboard expansion joints sealant at 50 m centres; on 100 mm hardcore blinded sand; edgings 150 × 50 mm to both sides; including foundations haunched one side in 11.5 N/mm² concrete with all necessary formwork		
Concrete 150 mm thick		
1.00 m wide	m	165.00
2.00 m wide	m	295.00
3.00 m wide	m	415.00
GRASS CONCRETE		
IN SITU GRASS CONCRETE TO CAR PARKS		
Grass Concrete Grasscrete in situ continuously reinforced cellular surfacing; fill with topsoil and peat (5:1) and fertilizer at 35 g/m²; seed with dwarf rye grass at 35 g/m²; excludes marking		
GC1; 100 mm thick for cars and light traffic; per bay; 5.0 × 2.4 m	each	1275.00
GC1; 100 mm thick for cars and light traffic; gangway	m²	105.00
GC2; 150 mm thick for HGV traffic including dust carts		
per bay; 5.0 × 2.4 m	each	1550.00
gangway	m²	130.00
interlocking concrete blocks 200 × 100 × 60 mm; grey; but mark out bays		
per bay; 5.80 × 3.25 ambulant	each	960.00
per bay; 6.55 × 3.80 wheelchair	each	1250.00
gangway	m²	50.00
MODULAR GRASS CONCRETE PAVING		
Reduce levels; lay 150 mm Type 1 granular fill compacted; supply and lay precast grass concrete blocks on 20 sand and level by hand; fill blocks with 3 mm sifted topsoil and pre-seeding fertilizer at 50 g/m²; sow with perennial ryegrass/chewings fescue seed at 35 g/m²		
Grass concrete		
GB103 406 × 406 × 103 mm	m²	78.00
GB83 406 × 406 × 83 mm	m²	76.00
extra for geotextile fabric underlayer	m²	1.00
Firepaths		
Excavate to reduce levels; lay 300 mm well rammed hardcore; blinded with 100 mm type 1; supply and lay Marshalls Mono Grassguard 180 precast grass concrete blocks on 50 mm sand and level by hand; fill blocks with sifted topsoil and pre-seeding fertilizer at 50 g/m²		
firepath 3.8 m wide	m	500.00
firepath 4.4 m wide	m	580.00
firepath 5.0 m wide	m	650.00
turning areas	m²	130.00

8.2.1 ROADS, PATHS AND PAVINGS

Item	Unit	Range £
RESIN BOUND AND RESIN BONDED SURFACES		
Resin bonded surfaces; excavate 230 mm; bring to grade; lay 150 mm Type 1 granular material; hand lay base course of 60 mm macadam and resin bonded wearing course; all disposal off site; edgings excluded		
Natratex resin bonded macadam wearing course		
general paved area	m^2	100.00
pathway 1.50 m wide	m	165.00
pathway 3.00 m wide	m	325.00
Addagrip; resin bonded aggregate; 1–3 mm golden pea gravel with buff adhesive		
general paved area	m^2	87.00
pathway 1.50 m wide	m	140.00
pathway 3.00 m wide	m	275.00
Addagrip; resin bonded aggregate; 6 mm chocolate gravel 18 mm depth		
general paved area	m^2	120.00
pathway 1.50 m wide	m	185.00
pathway 3.00 m wide	m	370.00
Addagrip; resin bonded aggregate; Cobalt blue recycled glass 15 mm depth		
general paved area	m^2	100.00
pathway 1.50 m wide	m	155.00
pathway 3.00 m wide	m	315.00
Resin bound surfaces; excavate 230 mm; bring to grade; lay 150 mm Type 1 granular material;hand lay base course of 60 mm macadam and resin bound wearing course; all disposal off site; edgings excluded		
Sureset; resin bound permeable surfacing; natural gravel 6 mm aggregate 18 mm thick		
paved area	m^2	105.00
pathway 1.50 m wide	m	165.00
pathway 3.00 m wide	m	325.00
Sureset; resin bound permeable surfacing; crushed rock 6 mm aggregate 18 mm thick		
paved area	m^2	140.00
pathway 1.50 m wide	m	165.00
pathway 3.00 m wide	m	550.00
Sureset; resin bound permeable surfacing; marble 6 mm aggregate 18 mm thick		
paved area	m^2	120.00
pathway 1.50 m wide	m	190.00
pathway 3.00 m wide	m	380.00
GRAVEL PATHWAYS		
Reduce levels and remove spoil to dump on site; lay 150 mm hardcore well rolled; lay geofabric; fix timber edge 150 × 30 mm to both sides of straight paths		
Lay Cedec gravel 50 mm thick; watered and rolled		
1.0 m wide	m	54.00
1.5 m wide	m	69.00
2.0 m wide	m	94.00
Lay Breedon gravel 50 mm thick; watered and rolled		
1.0 m wide	m	43.00
1.5 m wide	m	54.00
2.0 m wide	m	65.00

8.2.1 ROADS, PATHS AND PAVINGS

Item	Unit	Range £
GRAVEL PATHWAYS – cont		
Reduce levels and remove spoil to dump on site; lay 150 mm Type 1 well rolled; lay geofabric; fix timber edge 150 × 38 mm to both sides of straight paths		
Lay Cedec gravel 50 mm thick; watered and rolled		
1.0 m wide	m	62.00
1.5 m wide	m	81.00
2.0 m wide	m	110.00
Lay Breedon gravel 50 mm thick; watered and rolled		
1.0 m wide	m	50.00
1.5 m wide	m	65.00
2.0 m wide	m	79.00
Lay 20 mm shingle; 30 mm thick		
1.0 m wide	m	45.00
1.5 m wide	m	55.00
2.0 m wide	m	73.00
CELLULAR CONFINED SURFACES		
Excavate ground and reduce levels to receive cellular confined paving; dispose of excavated material off site		
Lay granular fill Type 1; limestone aggregate 150 mm thick laid to falls and compacted; lay confined cellular surface;		
Geoweb 100 thick	m^2	59.00
StablePATH honeycomb cellular gravel grid; 25 mm deep for foot traffic and bicycles filled with shingle	m^2	54.00
StablePATH honeycomb cellular gravel grid; 25 mm deep for foot traffic and bicycles filled with decorative aggregate	m^2	63.00
StablePATH honeycomb cellular gravel grid 40 mm deep for vehicular traffic; filled with shingle	m^2	57.00
StablePATH honeycomb cellular gravel grid 40 mm deep for vehicular traffic; filled with decorative aggregate	m^2	66.00
FLAG PAVINGS		
SLAB/BRICK PATHS		
Stepping stone path inclusive of hand excavation and mechanical disposal, 100 mm Type 1 and sand blinding; slabs laid 100 mm apart to existing turf		
600 mm wide; 600 × 600 × 50 mm slabs		
natural finish	m	90.00
coloured	m	93.00
900 mm wide; 600 × 900 × 50 mm slabs		
natural finish	m	99.00
coloured	m	105.00

8.2.1 ROADS, PATHS AND PAVINGS

Item	Unit	Range £
Pathway inclusive of mechanical excavation and disposal; 100 mm Type 1 and sand blinding; slabs close butted		
Straight butted path 900 mm wide; 600 × 900 × 50 mm slabs		
natural finish	m	40.00
coloured	m	44.00
Straight butted path 1200 mm wide; double row; 600 × 900 × 50 mm slabs; laid stretcher bond		
natural finish	m	69.00
coloured	m	79.00
Straight butted path 1200 mm wide; one row of 600 × 600 × 50 mm; one row 600 × 900 × 50 mm slabs		
natural finish	m	67.00
coloured	m	77.00
Straight butted path 1500 mm wide; slabs of 600 × 900 × 50 mm and 600 × 600 × 50 mm; laid to bond		
natural finish	m	83.00
coloured	m	94.00
Straight butted path 1800 mm wide; two rows of 600 × 900 × 50 mm slabs; laid bonded		
natural finish	m	92.00
coloured	m	115.00
Straight butted path 1800 mm wide; three rows of 600 × 900 × 50 mm slabs; laid stretcher bond		
natural finish	m	100.00
coloured	m	125.00
MARSHALLS FLAG PAVINGS		
Supply and lay precast concrete flags; excavate ground and reduce levels; treat substrate with total herbicide; supply and lay granular fill Type 1 150 mm thick laid to falls and compacted		
Marshalls Urbex paving		
450 × 450 × 35 mm	m²	94.00
600 × 300 × 35 mm	m²	95.00
Marshalls Modal paving		
450 × 450 × 50 mm	m²	105.00
600–900 × 600 × 50 mm	m²	105.00
Random lengths × 200/300/450/600/900 × 50 mm	m²	120.00
FLAG PAVING TO PEDESTRIAN AREAS		
Prices are inclusive of all mechanical excavation and disposal.		
Supply and lay precast concrete flags; excavate ground and reduce levels; treat substrate with total herbicide; supply and lay granular fill Type 1 150 mm thick laid to falls and compacted		
Standard precast concrete flags bedded and jointed in lime: sand mortar (1:3)		
450 × 450 × 50 mm; chamfered	m²	74.00
600 × 300 × 50 mm	m²	79.00
600 × 450 × 50 mm	m²	68.00
600 × 600 × 50 mm	m²	64.00
750 × 600 × 50 mm	m²	63.00
900 × 600 × 50 mm	m²	61.00

8.2.1 ROADS, PATHS AND PAVINGS

Item	Unit	Range £
FLAG PAVINGS – cont		
Supply and lay precast concrete flags – cont		
Coloured flags bedded and jointed in lime: sand mortar (1:3)		
600 × 600 × 50 mm	m^2	72.00
400 × 400 × 65 mm	m^2	74.00
900 × 600 × 50 mm	m^2	70.00
Marshalls Saxon; textured concrete flags; reconstituted Yorkstone in colours; butt jointed bedded in lime: sand mortar (1:3)		
300 × 300 × 35 mm	m^2	145.00
450 × 450 × 50 mm	m^2	105.00
600 × 300 × 35 mm	m^2	120.00
600 × 600 × 50 mm	m^2	95.00
Tactile flags; Marshalls blister tactile pavings; red or buff; for blind pedestrian guidance laid to designed pattern		
450 × 450 mm	m^2	105.00
400 × 400 mm	m^2	96.00
PEDESTRIAN DETERRENT PAVING		
Excavate ground and bring to levels; treat substrate with total herbicide; supply and lay granular fill Type 1 150 mm thick laid to falls and compacted; supply and lay precast deterrent paving units bedded in lime: sand mortar (1:3) and jointed in lime: sand mortar (1:3)		
Marshalls Mono		
Contour deterent paving; 600 × 600 × 75 mm	m^2	105.00
BARK PAVINGS		
Excavate to reduce levels; remove all topsoil to dump on site; treat area with herbicide; lay 150 mm clean hardcore; blind with sand; lay 0.7 mm geotextile filter fabric water flow 50 l/m²/sec; supply and fix treated softwood edging boards 50 × 150 mm to bark area on hardcore base extended 150 mm beyond the bark area; boards fixed with galvanized nails to treated softwood posts 750 × 50 × 75 mm driven into firm ground at 1.0 m centres; edging boards to finish 25 mm above finished bark surface; tops of posts to be flush with edging boards and once weathered		
Supply and lay 100 mm Melcourt conifer walk chips 10–40 mm size		
1.00 m wide	m	58.00
2.00 m wide	m	79.00
3.00 m wide	m	100.00
4.00 m wide	m	120.00
Extra over for wood fibre	m^2	−2.00
Extra over for hardwood chips	m^2	−2.00

8.2.1 ROADS, PATHS AND PAVINGS

Item	Unit	Range £
COBBLE PAVINGS		
BEACH COBBLE PAVING		
Excavate ground and bring to levels and fill with compacted Type 1 fill 100 mm thick; lay GEN 1 concrete base 100 mm thick; supply and lay cobbles individually laid by hand bedded in cement: sand mortar (1:3) 25 mm thick minimum; dry grout with 1:3 cement: sand grout; brush off surplus grout and water in; sponge off cobbles as work proceeds; all excluding formwork edgings or kerbs measured separately		
Scottish beach cobbles 50–75 mm	m^2	205.00
Scottish beach cobbles 100–200 mm	m^2	180.00
Kidney flint cobbles 75–100 mm	m^2	195.00
NATURAL STONE PAVING		
GRANITE SETT PAVING – PEDESTRIAN		
Excavate and grade; 100 mm hardcore; Type 1 50 mm thick; 100 mm concrete 1:2:4; supply and lay granite setts 100 × 100 × 100 mm bedded in cement: sand mortar (1:3) 25 mm thick minimum; jointed in mortar 1:3		
Setts laid to bonded pattern and jointed; bedded in cement: sand mortar 1:3 25 mm thick		
traditional cropped setts; 100 × 100 × 100 mm	m^2	380.00
diamond sawn setts;100 × 100 × 100 mm	m^2	415.00
Setts laid in curved pattern		
traditional cropped setts; 100 × 100 × 100 mm	m^2	390.00
diamond sawn setts;100 × 100 × 100 mm	m^2	425.00
GRANITE SETT PAVING – TRAFFICKED AREAS		
Excavate to reduce levels; lay 100 mm hardcore; compact; Type 1 50 mm thick; lay 150 mm site mixed concrete 1:2:4 reinforced with steel fabric; granite setts 100 × 100 × 100 mm bedded in cement: sand mortar (1:3) 25 mm thick minimum; jointed in mortar 1:3; all excluding edgings or kerbs measured separately		
Site mixed concrete		
traditional cropped setts; 100 × 100 × 100 mm	m^2	395.00
diamond sawn setts;100 × 100 × 100 mm	m^2	425.00
Ready mixed concrete		
traditional cropped setts; 100 × 100 × 100 mm	m^2	375.00
diamond sawn setts;100 × 100 × 100 mm	m^2	405.00

8.2.1 ROADS, PATHS AND PAVINGS

Item	Unit	Range £
NATURAL STONE SLAB PAVING		
WORKS BY MACHINE		
Excavate ground by machine and reduce levels; to receive 65 mm thick slab and 35 mm mortar bed; dispose of excavated material off site; lay granular fill Type 1 150 mm thick laid to falls and compacted; lay to random rectangular pattern on 35 mm mortar bed		
New riven slabs		
laid random rectangular on 35 mm 1:3 mortar bed	m^2	220.00
laid random rectangular on 30 mm thick Steintec mortar bed	m^2	315.00
New riven slabs; but to 150 mm plain concrete base		
laid random rectangular on 35 mm 1:3 mortar bed	m^2	245.00
laid random rectangular on 30 mm thick Steintec mortar bed	m^2	335.00
New riven slabs; disposal by grab		
laid random rectangular	m^2	225.00
Reclaimed Cathedral grade riven slabs		
laid random rectangular	m^2	335.00
Reclaimed Cathedral grade riven slabs; but to 150 mm plain concrete base		
laid random rectangular	m^2	360.00
Reclaimed Cathedral grade riven slabs; disposal by grab		
laid random rectangular	m^2	335.00
New slabs sawn 6 sides		
laid random rectangular	m^2	265.00
3 sizes; laid to coursed pattern	m^2	245.00
New slabs sawn 6 sides; but to 150 mm plain concrete base		
laid random rectangular	m^2	290.00
3 sizes; laid to coursed pattern	m^2	270.00
New slabs sawn 6 sides; disposal by grab		
laid random rectangular	m^2	265.00
3 sizes, laid to coursed pattern	m^2	245.00
Portland stone paving – Albion stone; inclusive of excavation and 150 mm reinforced blinded concrete base;		
Portland stone pavings 40 mm thick		
coursed 3 sizes	m^2	340.00
600 × 600 mm	m^2	280.00
Portland stone pavings 50 mm thick		
coursed 3 sizes	m^2	385.00
600 × 600 mm	m^2	325.00
1200 × 750 mm	m^2	430.00

8.2.1 ROADS, PATHS AND PAVINGS

Item	Unit	Range £
WORKS BY HAND		
Excavate ground by hand and reduce levels, to receive 65 mm thick slab and 35 mm mortar bed; barrow all materials and arisings 25 m; dispose of excavated material off site by grab; treat substrate with total herbicide; lay granular fill Type 1 150 thick laid to falls and compacted; lay to random rectangular pattern on 35 mm mortar bed		
New riven slabs		
laid random rectangular	m^2	285.00
New riven slabs laid random rectangular; but to 150 mm plain concrete base		
laid random rectangular	m^2	320.00
New riven slabs; but disposal to skip		
laid random rectangular	m^2	310.00
Reclaimed Cathedral grade riven slabs		
laid random rectangular	m^2	400.00
Reclaimed Cathedral grade riven slabs; but to 150 mm plain concrete base		
laid random rectangular	m^2	435.00
Reclaimed Cathedral grade riven slabs; but disposal to skip		
laid random rectangular	m^2	430.00
New slabs sawn 6 sides		
laid random rectangular	m^2	330.00
3 sizes; sawn 6 sides laid to coursed pattern	m^2	310.00
New slabs sawn 6 sides; but to 150 mm plain concrete base		
laid random rectangular	m^2	365.00
3 sizes; sawn 6 sides laid to coursed pattern	m^2	345.00
DECKING – TIMBER AND COMPOSITE		
HARDWOOD DECKS		
Timber decking; Exterior Decking Limited; timber decking to roof gardens; cranage or movement of materials to roof top not included; coated one coat with Seasonite		
Joist size 47 × 47 mm		
Ipe; Exterpark; 21 mm invisibly fixed	m^2	170.00
Ipe; pre-drilled and countersunk; screw fixed	m^2	205.00
Rustic Teak; 21 mm invisibly fixed	m^2	225.00
Timber decking to commercial applications; double beams at 300 mm centres; coated one coat with Seasonite; Exterior Decking Limited		
Joist size 47 × 150 mm		
Ipe; Exterpark; 21 mm invisibly fixed	m^2	170.00
Ipe; pre-drilled and countersunk; screw fixed	m^2	205.00
Rustic Teak; 21 mm invisibly fixed	m^2	225.00

8.2.1 ROADS, PATHS AND PAVINGS

Item	Unit	Range £
DECKING – TIMBER AND COMPOSITE – cont		
SOFTWOOD DECKS		
Timber decking; support structure of timber joists for decking laid laid at 400 mm centres on blinded base (measured separately); Exterior Decking Limited		
Yellow balau 125 mm × 21 mm		
joists 47 × 47 mm	m^2	145.00
joists 47 × 100 mm	m^2	145.00
joists 47 × 150 mm	m^2	150.00
Red cedar decking boards 90 mm wide × 25 mm thick		
joists 47 × 47 mm	m^2	145.00
joists 47 × 100 mm	m^2	145.00
joists 47 × 150 mm	m^2	150.00
Composite decking – Millboard; London Stone Ltd		
Decking type on Composite substructure; laid to blinded leveled ground; Average area 25–50 m^2		
Enhanced grain range	m^2	340.00
Weathered Oak range	m^2	360.00
Decking type on Composite substructure; average area 25–50 m^2; On support pedestals 28 mm high		
Enhanced grain range	m^2	350.00
Weathered Oak range	m^2	370.00
Decking type on Composite substructure; average area 25–50 m^2; On support pedestals 175–285 mm high		
Enhanced grain range	m^2	380.00
Weathered Oak range	m^2	400.00
Decking type on Composite substructure; average area 25–50 m^2; On support pedestals 285–400 mm high		
Enhanced grain range	m^2	400.00
Weathered Oak range	m^2	420.00
Handrails		
handrails fixed to posts 100 × 100 × 1370 mm high		
square balusters at 100 mm centres	m	130.00
square balusters at 300 mm centres	m	97.00
turned balusters at 100 mm centres	m	165.00
turned balusters at 300 mm centres	m	110.00
STEPS		
Excavate for new steps; grade to correct levels; remove excavated material from site and lay concrete base; Construct formwork; supply and lay site mixed concrete; Fix treads and risers to steps		
Stone clad steps; 1.20 m wide; tread depth 225 mm; Granite cut off-site		
1 risers	nr	460.00
2 risers	nr	770.00
3 risers	nr	1100.00
4 risers	nr	1500.00
6 risers	nr	2100.00

8.2.1 ROADS, PATHS AND PAVINGS

Item	Unit	Range £
Stone clad steps; 1.20 m wide; tread depth 225 mm; Riven Yorkstone cut on site		
1 risers	nr	410.00
2 risers	nr	680.00
3 risers	nr	950.00
4 risers	nr	1275.00
6 risers	nr	1825.00
Stone clad steps; 2.40 m wide; tread depth 225 mm; Granite cut off-site		
1 risers	nr	860.00
2 risers	nr	1475.00
3 risers	nr	2125.00
4 risers	nr	2900.00
6 risers	nr	4225.00
Steps; Portland stone; Albion stone; Cast bases of in situ concrete including excavation; Construct steps; risers of Portland stone 30 mm thick fixed to concrete step face		
tread depth 350 mm; 50 mm thick		
1 st riser	m	590.00
subsequent risers	m	540.00
tread depth 500 mm; 50 mm thick		
1 st riser	m	730.00
subsequent risers	m	710.00
Excavate for new steps; grade to correct levels; remove excavated material from site and lay concrete base; Construct formwork; supply and lay site mixed concrete; Fix treads and risers to steps		
Stone clad steps; tread depth 250 mm; Granite cut off-site; Marshalls Bega LED Lights at 2.00 m centres		
1 st riser	m	690.00
subsequent risers	m	670.00
Stone clad steps; tread depth 250 mm; Granite cut off-site; Marshalls Bega LED Lights at 1.00 m centres		
1 st riser	m	900.00
subsequent risers	m	870.00
Stone clad steps; tread depth 400 mm; Granite cut off-site; Marshalls Bega LED Lights at 2.00 m centres		
1 st riser	m	790.00
subsequent risers	m	790.00
Stone clad steps; tread depth 450 mm; Granite cut off-site; Marshalls Bega LED Lights at 1.00 m centres		
1 st riser	m	980.00
subsequent risers	m	970.00
Stone clad steps; tread depth 600 mm; Granite cut off-site; Marshalls Bega LED Lights at 1.00 m centres		
1 st riser	m	1125.00
subsequent risers	m	1100.00

8.2.1 ROADS, PATHS AND PAVINGS

Item	Unit	Range £
PLAY AREA FOR BALL GAMES		
Excavate for playground; bring to grade; lay 225 mm consolidated hardcore; blind with 100 mm type 1; lay macadam base of 28 mm size dense bitumen macadam to 75 mm thick; lay wearing course 30 mm thick		
Over 1000 m^2	100 m^2	7200.00
400–1000 m^2	100 m^2	7800.00
Excavate for playground; bring to grade; lay 100 mm consolidated hardcore; blind with 100 mm type 1; lay macadam base of dense bitumen macadam to 50 mm thick; lay wearing course 20 mm thick		
Over 1000 m^2	100 m^2	5400.00
400–1000 m^2	100 m^2	7100.00
Excavate for playground; bring to grade; lay 100 mm consolidated hardcore; blind with 100 mm type 1; lay macadam base of dense bitumen macadam to 50 mm thick; lay wearing course of Addagrip resin coated aggregate 3 mm dia. × 6 mm thick		
Over 1000 m^2	100 m^2	8600.00
400–1000 m^2	100 m^2	9800.00
SAFETY SURFACED PLAY AREAS		
Excavate ground and reduce levels to receive safety surface; dispose of excavated material off site; treat substrate with total herbicide; lay granular fill Type 1 150 mm thick laid to falls and compacted; lay macadam base 40 mm thick; supply and lay Rubaflex wet pour safety system to thicknesses as specified		
All by machine except macadam by hand		
black		
15 mm thick	m^2	98.00
35 mm thick	m^2	125.00
60 mm thick	m^2	140.00
coloured		
15 mm thick	m^2	120.00
35 mm thick	m^2	130.00
60 mm thick	m^2	165.00
All by hand except disposal by grab		
black		
15 mm thick	m^2	140.00
35 mm thick	m^2	170.00
60 mm thick	m^2	195.00
coloured		
15 mm thick	m^2	160.00
35 mm thick	m^2	200.00
60 mm thick	m^2	215.00
Excavate for playground; bring to grade; lay with 150 mm type 1; lay macadam base of dense bitumen macadam to 50 mm thick; lay safety surface		
Wetpour coloured; 0.50 m critical fall height	m	145.00
Wetpour black; 0.50 m critical fall height	m	99.00
Wetpour coloured; 1.50 m critical fall height	m	165.00
Wetpour black; 1.50 m critical fall height	m	140.00

8.2.1 ROADS, PATHS AND PAVINGS

Item	Unit	Range £
Excavate for playground; bring to grade; lay with 150 mm type 1; lay safety surface Meon UK		
Black; 10 m^2–50 m^2	m	140.00
Black; over 50 m^2	m	140.00
Colours; 10 m^2–50 m^2	m	165.00
Colours; over 50 m^2	m	165.00
BARK PLAY AREA		
Excavate playground area to 450 mm depth; lay 150 mm broken stone or clean hardcore; lay filter membrane; lay bark surface		
Melcourt Industries		
Playbark 10/50; 300 mm thick	m^2	62.00
Playbark 8/25; 300 mm thick	m^2	69.00

8.2.2 SPECIAL SURFACINGS AND PAVINGS

Item	Unit	Range £
SPORTS FIELD CONSTRUCTION		
Strip existing vegetation; grade to levels using laser guided equipment; cultivate ground; apply pre-seeding fertilizer at 900 kg/ha; apply pre-seeding selective weedkiller; seed in two operations with sports pitch type grass seed at 350 kg/ha; harrow and roll lightly; including initial cut; size; mark out and paint lines; erect goal posts		
New ground construction of existing field; greenfield or brownfield; existing lightweight or loam soils		
association football; senior 114 × 72 m	each	37000.00
association football; junior 106 × 58 m	each	28000.00
New ground construction of existing field; greenfield or brownfield; existing clay soil; replacement of rootzone 150 mm		
association football; senior 114 × 72 m	each	130000.00
association football; junior 106 × 58 m	each	100000.00
Plain sports pitches; site clearance, grading and drainage not included		
Agripower Ltd; cultivate ground and grade to levels; apply pre-seeding fertilizer at 900 kg/ha; apply pre-seeding selective weedkiller; seed in two operations with sports pitch type grass seed at 350 kg/ha; harrow and roll lightly; including initial cut; size		
Existing level and graded ground previously used for playing surface		
association football; senior 114 × 72 m	each	5100.00
association football; junior 106 × 58 m	each	4000.00
rugby union pitch; 156 × 81 m	each	6600.00
rugby league pitch; 134 × 60 m	each	6600.00
hockey pitch; 95 × 60 m	each	5100.00
Agripower Ltd; cricket square; excavate to depth of 100 mm; lay imported marl or clay loam; bring to accurate levels; apply pre-seeding fertilizer at 50 g/m; apply selective weedkiller; seed with cricket square type g grass seed at 50 g/m²; rake in and roll lightly; erect and remove temporary protective chestnut fencing; allow for initial cut and watering three times		
22.8 × 22.8 m	each	21000.00
JOGGING TRACK		
Excavate track 250 mm deep; lay filter membrane; lay 100 mm depth gravel waste or similar; lay 100 mm compacted gravel	100 m²	13000.00
Extra for treated softwood edging 50 × 150 mm on 50 × 50 mm posts		
both sides	m	21.00

8.3.1 SOFT LANDSCAPING SEEDING AND TURFING

Item	Unit	Range £
SPORTS GROUND FOR AMENITY SPORT		
To existing field; Spray to remove existing turf. Scarify top 25 mm; Grade and level using ripper power harrow cultivator and stone burier combination inclusive of grass seed and fertilizers		
Light to medium soils		
Existing area to 1:100 existing grade	6000 m^2	4625.00
Clay soils		
Existing area to 1:100 existing grade	6000 m^2	5800.00
SURFACE PREPARATION FOR SEEDING AND TURFING		
BY MACHINE		
Mechanized preparation; Stone burier and cultivator; Treat area with systemic herbicide in advance of operations (not included); rip subsoil using subsoiling machine to a depth of 500 mm at 1.20 m centres in light to medium soils; rotavate to 200 mm deep in two passes;		
Larger areas		
light to meduim soils	100 m^2	31.00
heavier soils	100 m^2	56.00
Rip area by tractor and subsoiler; soil preparation by pedestrian operated rotavator; clearance and raking by hand		
light to medium soils	100 m^2	120.00
heavier soils	100 m^2	130.00
Spread and lightly consolidate topsoil brought from spoil heap not exceeding 100 m; in layers not exceeding 150 mm; grade to specified levels; remove stones over 25 mm; all by machine		
100 mm thick	100 m^2	140.00
Extra to the above for imported topsoil		
topsoil to BS3882 PC £29.59 m^3 calculation allows for 20% settlement		
100 mm thick	100 m^2	405.00
Excavate existing topsoil and remove off site; fill with new good quality topsoil to BS3882; add 50 mm mushroom compost (delivered in 65 m^3 loads) and cultivate in; grade to level removing stones		
For turfing		
200 mm deep	m^2	24.00
300 mm deep	m^2	32.00

8.3.1 SOFT LANDSCAPING SEEDING AND TURFING

Item	Unit	Range £
SURFACE PREPARATION FOR SEEDING AND TURFING – cont		
SURFACE PREPARATION BY HAND		
To area previously cleared and sprayed; cultivate existing topsoil to receive new ornamental planting; rotavate using pedestrian rotavator; dig over by hand 1 spit deep; rake and grade to level to receive new planting		
For areas to be seeded or turfed	m^2	1.00
Spread only and lightly consolidate topsoil brought from spoil heap in layers not exceeding 150 mm; grade to specified levels; remove stones over 25 mm; all by hand		
100 mm thick	100 m^2	720.00
100 mm thick; but inclusive of loading to location by barrow maximum distance 25 m; finished topsoil depth	100 m^2	1550.00
100 mm thick but loading to location by barrow maximum distance 100 m; finished topsoil depth	100 m^2	1125.00
Extra to above for imported topsoil PC £29.59 m^3; Calculation allows for 20% settlement		
100 mm thick	100 m^2	405.00
SEEDING		
Cultivate topsoil 200 mm a fine tilth by tractor; remove stones over 25 mm by mechanical stone rake and bring to final tilth by harrow; apply pre-seeding fertilizer at 50 g/m^2 and work into top 50 mm during final cultivation; seed with certified grass seed in two operations; roll seedbed lightly after sowing		
General amenity grass areas; 35 g/m^2; BSH A3		
general grass seeded areas	100 m^2	94.00
lawn areas for public or domestic general use	100 m^2	115.00
high grade lawn areas	100 m^2	140.00
General amenity grass at 35 g/m^2; DLF Seeds Ltd Promaster 10		
general turfed areas	100 m^2	110.00
lawn areas for public or domestic general use	100 m^2	130.00
high grade lawn areas	100 m^2	155.00
Shaded areas; BSH A6 at 50 g/m^2	100 m^2	105.00
Motorway and road verges; DLF Seeds Ltd Promaster 120 at 25–35 g/m^2	100 m^2	100.00
Cultivate topsoil 200 mm a fine tilth by tractor; remove stones over 25 mm by mechanical stone rake and bring to final tilth by harrow; add 50 mm of imported topsoil over surface of area to be seeded; apply pre-seeding fertilizer at 50 g/m^2 and work into top 50 mm during final cultivation; seed with certified grass seed in two operations; roll seedbed lightly after sowing		
General amenity grass areas; 35 g/m^2; BSH A3		
general turfed areas	100 m^2	335.00
lawn areas for public or domestic general use	100 m^2	355.00
high grade lawn areas	100 m^2	385.00

8.3.1 SOFT LANDSCAPING SEEDING AND TURFING

Item	Unit	Range £
Cultivate topsoil 200 mm a fine tilth by tractor; remove stones over 25 mm by mechanical stone rake and bring to final tilth by harrow; add 50 mm of topsoil from site spoil heaps maximum distance 100 m over surface of area to be seeded; apply pre-seeding fertilizer at 50 g/m² and work into top 50 mm during final cultivation; seed with certified grass seed in two operations; roll seedbed lightly after sowing		
General amenity grass areas; 35 g/m²; BSH A3		
general turfed areas	100 m²	160.00
lawn areas for public or domestic general use	100 m²	185.00
high grade lawn areas	100 m²	210.00
Bring top 200 mm of topsoil to a fine tilth using pedestrian operated rotavator; remove stones over 25 mm; apply pre-seeding fertilizer at 50 mm g/m² and work into top 50 mm during final hand cultivation; seed with certified grass seed in two operations; rake and roll seedbed lightly after sowing		
General amenity grass areas; 35 g/m²; BSH A3		
general grass seeded areas	100 m²	210.00
lawn areas for public or domestic general use	100 m²	235.00
high grade lawn areas	100 m²	255.00
Extra to above for using imported topsoil spread by machine		
100 mm minimum depth	100 m²	465.00
150 mm minimum depth	100 m²	690.00
extra for slopes over 30°	50%	–
Extra to above for using imported topsoil spread by hand; maximum distance for transporting soil 100 m		
100 mm minimum depth	m²	1125.00
150 mm minimum depth	m²	1675.00
extra for slopes over 30°	50%	–
Extra for mechanically screening top 25 mm of topsoil through 6 mm screen and spreading on seedbed; debris carted to dump on site not exceeding 100 m	m³	9.00
Bring top 200 mm of topsoil to a fine tilth; remove stones over 50 mm; apply pre-seeding fertilizer at 50 g/m² and work into top 50 mm of topsoil during final cultivation; seed with certified grass seed in two operations; harrow and roll seedbed lightly after sowing		
Areas inclusive of fine levelling by specialist machinery		
outfield grass at 350 kg/ha; DLF Seeds Ltd Promaster 40	ha	11500.00
outfield grass at 350 kg/ha; DLF Seeds Ltd Promaster 70	ha	11000.00
sportsfield grass at 300 kg/ha; DLF Seeds Ltd J Pitch	ha	11000.00
Seeded areas prepared by chain harrow		
low maintenance grass at 350 kg/ha	ha	12000.00
verge mixture grass at 150 kg/ha	ha	10500.00
Extra for wild flora mixture at 30 kg/ha		
BSH WSF 75 kg/ha	ha	4925.00
extra for slopes over 30°	50%	–
Cut existing turf to 1.0 × 1.0 × 0.5 m turves, roll up and move to stack not exceeding 100 m		
by pedestrian operated machine; roll up and stack by hand	100 m²	130.00
all works by hand	100 m²	385.00
Extra for boxing and cutting turves	100 m²	7.00

8.3.1 SOFT LANDSCAPING SEEDING AND TURFING

Item	Unit	Range £
TURFING		
Cultivate topsoil in two passes; remove stones over 25 mm and bring to final tilth; apply pre-seeding fertilizer at 50 g/m² and work into top 50 mm during final cultivation; roll turf bed lightly		
Using tractor drawn implements and mechanical stone rake	100 m²	76.00
Cultivation by pedestrian rotavator; all other operations by hand		
light or medium soils	m²	2.00
heavy wet or clay soils	m²	2.00
Cultivate topsoil in two passes; remove stones over 25 mm and bring to final tilth; apply pre-seeding fertilizer at 50 g/m² and work into top 50 mm during final cultivation; lay imported turves maximum distance from turf stack 25 m		
using tractor drawn implements and mechanical stone rake	m²	11.00
cultivation by pedestrian rotavator; all other operations by hand	m²	11.00
Cultivate topsoil in two passes; remove stones over 25 mm and bring to final tilth; apply pre-seeding fertilizer at 50 g/m² and work into top 50 mm during final cultivation; roll turf bed supply and lay imported turf; Rolawn Medallion		
using tractor drawn implements and mechanical stone rake	m²	10.00
cultivation by pedestrian rotavator; all other operations by hand	m²	10.00
Turfed areas; watering on two occasions and carrying out initial cut		
by ride on triple mower	100 m²	2.00
by pedestrian mower	100 m²	9.00
by pedestrian mower; box cutting	100 m²	10.00
Extra for using imported topsoil spread and graded by machine		
25 mm minimum depth	m²	1.00
75 mm minimum depth	m²	4.00
100 mm minimum depth	m²	5.00
150 mm minimum depth	m²	7.00
Extra for using imported topsoil spread and graded by hand; distance of barrow run 25 m		
25 mm minimum depth	m²	3.00
75 mm minimum depth	m²	7.00
100 mm minimum depth	m²	9.00
150 mm minimum depth	m²	14.00
Extra for work on slopes over 30° including pegging with 200 galvanized wire pins	m²	3.00
INTURF BIG ROLL		
Supply, deliver in one consignment, fully prepare the area and install in Big Roll format Inturf 553, a turfgrass comprising dwarf perennial ryegrass, smooth stalked rneadowgrass and fescues; installation by tracked machine		
Preparation by tractor drawn rotavator	m²	10.00
Cultivation by pedestrian rotavator; all other operations by hand	m²	11.00

8.3.1 SOFT LANDSCAPING SEEDING AND TURFING

Item	Unit	Range £
EROSION CONTROL		
On ground previously cultivated, bring area to level, treat with herbicide; lay 20 mm thick open texture erosion control mat with 100 mm laps, fixed with 8 × 400 mm steel pegs at 1.0 m centres; sow with low maintenance grass suitable for erosion control on slopes at 35 g/m²; spread imported topsoil 25 mm thick and fertilizer at 35 g/m²; water lightly using sprinklers on two occasions	m²	11.00
hand-watering by hose pipe instead of by sprinkler maximum distance from mains supply 50 m	m²	11.00
ARTIFICIAL TURF		
Excavate ground and reduce levels to receive 20 mm thick artifial grass surface on a compacted 150 thick Type 1 base; dispose of excavated material off site		
Lay artificial turf; by machine		
Trulawn Value; 16 mm	m²	62.00
Trulawn Play; 20 mm; entry level; lawn and play areas	m²	55.00
Trulawn Continental; 20 mm; realistic appearance; lawns and patios	m²	57.00
Trulawn Optimum; 26 mm; softest pile; lush green; lawns and patios	m²	60.00
Trulawn Luxury; 30 mm; deep pile; realistic lawn	m²	66.00
All by hand except disposal by grab		
Trulawn Value; 16 mm	m²	110.00
Trulawn Play; 20 mm; entry level; lawn and play areas	m²	105.00
Trulawn Continental; 20 mm; realistic appearance; lawns and patios	m²	105.00
Trulawn Optimum; 26 mm; softest pile; lush green; lawns and patios	m²	105.00
Trulawn Luxury; 30 mm; deep pile; realistic lawn	m²	120.00
MAINTENANCE OF TURFED/SEEDED AREAS		
Maintenance executed as part of a landscape construction contract		
Grass cutting		
Grass cutting; fine turf; using pedestrian guided machinery; arisings boxed and disposed of off site		
per occasion	100 m²	11.00
per annum 26 cuts	100 m²	275.00
per annum 18 cuts	100 m²	200.00
Grass cutting; standard turf; using self propelled three gang machinery		
per occasion	100 m²	1.00
per annum 26 cuts	100 m²	18.00
per annum 18 cuts	100 m²	12.00
Grass cutting for one year; recreation areas, parks, amenity grass areas; using tractor drawn machinery		
per occasion	ha	45.00
per annum 26 cuts	ha	1150.00
per annum 18 cuts	ha	810.00

8.3.1 SOFT LANDSCAPING SEEDING AND TURFING

Item	Unit	Range £
MAINTENANCE OF TURFED/SEEDED AREAS – cont		
Maintain grass area inclusive of grass cutting; fertilization twice per annum; leaf and litter clearance		
using pedestrian guided machinery; arisings boxed and disposed of off site		
26 cuts; public areas	100 m^2	315.00
26 cuts; private areas	100 m^2	295.00
18 cuts; public areas	100 m^2	240.00
18 cuts; private areas	100 m^2	220.00
Maintain grass area inclusive of grass cutting; fertilization twice per annum; leaf and litter clearance		
using ride-on three cylinder machinery; arisings left on site		
26 cuts; public areas	100 m^2	52.00
26 cuts; private areas	100 m^2	28.00
18 cuts; public areas	100 m^2	48.00
18 cuts; private areas	100 m^2	24.00
Aeration of turfed areas		
Aerate ground with spiked aerator; apply spring/summer fertilizer once; apply autumn/winter fertilizer once; cut grass, 16 cuts; sweep up leaves twice		
as part of a landscape contract, defects liability	ha	2750.00
as part of a long term maintenance contract	ha	1500.00

8.3.2 EXTERNAL PLANTING

Item	Unit	Range £
SURFACE PREPARATIONS FOR PLANTING		
SURFACE PREPARATION BY MACHINE		
Rip subsoil using subsoiling machine to a depth of 250 mm below topsoil at 1.20 m centres in light to medium soils; rotavate to 200 mm deep in two passes; cultivate with chain harrow; roll lightly; clear stones over 50 mm		
By tractor	100 m^2	36.00
by tractor but carrying out operations in clay or compacted gravel	100 m^2	40.00
ripping by tractor rotavation by pedestrian operated rotavator; clearance and raking by hand; herbicide application by knapsack sprayer	100 m^2	83.00
ripping by tractor; rotavation by pedestrian operated rotavator; clearance and raking by hand; herbicide application by knapsack sprayer but carrying out operations in clay or compacted gravel	100 m^2	86.00
Spread and lightly consolidate topsoil brought from spoil heap not exceeding 100 m; in layers not exceeding 150 mm; grade to specified levels; remove stones over 25 mm; all by machine		
100 mm thick	m^2	1.00
150 mm thick	m^2	2.00
300 mm thick	m^2	4.00
450 mm thick	m^2	6.00
Extra to the above for imported topsoil		
topsoil PC £29.59 / m^3; calculation allows for 20% settlement		
100 mm thick	m^2	4.00
150 mm thick	m^2	6.00
300 mm thick	m^2	12.00
450 mm thick	m^2	18.00
500 mm thick	m^2	20.00
600 mm thick	m^2	24.00
750 mm thick	m^2	30.00
1.00 m thick	m^2	41.00
topsoil to BS3882 topsoil PC £29.59 m^3; calculation allows for 20% settlement		
100 mm thick	m^2	4.00
150 mm thick	m^2	6.00
300 mm thick	m^2	12.00
450 mm thick	m^2	18.00
500 mm thick	m^2	20.00
600 mm thick	m^2	24.00
750 mm thick	m^2	30.00
1.00 m thick	m^2	41.00
Excavate existing topsoil and remove off site; fill with new good quality topsoil to BS3882; add 50 mm mushroom compost (delivered in 65 m^3 loads) and cultivate in; grade to level removing stones		
For planting		
300 mm deep	m^2	31.00
400 mm deep	m^2	44.00
500 mm deep	m^2	52.00

8.3.2 EXTERNAL PLANTING

Item	Unit	Range £
SURFACE PREPARATIONS FOR PLANTING – cont		
Excavate existing topsoil and remove off site – cont		
Extra for incorporating mushroom compost at 50 mm/m² into the top 150 mm of topsoil (compost delivered in 20 m³) loads		
manually spread; mechanically rotavated	m²	3.00
mechanically spread and rotavated	m²	2.00
Extra for incorporating manure at 50 mm/m² into the top 150 mm of topsoil loads		
manually spread; mechanically rotavated; 20 m³ loads	m²	3.00
mechanically spread and rotavated; 60 m³ loads	m²	2.00
SURFACE PREPARATION BY HAND		
To area previously cleared and sprayed; cultivate existing topsoil to receive new ornamental planting; rotavate using pedestrian rotavator. dig over by hand 1 spit deep; rake and grade to level to receive new planting		
For ornamental planting beds	m²	1.00
Spread only and lightly consolidate topsoil brought from spoil heap in layers not exceeding 150 mm; grade to specified levels; remove stones over 25 mm; all by hand		
excavated material alongside or within area to be treated		
100 mm thick	100 m²	720.00
150 mm thick	100 m²	1075.00
300 mm thick	100 m²	2125.00
450 mm thick	100 m²	3200.00
loading to location by barrow maximum distance 25 m; finished topsoil depth		
100 mm thick	100 m²	1550.00
150 mm thick	100 m²	2350.00
300 mm thick	100 m²	4725.00
450 mm thick	100 m²	7100.00
500 mm thick	100 m²	7800.00
600 mm thick	100 m²	9400.00
loading to location by barrow maximum distance 100 m; finished topsoil depth		
100 mm thick	100 m²	1125.00
150 mm thick	100 m²	1700.00
300 mm thick	100 m²	3425.00
450 mm thick	100 m²	5200.00
500 mm thick	100 m²	5700.00
600 mm thick	100 m²	6800.00
Extra to the above for incorporating mushroom compost at 50 mm/m² into the top 150 mm of topsoil; by hand	100 m²	340.00
Extra to above for imported topsoil topsoil PC £29.59 m³; calculation allows for 20% settlement		
100 mm thick	100 m²	405.00
150 mm thick	100 m²	620.00
300 mm thick	100 m²	1225.00
450 mm thick	100 m²	1825.00
500 mm thick	100 m²	2025.00
600 mm thick	100 m²	2425.00
750 mm thick	100 m²	3025.00
1.00 m thick	100 m²	4050.00

8.3.2 EXTERNAL PLANTING

Item	Unit	Range £
TREE PLANTING		
Specification Corresponding Tree Girth Size		
Excavate tree pit by hand; fork over bottom of pit; plant tree with roots well spread out; backfill with excavated material incorporating treeplanting compost at 1 m³ per 3 m³ of soil; one tree stake and two ties; tree pits square in sizes shown		
Standard bare root tree in pit; PC £30.00		
600 × 600 mm deep	each	110.00
900 × 600 mm deep	each	130.00
Standard root balled tree in pit; PC £40.47		
600 × 600 mm deep	each	130.00
900 × 600 mm deep	each	140.00
1.00 × 1.00 deep	each	185.00
Selected standard bare root tree in pit; PC £53.00		
900 × 900 mm deep	each	160.00
1.00 × 1.00 deep	each	215.00
Selected standard root ball tree in pit; PC £67.02		
900 × 900 mm deep	each	210.00
1.00 × 1.00 deep	each	300.00
Heavy standard bare root tree in pit; PC £83.00		
900 × 900 mm deep	each	245.00
1.00 × 1.00 deep	each	270.00
1.50 m × 750 mm deep	each	350.00
Heavy standard root ball tree in pit; PC £100.68		
900 × 900 mm deep	each	265.00
1.00 × 1.00 deep	each	300.00
1.50 m × 750 mm deep	each	370.00
Extra heavy standard bare root tree in pit; PC £103.00		
1.00 × 1.00 deep	each	315.00
1.20 × 1.00 deep	each	320.00
1.50 m × 750 mm deep	each	420.00
Extra heavy standard root ball tree in pit; PC £125.00		
1.00 × 1.00 deep	each	350.00
1.50 m × 750 mm deep	each	415.00
1.50 × 1.00 deep	each	495.00

8.3.2 EXTERNAL PLANTING

Item	Unit	Range £
SEMI-MATURE TREE PLANTING		
Trees in soft landscape environments		
Excavate tree pit deep by machine; fork over bottom of pit; plant rootballed tree Acer platanoides Emerald Queen using telehandler where necessary; backfill with excavated material, incorporating Melcourt Topgrow bark/ manure mixture at 1 m^3 per 3 m^3 of soil; Platipus underground guying system; tree pits 1500 × 1500 × 1500 mm deep inclusive of Platimats; excavated material not backfilled to treepit spread to surrounding area		
treepits 1500 × 1500 × 1500 mm		
16–18 cm girth; PC £110.00	each	530.00
18–20 cm girth; PC £150.00	each	600.00
20–25 cm girth; PC £180.00	each	780.00
25–30 cm girth; PC £250.00	each	1000.00
30–35 cm girth; PC £380.00	each	1300.00
treepits 2.00 × 2.00 × 1.5 m deep		
40–45 cm girth; PC £1000.00	each	2500.00
45–50 cm girth; PC £2000.00	each	4100.00
55–60 cm girth; PC £3000.00	each	5400.00
60–70 cm girth; PC £4000.00	each	7700.00
80–90 cm girth; PC £7500.00	each	13000.00
Excavate tree pit deep by machine; fork over bottom of pit; plant rootballed tree Acer platanoides Emerald Queen using telehandler where necessary; backfill with excavated material, incorporating Melcourt Topgrow bark/ manure mixture at 1 m^3 per 3 m^3 of soil; Double timber stake and crossbar staking; tree pits 1500 × 1500 × 1500 mm deep; excavated material not backfilled to treepit spread to surrounding area		
16–18 cm girth; PC £110.00	each	470.00
18–20 cm girth; PC £150.00	each	540.00
20–25 cm girth; PC £180.00	each	730.00
Extra to the above for imported topsoil moved 25 m from tipping area and disposal off site of excavated material		
Tree pits 1500 × 1500 × 1500 mm deep		
16–18 cm girth	each	345.00
18–20 cm girth	each	330.00
20–25 cm girth	each	315.00
25–30 cm girth	each	305.00
30–35 cm girth	each	290.00
Tree pit in Trafficked environments		
Excavate tree pit; remove excavated material; break up bottom of pit; line pit with root barrier; Lay drainage layer 200 mm thick; backfill treepit with cellular load supporting cells to 7.5 tonne traffic weight limit; Plant tree with underground guying system; inclusive of root director Add rootball irrigation system; backfill over rootball with topsoil enriched with composts and fertilizers		
Tree pit tree size		
2.00 × 2.00 × 1.00 deep; 30–35 cm tree	nr	3100.00
2.50 × 2.50 × 1.00 deep; 40–45 cm tree	nr	5100.00

8.3.2 EXTERNAL PLANTING

Item	Unit	Range £
Excavate tree pit; remove excavated material; break up bottom of pit; line pit with root barrier; Lay drainage layer 200 mm thick; backfill treepit with cellular load supporting cells to 40 tonne traffic weight limit; Plant tree with underground guying system; inclusive of root director. Add rootball irrigation system; backfill over rootball with topsoil enriched with composts and fertilizers		
Tree pit tree size		
2.00 × 2.00 × 1.00 deep; 30–35 cm tree	nr	12000.00
2.50 × 2.50 × 1.00 deep; 40–45 cm tree	nr	19000.00
TREE PLANTING WITH MOBILE CRANES		
Excavate treepit 1.50 × 1.50 × 1.00 deep; supply and plant semi mature trees delivered in full loads; trees lifted by crane; inclusive of backfilling tree pit with imported topsoil, compost, fertilizers and underground guying using Platipus anchors		
Self managed lift; local authority applications, health and safety, traffic management or road closures not included; tree size and distance of lift; 35 tonne crane		
25–30 cm; maximum 25 m distance	each	1250.00
30–35 cm; maximum 25 m distance	each	1575.00
35–40 cm; maximum 25 m distance	each	2450.00
55–60 cm; maximum 15 m distance	each	5400.00
80–90 cm; maximum 10 m distance	each	16000.00
Managed lift; inclusive of all local authority applications, health and safety, traffic management or road closures all by crane hire company; tree size and distance of lift; 35 tonne crane		
25–30 cm; maximum 25 m distance	each	1275.00
30–35 cm; maximum 25 m distance	each	1575.00
35–40 cm; maximum 25 m distance	each	2450.00
55–60 cm; maximum 15 m distance	each	5500.00
80–90 cm; maximum 10 m distance	each	16000.00
Self managed lift; local authority applications, health and safety, traffic management or road closures not included; tree size and distance of lift; 80 tonne crane		
25–30 cm; maximum 40 m distance	each	1275.00
30–35 cm; maximum 40 m distance	each	1575.00
35–40 cm; maximum 40 m distance	each	2450.00
55–60 cm; maximum 33 m distance	each	5400.00
80–90 cm; maximum 23 m distance	each	16000.00
Managed lift; inclusive of all local authority applications, health and safety, traffic management or road closures all by crane hire company; tree size and distance of lift; 35 tonne crane		
25–30 cm; maximum 40 m distance	each	1275.00
30–35 cm; maximum 40 m distance	each	1600.00
35–40 cm; maximum 40 m distance	each	2500.00
55–60 cm; maximum 33 m distance	each	5500.00
80–90 cm; maximum 23 m distance	each	16000.00

8.3.2 EXTERNAL PLANTING

Item	Unit	Range £
PLANTING SHRUBS AND GROUNDCOVERS		
SHRUBS		
Excavate planting holes 250 × 250 × 300 mm deep to area previously ripped and rotavated; excavated material left alongside planting hole		
By mechanical auger		
250 mm centres (16 plants per m^2)	m^2	20.00
300 mm centres (11.11 plants per m^2)	m^2	14.00
400 mm centres (6.26 plants per m^2)	m^2	8.00
450 mm centres (4.93 plants per m^2)	m^2	6.00
500 mm centres (4 plants per m^2)	m^2	5.00
600 mm centres (2.77 plants per m^2)	m^2	3.00
750 mm centres (1.77 plants per m^2)	m^2	2.00
900 mm centres (1.23 plants per m^2)	m^2	2.00
1.00 m centres (1 plant per m^2)	m^2	1.00
1.50 m centres (0.44 plants per m^2)	m^2	1.00
Excavation by hand		
250 mm centres (16 plants per m^2)	m^2	20.00
300 mm centres (11.11 plants per m^2)	m^2	14.00
400 mm centres (6.26 plants per m^2)	m^2	8.00
450 mm centres (4.93 plants per m^2)	m^2	6.00
500 mm centres (4 plants per m^2)	m^2	5.00
600 mm centres (2.77 plants per m^2)	m^2	3.00
750 mm centres (1.77 plants per m^2)	m^2	2.00
900 mm centres (1.23 plants per m^2)	m^2	2.00
1.00 m centres (1 plant per m^2)	m^2	1.00
1.50 m centres (0.44 plants per m^2)	m^2	1.00
Clear light vegetation from planting area and remove to dump on site; dig planting holes; plant whips with roots well spread out; backfill with excavated topsoil; including one 38 × 38 mm treated softwood stake, two tree ties and mesh guard 1.20 m high; allow for beating up once at 10% of original planting, cleaning and weeding round whips once, applying fertilizer once at 35 gm/m²; using the following mix of whips, bare rooted		
Plant bare root plants average PC £0.51 each to matrix		
1500 × 1500 mm	100 m^2	570.00
Cultivate and grade shrub bed; bring top 300 mm of topsoil to a fine tilth, incorporating mushroom compost at 50 mm and Enmag slow release fertilizer; rake and bring to given levels; remove all stones and debris over 50 mm; dig planting holes average 300 × 300 × 300 mm deep; supply and plant specified shrubs in quantities as shown below; backfill with excavated material; water to field capacity and mulch 50 mm bark chips 20–40 mm size; water and weed regularly for 12 months and replace failed plants		
Shrubs 3 l PC £3.68; ground covers 9 cm PC £1.60		
100% shrub area		
300 mm centres	100 m^2	6200.00
400 mm centres	100 m^2	3900.00
500 mm centres	100 m^2	2775.00
600 mm centres	100 m^2	2225.00

8.3.2 EXTERNAL PLANTING

Item	Unit	Range £
100% groundcovers		
200 mm centres	100 m²	6900.00
300 mm centres	100 m²	3500.00
400 mm centres	100 m²	2350.00
500 mm centres	100 m²	1825.00
groundcover 30%/shrubs 70% at the distances shown below		
200/300 mm	100 m²	6300.00
300/400 mm	100 m²	3700.00
300/500 mm	100 m²	2975.00
400/500 mm	100 m²	2625.00
groundcover 50%/shrubs 50% at the distances shown below		
200/300 mm	100 m²	6500.00
300/400 mm	100 m²	3700.00
300/500 mm	100 m²	3200.00
400/500 mm	100 m²	2550.00
BULB PLANTING		
Cultivate ground by machine and rake to level; plant bulbs as shown; bulbs PC £27.00/100		
15 bulbs per m²	100 m²	970.00
25 bulbs per m²	100 m²	1575.00
50 bulbs per m²	100 m²	3075.00
Cultivate ground by machine and rake to level; plant bulbs as shown; bulbs PC £14.00/100		
15 bulbs per m²	100 m²	490.00
25 bulbs per m²	100 m²	790.00
50 bulbs per m²	100 m²	1525.00
Form holes in grass areas and plant bulbs using bulb planter, backfill with organic manure and turf plug; bulbs PC £14.00/100		
15 bulbs per m²	100 m²	900.00
25 bulbs per m²	100 m²	1475.00
50 bulbs per m²	100 m²	2950.00
HEDGE PLANTING		
Works by machine; excavate trench for hedge 300 mm wide × 300 mm deep; deposit spoil alongside and plant hedging plants in single row at 200 mm centres; backfill with excavated material		
Trench 300 mm × 300 mm; bare root hedging plants PC £0.60 per plant; single row		
200 mm centres	m	9.00
300 mm centres	m	8.00
400 mm centres	m	7.00
600 mm centres	m	6.00
800 mm centres	m	5.00

8.3.2 EXTERNAL PLANTING

Item	Unit	Range £
HEDGE PLANTING – cont		
Works by machine – cont		
Trench 600 mm × 300 mm deep; bare root hedging plants PC £0.60 per plant; double staggered row		
300 mm centres	m	15.00
400 mm centres	m	6.00
500 mm centres	m	11.00
600 mm centres	m	10.00
800 mm centres	m	9.00
1.00 m centres	m	8.00
Works by hand; excavate trench for hedge 300 mm wide × 450 mm deep; deposit spoil alongside and plant hedging plants in single row at 200 mm centres; backfill with excavated material incorporating organic manure at 1 m³ per 5 m³; carry out initial cut; including delivery of plants from nursery		
Trench 300 mm × 300 mm; bare root hedging plants PC £0.60 per plant; single row		
200 mm centres	m	14.00
300 mm centres	m	13.00
400 mm centres	m	12.00
600 mm centres	m	11.00
800 mm centres	m	10.00
Trench 600 mm × 300 mm deep; bare root hedging plants PC £0.60 per plant; double staggered row		
300 mm centres	m	27.00
400 mm centres	m	24.00
500 mm centres	m	23.00
600 mm centres	m	22.00
800 mm centres	m	21.00
1.00 m centres	m	20.00
Readyhedge Ltd; Fully preformed hedge planting excavate trench by machine 700 mm wide × 500 mm deep; add compost at 100 mm/m² mixed to excavated material; plant mature hedge plants; backfill with excavated material and compost; allow for disposal of 50% of excavated material to spoil heaps 50 m distant		
Beech or hornbeam hedging		
1.75 m × 400 mm wide at 500 mm centres	m	100.00
1.75 m × 500 mm wide at 750 mm centres	m	65.00
Yew (Taxus) hedging		
1.50 m × 500 mm wide at 500 mm centres	m	225.00
1.50 m × 500 mm wide at 750 mm centres	m	140.00
1.50 m × 500 mm wide at 1.00 m centres	m	125.00
1.75 m × 500 mm wide at 500 mm centres	m	255.00
1.75 m × 500 mm wide at 750 mm centres	m	155.00
1.75 m × 500 mm wide at 1.00 m centres	m	135.00
Laurel (Prunus) hedging		
2.00 m × 500 mm wide at 1.00 m centres	m	16.00

8.3.2 EXTERNAL PLANTING

Item	Unit	Range £
Box (Buxus) hedging		
800 mm/1.00 m high × 300 mm wide	m	86.00
1.00 m/1.20 m high × 300 mm wide	m	175.00
Readyhedge Ltd; pre-clipped and preformed hedges planting; excavate trench by machine 500 mm wide × 500 mm deep; add compost at 100 mm/m^2 mixed to excavated material; plant mature hedge plants; backfill with excavated material and compost; allow for disposal of 50% of excavated material to spoil heaps 50 m distant		
Beech or Hornbeam hedging		
1.75 m high × 300 mm wide at 400 mm centres	m	94.00
1.75 m high × 300 mm wide at 500 mm centres	m	78.00
1.75 m high × 300 mm wide at 750 mm centres	m	58.00
1.75 m high × 300 mm wide at 900 mm centres	m	48.00
2.00 m high × 300 mm wide at 400 mm centres	m	205.00
2.00 m high × 300 mm wide at 500 mm centres	m	165.00
Yew (Taxus) hedging		
700 mm high × 250 mm wide	m	225.00
1.60 −1.80 m high × 350 mm wide	m	255.00
Readyhedge; strip hedge planting; excavate trench 700 mm wide × 500 mm deep by machine; add compost at 100 mm/m^2 mixed to excavated material; plant instant strip hedging; backfill with excavated material and compost; allow for disposal of 50% of excavated material to spoil heaps 50 m distant		
Beech hedging		
1.00 m high × 400 mm wide	m	95.00
1.40/1.60 m high × 500 mm wide	m	210.00
Yew (Taxus) hedging		
700 m high × 250 mm wide	m	225.00
Laurel (Prunus) hedging		
1.80/2.00 m high × 500 mm wide	m	220.00
Box (Buxus) hedging		
800 mm/1.00 m high × 300 mm wide	m	86.00
Hedges direct nurseries; individual feathered hedge planting; excavate trench by machine 500 mm wide × 500 mm deep; add compost at 100 mm/m^2 mixed to excavated material; plant mature hedge plants; backfill with excavated material and compost; allow for disposal of 50% of excavated material to spoil heaps 50 m distant		
Yew hedging		
1.0–1.20 m high × 300 mm wide at 2 / m	m	85.00
1.25–1.50 m high × 300 mm wide at 3 / m	m	160.00
1.50–1.75 m high at 450 mm centres	m	225.00
1.50–1.75 m high at 600 mm centres	m	175.00

8.3.2 EXTERNAL PLANTING

Item	Unit	Range £
HEDGE PLANTING – cont		
Hedges direct nurseries; Works by machine; excavate trench for hedge 300 mm wide × 300 mm deep; deposit spoil alongside and plant hedging plants in single row; backfill with excavated material		
Box hedging; rootballed		
400–500 mm high; 200 mm centres	m	61.00
400–500 mm high; 300 mm centres	m	42.00
400–500 mm high; 500 mm centres	m	27.00
500–600 mm mm high; 200 mm centres	m	82.00
500–600 mm mm high; 300 mm centres	m	56.00
500–600 mm mm high; 600 mm centres	m	36.00
600–800 mm high; 250 mm centres	m	91.00
600–800 mm high; 400 mm centres	m	59.00
800 mm–1.0 mm high; 300 mm centres	m	110.00
800 mm–1.0 m high; 500 mm centres	m	68.00
1.0–1.25 m high; 600 mm centres	m	86.00
1.0–1.25 m high; 900 mm centres	m	59.00
ANNUAL BEDDING		
BEDDING		
Spray surface with glyphosate; lift and dispose of turf when herbicide action is complete; cultivate new area for bedding plants to 400 mm deep; spread compost 100 mm deep and chemical fertilizer Enmag and rake to fine tilth to receive new bedding plants; remove all arisings		
Existing turf area		
disposal to skip	m^2	9.00
disposal to compost area on site; distance 25 m	m^2	10.00
Plant bedding to existing planting area; bedding planting PC £0.27 each		
Clear existing bedding; cultivate soil to 230 mm deep; incorporate compost 75 mm and rake to fine tilth; collect bedding from nursery and plant at 100 mm centres; irrigate on completion; maintain weekly for 12 weeks		
mass planted; 100 mm centres	m^2	39.00
to patterns; 100 mm centres	m^2	45.00
mass planted; 150 mm centres	m^2	25.00
to patterns; 150 mm centres	m^2	31.00
mass planted; 200 mm centres	m^2	25.00
to patterns; 200 mm centres	m^2	31.00
Watering of bedding by hand held hose pipe; The prices shown here are per occasion; bedding may require up to 40 visits during a dry season		
Flow rate 25 litres/minute; per occasion		
10 litres/m^2	100 m^2	–
15 litres/m^2	100 m^2	–
20 litres/m^2	100 m^2	–
25 litres/m^2	100 m^2	1.00

8.3.2 EXTERNAL PLANTING

Item	Unit	Range £
Flow rate 40 litres/minute		
10 litres/m²	100 m²	–
15 litres/m²	100 m²	–
20 litres/m²	100 m²	–
25 litres/m²	100 m²	–
LIVING WALL		
Living wall; Scotscape Ltd; design and installation of planted modules with automatic irrigation systems		
Fabric based systems; indicative area rates as shown		
Wall 20 m²	m²	700.00
Wall 50 m²	m²	600.00
Wall 100 m²	m²	570.00
Wall 150 m²	m²	560.00
Annual maintenance costs for living walls		
Walls over 30–70 m²	item	5600.00
Walls under over 150 m²	item	13000.00
PLANTERS		
To brick planter; coat insides with 2 coats RIW liquid asphaltic composition; fill with 50 mm shingle and cover with geofabric; fill with screened topsoil incorporating 25% Topgrow compost and Enmag		
Planters 1.00 m deep		
1.00 × 1.00	each	180.00
1.00 × 2.00	each	310.00
1.00 × 3.00	each	445.00
Planters 1.50 m deep		
1.00 × 1.00	each	265.00
1.00 × 2.00	each	410.00
1.00 × 3.00	each	660.00
Container planting; fill with 50 mm shingle and cover with geofabric; fill with screened topsoil incorporating 25% Topgrow compost and Enmag		
Planters 1.00 m deep		
400 × 400 × 400 mm deep	each	19.00
400 × 400 × 600 mm deep	each	21.00
1.00 m × 400 mm wide × 400 mm deep	each	37.00
1.00 m × 600 mm wide × 600 mm deep	each	44.00
1.00 m × 100 mm wide × 400 mm deep	each	45.00
1.00 m × 100 mm wide × 600 mm deep	each	66.00
1.00 m × 100 mm wide × 1.00 m deep	each	105.00
1.00 m dia. × 400 mm deep	each	36.00
1.00 m dia. × 1.00 m deep	each	85.00
2.00 m dia. × 1.00 m deep	each	330.00

8.3.2 EXTERNAL PLANTING

Item	Unit	Range £
PLANTERS – cont		
STREET PLANTERS		
Supply and locate in position precast concrete planters; fill with topsoil placed over 50 mm shingle and terram; plant with assorted 5 litre and 3 litre shrubs at to provide instant effect		
700 mm dia. × 470 mm high; white exposed aggregate finish; Marshalls Boulevard	each	1225.00
LANDSCAPE MAINTENANCE		
AS PART OF A TERM MAINTENANCE CONTRACT		
MAINTENANCE OF ORNAMENTAL SHRUB BEDS		
Ornamental shrub beds; private gardens or business parks and restricted commercial environments		
Hand weed ornamental shrub bed during the growing season; clear leaves and litter; edge beds when necessary; prune shrubs or hedges and remove replace mulch to 50 mm at end of each year; (mulched areas only) planting less than 2 years old		
Mulched beds; weekly visits; 12 month period planting centres shown; public areas; shopping areas		
600 mm centres	100 m^2	225.00
400 mm centres	100 m^2	200.00
300 mm centres	100 m^2	175.00
ground covers	100 m^2	155.00
Mulched beds; weekly visits; 12 month period planting centres shown; private areas ; business or office parks or similar		
600 mm centres	100 m^2	205.00
400 mm centres	100 m^2	160.00
300 mm centres	100 m^2	140.00
ground covers	100 m^2	110.00
Non-mulched beds; weekly visits; planting centres		
600 mm centres	100 m^2	185.00
400 mm centres	100 m^2	175.00
300 mm centres	100 m^2	170.00
ground covers	100 m^2	165.00
Non-mulched beds; monthly visits; planting centres		
600 mm centres	100 m^2	145.00
400 mm centres	100 m^2	135.00
300 mm centres	100 m^2	130.00
ground covers	100 m^2	125.00
Maintain rose bed		
Weed around base of plants; dead head regularly; fertilize once per annum with Enmag; prune once per annum; edge with half moon edging tool; add compost to 50 mm once per annum; clear of litter and leaves;		
private or restricted access beds	100 m^2	390.00
public areas	100 m^2	440.00

8.3.2 EXTERNAL PLANTING

Item	Unit	Range £
Remulch planting bed at the start of the planting season; top up mulch 25 mm thick; Melcourt Ltd		
Larger areas maximum distance 25 m; 80 m^3 loads		
Ornamental bark mulch	100 m^2	225.00
Melcourt Bark Nuggets	100 m^2	220.00
Amenity Bark	100 m^2	170.00
Forest biomulch	100 m^2	160.00
Smaller areas; maximum distance 25 m; 25 m^3 loads		
Ornamental bark mulch	100 m^2	270.00
Melcourt Bark Nuggets	100 m^2	260.00
Amenity Bark	100 m^2	215.00
Forest biomulch	100 m^2	205.00
Shrub beds in ornamental gardens		
Maintain mature shrub bed; shrubs average 1.00 m–3.00 m high with occasional mature trees in bed or in proximity; prune one third of all planting by 30% every year in rotation; fertilize; add compost; annually; add fertilizers; clear leaves; edge bed with half moon tool; water with hand held hose on 6 occasions		
massed plants; 1.00 high	m^2	630.00
plants up to 2 m high;	m^2	710.00
plants 2 m–3 m high;	m^2	910.00
VEGETATION CONTROL		
Native planting; roadside railway or forestry planted areas.		
Post planting maintenance; control of weeds and grass; herbicide spray applications; maintain weed free circles 1.00 m dia. to planting less than 5 years old in roadside, rail or forestry planting environments and the like; strim grass to 50–75 mm; prices per occasion (three applications of each operation normally required)		
Knapsack spray application; glyphosate; planting at		
1.50 m centres	ha	1725.00
1.75 m centres	ha	1125.00
2.00 m centres	ha	1225.00
Maintain planted areas; control of weeds and grass; maintain weed free circles 1.00 m dia. to planting less than 5 years old in roadside, rail or forestry planting environments and the like; strim surrounding grass to 50–75 mm; prices per occasion (three applications of each operation normally required)		
Herbicide spray applications; CDA (controlled droplet application) glyphosate and strimming; plants planted at the following centres		
1.50 m centres	ha	990.00
1.75 m centres	ha	1125.00
2.00 m centres	ha	1225.00

8.3.2 EXTERNAL PLANTING

Item	Unit	Range £
LANDSCAPE MAINTENANCE – cont		
Post planting maintenance; control of weeds and grass; herbicide spray applications; CDA (controlled droplet application); Xanadu glyphosate/ diuron; maintain weed free circles 1.00 m dia. to planting less than 5 years old in roadside, rail or forestry planting environments and the like; strim grass to 50–75 mm; prices per occasion (1.5 applications of herbicide and three strim operations normally required)		
Plants planted at the following centres		
1.50 m centres	ha	1050.00
1.75 m centres	ha	1150.00
2.00 m centres	ha	1225.00
MAINTENANCE AS DEFECTS LIABILITY AS PART OF A LANDSCAPE CONSTRUCTION PROJECT		
Clean and weed plant beds; fork over as necessary; water with full liability for the duration of the first growing season using moveable sprinklers; clear leaf fall; top up mulch at the end of the 12 month period		
1 st year after planting		
weekly visits to high profile areas new planting areas	100 m^2	570.00
monthly visits to new planting areas	100 m^2	510.00

8.3.3 IRRIGATION SYSTEMS

Item	Unit	Range £
IRRIGATION SYSTEMS		
LANDSCAPE IRRIGATION		
Automatic irrigation; TK Solutions Ltd		
Large garden consisting of 24 stations; 7000 m^2 irrigated area		
turf only	nr	16000.00
70/30% turf/shrub beds	nr	20000.00
Medium sized garden consisting of 12 stations; 3500 m^2 irrigated area		
turf only	nr	10500.00
70/30% turf/shrub beds	nr	15500.00
Medium sized garden consisting of 6 stations of irrigated area; 1000 m^2		
turf only	nr	7000.00
70/30% turf/shrub beds	nr	8000.00
50/50% turf/shrub beds	nr	8900.00
Leaky pipe irrigation; Leaky Pipe Ltd		
Works by machine; main supply and connection to laterals; excavate trench for main or ring main 450 mm deep; supply and lay pipe; backfill and lightly compact trench		
20 mm LDPE	100 m	335.00
16 mm LDPE	100 m	320.00
Works by hand; main supply and connection to laterals; excavate trench for main or ring main 450 mm deep; supply and lay pipe; backfill and lightly compact trench		
20 mm LDPE	100 m	2400.00
16 mm LDPE	100 m	2375.00
Turf area irrigation; laterals to mains; to cultivated soil; excavate trench 150 mm deep using hoe or mattock; lay moisture leaking pipe laid 150 mm below ground at centres of 350 mm		
low leak	100 m^2	710.00
high leak	100 m^2	730.00
Landscape area irrigation; laterals to mains; moisture leaking pipe laid to the surface of irrigated areas at 600 mm centres		
low leak	100 m^2	345.00
high leak	100 m^2	355.00
Landscape area irrigation; laterals to mains; moisture leaking pipe laid to the surface of irrigated areas at 900 mm centres		
low leak	100 m^2	225.00
high leak	100 m^2	235.00
multistation controller	each	405.00
solenoid valves connected to automatic controller	each	660.00

8.3.3 IRRIGATION SYSTEMS

Item	Unit	Range £
RAINWATER HARVESTING		
Rainwater Harvesting Ltd; rainwater tanks; soft landscape areas installed below ground		
Excavate in soft landscape area by machine to install rainwater harvesting tank; grade base and lay compacted granular base to level; connect filters to rainwater down pipe of building 10 m distance and lay 110 mm uPVC pipe in trench. Install rainwater tank complete with filters for irrigation and submersible pump; cover with geofabric and backfill with excavated material compacting lightly; remove surplus material off -site		
3600 litre	nr	3150.00
4500 litre	nr	3825.00
10000 litre	nr	6300.00
Excavate in soft landscape area by hand to install rainwater harvesting tank; grade base and lay compacted granular base to level; connect filters to rainwater down pipe of building 10 m distance and lay 110 mm uPVC pipe in trench. Install low profile rainwater tank complete with filters for irrigation and submersible pump; cover with geofabric and backfill with excavated material compacting lightly; remove surplus material off-site		
Flat tank 1500 litre	nr	5400.00
Flat tank 3000 litre	nr	6800.00

8.4.1 FENCING AND RAILINGS

Item	Unit	Range £
TEMPORARY FENCING		
Site fencing; supply and erect temporary protective fencing and remove at completion of works		
Cleft chestnut paling		
1.20 m high 75 mm larch posts at 3 m centres	100 m	2250.00
Heras fencing		
2 week hire period	100 m	1200.00
4 week hire period	100 m	1575.00
8 week hire period	100 m	2350.00
12 week hire period	100 m	3125.00
16 week hire period	100 m	3950.00
24 week hire period	100 m	5500.00
CHAIL LINK AND WIRE FENCING		
Chain link fencing; supply and erect chain link fencing; form post holes and erect concrete posts and straining posts with struts at 50 m centres all set in 1:3:6 concrete; fix line wires		
3 mm galvanized wire 50 mm chainlink fencing		
1800 mm high	m	62.00
Plastic coated 3.15 gauge galvanized wire mesh		
900 mm high	m	42.00
1200 mm high	m	45.00
1800 mm high	m	60.00
Extra for additional concrete straining posts with 1 strut set in concrete		
900 mm high	each	155.00
1200 mm high	each	160.00
1800 mm high	each	170.00
Extra for additional concrete straining posts with 2 struts set in concrete		
900 mm high	each	220.00
1200 mm high	each	225.00
1800 mm high	each	255.00
Extra for additional angle iron straining posts with 2 struts set in concrete		
900 mm high	each	120.00
1200 mm high	each	86.00
1400 mm high	each	140.00
1800 mm high	each	165.00
AGRICULTURAL FENCING		
RABBIT FENCING		
Construct rabbit-stop fencing; erect galvanized wire netting; mesh 31 mm; 900 mm above ground, 150 mm below ground turned out and buried; on 75 mm dia. treated timber posts 1.8 m long driven 700 mm into firm ground at 4.0 m centres; netting clipped to top and bottom straining wires 2.63 mm dia.; straining post 150 mm dia. × 2.3 m long driven into firm ground at 50 m intervals		
turned in 150 mm	100 m	2000.00
buried 150 mm	100 m	2150.00

8.4.1 FENCING AND RAILINGS

Item	Unit	Range £
AGRICULTURAL FENCING – cont		
Deer fence		
Construct deer-stop fencing; erect five 4 mm dia. plain galvanized wires and five 2 ply galvanized barbed wires at 150 mm spacing on 45 × 45 × 5 mm angle iron posts 2.4 m long driven into firm ground at 3.0 m centres, with timber droppers		
25 × 38 mm × 1.0 m long at 1.5 m centres	100 m	2425.00
Forestry fencing		
Supply and erect forestry fencing of three lines of 3 mm plain galvanized wire tied to 1700 × 65 mm dia. angle iron posts at 2750 m centres with 1850 × 100 mm dia. straining posts and 1600 × 80 mm dia. struts at 50.0 m centres driven into firm ground		
1800 mm high; three wires	100 m	1575.00
1800 mm high; three wires including cattle fencing	100 m	2150.00
TIMBER FENCING		
Clear fence line of existing evergreen shrubs 3 m high average; grub out roots by machine chip on site and remove off site; erect closeboarded timber fence in treated softwood, pales 100 × 22 mm lapped, 150 × 22 mm gravel boards		
Concrete posts 100 × 100 mm at 3.0 m centres set into ground in 1:3:6 concrete		
1500 mm high	m	145.00
1800 mm high	m	155.00
softwood posts; three arris rails		
1350 mm high	m	145.00
1650 mm high	m	155.00
1800 mm high	m	155.00
Erect chestnut pale fencing; cleft chestnut pales; two lines galvanized wire, galvanized tying wire, treated softwood posts at 3.0 m centres and straining posts and struts at 50 m centres driven into firm ground		
900 mm high; posts 75 mm dia. × 1200 mm long	m	34.00
1200 mm high; posts 75 mm dia. × 1500 mm long	m	22.00
Construct timber rail; horizontal hit and miss type; rails 150 × 25 mm; posts 100 × 100 mm at 1.8 m centres; twice stained with coloured wood preservative; including excavation for posts and concreting into ground (C7P)		
In treated softwood		
1800 mm high	m	125.00
Construct cleft oak rail fence with rails 300 mm minimum girth tennoned both ends; 125 × 100 mm treated softwood posts double mortised for rails; corner posts 125 × 125 mm, driven into firm ground at 2.5 m centres		
3 rails	m	40.00
4 rails	m	50.00

8.4.1 FENCING AND RAILINGS

Item	Unit	Range £
Birdsmouth fencing; timber		
600 mm high	m	41.00
900 mm high	m	44.00
SECURITY FENCING		
Supply and erect chainlink fence; 51 × 3 mm mesh; with line wires and stretcher bars bolted to concrete posts at 3.0 m centres and straining posts at 10 m centres; posts set in concrete 450 × 450 mm × 33% of height of post deep; fit straight extension arms of 45 × 45 × 5 mm steel angle with three lines of barbed wire and droppers; all metalwork to be factory hot-dip galvanized for painting on site		
Galvanized 3 mm mesh		
1800 mm high	m	76.00
Galvanized mesh but with PVC coated 3.15 mm mesh (dia. of wire 2.5 mm)		
900 mm high	m	53.00
1200 mm high	m	58.00
1800 mm high	m	74.00
Add to fences above for base of fence to be fixed with hairpin staples cast into concrete ground beam 1:3:6 site mixed concrete; mechanical excavation disposal to on site spoil heaps		
125 × 225 mm deep	m	11.00
Add to fences above for straight extension arms of 45 × 45 × 5 mm steel angle with three lines of barbed wire and droppers	m	6.00
Supply and erect palisade security fence Jacksons Barbican 2500 mm high with rectangular hollow section steel pales at 150 mm centres on three 50 × 50 × 6 mm rails; rails bolted to 80 × 60 mm posts set in concrete 450 × 450 × 750 mm deep at 2750 mm centres; tops of pales to points and set at 45° angle; all metalwork to be hot-dip factory galvanized for painting on site	m	135.00
Supply and erect single gate to match above complete with welded hinges and lock		
1.0 m wide	each	930.00
4.0 m wide	each	800.00
8.0 m wide	pair	1825.00
BALL STOP FENCING		
Supply and erect plastic coated 30 × 30 mm netting fixed to 60.3 mm dia. 12 mm solid bar lattice galvanized dual posts; top, middle and bottom rails with 3 horizontal rails on 60.3 mm dia. nylon coated tubular steel posts at 3.0 m centres and 60.3 mm dia. straining posts with struts at 50 m centres set 750 mm into FND2 concrete footings 300 × 300 × 600 mm deep; include framed chain link gate 900 × 1800 mm high to match complete with hinges and locking latch		
4500 mm high	100 m	24500.00
5000 mm high	100 m	25500.00
6000 mm high	100 m	29500.00

8.4.1 FENCING AND RAILINGS

Item	Unit	Range £
GATES		
STOCK GATE		
Erect stock gate, ms tubular field gate, diamond braced 1.8 m high hung on tubular steel posts set in concrete (C7P); complete with ironmongery; all galvanized		
Width 3.00 m	each	600.00
Width 4.20 m	each	650.00

8.4.2 WALLS AND SCREENS

Item	Unit	Range £
FOUNDATIONS FOR WALLS		
IN SITU CONCRETE FOUNDATIONS		
Excavate foundation trench mechanically; remove spoil off site; lay 1:3:6 site mixed concrete foundations; distance from mixer 25 m; depth of trench to be 225 mm deeper than foundation to allow for three underground brick courses priced separately		
Concrete poured to blinded exposed ground		
200 mm deep × 400 mm wide	m	46.00
300 mm deep × 500 mm wide	m	78.00
400 mm deep × 400 mm wide	m	77.00
400 mm deep × 600 mm wide	m	115.00
600 mm deep × 600 mm wide	m	165.00
Concrete poured to formwork		
200 mm deep × 400 mm wide	m	92.00
300 mm deep × 500 mm wide	m	125.00
300 mm deep × 600 mm wide	m	135.00
300 mm deep × 800 mm wide	m	170.00
400 mm deep × 400 mm wide	m	120.00
400 mm deep × 600 mm wide	m	195.00
600 mm deep × 600 mm wide	m	235.00
Excavate foundation trench by hand; remove spoil off site; lay 1:3:6 site mixed concrete foundations; distance from mixer 25 m; depth of trench to be 225 mm deeper than foundation to allow for three underground brick courses priced separately		
Disposal to spoil heap 25 m by barrow and off site by grab vehicle		
200 mm deep × 400 mm wide	m	125.00
300 mm deep × 500 mm wide	m	175.00
400 mm deep × 400 mm wide	m	175.00
400 mm deep × 600 mm wide	m	255.00
600 mm deep × 600 mm wide	m	355.00
Excavate foundation trench mechanically; remove spoil off site; lay ready mixed concrete C10 discharged directly from delivery vehicle to location; depth of trench to be 225 mm deeper than foundation to allow for three underground brick courses priced separately; foundation size		
200 mm deep × 400 mm wide	m	28.00
300 mm deep × 500 mm wide	m	43.00
400 mm deep × 400 mm wide	m	41.00
400 mm deep × 600 mm wide	m	61.00
400 mm deep × 800 mm wide	m	76.00
400 mm deep × 1000 mm wide	m	96.00
400 mm deep × 1200 mm wide	m	115.00
600 mm deep × 600 mm wide	m	88.00
Reinforced concrete wall to foundations above (site mixed concrete)		
Up to 1.00 m high × 200 mm thick	m^2	135.00
Up to 1.00 m high × 300 mm thick	m^2	135.00

8.4.2 WALLS AND SCREENS

Item	Unit	Range £
FOUNDATIONS FOR WALLS – cont		
Reinforced concrete wall to foundations above (ready mix concrete RC35)		
1.00 m high × 200 mm thick	m^2	205.00
1.00 m high × 300 mm thick	m^2	235.00
Grading; excavate to reduce levels for concrete slab; lay waterproof membrane, 100 mm hardcore and form concrete slab reinforced with A142 mesh in 1:2:4 site mixed concrete to thickness; remove excavated material off site		
Concrete 1:2:4 site mixed		
100 mm thick	m^2	89.00
150 mm thick	m^2	105.00
250 mm thick	m^2	140.00
300 mm thick	m^2	160.00
Concrete ready mixed GEN 2		
100 mm thick	m^2	73.00
150 mm thick	m^2	83.00
250 mm thick	m^2	100.00
300 mm thick	m^2	110.00
BLOCK WALLING		
Note: Measurements allow for works above ground only		
Concrete block walls; including excavation of foundation trench 450 mm deep; remove spoil off site; lay site mixed concrete foundations 600 × 300 mm thick; 1 block below ground		
Solid blocks 7 N/mm^2; 100 mm thick		
500 mm high	m	115.00
750 mm high	m	130.00
1.00 m high	m	145.00
1.25 m high	m	165.00
1.50 m high	m	180.00
1.80 m high	m	200.00
140 mm thick		
generally	m	170.00
100 mm blocks; laid on flat		
500 mm high	m	165.00
750 mm high	m	205.00
100 mm blocks laid flat; 2 courses underground		
1.00 m high	m	260.00
1.20 m high	m	295.00
1.50 m high	m	350.00
1.80 m high	m	395.00
Hollow blocks filled with concrete; 440 × 215 × 215 mm		
500 mm high	m	170.00
750 mm high	m	205.00
1.00 m high	m	245.00
1.25 m high	m	275.00
1.50 m high	m	320.00
1.80 m high	m	365.00

8.4.2 WALLS AND SCREENS

Item	Unit	Range £
Concrete block retaining walls; 2 courses underground; including excavation of foundation trench 450 mm deep; remove spoil off site; lay ready mixed concrete foundations 1200 × 400 mm thick		
Solid blocks 7 N/mm²; 100 mm thick		
1.00 m high	m	200.00
1.50 m high	m	245.00
1.80 m high	m	270.00
Double skin overall 300 mm thick; Solid blocks 7 N/mm²; 100 mm thick with 100 cavity filled with concrete; blockwork skins tied with 200 mm stainless steel ties at 400 mm centres		
1.00 m high	m	340.00
1.50 m high	m	425.00
1.80 m high	m	480.00
100 mm blocks; laid on flat; 225 mm thick		
1.00 m high	m²	325.00
1.50 m high	m²	410.00
1.80 m high	m²	460.00
Hollow blocks filled with concrete; 215 mm thick		
1.00 m high	m²	355.00
1.50 m high	m²	455.00
1.80 m high	m²	490.00
Hollow blocks with steel bar cast into the foundation		
1.00 m high	m²	520.00
1.50 m high	m²	630.00
1.80 m high	m²	680.00
Concrete block wall with brick face 112.5 mm thick to stretcher bond; including excavation of foundation trench 450 mm deep; remove spoil off site; lay GEN 1 concrete foundations 600 × 300 mm thick; place stainless steel ties at 4 nr/m² of wall face		
Solid blocks 7 N/mm² with stretcher bond brick face; reclaimed bricks PC £866.00/1000		
100 mm thick	m²	295.00
140 mm thick	m²	315.00
Concrete block wall Rendered with K-Rend; including excavation of foundation trench 450 mm deep; remove spoil off site; lay GEN 1 concrete foundations 600 × 300 mm thick; place stainless steel ties at 4 nr/m² of wall face		
Solid blocks 7 N/mm² with stretcher bond brick face; 2 coats K-Rend Silicone FT		
100 mm thick	m²	195.00
140 mm thick	m²	220.00

8.4.2 WALLS AND SCREENS

Item	Unit	Range £
BRICK WALLING		
One brick thick wall (225 mm thick); excavation 400 mm deep; remove arisings off site; lay GEN 1 concrete foundations 450 mm wide × 250 mm thick; all in English Garden Wall bond; laid in cement: lime: sand (1:1:6) mortar with flush joints, fair face one side; DPC two courses engineering brick in cement: sand (1:3) mortar; brick on edge coping		
Wall 900 mm high above DPC		
engineering brick (class B)	m	315.00
brick PC £500.00/1000	m	330.00
brick PC £800.00/1000	m	365.00
Curved wall; 6.00 m radius 900 mm high above DPC		
engineering brick (class B)	m	370.00
brick PC £500.00/1000	m	480.00
brick PC £800.00/1000	m	485.00
Wall 1200 mm high above DPC		
engineering brick (class B)	m	405.00
brick PC £500.00/1000	m	400.00
brick PC £800.00/1000	m	450.00
Curved wall; 6.00 m radius; 1200 mm high above DPC		
engineering brick (class B)	m	475.00
brick PC £500.00/1000	m	630.00
brick PC £800.00/1000	m	900.00
Wall 1800 mm high above DPC		
engineering brick (class B)	m	540.00
brick PC £500.00/1000	m	580.00
brick PC £800.00/1000	m	650.00
Curved wall; 6.00 m radius; 1800 mm high above DPC		
engineering brick (class B)	m	670.00
brick PC £500.00/1000	m	890.00
brick PC £800.00/1000	m	880.00
One and a half brick wall; excavate 450 mm deep; remove arisings off site; lay GEN 1 concrete foundations 600 × 300 mm thick; two thick brick piers at 3.0 m centres; all in English Garden Wall bond; laid in cement: lime: sand (1:1:6) mortar with flush joints; fair face one side; DPC two courses engineering brick in cement: sand (1:3) mortar; coping of headers on edge		
Wall 900 mm high above DPC		
engineering brick (class B)	m	510.00
brick PC £500.00/1000	m	510.00
Wall 1200 mm high above DPC		
engineering brick (class B)	m	700.00
brick PC £500.00/1000	m	435.00
Wall 1800 mm high above DPC		
engineering brick (class B)	m	870.00
brick PC £500.00/1000	m	660.00
brick PC £800.00/1000	m	1825.00

8.4.2 WALLS AND SCREENS

Item	Unit	Range £
BRICK WALLING WITH PIERS		
Half brick thick wall; 102.5 mm thick; with one brick piers at 2.00 m centres; excavate foundation trench 500 mm deep; remove spoil off site; lay site mixed concrete foundations 1:3:6 350 × 150 mm thick; laid in cement: lime: sand (1:1:6) mortar with flush joints; fair face one side; DPC two courses underground; engineering brick in cement: sand (1:3) mortar; coping of headers on end		
Wall 900 mm high above DPC		
engineering brick (class B)	m	360.00
brick PC £500.00/1000	m	225.00
brick PC £800.00/1000	m	270.00
One brick thick wall; 225 mm thick; with one and a half brick piers at 3.0 m centres; excavation 400 mm deep; remove arisings off site; lay GEN 1 concrete foundations 450 mm wide × 250 mm thick; all in English Garden Wall bond; laid in cement: lime: sand (1:1:6) mortar with flush joints; fair face one side; DPC two courses engineering brick in cement: sand (1:3) mortar; precast concrete coping 152 × 75 mm		
Wall 900 mm high above DPC		
engineering brick (class B)	m	620.00
brick PC £500.00/1000	m	335.00
brick PC £800.00/1000	m	850.00
Wall 1200 mm high above DPC		
engineering brick (class B)	m	680.00
brick PC £500.00/1000	m	410.00
brick PC £800.00/1000	m	4400.00
Wall 1800 mm high above DPC		
engineering brick (class B)	m	990.00
brick PC £500.00/1000	m	580.00
brick PC £800.00/1000	m	6300.00
CLADDING TO WALLS		
Concrete block wall with brick face 112.5 mm thick to bond pattern using snap headers; including excavation of foundation trench 450 mm deep; remove spoil off site; lay GEN 1 concrete foundations 600 × 300 mm thick; place stainless steel ties at 4 nr/m^2 of wall face; lay coping of brick on edge in engineering brick		
Solid blocks 7 N/mm^2 with snap header brick face; bricks PC £866.00/1000		
500 mm high	m^2	230.00
750 mm high	m^2	285.00
1.00 m high	m^2	320.00
1.50 m high	m^2	430.00
1.80 m high	m^2	530.00

8.4.2 WALLS AND SCREENS

Item	Unit	Range £
CLADDING TO WALLS – cont		
IN SITU CONCRETE CLAD WALLS		
Excavate foundation trench mechanically; remove spoil off site; fix reinforcement starter bars 12 mm at 200 mm centres; lay concrete foundations; depth of trench to be 225 mm deeper than foundation; cast in situ concrete walls 250 mm thick inclusive of bar reinforcement 12 mm; footings cast to blinded exposed ground; Cladding single side		
Cladding with Yorkstone – Johnsons Wellfield Quarries		
500 mm high	m	570.00
1.00 m high	m	840.00
1.50 m high	m	1400.00
1.80 m high	m	1650.00
Cladding with single size Portland stone and Portland stone coping – Albion stone		
500 mm high	m	670.00
1.00 m high	m	950.00
1.50 m high	m	1500.00
1.80 m high	m	1775.00
Cladding with silver grey granite and granite coping – CED Ltd		
500 mm high	m	455.00
1.00 m high	m	660.00
1.50 m high	m	1150.00
1.80 m high	m	1400.00
Cladding with Tier system adhesive stone panelling and granite coping – CED Ltd		
500 mm high	m	450.00
1.00 m high	m	660.00
1.50 m high	m	1150.00
1.80 m high	m	1400.00
Cladding and coping with reconstituted stone – Haddonstone Ltd		
500 mm high	m	450.00
1.00 m high	m	590.00
1.50 m high	m	1025.00
1.80 m high	m	1200.00
NATURAL STONE WALLING		
Dry stone wall; wall on concrete foundation 800 × 300 mm; dry stone coursed wall inclusive of locking stones and filling to wall with broken stone or rubble; walls up to 1.20 m high; battered; 2 sides fair faced; Note that dry-stone walls are not traditionally built on concrete footings.		
2 sides fair faced; 1.00 m high		
Yorkstone	m	760.00
Purbeck	m	700.00
Cotswold stone	m	590.00
2 sides fair faced; 1.50 m high		
Yorkstone	m	1000.00
Purbeck	m	920.00
Cotswold stone	m	760.00

8.4.2 WALLS AND SCREENS

Item	Unit	Range £
Dry stone or dry stone effect wall; to concrete block wall; Solid blocks 7 N/mm² 100 mm thick; including excavation of foundation trench 450 mm deep; remove spoil off site; lay GEN 1 concrete foundations 600 × 300 mm thick; place stainless steel ties at 4 nr/m² of wall face to receive stone walling treatments. Coping of 40 mm thick sandstone with chamfer and drip		
Walls faced one side only 1.00 m high		
Yorkstone	m	730.00
Purbeck	m	670.00
Cotswold stone	m	550.00
Walls faced one side only 1.50 m high		
Yorkstone	m	940.00
Purbeck	m	860.00
Cotswold stone	m	700.00
Dry stone or dry stone effect wall 140 mm thick; to concrete block wall; Solid blocks 7 N/mm²; including excavation of foundation trench 450 mm deep; remove spoil off site; lay GEN 1 concrete foundations 900 × 300 mm thick; place stainless steel ties at 4 nr/m² of wall face to receive stone walling treatments. Coping of 40 mm thick sandstone with chamfer and drip		
Walls faced two sides; 1.00 m high		
Yorkstone	m	1325.00
Purbeck	m	1200.00
Cotswold stone	m	970.00
Walls faced two sides; 1.50 m high		
Yorkstone	m	1775.00
Purbeck	m	1600.00
Cotswold stone	m	1250.00
Portland stone wall with white cememnt mortar on concrete footing; including excavations and disposal; concrete footing 600 × 300 mm		
Portland stone with white cement mortar 300 × 100 × 250 mm thick; 1.00 m high		
fair-faced one side	m	920.00
fair-faced two sides	m	960.00
Portland stone with white cement mortar 300 × 100 × 250 mm thick; 1.50 m high		
fair-faced one side	m	1150.00
fair-faced two sides	m	1225.00
Portland stone with white cement mortar 440 × 300 × 225 mm thick; 1.00 m high		
fair-faced one side	m	1575.00
fair-faced two sides	m	1625.00
Portland stone with white cement mortar 440 × 300 × 225 mm thick; 1.50 m high		
fair-faced one side	m	2100.00
fair-faced two sides	m	2150.00

8.4.2 WALLS AND SCREENS

Item	Unit	Range £
PIERS		
BLOCK PIERS		
Block piers; blocks 440 × 215 × 100 mm thick laid on flat piers on concrete footings 600 × 600 × 250 mm thick; inclusive of excavations and disposal off site; coping of engineering brick on edge; 2 courses underground		
Pier 450 × 450 mm; to receive cladding treatment priced separately		
500 mm high	nr	92.00
750 mm high	nr	120.00
1.00 m high	nr	150.00
1.25 m high	nr	180.00
1.50 m high	nr	205.00
1.80 m high	nr	250.00
Block piers; Hollow blocks 440 × 215 × 215 mm thick on concrete footings 600 × 600 × 250 mm thick; inclusive of excavations and disposal off site; coping of engineering brick on edge; 2 courses underground		
Pier 450 × 450 mm; to receive cladding treatment priced separately		
500 mm high	nr	88.00
750 mm high	nr	120.00
1.00 m high	nr	150.00
1.25 m high	nr	180.00
1.50 m high	nr	205.00
1.80 m high	nr	250.00
BRICK PIERS		
Brick piers on concrete footings 500 × 500 × 250 mm thick; inclusive of excavations and disposal off site; coping of engineering brick on edge; 2 courses underground		
Pier 215 × 215 mm; one brick thick; brick PC £541.00/1000		
500 mm high	nr	95.00
750 mm high	nr	120.00
1.00 m high	nr	140.00
1.25 m high	nr	160.00
1.50 m high	nr	185.00
1.80 m high	nr	205.00
Pier 215 × 215 mm; one brick thick; brick PC £866.00/1000		
500 mm high	nr	100.00
750 mm high	nr	125.00
1.00 m high	nr	150.00
1.25 m high	nr	170.00
1.50 m high	nr	200.00
1.80 m high	nr	225.00
Pier 337.5 × 337.5 mm; one and a half brick thick; brick PC £541.00/1000		
500 mm high	nr	160.00
750 mm high	nr	205.00
1.00 m high	nr	255.00
1.25 m high	nr	295.00
1.50 m high	nr	345.00
1.80 m high	nr	400.00

8.4.2 WALLS AND SCREENS

Item	Unit	Range £
Pier 337.5 × 337.5 mm; one and a half brick thick; brick PC £866.00/1000		
500 mm high	nr	410.00
750 mm high	nr	560.00
1.00 m high	nr	700.00
1.25 m high	nr	840.00
1.50 m high	nr	980.00
1.80 m high	nr	1150.00
Brick piers on concrete footings 800 × 800 × 600 mm thick; inclusive of excavations and disposal off site; coping of Haddonstone S150C classically moulded corbelled piercap PC £139.06; 120 mm thick overall		
Pier 450 × 450 mm; two bricks thick; brick PC £541.00/1000; height of brickwork above ground		
500 mm high	nr	510.00
750 mm high	nr	580.00
1.00 m high	nr	660.00
1.25 m high	nr	720.00
1.50 m high	nr	800.00
1.80 m high	nr	880.00
2.00 m high	nr	930.00
Pier 450 × 450 mm; two bricks thick; brick PC £866.00/1000; height of brickwork above ground		
500 mm high	nr	530.00
750 mm high	nr	610.00
1.00 m high	nr	700.00
1.25 m high	nr	780.00
1.50 m high	nr	860.00
1.80 m high	nr	960.00
2.00 m high	nr	1025.00
Brick piers on concrete footings 1000 × 1000 × 600 mm thick; inclusive of excavations and disposal off site; coping Haddonstone S215C classically moulded corbelled pyramidal piercap PC £257.32		
Pier 600 × 600 mm; three bricks thick; brick PC £866.00/1000; height of brickwork above ground		
1.50 m high	nr	1650.00
1.80 m high	nr	1825.00
2.00 m high	nr	1975.00
BLOCK PIERS		
Blockwork piers with Haddonstone cast stone cladding; internal blockwork 450 × 450 mm on concrete footings 800 × 800 × 600 mm thick; inclusive of excavations and disposal off site; coping of Haddonstone S150C classically moulded corbelled piercap PC £139.06; 120 mm thick.		
Pier 495 × 495 mm overall;		
500 mm high	nr	630.00
1.00 mm high	nr	880.00
1.50 mm high	nr	1150.00
2.00 mm high	nr	1375.00

8.4.2 WALLS AND SCREENS

Item	Unit	Range £
PIERS – cont		
Blockwork piers with Haddonstone cast stone cladding; internal blockwork of 655 × 655 mm on concrete footings 1.00 × 1.00 × 600 mm thick; inclusive of excavations and disposal off site; coping of Haddonstone S120C weathered piercap PC £165.00; 114 mm thick.		
Pier 695 × 695 mm overall		
1.00 mm high	nr	1300.00
1.50 mm high	nr	1725.00
2.00 mm high	nr	2125.00
Gate Piers Haddonstone; Cast stone piers. On concrete footings, inclusive of all excavations and disposal		
S120 cast stone pier; 699 × 699 mm at base with 533 × 533 shaft. Weathered pier cap 737 × 737 mm with solid concrete core filled incrementally during construction.		
1.86 m high	nr	2000.00
2.10 m high	nr	2200.00
IN SITU CONCRETE WALLS		
IN SITU CONCRETE FOOTINGS		
Excavate trench mechanically and dispose excavated material off site; Fix starter bars of T12 at 300 mm centres and horizontal bars of T25 in two lengths through the width of the trench. Pour ready mixed concrete C25 to 225 mm below ground level		
Size of footing		
600 × 350 mm deep	m	135.00
800 × 450 mm deep	m	175.00
800 × 600 mm deep	m	195.00
1.00 × 500 mm deep	m	205.00
1.20 × 600 mm deep	m	255.00
IN SITU CONCRETE WALLS		
Excavate foundation trench mechanically; remove spoil off site; fix reinforcement starter bars 12 mm at 200 mm centres; lay concrete foundations; depth of trench to be 225 mm deeper than foundation; cast in situ concrete walls inclusive of bar reinforcement 12 mm; footings cast to blinded exposed ground		
Wall height 500 mm above ground; on foundation 400 mm × 200 mm thick		
150 thick wall; site mixed concrete 20 tonne aggregate loads	m	190.00
150 thick wall; ready mixed concrete C25	m	160.00
250 thick wall; site mixed concrete; 20 tonne aggregate loads	m	295.00
250 thick wall; ready mixed concrete C25	m	250.00
Wall 1.00 m high on foundation 500 mm wide × 300 mm deep		
150 thick wall; site mixed concrete; 20 tonne aggregate loads	m	335.00
150 thick wall; ready mixed concrete C25	m	275.00
250 thick wall; site mixed concrete; 20 tonne aggregate loads	m	380.00
250 thick wall; ready mixed concrete C25	m	295.00

8.4.2 WALLS AND SCREENS

Item	Unit	Range £
Wall 1.50 m high on foundation 500 mm wide × 300 mm deep		
250 thick wall; site mixed concrete; 20 tonne aggregate loads	m	730.00
250 thick wall; ready mixed concrete C25	m	620.00
Wall 1.80 m high on foundation 600 mm wide × 300 mm deep		
250 thick wall; site mixed concrete; 20 tonne aggregate loads	m	910.00
250 thick wall; ready mixed concrete C25	m	780.00
Wall 1.80 m high on foundation 800 mm wide × 300 mm deep		
250 thick wall; site mixed concrete; 20 tonne aggregate loads	m	940.00
250 thick wall; ready mixed concrete C25	m	840.00
Curved walls; 6.00 m radius		
Wall 1.50 m high on foundation 500 mm wide × 300 mm deep		
250 thick wall; site mixed concrete; 20 tonne aggregate loads	m	770.00
250 thick wall; ready mixed concrete C25	m	670.00
Wall 1.80 m high on foundation 600 mm wide × 300 mm deep		
250 thick wall; site mixed concrete; 20 tonne aggregate loads	m	960.00
250 thick wall; ready mixed concrete C25	m	830.00
Curved walls; 3.00 m radius		
Wall 1.50 m high on foundation 500 mm wide × 300 mm deep		
250 thick wall; site mixed concrete; 20 tonne aggregate loads	m	830.00
250 thick wall; ready mixed concrete C25	m	720.00
Wall 1.80 m high on foundation 600 mm wide × 300 mm deep		
250 thick wall; site mixed concrete; 20 tonne aggregate loads	m	1025.00
250 thick wall; ready mixed concrete C25	m	900.00
SITE MIXED CONCRETE		
Mix concrete on site; aggregates delivered in 20 tonne loads; deliver mixed concrete to location by mechanical dumper distance 25 m		
1:3:6	m^3	235.00
1:2:4	m^3	255.00
Ready mixed concrete; deliver mixed concrete to location by mechanical dumper distance 25 m		
10 N/mm^2	m^3	135.00
15 N/mm^2	m^3	135.00
Mix concrete on site; aggregates delivered in 20 tonne loads; deliver mixed concrete to location by barrow distance 25 m		
Aggregates delivered in 20 t loads to site		
1:3:6	m^3	315.00
1:2:4	m^3	335.00
aggregates delivered in 850 kg bulk bags		
1:3:6	m^3	370.00
1:2:4	m^3	425.00
ready mixed concrete		
10 N/mm^2	m^3	215.00
15 N/mm^2	m^3	215.00

8.4.2 WALLS AND SCREENS

Item	Unit	Range £
BALUSTRADES		
Bottle balustrades reconstituted stone; Haddonstone Ltd; to prepared brickwork walls or bases (not included).		
Top and bottom rails to support balustrades; laid to builders work of brick block or other stonework supports (not included); under coping; pier shafts; Piers 798 mm high		
Overall 1519 mm high Balusters 457 mm high	m	660.00
Overall 1672 mm high; Balusters 610 mm high	m	740.00
Top and bottom rails to support balustrades; laid to builders work of brick block or other stonework supports (not included); under coping; pier shafts; Piers 898 mm high		
Overall 1619 mm high; Balusters 457 mm high	m	680.00
Overall 1712 mm high; Balusters 610 mm high	m	760.00

8.4.3 RETAINING WALLS

Item	Unit	Range £
FOOTINGS FOR RETAINING WALLS		
Excavate trench for retaining wall; Cart material off site; pour concrete footings with L shaped T20 starter bars; total length 2.00 m at 200 mm centres; Concrete poured to 225 mm below ground level		
Ready mixed concrete		
600 mm wide × 300 mm deep	m	89.00
800 mm wide × 300 mm deep	m	100.00
1.00 m wide × 300 mm deep	m	115.00
1.20 m wide × 300 mm deep	m	150.00
1.00 m wide × 400 mm deep	m	155.00
1.50 m wide × 300 mm deep	m	175.00
1.50 m wide × 400 mm deep	m	230.00
Site mixed concrete; Aggregates in 20 tonne loads		
600 mm wide × 300 mm deep	m	115.00
800 mm wide × 300 mm deep	m	140.00
1.00 m wide × 300 mm deep	m	165.00
1.20 m wide × 300 mm deep	m	215.00
1.00 m wide × 400 mm deep	m	225.00
1.50 m wide × 300 mm deep	m	245.00
1.50 m wide × 400 mm deep	m	305.00
BLOCK RETAINING WALLS		
Note: The following models do not allow for excavation to the existing bank in preparation of the retaining system. Please amalgamate with the Grading and preparation section above.		
Concrete block retaining walls; Stepoc		
Excavate trench 750 mm deep and lay concrete foundation 750 mm wide × 400 mm deep; construct Forticrete precast hollow concrete block wall with 450 mm below ground laid all in accordance with manufacturer's instructions; fix reinforcing bar 12 mm as work proceeds; fill blocks with ready mixed concrete; Allow for two coats of bitumen based waterproofing to back of wall; backfilling against face of wall with drainage trench perforated pipe filled with gravel rejects 600 mm deep		
Walls 1.00 m high		
type 200; 400 × 225 × 200 mm	m	335.00
type 256; 400 × 225 × 256 mm	m	365.00
Walls 1.50 m high		
type 200; 400 × 225 × 200 mm	m	405.00
type 256; 400 × 225 × 256 mm	m	450.00
Excavate trench 750 mm deep and lay concrete foundation 1.80 m wide × 450 mm deep; construct Forticrete precast hollow concrete block wall with 450 mm below ground laid all in accordance with manufacturer's instructions; fix reinforcing bar 12 mm as work proceeds; fill blocks with ready mixed concrete; Allow for two coats of bitumen based waterproofing to back of wall; backfilling against face of wall with drainage trench perforated pipe filled with gravel rejects 600 mm deep		
Walls 1.60 m high		
type 200; 400 × 225 × 200 mm	m	630.00
type 256; 400 × 225 × 256 mm	m	670.00
Walls 1.80 m high		
type 200; 400 × 225 × 200 mm	m	670.00
type 256; 400 × 225 × 256 mm	m	720.00

8.4.3 RETAINING WALLS

Item	Unit	Range £
BLOCK RETAINING WALLS – cont		
Hollow concrete block retaining wall		
Excavate trench 750 mm deep and lay concrete foundation 750 mm wide × 400 mm deep; construct hollow concrete block wall with 450 mm below ground laid all in accordance with manufacturer's instructions; fix reinforcing bar 12 mm as work proceeds; fill blocks with concrete 1:3:6 as work proceeds; Allow for two coats of bitumen based waterproofing to back of wall; backfilling against face of wall with drainage trench perforated pipe filled with gravel rejects 600 mm deep		
Walls 1.00 m high	m	315.00
Walls 1.25 m high	m	350.00
Walls 1.50 m high	m	385.00
Excavate trench 750 mm deep and lay concrete foundation 1.80 m wide × 400 mm deep; construct hollow concrete block wall with 225 mm below ground; fix reinforcing bar 12 mm as work proceeds; fill blocks with concrete 1:3:6 as work proceeds; Allow for two coats of bitumen based waterproofing to back of wall; backfilling against face of wall with drainage trench perforated pipe filled with gravel rejects 600 mm deep		
Walls 1.50 m high	m	620.00
Walls 1.80 m high	m	660.00
PRECAST CONCRETE RETAINING WALLS		
Excavate trench and lay concrete foundation 600 × 300; 1:3:6: plain concrete, supply and install Milton Precast Concrete precast concrete L shaped units, constructed all in accordance with manufacturer's instructions; backfill with approved excavated material compacted as the work proceeds		
1500 mm high × 1000 mm wide	m	405.00
2500 mm high × 1000 mm wide	m	780.00
3000 mm high × 1000 mm wide	m	900.00
GABION WALLS		
Gabion walls; excavation costs excluded; see Grading and preparation of banks above		
Gabions 500 mm high; allowance of 1.00 m working space at top of bank; backfill working space and remove balance of arisings off site; lay concrete footing; 200 mm deep × 1.50 m wide		
Height retained		
500 mm	m	195.00
1.00 m	m	315.00
1.50 m	m	460.00
2.00 m	m	580.00

8.4.3 RETAINING WALLS

Item	Unit	Range £
TIMBER RETAINING WALLS		
Timber log retaining walls		
Excavate trench 300 mm wide to one third of the finished height of the retaining walls below; lay 100 mm hardcore; fix machine rounded logs set in concrete 1:3:6; remove excavated material from site; fix geofabric to rear of timber logs; backfill with previously excavated material set aside in position; all works by machine		
100 mm dia. logs		
500 mm high (constructed from 1.80 m lengths)	m	170.00
1.20 m high (constructed from 1.80 m lengths)	m	285.00
1.60 m high (constructed from 2.40 m lengths)	m	450.00
2.00 m high (constructed from 3.00 m lengths)	m	520.00
150 mm dia. logs		
1.20 m high (constructed from 1.80 m lengths)	m	490.00
1.60 m high (constructed from 2.40 m lengths)	m	550.00
Timber crib wall		
Excavate trench to receive foundation 300 mm deep; place plain concrete foundation 150 mm thick in 11.50 N/mm² concrete (sulphate-resisting cement); construct timber crib retaining wall and backfill with excavated spoil behind units		
timber crib wall system; average 5.0 m high	m	1725.00
timber crib wall system; average 4.0 m high	m	1300.00
timber crib wall system; average 2.0 m high	m	610.00
timber crib wall system; average 1.00 m high	m	305.00
Railway sleeper wall; Excavate and remove spoil offsite; Pour concrete footing 500 × 250 deep; lay sleeper retaining wall;		
Grade 1 hardwood sleeper 2 450 m × 250 mm wide × 150 mm high; Sleepers set to galvanized pins set in the concrete		
150 high	m	43.00
300 high	m	65.00
450 high	m	87.00
600 high	m	110.00
Grade 1 hardwood; 2 galvanized angle iron stakes per sleeper length set into concrete		
750 high	m	185.00
900 high	m	205.00
CORTEN RETAINING WALLS		
Corten Retaining Walls – Adezz; Cast strip footings 200–500 wide as appropriate × 250 deep; Fix Corten retaining wall to footings. Excavation of banks or backfilling to planters not included		
200 mm high		
Straight	m	68.00
Internal or external corners	m	79.00
curved – short lengths	m	175.00
curved – longer lenghts of up to 2.00 m each	m	140.00

8.4.3 RETAINING WALLS

Item	Unit	Range £
CORTEN RETAINING WALLS – cont		
Corten Retaining Walls – Adezz – cont		
400 mm high		
Straight	m	135.00
Internal or external corners	m	165.00
curved – short lengths	m	265.00
curved – longer lenghts of up to 2.00 m each	m	255.00
600 mm high		
Straight	m	210.00
Internal or external corners	m	245.00
curved – short lengths	m	425.00
curved – longer lenghts of up to 2.00 m each	m	360.00
RECYCLED PLASTIC RETAINING WALLS		
Hahn Recycled plastic retaining wall systems		
Excavate trench 300 mm wide to one third of the finished height of the retaining walls below; lay 100 mm hardcore; fix recycled retaining system; concrete; 1:3:6; remove excavated material from site; fix geofabric to rear of units; backfill with previously excavated material set aside in position; all works by machine		
Ogee 'C' Section interlocking units; 100 mm dia.		
300 mm high	m	145.00
600 mm high	m	245.00
1.20 m high	m	395.00
Round palisades; 100 mm dia.; solid profile; retained height		
1.20 m high	m	330.00
1.60 m high (constructed from 2.40 m lengths)	m	580.00

8.5 EXTERNAL FIXTURES

Item	Unit	Range £
8.5.1 SITE/STREET FURNITURE AND EQUIPMENT		
Bollards and access restriction		
Supply and install 10 nr cast iron Doric bollards 920 mm high above ground × 170 mm dia. bedded in concrete base 400 mm dia. × 400 mm deep	10 nr	3500.00
Benches and seating		
In grassed area excavate for base 2500 × 1575 mm and lay 100 mm hardcore, 100 mm concrete, brick pavers in stack bond bedded in 25 mm cement: lime: sand mortar; supply and fix where shown on drawing proprietary seat, hardwood slats on black powder coated steel frame, bolted down with 4 nr 24 × 90 mm recessed hex-head stainless steel anchor bolts set into concrete	set	2150.00
Cycle stand		
Supply and fix cycle stand 1250 m × 550 mm of 60.3 mm black powder coated hollow steel sections, one-piece with rounded top corners; set 250 mm into paving	each	570.00
8.5.2 ORNAMENTAL WATER FEATURES		
FAIRWATER LTD		
Water feature; Fairwater Water Gardens Ltd; Chemically or biologically pumped feature; Inclusive of design; excavation concrete base and blockwork retaining walls; Waterproofing GRP; overflow to small soakaway located locally		
Constructional elements inclusive of excavation; mesh reinforced bases; 100 mm blockwork walls; and coping of 500 mm wide × 50 thick granite		
5.00 m × 1.00 m	nr	3150.00
10.00 m × 1.50 m	nr	6400.00
Biologically filtered system components of pump housing adjacent to water feature; pvc spider filters in gravel beds; pumping overflows and automatic top-up.		
5.00 × 1.00	nr	17500.00
10.00 × 1.50	nr	21000.00
Chemical filtered system components of pump housing adjacent to water feature; pumping overflows and automatic top-up.		
5.00 × 1.00	nr	12000.00
10.00 × 1.50	nr	15500.00
Water feature in public realm; Fairwater Water Gardens Ltd; Blockwork formed sides edged with non-slip concrete paving. with Firestone pond liner boulders knife edge to 60% of circumference; excludes aquatic planting; locally installed pump chambers and balancing tanks.		
Construction elements including excavation; disposal; shelved construction to support plants maximum depth graded to 1000 mm; textured concrete slab paving 2.00 m wide to perimeter		
surface area 930 m^2; perimeter 230 m; water volume; 400 m^3–450 m^3	item	60000.00
Specialist water feature elements inclusive of all pipework; pumping and UV filters and local balancing tank and pump chamber; spider filters		
surface area 930 m^2; perimeter 230 m; water volume; 400 m^3–450 m^3	item	57000.00

8.5 EXTERNAL FIXTURES

Item	Unit	Range £
8.5.2 ORNAMENTAL WATER FEATURES – cont		
Natural Pond; Fairwater Water Gardens Ltd; Natural lake with regular natural curvilinear shape; 200 m perimeeter; 2100 m^2 plan area maximum graded depth 4.00 m; EDPM liner overlay and underlays of geofabric; anchor trenches with mixed submerged formal edge;		
Excavations and grading to profile; excavated material onsite to form retaining bunds graded to 1:8.5; excavations of edge for recessed edge batters; total excavation volume based on 4600 m^3	item	51000.00
Grading of retaining bunds to required landforms 4600 m^3	item	46000.00
Laying of liners overlays and underlays inclusive of retention to anchor trench; EPDM liner 1.02 mm Firestone 2600 m^2	item	45500.00
Excavate for small pond or lake maximum depth 1.00 m; remove arisings off site; grade and trim to shape; lay 75 mm sharp sand; line with 0.75 mm butyl liner 75 mm sharp sand and geofabric; cover over with sifted topsoil; anchor liner to anchor trench; install balancing tank and automatic top-up system		
Pond or lake of organic shape		
100 m^2; perimeter 50 m	each	10500.00
250 m^2; perimeter 90 m	each	16000.00
500 m^2; perimeter 130 m	each	38500.00
1000 m^2; perimeter 175 m	each	78000.00
Excavate for lake average depth 1.0 m, allow for bringing to specified levels; reserve topsoil; remove spoil to approved dump on site; remove all stones and debris over 75 mm; lay polythene sheet including welding all joints and seams by specialist; screen and replace topsoil 200 mm thick		
Prices are for lakes of regular shape		
750 micron sheet	1000 m^2	22000.00
1000 micron sheet	1000 m^2	33500.00
Extra for removing spoil to tip	m^3	48.00
Extra for 25 mm sand blinding to lake bed	100 m^2	225.00
Extra for screening topsoil	m^2	2.00
Extra for spreading imported topsoil	100 m^2	1125.00
Plant aquatic plants in lake topsoil		
Representative plant prices are shown below and users should add these to to the labour rates shown		
Apongeton distachyos	nr	6.00
Acorus calamus	nr	8.00
Butomus umbellatus	nr	4.00
Typha latifolia	nr	2.00
Planting aquatics labours only	100	140.00

8.5 EXTERNAL FIXTURES

Item	Unit	Range £
FORMAL WATER FEATURES		
Excavate and construct water feature of regular shape; lay 100 mm hardcore base and 150 mm concrete 1:2:4 site mixed; line base and vertical face with butyl liner 0.75 micron and construct vertical sides of reinforced blockwork; rendering two coats; anchor the liner behind blockwork; install pumps balancing tanks and all connections to mains supply		
1.00 × 1.00 × 1.00 m deep	nr	3975.00
2.00 × 1.00 × 1.00 m deep	nr	5100.00
Excavate and construct water feature of circular shape; lay 100 mm hardcore base and 150 mm concrete 1:2:4 site mixed; line base and vertical face with butyl liner 0.75 micron and construct vertical sides of reinforced blockwork; rendering two coats; anchor the liner behind blockwork; install pumps balancing tanks and all connections to mains supply		
Surround water body with ornamental fountain surround		
Architectural Heritage; Great Westwood; 6.00 m dia. 350 mm high	nr	18500.00
Haddonstone; circular pool surround; internal dia. 1780 mm; kerb features continuous moulding enriched with ovolvo and palmette designs; inclusive of plinth and integral conch shell vases flanked by dolphins	nr	9200.00
8.5.3 ORNAMENTAL BUILDINGS		
Excavate and dispose excavated material; lay concrete base 150 thick ; erect garden building		
Pavillion; Cast stone; Haddonstone; floor of Yorkstone diamond sawn 6 sides.		
Venetian Folly L9400; Tuscan columns, pedimented arch, quoins and optional balustrading; 4.70 × 3.10	nr	24000.00
Cast stone; Haddonstone temples with domed roof and stepped floor		
Small classical; 2.54 m dia.	nr	18000.00
Large classical 3.19 m dia.	nr	27500.00
Temple; Soild limestone; Architectural Heritage		
Estate temple 2.70 m dia.; floor of Yorkstone; diamond sawn 6 sides	nr	27000.00
Pergola; Solid limestone columns with oak beams; Architectural Heritage		
Estate temple 2.70 m dia.; floor of Yorkstone; diamond sawn 6 sides	nr	22000.00

8.6 EXTERNAL DRAINAGE

Item	Unit	Range £
8.6.1 SURFACE WATER AND FOUL WATER DRAINAGE		
PIPE LAYING		
PVC-u pipe laying; excavate trench 300 mm wide by excavator; lay bedding; lay surface water drain pipe to depth shown; backfill to 150 mm above pipe with gravel or sand rejects; backfill with excavated material to ground level		
Trench 750 mm deep; pipe 640–590 mm deep		
110 mm	m	18.00
110 mm; short lengths	m	20.00
160 mm	m	21.00
160 mm; short lengths	m	24.00
Trench 900 mm deep; pipe 640–590 mm deep		
110 mm	m	33.00
110 mm; short lengths	m	35.00
160 mm	m	36.00
160 mm; short lengths	m	39.00
Trench 1.20 m deep; pipe 940–890 mm deep		
110 mm	m	36.00
110 mm; short lengths	m	39.00
160 mm	m	40.00
160 mm; short lengths	m	43.00
Trench 1.50 m deep; pipe 940–890 mm deep		
110 mm	m	42.00
110 mm; short lengths	m	53.00
160 mm	m	45.00
160 mm; short lengths	m	48.00
Vitrified clay pipe laying; excavate trench by excavator; backfill with excavated material to ground level; lay surface water drain pipe to depth shown; note that the use of clay pipes precludes the use of bedding materials		
Trench 750 mm deep; pipe 640–590 mm deep		
100 mm	m	34.00
100 mm; short lengths	m	36.00
160 mm	m	58.00
160 mm; short lengths	m	69.00
Trench 1.20 m deep; pipe 940–890 mm deep; bedding on reject sand		
110 mm	m	53.00
110 mm; short lengths	m	55.00
160 mm	m	75.00
160 mm; short lengths	m	88.00
ACCESS CHAMBERS		
Excavate inspection chamber by machine; lay base of concrete 150 thick; allow for half section pipework and benchings		
Precast concrete inspection chamber		
600 × 400 × 600 mm deep	nr	790.00
600 × 400 × 900 mm deep	nr	830.00

8.6 EXTERNAL DRAINAGE

Item	Unit	Range £
Polypropylene inspection chamber		
Mini access chamber 600 mm deep	nr	425.00
475 mm dia. × 900 mm deep; polymer cover	nr	1125.00
475 mm dia. × 900 mm deep; ductile iron cover with screw down lid	nr	1125.00
GULLIES		
Gullies; vitrified clay; excavate by hand; supply and set gully in concrete (C10P); connect to drainage system with flexible joints; backfilling with excavated material to 250 mm below finished level. Lay 150 mm type 1 to receive surface treatments (not included)		
Trapped mud (dirt) gully; complete with galvanized bucket and cast iron hinged locking grate and frame		
loading to 1 tonne; 100 or 150 mm outlet	each	720.00
loading to 5 tonne; 100 mm outlet	each	840.00
loading to 5 tonne; 150 mm outlet	each	880.00
Trapped mud (dirt) gully with rodding eye; complete with galvanized bucket and cast iron hinged locking grate and frame		
100 outlet; 300 mm internal dia.; 300 mm internal depth	each	710.00
150 outlet; 400 mm internal dia.; 750 mm internal depth	each	760.00
Concrete road gully		
Excavate and lay 100 mm concrete base (1:3:6) 150 x150 mm to suit given invert level of drain; supply and connect trapped precast concrete road gully set in concrete surround; connect to vitrified clay or uPVC drainage system with flexible joints; supply and fix straight bar dished top cast iron grating and frame; bedded in cement: sand mortar (1:3)		
375 mm dia. × 750 m deep	each	860.00
450 mm dia. × 750 m deep	each	860.00
450 mm dia. × 1.05 m deep	each	700.00
Gullies PVC-u		
Excavate and lay 100 mm concrete (C20P) base 150 × 150 mm to suit given invert level of drain; connect to drainage system; backfill with DoT Type 1 granular fill; install gully; complete with cast iron grate and frame		
yard gully 300 mm dia. × 600 mm deep	each	305.00
trapped PVC-u gully; 110 mm dia. × 215 mm deep; ductile iron frame	each	300.00
bottle gully 228 × 228 × 642 mm deep	each	235.00
bottle gully 228 × 228 × 317 mm deep	each	185.00
LINEAR DRAINAGE		
Linear drainage to vehicular areas D400 loadings; Excavate trench by machine; lay Aco Brickslot channel drain on concrete base and surround to falls; all to manufacturers specifications; paving surround to both sides of channel		
Aco M100; internal width 100 mm; on concrete base and surround to falls; all to manufacturers specifications; paving surround to both sides of channel		
Galvanized	m	225.00
Stainless steel	m	380.00
Slotted; stainless steel	m	195.00
Heelguard composite black	m	165.00

8.6 EXTERNAL DRAINAGE

Item	Unit	Range £
8.6.1 SURFACE WATER AND FOUL WATER DRAINAGE – cont		
Linear drainage to vehicular areas D400 loadings – cont		
Aco M150; internal width 150 mm; on concrete base and surround to falls; all to manufacturers specifications; paving surround to both sides of channel		
Galvanized	m	225.00
Stainless steel	m	380.00
Heelguard ductile	m	470.00
Heelguard mesh stainless steel	m	315.00
Slotted ductile	m	210.00
Linear drainage to pedestrian areas C250 loadings; Excavate trench by machine; lay Aco Brickslot channel drain on concrete base and surround to falls; all to manufacturers specifications; paving surround to both sides of channel		
100 mm internal width C250 loading		
Brickslot galvanized grating	m	225.00
Brickslot stainless steel grating	m	380.00
Heelguard mesh stainless steel grating	m	220.00
Brickslot twinslot offset	m	440.00
MANHOLES		
Brick manhole; excavate pit including earthwork support and working space disposal of surplus spoil to dump on site not exceeding 100 m; lay concrete (1:2:4) base 1500 mm dia. × 200 mm thick; reinforced with mesh reinforcement; 110 mm vitrified clay channels; benching in concrete (1:3:6) allowing one outlet and two inlets for 110 mm dia. pipe; construct inspection chamber 1 brick thick walls of engineering brick Class B; backfill with excavated material; complete with 2 nr cast iron step irons		
1200 × 1200 × 1200 mm		
cover slab of precast concrete	each	2325.00
access cover; Group 2; 600 × 450 mm	each	2425.00
1200 × 1200 × 1500 mm		
access cover; Group 2; 600 × 450 mm	each	3000.00
recessed cover 5 tonne load; 600 × 450 mm; filled with block paviors	each	3050.00
1200 × 1200 × 2000 mm; Walls 327.5 thick		
access cover; Group 2; 600 × 450 mm	each	4400.00
recessed cover 5 tonne load; 600 × 450 mm; filled with block paviors	each	4450.00
1500 × 1500 × 2500 mm; Walls 327.5 thick		
access cover; Group 2; 600 × 450 mm	each	6400.00
recessed cover 5 tonne load; 600 × 450 mm; filled with block paviors	each	6500.00
Concrete manholes; excavate pit including earthwork support and working space disposal of surplus spoil to dump on site not exceeding 100 m; lay concrete (1:2:4) base 1500 mm dia. × 200 mm thick; reinforced with mesh reinforcement; 110 mm vitrified clay channels; benching in concrete (1:3:6) allowing one outlet and two inlets for 110 mm dia. pipe; construct inspection chamber pre-cast concrete rings; surround in concrete 150 mm thick backfill with excavated material; complete with step irons		

8.6 EXTERNAL DRAINAGE

Item	Unit	Range £
Manhole dia. 900 mm		
1.00 m deep	each	1700.00
Manhole dia. 1.05 m		
1.00 m deep	each	1950.00
2.00 m deep	each	2600.00
3.00 m deep	each	3250.00
Manhole dia. 1.20 m		
1.00 m deep	each	2325.00
2.00 m deep	each	3050.00
3.00 m deep	each	3775.00
Manhole dia. 1.50 m		
1.00 m deep	each	3025.00
2.00 m deep	each	3975.00
3.00 m deep	each	4850.00
8.6.4 LAND DRAINAGE		
SUBSOIL DRAINAGE – BY MACHINE		
Main drain; remove 150 mm topsoil and deposit alongside trench, excavate drain trench by machine and lay flexible perforated drain; lay bed of gravel rejects 100 mm; backfill with gravel rejects or similar to within 150 mm of finished ground level; complete fill with topsoil; remove surplus spoil to approved dump on site		
Main drain 160 mm supplied in 35 m lengths		
450 mm deep	100 m	910.00
600 mm deep	100 m	1075.00
900 mm deep	100 m	1800.00
Extra for couplings	each	3.00
100 mm main drain supplied in 100 m lengths		
450 mm deep	100 m	1550.00
600 mm deep	100 m	2050.00
900 mm deep	100 m	3350.00
Extra for couplings	each	3.00
Laterals to main drains; herringbone pattern; excavation and backfilling; inclusive of connecting lateral to main drain		
100 mm pipe to 450 mm deep trench		
laterals at 1.0 m centres	100 m^2	1550.00
laterals at 2.0 m centres	100 m^2	780.00
laterals at 3.0 m centres	100 m^2	510.00
laterals at 5.0 m centres	100 m^2	310.00
100 mm pipe to 600 mm deep trench		
laterals at 1.0 m centres	100 m^2	2050.00
laterals at 2.0 m centres	100 m^2	1025.00
laterals at 3.0 m centres	100 m^2	680.00
laterals at 5.0 m centres	100 m^2	410.00

8.6 EXTERNAL DRAINAGE

Item	Unit	Range £
8.6.4 LAND DRAINAGE – cont		
Laterals to main drains – cont		
100 mm pipe to 900 mm deep trench		
laterals at 1.0 m centres	100 m^2	3350.00
laterals at 2.0 m centres	100 m^2	1700.00
laterals at 3.0 m centres	100 m^2	1100.00
laterals at 5.0 m centres	100 m^2	680.00
80 mm pipe to 450 mm deep trench		
laterals at 1.0 m centres	100 m^2	1475.00
laterals at 2.0 m centres	100 m^2	750.00
laterals at 3.0 m centres	100 m^2	495.00
laterals at 5.0 m centres	100 m^2	300.00
80 mm pipe to 600 mm deep trench		
laterals at 1.0 m centres	100 m^2	2025.00
laterals at 2.0 m centres	100 m^2	1000.00
laterals at 3.0 m centres	100 m^2	660.00
laterals at 5.0 m centres	100 m^2	400.00
80 mm pipe to 900 mm deep trench		
laterals at 1.0 m centres	100 m^2	3350.00
laterals at 2.0 m centres	100 m^2	1675.00
laterals at 3.0 m centres	100 m^2	1100.00
laterals at 5.0 m centres	100 m^2	670.00
Extra for 100/80 mm junctions connecting laterals to main drain		
laterals at 1.0 m centres	10 m	81.00
laterals at 2.0 m centres	10 m	40.00
laterals at 3.0 m centres	10 m	27.00
laterals at 5.0 m centres	10 m	16.00
SUBSOIL DRAINAGE – BY HAND		
Main drain; remove 150 mm topsoil and deposit alongside trench; excavate drain trench by machine and lay flexible perforated drain; lay bed of gravel rejects 100 mm; backfill with gravel rejects or similar to within 150 mm of finished ground level; complete fill with topsoil; remove surplus spoil to approved dump on site		
Main drain 160 mm in supplied in 35 m lengths		
450 mm deep	m	43.00
600 mm deep	m	56.00
900 mm deep	m	79.00
Extra for couplings	each	3.00
100 mm main drain supplied in 100 m lengths		
450 mm deep	m	30.00
600 mm deep	m	38.00
900 mm deep	m	54.00
Extra for couplings	each	3.00

8.6 EXTERNAL DRAINAGE

Item	Unit	Range £
Laterals to main drains; herringbone pattern; excavation and backfilling; inclusive of connecting lateral to main drain		
160 mm pipe to 450 mm deep trench		
laterals at 1.0 m centres	m^2	43.00
laterals at 2.0 m centres	m^2	32.00
160 mm pipe to 600 mm deep trench		
laterals at 1.0 m centres	m^2	56.00
laterals at 2.0 m centres	m^2	40.00
100 mm pipe to 450 mm deep trench		
laterals at 1.0 m centres	m^2	21.00
laterals at 2.0 m centres	m^2	13.00
100 mm pipe to 600 mm deep trench		
laterals at 1.0 m centres	m^2	38.00
laterals at 2.0 m centres	m^2	28.00
80 mm pipe to 450 mm deep trench		
laterals at 1.0 m centres	m^2	20.00
laterals at 2.0 m centres	m^2	12.00
80 mm pipe to 600 mm deep trench		
laterals at 1.0 m centres	m^2	38.00
laterals at 2.0 m centres	m^2	27.00
Extra for 100/80 mm couplings connecting laterals to main drain		
laterals at 1.0 m centres	10 m	81.00
laterals at 2.0 m centres	10 m	40.00
DRAINAGE DITCHES		
Excavate and form ditch and bank with 45° sides in light to medium soils; all widths taken at bottom of ditch		
300 mm wide × 600 mm deep	100 m	180.00
600 mm wide × 900 mm deep	100 m	620.00
1.20 m wide × 900 mm deep	100 m	1500.00
1.50 m wide × 1.20 m deep	100 m	2525.00
Clear and bottom existing ditch average 1.50 m deep, trim back vegetation and remove debris to licensed tip, lay jointed concrete pipes; including bedding, haunching and topping with 150 mm concrete; 11.50 N/mm^2–40 mm aggregate; backfill with approved spoil from site		
Pipes 300 mm dia.	100 m	13500.00
Pipes 450 mm dia.	100 m	18000.00
Pipes 600 mm dia.	100 m	24000.00
LAND DRAINS		
Excavate trench by excavator; lay Type 2 bedding; backfill to 150 mm above pipe with gravel rejects; lay non woven geofabric and fill with topsoil to ground level		
Trench 600 deep		
160 mm PVC-u drainpipe	100 m	2425.00
110 mm PVC-u drainpipe	100 m	2125.00
150 mm vitrified clay	100 m	6200.00
100 mm vitrified clay	100 m	3750.00

8.6 EXTERNAL DRAINAGE

Item	Unit	Range £
8.6.4 LAND DRAINAGE – cont		
SOAKAWAYS		
Construct soakaway from perforated concrete rings; excavation, casting in situ concrete ring beam base; filling with gravel 250 mm deep; placing perforated concrete rings; surrounding with geofabric and backfilling to external surround of soakaway with 250 mm wide granular surround and excavated material; step irons and cover slab; inclusive of all earthwork retention and disposal off site of surplus material		
900 mm dia.		
1.00 m deep	nr	1175.00
2.00 m deep	nr	2325.00
1200 mm dia.		
1.00 m deep	nr	1575.00
2.00 m deep	nr	2925.00
2400 mm dia.		
1.00 m deep	nr	4950.00
2.00 m deep	nr	8400.00
Aquacell Infiltration unit soakaway; 1.00 m × 500 × 400 mm; 200 litre volume each in trench		
Excavate for new soakaway; grade bottom of excavation; Lay bedding layer of 100 mm sharp sand; Install geotextile to line excavation and bedding; Install Aquacell units in bonded formation; wrap entire installation in geotextile; back filling to surround in MOT Type 2 150 mm thick. Place sharp sand over infiltration units; inclusive of connection only of in-flow pipework; Cover over 500 deep with Type 2 granular material.		
8 units in trench; 1520 litres	nr	970.00
12 units in trench; 2280 litres	nr	1550.00
16 units in trench; 3040 litres	nr	1975.00
20 units in trench; 3800 litres	nr	2475.00
30 units; in trench 5700 litres	nr	4000.00
60 units; in trench 11400 litres	nr	6800.00

8.7 EXTERNAL SERVICES

Item	Unit	Range £
8.7.1 WATER MAINS SUPPLY		
COLD WATER		
Excavate trench for cold water pipe lay bed of sharp sand 100 mm thick; lay pipe		
1.20 m deep		
20 mm MDPE pipe	m	55.00
25 mm MDPE pipe	m	55.00
32 mm MDPE pipe	m	55.00
50 mm MDPE pipe	m	58.00
60 mm MDPE pipe	m	59.00
750 mm deep		
20 mm MDPE pipe	m	35.00
25 mm MDPE pipe	m	35.00
32 mm MDPE pipe	m	35.00
50 mm MDPE pipe	m	37.00
60 mm MDPE pipe	m	39.00
8.7.2 ELECTRICITY MAINS SUPPLY		
Power supply; cable installation for external services; trenching and backfilling; supply and place ring main armoured cable for supply of services for external elctrical systems; straight clear runs; 600 mm deep		
Armoured cable		
Single circuit	m	10.00
Dual circuit	m	14.00
Three circuit	m	17.00
Four circuits	m	21.00
External grade cable in ducts; laid on sand bed covered with sand and tape		
Single circuit	m	12.00
Dual circuit	m	16.00
Three circuit	m	21.00
Four circuits	m	25.00
Pit housing for in-ground power by Kent Stainless Ltd; excavation, disposal to form housing of concrete blockwork 140 mm thick with waterproofing to rear face on 100 mm mesh reinforced concrete base; allow for services cable entry and drainage outlet ducts of 100 mm uPVC pipe. Install pop-up power unit. Inclusive of cable ducts and armoured cable inclusive of trenching and drainage connection to main drainage system 10 m distance. Includes connection of unit; excludes connections to distribution board.		
In-ground unit Type 2		
400 × 600 with fully recessed FACTA graded cover	nr	5200.00
8.7.9 EXTERNAL STREET LIGHTING SYSTEMS		
Streetlighting Marshalls Bega; to parking area or precinct; Installation of lighting poles and luminaries; excludes excavation of trenches for supply cable; ducted cable		
Lighting poles to 6.00 m		
Assymetrical lighting Bega M99 515	–	2125.00
Traditional bow top pole and luminary M70–991 – single luminary	–	3025.00
Traditional bow top pole and luminary M70–910 – two luminaries	–	4650.00

PART 3

Prices for Measured Works

INTRODUCTION

Typical Project Profile

Contract value	£100,000.00–£3,000,000.00
Labour rate (see page 7)	£31.00 per hour
Bricklayer	£37.55 per hour
Labour rate for maintenance contracts	£25.20 per hour
Number of site staff	5–35
Project area	6000 m^2
Project location	Outer London
Project components	50% hard landscape 50% soft landscape and planting
Access to works areas	Very good
Contract	Main contract
Delivery of materials	Full loads
Exchange rate for goods imported from the EU	We have used an exchange rate of £0.90 = €1.00

Rates shown include O&P as indicated below

Labour Type	Unit	Cost incl. Overhead	Mark-up %	Rate incl. Overhead & Profit
General contracting	/man-hour	£31.00	15%	£35.65
Brickwork 1+1 man gang	/team-hour	£71.00	15%	£81.65
Brickwork 1+2 man gang	/team-hour	£102.11	15%	£117.43
Maintenance contracting	/man-hour	£25.20	15%	£28.98

Other resource categories

Materials generally	15%
Plant	15%
Subcontractors generally	10%

Some high cost material or subcontract items may be at lower rates of 5%.

MARKET PRICES OF PLANT USED IN THIS EDITION

Item	PC £	Labour hours	Labour £	Plant £	Material £	Unit	**Total rate £**
AGGREGATES							
Aggregates MCM (SE) Ltd							
Aggregates in load sizes as shown; Bags in 20 tonne deliveries comprising 18–20 850 kg bags unloaded by truck mounted Hiab; Measured as compacted or settled volumes; Prices shown here include 15% profit mark-up							
Type 1 primary							
20 tonne loose tipped	49.39		–	–	49.39	m^3	**49.39**
in bulk bags	110.86		–	–	110.86	m^3	**110.86**
Type 1 recycled							
20 tonne loose tipped	24.64		–	–	24.64	m^3	**24.64**
in bulk bags	79.74		–	–	79.74	m^3	**79.74**
Type 1 crushed concrete							
20 tonne loose tipped	24.64		–	–	24.64	m^3	**24.64**
Type 3 primary							
20 tonne loose tipped	58.21		–	–	58.21	m^3	**58.21**
in bulk bags	113.68		–	–	113.68	m^3	**113.68**
Type 3 recycled							
20 tonne loose tipped	55.44		–	–	55.44	m^3	**55.44**
Ballast – all in							
20 tonne loose tipped	49.39		–	–	49.39	m^3	**49.39**
in bulk bags	108.49		–	–	108.49	m^3	**108.49**
Sand – Sharp							
20 tonne loose tipped	45.36		–	–	45.36	m^3	**45.36**
in bulk bags	108.43		–	–	108.43	m^3	**108.43**
Sand Soft building							
20 tonne loose tipped	49.28		–	–	49.28	m^3	**49.28**
in bulk bags	108.43		–	–	108.43	m^3	**108.43**
6F2 75 mm down crusher run capping material							
20 tonne loose tipped	15.12		–	–	15.12	m^3	**15.12**
in bulk bags	66.08		–	–	66.08	m^3	**66.08**
Shingle 20 mm							
20 tonne loose tipped	55.44		–	–	55.44	m^3	**55.44**
in bulk bags	99.57		–	–	99.57	m^3	**99.57**
Shingle – pea							
20 tonne loose tipped	55.44		–	–	55.44	m^3	**55.44**
in bulk bags	99.57		–	–	99.57	m^3	**99.57**
Crushed rock							
20 tonne loose tipped	125.83		–	–	125.83	m^3	**125.83**
Stone angular 40–20 or 20/4							
20 tonne loose tipped	73.80		–	–	73.80	m^3	**73.80**
in bulk bags	112.56		–	–	112.56	m^3	**112.56**
Grit 6–2 mm for permeable paving							
20 tonne loose tipped	–		–	–	73.80	m^3	**73.80**

MARKET PRICES OF PLANT USED IN THIS EDITION

Item	PC £	Labour hours	Labour £	Plant £	Material £	Unit	Total rate £
HIRED IN PLANT							
Hired Plant; Prices exclusive of Overhead & Profit							
All equipment below is based on a productivity rate of 32 hours per week and include costs per hour for an operator and fuel							
Dumper 3 tonne	–		–	39.32	–	hr	**39.32**
Dumper 5 tonne Thwaites self drive	–		–	43.89	–	hr	**43.89**
Dumper 6 tonne Thwaites self drive	–		–	43.74	–	hr	**43.74**
Excavator 360 Tracked 21 tonne operated	–		–	71.88	–	hr	**71.88**
Excavator 360 Tracked 8 tonne	–		–	53.01	–	hr	**53.01**
Excavator Tracked 5 tonne	–		–	47.70	–	hr	**47.70**
Fork lift telehandler (24 hours/ week)	–		–	44.18	–	hr	**44.18**
The items below exclude operators							
Mini Excavator 1.5 tonne	–		–	6.46	–	hr	**6.46**
Mini Excavator JCB 803 Rubber tracks	–		–	9.14	–	hr	**9.14**
Mini Excavator JCB 803 Steel tracks	–		–	9.14	–	hr	**9.14**
Skip loader 1 tonne	–		–	3.72	–	hr	**3.72**
Fuel charge Red diesel	–		–	0.76	–	l	**0.76**
HSS Ltd							
Access tower; alloy; 5.2 m	–		–	25.99	–	day	**25.99**
Cultivator; 110 kg	–		–	5.29	–	hr	**5.29**
Diamond blade consumable; 450 mm; concrete	–		–	13.54	–	mm	**13.54**
Heavy-duty breaker; 110 v; 2200 w	–		–	4.07	–	hr	**4.07**
Heavy-duty breaker; petrol; 5 hrs/ day; single tool	–		–	8.40	–	hr	**8.40**
Mesh fence; temporary security fencing; 2.85 × 2.0 m high	–		–	1.70	–	1 m/ wk	**1.70**
Petrol masonry saw bench; 350 mm	–		–	7.58	–	hr	**7.58**
Petrol poker vibrator + 50 mm head	–		–	3.81	–	hr	**3.81**
Post hole borer; 1 man; weekly rate	–		–	6.93	–	hr	**6.93**
Vibrating plate compactor	–		–	4.12	–	hr	**4.12**
Vibrating roller; 136 kg; 10.1 kN; 4 hrs/day	–		–	4.98	–	hr	**4.98**
Vibration damped breaker (light); 1600 w; 110 v	–		–	1.69	–	hr	**1.69**

1 PRELIMINARIES

Item	PC £	Labour hours	Labour £	Plant £	Material £	Unit	Total rate £
1.2 EMPLOYER'S REQUIREMENTS: TENDERING/ SUBLETTING/SUPPLY							
Tendering costs for an employed estimator for the acquisition of a landscape main or subcontract; inclusive of measurement, sourcing of materials and suppliers, cost and area calculations, pre and post tender meetings, bid compliance method statements and submission documents							
Remeasureable contract from prepared bills; contract value							
£25,000	–		–	–	–	nr	**635.18**
£50,000	–		–	–	–	nr	**1058.63**
£100,000	–		–	–	–	nr	**2540.72**
£200,000	–		–	–	–	nr	**4234.53**
£500,000	–		–	–	–	nr	**5504.89**
£1,000,000	–		–	–	–	nr	**8469.06**
Lump sum contract from specifications and drawings only; contract value							
£25,000	–		–	–	–	nr	**1693.81**
£50,000	–		–	–	–	nr	**2117.27**
£100,000	–		–	–	–	nr	**4234.53**
£200,000	–		–	–	–	nr	**7622.15**
£500,000	–		–	–	–	nr	**10586.33**
£1,000,000	–		–	–	–	nr	**16938.12**
Material costs for tendering							
Drawing and plan printing and distribution costs; plans issued by employer on CD; project value							
£30,000	–		–	–	49.82	nr	**49.82**
£30,000 to £80,000	–		–	–	124.55	nr	**124.55**
£80,000 to £150,000	–		–	–	249.09	nr	**249.09**
£150,000 to £300,000	–		–	–	348.73	nr	**348.73**
£300,000 to £1,000,000	–		–	–	597.82	nr	**597.82**
Programmes; tender stage programmes							
Allow for production of an outline works programmes for submission with the tenders; project value							
£30,000	–	1.50	75.53	–	–	nr	**75.53**
£50,000	–	6.00	302.12	–	–	nr	**100.71**
£75,000	–	6.00	302.12	–	–	nr	**125.88**
£100,000	–	3.00	151.06	–	–	nr	**151.06**
£200,000	–	6.00	302.12	–	–	nr	**201.41**
£500,000	–	6.00	302.12	–	–	nr	**226.59**
£1,000,000	–	6.00	302.12	–	–	nr	**302.12**

1 PRELIMINARIES

Item	PC £	Labour hours	Labour £	Plant £	Material £	Unit	Total rate £
1.2 EMPLOYER'S REQUIREMENTS: TENDERING/ SUBLETTING/SUPPLY – cont							
Method statements							
Method statements for private or commercial projects where award is not points based; provide detailed method statements on all aspects of the works; project value							
£30,000	–	2.50	125.88	–	–	nr	**125.88**
£50,000	–	3.50	176.23	–	–	nr	**176.23**
£75,000	–	4.00	201.41	–	–	nr	**201.41**
£100,000	–	5.00	251.76	–	–	nr	**251.76**
£200,000	–	6.00	302.12	–	–	nr	**302.12**
£500,000	–	7.00	352.47	–	–	nr	**352.47**
£1,000,000	–	8.00	402.82	–	–	nr	**402.82**
Method statements for local authority type projects where award is points rated; provide detailed method statements on all aspects of the works; project value							
£30,000	–	5.00	251.76	–	–	nr	**251.76**
£50,000	–	5.00	251.76	–	–	nr	**251.76**
£75,000	–	6.00	302.12	–	–	nr	**302.12**
£100,000	–	7.00	352.47	–	–	nr	**352.47**
£200,000	–	8.00	402.82	–	–	nr	**402.82**
£500,000	–	9.00	453.17	–	–	nr	**453.17**
£1,000,000	–	16.00	805.64	–	–	nr	**805.64**
Health and safety							
Produce health and safety file including preliminary meeting and subsequent progress meetings with external planning officer in connection with health and safety; project value							
£35,000	–	8.00	402.82	–	–	nr	**402.82**
£75,000	–	12.00	604.23	–	–	nr	**604.23**
£100,000	–	16.00	805.64	–	–	nr	**805.64**
£200,000 to £500,000	–	40.00	2014.11	–	–	nr	**2014.11**
Maintain health and safety file for project duration; project value							
£35,000	–	4.00	201.41	–	–	week	**201.41**
£75,000	–	4.00	201.41	–	–	week	**201.41**
£100,000	–	8.00	402.82	–	–	week	**402.82**
£200,000 to £500,000	–	8.00	402.82	–	–	week	**402.82**

1 PRELIMINARIES

Item	PC £	Labour hours	Labour £	Plant £	Material £	Unit	Total rate £
Produce written risk assessments on all areas of operations within the scope of works of the contract; project value							
£35,000	–	2.00	100.71	–	–	nr	**100.71**
£75,000	–	3.00	151.06	–	–	nr	**151.06**
£100,000	–	5.00	251.76	–	–	nr	**251.76**
£200,000 to £500,000	–	8.00	402.82	–	–	nr	**402.82**
Produce Coshh assessments on all substances to be used in connection with the contract; project value							
£35,000	–	1.50	75.53	–	–	nr	**75.53**
£75,000	–	2.00	100.71	–	–	nr	**100.71**
£100,000	–	3.00	151.06	–	–	nr	**151.06**
£200,000 to £500,000	–	3.00	151.06	–	–	nr	**151.06**
1.2.4 EMPLOYER'S REQUIREMENTS: SECURITY/ SAFETY/PROTECTION							
Temporary security fence; HSS Hire; mesh framed unclimbable fencing; including precast concrete supports and couplings							
Weekly hire; 2.85 × 2.00high							
weekly hire rate	–		–	1.96	–	m	**1.96**
erection of fencing; labour only	–	0.10	3.57	–	–	m	**3.57**
removal of fencing loading to collection vehicle	–	0.07	2.38	–	–	m	**2.38**
delivery charge	–		–	1.25	–	m	**1.25**
return haulage charge	–		–	0.82	–	m	**0.82**
Tree Root zone protection; Greenfix Ltd Carefully hand dig around base of tree but away from trunk to required depth; Lay Treetex geofabric to subgrade; Lay Geoweb cellular root protection to required depth and backfill with angular material							
To graded and compacted substrate (not included), on 50 mm sharp sand base filled with angular drainage aggregate 20–5 mm; not mechanical traffic on the area of the rootzone permitted; excavated material loaded outside the root protection zone to spoil heaps							
Geoweb 75 mm depth	8.99	0.07	7.13	1.33	19.88	m²	**28.34**
Geoweb 100 mm depth	10.35	0.08	8.02	1.54	23.61	m²	**33.18**
Geoweb 150 mm depth	12.94	0.10	10.16	1.67	30.51	m²	**42.34**
Geoweb 200 mm depth	18.23	0.10	11.08	2.50	40.50	m²	**54.09**
Geoweb 300 mm depth	20.90	0.13	14.09	4.00	51.41	m²	**69.50**

1 PRELIMINARIES

Item	PC £	Labour hours	Labour £	Plant £	Material £	Unit	Total rate £
1.2.4 EMPLOYER'S REQUIREMENTS: SECURITY/ SAFETY/PROTECTION – cont							
Lay ply boards to protect surfaces							
To horizontal surfaces	–	0.01	0.89	–	18.72	m^2	**19.61**
To vertical surfaces	–	0.07	4.76	–	18.72	m^2	**23.47**
1.2.6 SUBMISSION OF SAMPLES							
Set up sample panels for the works inclusive of arranging delivery of sample materials; paving samples on 100 mm base							
Costs of labours only							
brick paving panel (pointed)	–	6.00	186.00	–	–	m^2	**186.00**
brick paving panel (butt jointed)	–	4.00	124.00	–	–	m^2	**124.00**
block paving panel	–	4.00	124.00	–	–	m^2	**124.00**
stone slab paving panel	–	6.00	186.00	–	–	m^2	**186.00**
cladding panel	–	7.00	217.00	–	–	m^2	**217.00**
brick wall panel	–	5.00	155.00	–	–	m^2	**155.00**
render panel	–	3.50	108.50	–	–	m^2	**108.50**
paint panel	–	3.00	93.00	–	–	m^2	**93.00**
1.2.7 CONTRACTOR'S GENERAL COST ITEMS: MECHANICAL PLANT							
Coveya; moving only of granular material by belt conveyor; conveyors fitted with troughed belts, receiving hopper, front and rear undercarriages and driven by electrical and air motors; support work for conveyor installation (scaffolding), delivery, collection and installation all excluded							
Delivery collection haulage and setup charges							
Conveyors up to 10 m long	–		–	622.73	–	nr	**622.73**
Conveyors up to 20 m long	–		–	1120.91	–	nr	**1120.91**
Conveyors up to 50 m long	–		–	3113.63	–	nr	**3113.63**
Conveyor belt width 400 mm; mechanically loaded and removed at offload point; conveyor length							
up to 5 m	–	1.50	53.48	22.18	–	m^3	**75.65**
10 m	–	1.50	53.48	22.82	–	m^3	**76.29**
12.5 m	–	1.50	53.48	23.57	–	m^3	**77.04**
15 m	–	1.50	53.48	24.10	–	m^3	**77.58**
20 m	–	1.50	53.48	25.39	–	m^3	**78.86**
25 m	–	1.50	53.48	26.99	–	m^3	**80.47**
30 m	–	1.50	53.48	28.60	–	m^3	**82.07**

1 PRELIMINARIES

Item	PC £	Labour hours	Labour £	Plant £	Material £	Unit	Total rate £
Conveyor belt width 600 mm; mechanically loaded and removed at offload point; conveyor length							
up to 5 m	–	1.00	35.66	15.14	–	m^3	**50.80**
10 m	–	1.00	35.66	15.86	–	m^3	**51.52**
12.5 m	–	1.00	35.66	16.37	–	m^3	**52.03**
15 m	–	1.00	35.66	16.93	–	m^3	**52.59**
20 m	–	1.00	35.66	18.00	–	m^3	**53.66**
25 m	–	1.00	35.66	19.78	–	m^3	**55.44**
30 m	–	1.00	35.66	21.21	–	m^3	**56.87**
Conveyor belt width 400 mm; mechanically loaded and removed by hand at offload point; conveyor length							
up to 5 m	–	3.19	113.74	7.78	–	m^3	**121.52**
10 m	–	3.19	113.74	8.64	–	m^3	**122.38**
12.5 m	–	3.19	113.74	9.64	–	m^3	**123.38**
15 m	–	3.19	113.74	10.35	–	m^3	**124.09**
20 m	–	3.19	113.74	12.06	–	m^3	**125.80**
25 m	–	3.19	113.74	14.20	–	m^3	**127.94**
30 m	–	3.19	113.74	16.34	–	m^3	**130.08**
Terranova Cranes Ltd; crane hire; materials handling and lifting; telescopic cranes supply and management; exclusive of roadway management or planning applications; prices below illustrate lifts of 1 tonne at maximum crane reach; lift cycle of 0.25 hours; mechanical filling of material skip if appropriate							
35 tonne mobile crane; lifting capacity of 1 tonne at 26 m; C.P.A hire (all management by hirer)							
granular materials including concrete	–	0.75	26.74	26.37	–	tonne	**53.11**
palletized or packed materials	–	0.50	17.83	23.35	–	tonne	**41.18**
5 tonne mobile crane; lifting capacity of 1 tonne at 26 m; contract lift							
granular materials or concrete	–	0.75	26.74	62.18	–	tonne	**88.91**
palletized or packed materials	–	0.50	17.83	59.16	–	tonne	**76.98**
50 tonne mobile crane; lifting capacity of 1 tonne at 34 m; C.P.A hire (all management by hirer)							
granular materials including concrete	–	0.75	26.74	29.48	–	tonne	**56.22**
palletized or packed materials		1.00	35.65	26.47	–	tonne	**62.12**

1 PRELIMINARIES

Item	PC £	Labour hours	Labour £	Plant £	Material £	Unit	Total rate £
1.2.7 CONTRACTOR'S GENERAL COST ITEMS: MECHANICAL PLANT – cont							
Terranova Cranes Ltd – cont							
50 tonne mobile crane; lifting capacity of 1 tonne at 34 m; contract lift							
granular materials or concrete	–	0.75	26.74	74.63	–	tonne	**101.37**
palletized or packed materials	–	0.50	17.83	71.61	–	tonne	**89.44**
80 tonne mobile crane; lifting capacity of 1 tonne at 44 m; C.P.A. hire (all management by hirer)							
granular materials including concrete	–	0.75	26.74	46.61	–	tonne	**73.34**
palletized or packed materials	–	0.50	17.83	43.59	–	tonne	**61.42**
80 tonne mobile crane; lifting capacity of 1 tonne at 44 m; contract lift							
granular materials or concrete	–	0.75	26.74	87.08	–	tonne	**113.82**
palletized or packed materials	–	0.50	17.83	84.07	–	tonne	**101.89**
Other mechanical or hired plant required to manage a landscape contract							
Hired in plant							
General purpose Excavator and dumper – non productive time, or not task allocated; operated	–		–	91.59	–	hour	**91.59**
Telehandller; operated	–		–	44.18	–	week	**44.18**
Site skips	–		–	259.92	–	nr	**259.92**
Fuel bowsers	–		–	79.06	–	week	**79.06**

1 PRELIMINARIES

Item	PC £	Labour hours	Labour £	Plant £	Material £	Unit	**Total rate £**
1.2.8 CONTRACTOR'S GENERAL COST ITEMS: TEMPORARY WORKS							
LiveTrakway; portable roadway systems							
Temporary roadway system laid directly onto existing surface or onto PVC matting to protect existing surface; most systems are based on a weekly hire charge with transportation, installation and recovery charges included							
Heavy Duty Trakpanel; per panel (3.05 × 2.59 per week	–		–	–	–	m^2	**6.64**
Outrigger Mats for use in conjunction with Heavy Duty Trakpanels; per set of 4 mats per week	–		–	–	–	set	**111.98**
Medium Duty Trakpanel; per panel (2.44 × 3.00 per week	–		–	–	–	m^2	**7.16**
LD20 Evolution; Light Duty Trakway; roll out system minimum delivery 50 m	–		–	–	–	m^2	**8.34**
Terraplas Walkways; turf protection system; per section (1 × 1 m); per week	–		–	–	–	m^2	**6.55**

1 PRELIMINARIES

Item	PC £	Labour hours	Labour £	Plant £	Material £	Unit	Total rate £
1.5 TENDER AND CONTRACT DOCUMENTS							
Note: The preliminary requirements are different for all projects. Many are time related and many are set costs. Others vary depending on the project requirements. We have attempted to provide a guide to the cost related items in Section A of the National Building Specification. These costs are a guide only. Users of the book should evaluate each project separately. There are other preliminary cost items which may be added in to the preliminary costs of any project which may not be featured here.							
Method statements							
Provide detailed method statements on all aspects of the works; project value							
£30,000	–	2.50	109.46	–	–	nr	**109.46**
£50,000	–	3.50	153.25	–	–	nr	**153.25**
£75,000	–	4.00	175.14	–	–	nr	**175.14**
£100,000	–	5.00	218.93	–	–	nr	**218.93**
£200,000	–	6.00	262.71	–	–	nr	**262.71**
Tender and contract documents							
Drawing and plan printing and distributioncosts; plans issued by employer on CD; project value							
£30,000	–		–	–	43.32	nr	**43.32**
£30,000 to £80,000	–		–	–	108.30	nr	**108.30**
£80,000 to £150,000	–		–	–	216.60	nr	**216.60**
£150,000 to £300,000	–		–	–	303.24	nr	**303.24**
£300,000 to £1,000,000	–		–	–	519.84	nr	**519.84**
Contract drawings for project management and distribution to suppliers, subcontractors and site staff							
£30,000	–		–	–	86.64	nr	**86.64**
£30,000 to £80,000	–		–	–	216.60	nr	**216.60**
£80,000 to £150,000	–		–	–	433.20	nr	**433.20**
£150,000 to £300,000	–		–	–	606.48	nr	**606.48**
£300,000 to £1,000,000	–		–	–	1039.68	nr	**1039.68**

1 PRELIMINARIES

Item	PC £	Labour hours	Labour £	Plant £	Material £	Unit	Total rate £
1.5 THE CONTRACT/ SUBCONTRACT							
Contract/subcontract evaluation							
Due diligence on evaluation or examination of clauses in contract documents; evaluation and report by a suitably qualified quantity surveyor or legal advisor where necessary							
minor works contract	–	1.00	43.79	–	–	nr	**43.79**
JCLI	–		–	–	86.64	nr	**86.64**
JCT intermediate contract or subcontract	–	1.00	43.79	–	173.28	nr	**217.07**
1.6 DESIGN COSTS FOR LANDSCAPE WORKS							
Landscape design; Landscape architects; mixed hard and soft landscape; indicative prices for design project stages; brief; planning; detailed design and supervision during construction stage; the complexity ratings shown refer to the Landscape Institute tables published in this book							
Commercial urban development; complexity rating 2; landscape construction value							
£100,000.00	–		–	–	–	nr	**10000.00**
£250,000.00	–		–	–	–	nr	**20000.00**
£500,000.00	–		–	–	–	nr	**35000.00**
£1,000,000.00	–		–	–	–	nr	**65000.00**
Hospitals and education; complexity rating 3; landscape construction value							
£100,000.00	–		–	–	–	nr	**11000.00**
£250,000.00	–		–	–	–	nr	**23000.00**
£500,000.00	–		–	–	–	nr	**40000.00**
£1,000,000.00	–		–	–	–	nr	**74000.00**
Public urban open space; complexity rating 4; landscape construction value							
£100,000.00	–		–	–	–	nr	**12000.00**
£250,000.00	–		–	–	–	nr	**26000.00**
£500,000.00	–		–	–	–	nr	**47500.00**
£1,000,000.00	–		–	–		nr	**85000.00**

1 PRELIMINARIES

Item	PC £	Labour hours	Labour £	Plant £	Material £	Unit	Total rate £
1.6 DESIGN COSTS FOR LANDSCAPE WORKS – cont							
Landscape design – cont							
Countryside or leisure open space; complexity rating 1							
£100,000.00	–		–	–	–	nr	**9000.00**
£250,000.00	–		–	–	–	nr	**18000.00**
£500,000.00	–		–	–	–	nr	**32000.00**
£1,000,000.00	–		–	–	–	nr	**60000.00**
Reclamation project; complexity rating 4							
£100,000.00	–		–	–	–	nr	**12000.00**
£250,000.00	–		–	–	–	nr	**26000.00**
£500,000.00	–		–	–	–	nr	**45000.00**
£1,000,000.00	–		–	–	–	nr	**80000.00**
Housing; complexity rating 3							
£100,000.00	–		–	–	–	nr	**11000.00**
£250,000.00	–		–	–	–	nr	**23000.00**
£500,000.00	–		–	–	–	nr	**40000.00**
£1,000,000.00	–		–	–	–	nr	**74000.00**
1.7 EMPLOYERS REQUIREMENTS: PROVISION CONTENT AND USE OF DOCUMENTS							
Provision, content and use of documents							
Supply as built drawings for elements of the project that may have carried from the original design drawings							
£35,000	–	2.00	91.77	–	–	nr	**91.77**
£75,000	–	2.00	91.77	–	–	nr	**91.77**
£100,000	–	4.00	183.54	–	–	nr	**183.54**
£200,000 to £500,000	–	5.00	229.43	–	–	nr	**229.43**
Fees for Considerate Contractors Scheme (CCS)							
Project value							
up to £100,000	–		–	–	124.55	nr	**124.55**
£100,000–£500000	–		–	–	292.68	nr	**292.68**
£500,000–£5,000,000	–		–	–	585.36	nr	**585.36**
over £5,000,000	–		–	–	878.04	nr	**878.04**
Climatic conditions – keep records of temperature and rainfall; delays due to weather including descriptions of weather							
daily cost	–	0.08	4.19	–	–	day	**4.19**
weekly cost	–	0.42	20.97	–	–	day	**20.97**

1 PRELIMINARIES

Item	PC £	Labour hours	Labour £	Plant £	Material £	Unit	Total rate £
Programmes; master programme for the contract works							
Allow for production of works programmes prior to the start of the works; project value							
£30,000	–	3.00	151.06	–	–	nr	**151.06**
£50,000	–	6.00	302.12	–	–	nr	**302.12**
£75,000	–	8.00	402.82	–	–	nr	**402.82**
£100,000	–	10.00	503.53	–	–	nr	**503.53**
£200,000	–	14.00	704.94	–	–	nr	**704.94**
£500,000	–	15.00	755.29	–	–	nr	**755.29**
£1,000,000	–	18.00	906.35	–	–	nr	**906.35**
Indicative staff resource chart							
Project value							
up to £100000	–	1.00	50.35	–	–	nr	**50.35**
up to £500000	–	1.50	75.53	–	–	nr	**75.53**
up to £1,000,000	–	2.00	100.71	–	–	nr	**100.71**
Curriculum vitae of staff							
Prepare and submit curriculum vitae of pre-construction and construction phase staff							
per staff member	–	0.75	37.76	–	–	nr	**37.76**
1.7 EMPLOYER'S REQUIREMENTS: MANAGEMENT OF THE WORKS							
Allow for updating the works programme during the course of the works; project value							
£30,000	–	1.00	50.35	–	–	nr	**50.35**
£50,000	–	1.50	75.53	–	–	nr	**75.53**
£75,000	–	2.00	100.71	–	–	nr	**100.71**
£100,000	–	3.00	151.06	–	–	nr	**151.06**
£200,000	–	5.00	251.76	–	–	nr	**251.76**
Setting out							
Setting out for external works operations comprising hard and soft works elements; placing of pegs and string lines to Landscape Architect's drawings; surveying levels and placing level pegs; obtaining approval from the Landscape Architect to commence works; areas of entire site							
1,000 m^2	–	5.00	178.25	–	10.65	nr	**188.90**
2,500 m^2	–	8.00	285.20	–	21.29	nr	**306.49**
5,000 m^2	–	8.00	285.20	–	21.29	nr	**306.49**
10,000 m^2	–	32.00	1140.80	–	53.23	nr	**1194.03**

1 PRELIMINARIES

Item	PC £	Labour hours	Labour £	Plant £	Material £	Unit	Total rate £
1.7 EMPLOYER'S REQUIREMENTS: MANAGEMENT OF THE WORKS – cont							
Setting out – cont							
Setting out engineer							
per half day	–		–	–	–	nr	**227.43**
per day	–		–	–	–	nr	**379.05**
1.11 CONTRACTOR'S GENERAL COST ITEMS: SITE ACCOMMODATION							
General							
The following items are instances of the commonly found preliminary costs associated with external works contracts. The assumption is made that the external works contractor is subcontracted to a main contractor.							
Elliot Hire; erect temporary accommodation and storage on concrete base measured separately							
Prefabricated office hire; jackleg; open plan							
3.6 × 2.4 m	–		–	41.10	–	week	**41.10**
4.8 × 2.4 m	–		–	42.35	–	week	**42.35**
Armoured store; erect temporary secure storage container for tools and equipment							
3.0 × 2.4 m	–		–	15.57	–	week	**15.57**
3.6 × 2.4 m	–		–	15.57	–	week	**15.57**
Delivery and collection charges on site offices							
delivery charge	–		–	259.05	–	load	**259.05**
collection charge	–		–	259.05	–	load	**259.05**
Toilet facilities							
HSS Hire; serviced self-contained toilet delivered to and collected from site; maintained by toilet supply company							
single chemical toilet including wash hand basin and water	–		–	49.82	–	week	**49.82**
delivery and collection; each way	–		–	47.33	–	nr	**47.33**

1 PRELIMINARIES

Item	PC £	Labour hours	Labour £	Plant £	Material £	Unit	Total rate £
1.11 THE SITE/EXISTING BUILDINGS							
Existing mains and services; mark positions of existing mains and services; locate and mark; site area							
less than 500 m²	–	1.50	53.48	–	–	nr	**53.48**
up to 1000 m²	–	3.00	106.95	–	–	nr	**106.95**
up to 2000 m²	–	8.00	285.20	–	–	nr	**285.20**
up to 4000 m²	–	16.00	570.40	–	–	nr	**570.40**
Access to the site							
For pedestrians							
security kiosk	–		–	236.64	–	week	**236.64**
security guard	–	40.00	579.60	–	–	week	**579.60**
protected walkways; Heras fencing on 2 sides	–		–	3.91	–	m	**3.91**
Parking							
Parking expenses where vehicles do not park on the site area; per vehicle							
metropolitan area; city centre	–		–	–	–	week	**230.00**
metropolitan area; outer areas	–		–	–	–	week	**184.00**
suburban restricted parking areas	–		–	–	–	week	**46.00**
Congestion charging							
London only	–		–	–	–	week	**66.13**
Site visit; pre-tender for purposes of understanding site and tender requirements; prices below are based on a single senior manager attending site							
City location; average distance of travel 20 miles inclusive of travel and parking costs							
less than 500 m²	–	4.00	201.41	–	63.52	nr	**264.93**
up to 1000 m²	–	5.00	251.76	–	67.26	nr	**319.02**
up to 2000 m²	–	6.00	302.12	–	78.46	nr	**380.58**
up to 4000 m²	–	8.00	402.82	–	93.41	nr	**496.23**
Town or rural location; average distance of travel 20 miles inclusive of travel and parking costs							
less than 500 m²	–	3.00	151.06	–	37.36	nr	**188.42**
up to 1000 m²	–	4.00	201.41	–	41.10	nr	**242.51**
up to 2000 m²	–	5.00	251.76	–	44.84	nr	**296.60**
up to 4000 m²	–	6.00	302.12	–	44.84	nr	**346.95**

2 OFF-SITE MANUFACTURED MATERIALS, COMPONENTS AND BUILDINGS

Item	PC £	Labour hours	Labour £	Plant £	Material £	Unit	Total rate £
PREFABRICATED BUILDINGS/ STRUCTURES/UNITS							
The labour in this section is calculated on a 3 person team. The labour time below should be multiplied by 3 to calculate the cost as shown.							
Cast stone buildings; Haddonstone Ltd; ornamental garden buildings in Portland Bath or Terracotta finished cast stone; prices for stonework and facades only; excavations, foundations, reinforcement, concrete infill, roofing and floors all priced separately							
Pavilion Venetian Folly L9400; Tuscan columns, pedimented arch, quoins and optional balustrading							
4184 mm high × 4728 mm wide × 3147 mm deep	13422.00	175.00	6238.75	132.02	15716.74	nr	**22087.51**
Pavilion L9300; Tuscan columns							
3496 mm high × 3634 mm wide	8142.00	144.00	5133.60	132.02	9644.74	nr	**14910.36**
Small Classical Temple L9250; 6 column with fibreglass lead effect finish dome roof							
overall height 3610 mm; dia. 2540 mm	7070.40	130.00	4634.50	66.01	8412.40	nr	**13112.91**
Large Classical Temple L9100; 8 column with fibreglass lead effect finish dome roof							
overall height 4664 mm; dia. 3190 mm	15576.00	165.00	5882.25	66.01	18193.84	nr	**24142.10**
Stepped floors to temples							
single step; Small Classical Temple	1434.00	24.00	855.60	–	1761.23	nr	**2616.83**
single step; Large Classical Temple	1878.00	26.00	926.90	–	2294.25	nr	**3221.15**

2 OFF-SITE MANUFACTURED MATERIALS, COMPONENTS AND BUILDINGS

Item	PC £	Labour hours	Labour £	Plant £	Material £	Unit	**Total rate £**
Stone structures; Architectural Heritage Ltd; hand carved from solid natural limestone with wrought iron domed roof, decorated frieze and base and integral seats; supply and erect only; excavations and concrete bases priced separately							
The Park Temple; 5 columns; 3500 mm high × 1650 mm dia.	12000.00	96.00	3422.40	121.93	13800.00	nr	**17344.33**
The Estate Temple; 6 columns; 4000 mm high × 2700 mm dia.	18000.00	120.00	4278.00	243.86	20700.00	nr	**25221.86**
Stone structures; Architectural Heritage Ltd; hand carved from solid natural limestone with solid oak trelliage; prices for stonework and facades only; supply and erect only; excavations and concrete bases priced separately							
The Pergola; 2240 mm high × 2640 mm wide × 6990 mm long	12000.00	96.00	3422.40	487.72	13887.48	nr	**17797.60**
Ornamental stone structures; Architectural Heritage Ltd; setting to bases or plinths (not included)							
The Small Obelisk; classic natural stone obelisk; tapering square form on panelled square base; 1860 mm high × 360 mm square	2400.00	2.00	71.30	–	2760.00	nr	**2831.30**
The Inverted Sundial Pedestal with Griffin Armillary Sphere	2600.00	2.00	71.30	–	2990.00	nr	**3061.30**
Large Obelisk; 2.40 m high × 460 mm square; hand carved natural limestone obelisk of tapering square form raised upon a panelled square base supported by four spheres. Reproduction.	3200.00	1.00	35.65	–	3680.00	nr	**3715.65**

3 DEMOLITIONS

Item	PC £	Labour hours	Labour £	Plant £	Material £	Unit	Total rate £
DEMOLITIONS/CLEARANCE							
The labour in this section is calculated on a 3 person team. The labour time below should be multiplied by 3 to calculate the cost as shown.							
Demolish existing structures; disposal off site; mechanical demolition; with 3 tonne excavator and dumper							
Brick wall							
112.5 mm thick	–	0.07	2.32	3.63	–	m^2	**12.40**
225 mm thick	–	0.08	2.89	4.51	–	m^2	**20.23**
337.5 mm thick	–		–	12.98	–	m^2	**34.51**
450 mm thick	–		–	17.33	–	m^2	**43.02**
Demolish existing structures; disposal off site mechanically loaded; all other works by hand							
Brick wall							
112.5 mm thick	–	0.33	11.70	–	–	m^2	**18.20**
225 mm thick	–	0.50	17.50	–	–	m^2	**30.45**
337.5 mm thick	–	0.67	23.28	–	–	m^2	**44.95**
450 mm thick	–	1.00	35.00	–	–	m^2	**60.90**
Demolish existing structures; disposal off site mechanically loaded; by diesel or electric breaker; all other works by hand							
Brick wall							
112.5 mm thick	–	0.17	5.81	0.32	–	m^2	**12.58**
225 mm thick	–	0.20	6.93	0.38	–	m^2	**20.14**
337.5 mm thick	–	0.25	8.65	0.47	–	m^2	**28.11**
450 mm thick	–	0.33	11.53	0.63	–	m^2	**37.77**
Break out concrete footings associated with free-standing walls; inclusive of all excavation and backfilling with excavated material; disposal off site mechanically loaded							
By mechanical breaker; diesel or electric							
plain concrete	–	1.50	52.60	10.06	–	m^3	**101.16**
reinforced concrete	–	2.50	87.75	14.68	–	m^3	**160.15**

3 DEMOLITIONS

Item	PC £	Labour hours	Labour £	Plant £	Material £	Unit	Total rate £
Remove existing free-standing buildings; demolition by hand							
Timber building with suspended timber floor; hardstanding or concrete base not included; disposal off site							
shed 6.0 m^2	–	2.00	69.89	–	–	nr	**129.75**
shed 10.0 m^2	–	3.00	104.64	–	–	nr	**212.15**
shed 15.0 m^2	–	3.50	121.68	–	–	nr	**288.39**
Timber building; insulated; with timber or concrete posts set in concrete, felt covered timber or tiled roof; internal walls cladding with timber or plasterboard; load arisings to self loaded muck-away							
timber structure 6.0 m^2	–	2.50	86.25	–	–	nr	**341.51**
timber structure 12.0 m^2	–	3.50	120.76	–	–	nr	**477.65**
timber structure 20.0 m^2	–	8.00	278.11	–	–	nr	**659.16**
Demolition of free-standing brick buildings with tiled or sheet roof; concrete foundations measured separately; mechanical demolition; maximum distance to stockpile 25 m; inclusive for all access scaffolding and the like; maximum height of roof 4.0 m; inclusive of all doors, windows, guttering and down pipes; including disposal by grab							
Half brick thick							
10 m^2	–	8.00	281.16	202.16	–	nr	**719.97**
20 m^2	–	16.00	564.40	376.16	–	nr	**1246.04**
1 brick thick							
10 m^2	–	10.00	350.27	344.14	–	nr	**1166.12**
20 m^2	–	16.00	561.57	517.40	–	nr	**1686.87**
Cavity wall with blockwork inner skin and brick outer skin; insulated							
10 m^2	–	12.00	420.23	486.67	–	nr	**1561.28**
20 m^2	–	20.00	701.22	659.91	–	nr	**2205.52**
Extra over to the above for disconnection of services							
Electrical							
disconnection	–	2.00	90.68	–	–	nr	**90.68**
grub out cables and dispose; backfilling; by machine	–		–	0.44	–	m	**1.39**
grub out cables and dispose; backfilling; by hand	–	0.50	17.79	–	–	m	**18.76**

3 DEMOLITIONS

Item	PC £	Labour hours	Labour £	Plant £	Material £	Unit	Total rate £
DEMOLITIONS/CLEARANCE – cont							
Extra over to the above for disconnection of services – cont							
Water supply, foul or surface water drainage							
disconnection; capping off	–	1.00	34.54	–	–	nr	**125.15**
grub out pipes and dispose; backfilling; by machine	–		–	0.44	–	m	**1.39**
grub out pipes and dispose; backfilling; by hand	–	0.50	17.79	–	–	m	**18.76**
The labour in this section is calculated on a 2 person team. The labour time below should be multiplied by 2 to calculate the cost as shown.							
Demolish existing fence; remove stakes or grub out posts as appropriate; remove to stock pile for removal off site; (not included)							
Mechanical demolition; clear fence line							
chain link fence 1.20–1.50 m high	–	0.01	0.71	0.80	–	m	**1.51**
closeboard or timber panel fence 1.80 m high	–	0.03	1.78	1.84	–	m	**3.62**
Mechanical demolition; light shrubs or creepers in chainlinks							
chain link fence 1.20–1.50 m high	–	0.01	0.89	1.02	–	m	**1.92**
closeboard or timber panel fence 1.80 m high	–	0.03	2.37	2.47	–	m	**4.84**
Mechanical demolition; heavy shrubs or creepers bramble and the like requiring clearance to enable removal							
fence 1.20–1.50 m high	–	0.05	3.57	2.58	–	m	**6.14**
closeboard or timber panel fence 1.80 m high	–	0.05	3.57	3.90	–	m	**7.46**
Demolition and site transport by hand; clear fence lines							
fence 1.20–1.50 m high	–	0.03	2.37	–	–	m	**2.37**
closeboard or timber panel fence 1.80 m high	–	0.07	4.76	–	–	m	**4.76**
Demolition and site transport by hand; light vegetation and creepers							
fence 1.20–1.50 m high	–	0.05	3.57	–	–	m	**3.57**
closeboard or timber panel fence 1.80 m high	–	0.13	8.91	–	–	m	**8.91**

3 DEMOLITIONS

Item	PC £	Labour hours	Labour £	Plant £	Material £	Unit	**Total rate £**
Demolition and site transport by hand; heavy shrubs or creepers bramble and the like requiring clearance to enable removal							
fence 1.20–1.50 m high	–	0.13	8.91	–	–	m	**8.91**
closeboard or timber panel fence 1.80 m high	–	0.17	11.89	–	–	m	**11.89**
Demolition of posts and straining posts							
Break out straining post; grub out concrete footings; remove to stockpile on site							
single straining post by machine	–	0.25	8.91	18.76	–	nr	**27.67**
double straining post by machine	–	0.30	10.70	21.50	–	nr	**32.20**
single straining post by hand	–	1.00	71.30	–	–	nr	**71.30**
double straining post by hand	–	1.25	89.13	–	–	nr	**89.13**
Break out gatepost; concrete steel or timber; girth not exceeding 150 mm; grub out concrete footings							
by machine	–	0.25	8.91	18.76	–	nr	**27.67**
by hand	–	1.00	35.65	–	–	nr	**35.65**
SITE PREPARATION							
Topsoil stripping for preservation; stripping to subsoil layer 300 mm deep							
larger sites with 21 tonne excavators							
moving to stockpile	–		–	2.79	–	m^2	**2.79**
spreading locally	–		–	3.80	–	m^2	**3.80**
The labour in this section is calculated on a 3 person team. The labour time below should be multiplied by 3 to calculate the cost as shown.							
Tree felling							
Felling and leaving on site							
girth 600 mm–1.50 m (95–240 mm trunk dia.)	–	1.00	106.95	–	–	nr	**106.95**
girth 1.50–3.00 m (240–475 mm trunk dia.)	–	4.00	427.80	–	–	nr	**427.80**
girth 3.00–4.00 m (475–630 mm trunk dia.)	–	6.00	641.70	–	–	nr	**641.70**

3 DEMOLITIONS

Item	PC £	Labour hours	Labour £	Plant £	Material £	Unit	Total rate £
SITE PREPARATION – cont							
Tree felling – cont							
Felling and removing trees off site; logging to arboriculturalists store; chipped material to free green waste site							
girth 600 mm–1.50 m (95–240 mm trunk dia.)	–	2.50	267.38	58.38	–	nr	**325.76**
girth 1.50–3.00 m (240–475 mm trunk dia.)	–	6.00	641.70	116.76	–	nr	**758.46**
girth 3.00–4.00 m (475–630 mm trunk dia.)	–	8.00	855.60	186.82	–	nr	**1042.42**
Removing tree stumps							
girth 600 mm–1.50 m	–	2.00	71.30	134.12	–	nr	**205.42**
girth 1.50–3.00 m	–	7.00	249.55	469.43	–	nr	**718.98**
girth over 3.00 m	–	12.00	427.80	804.74	–	nr	**1232.54**
Stump grinding; disposing to spoil heaps							
girth 600 mm–1.50 m	–	2.00	71.30	137.00	–	nr	**208.30**
girth 1.50–3.00 m	–	2.50	89.13	274.00	–	nr	**363.12**
girth over 3.00 m	–	4.00	142.60	239.75	–	nr	**382.35**
Clearing site vegetation							
mechanical clearance	–	0.25	8.91	16.91	–	100 m^2	**25.82**
hand clearance	–	2.00	71.30	–	–	100 m^2	**71.30**
Lifting turf for preservation							
sod cutter machine lift and stack	–	0.75	26.74	17.44	–	100 m^2	**44.17**
hand lift and stack	–	8.33	297.07	–	–	100 m^2	**297.07**
Lifting turf for disposal (disposal not included)							
by excavator; 5 tonne	–		–	50.04	–	100 m^2	**50.04**
by excavator; 8 tonne	–		–	35.39	–	100 m^2	**35.39**
by excavator; 8 tonne; working with two 3 tonne dumpers	–		–	40.31	–	100 m^2	**40.31**
by excavator; 8 tonne; working with two 5 tonne dumpers	–		–	32.86	–	100 m^2	**32.86**
hand lift and stack	–	8.33	297.07	–	–	100 m^2	**297.07**
Site clearance; by machine; clear site of mature shrubs from existing cultivated beds; dig out roots by machine							
Mixed shrubs in beds; planting centres 500 mm average							
height less than 1 m	–	0.03	0.89	1.27	–	m^2	**2.16**
1.00–1.50 m	–	0.04	1.43	2.03	–	m^2	**3.45**
1.50–2.00 m; pruning to ground level by hand	–	0.10	3.57	5.07	–	m^2	**8.63**
2.00–3.00 m; pruning to ground level by hand	–	0.10	3.57	10.14	–	m^2	**13.70**
3.00–4.00 m; pruning to ground level by hand	–	0.20	7.13	16.89	–	m^2	**24.02**

3 DEMOLITIONS

Item	PC £	Labour hours	Labour £	Plant £	Material £	Unit	Total rate £
Site clearance; by hand; clear site of mature shrubs from existing cultivated beds; dig out roots							
Mixed shrubs in beds; planting centres 500 mm average							
height less than 1 m	–	0.33	11.88	–	–	m^2	**11.88**
1.00–1.50 m	–	0.50	17.83	–	–	m^2	**17.83**
1.50–2.00 m	–	1.00	35.65	–	–	m^2	**35.65**
2.00–3.00 m	–	2.00	71.30	–	–	m^2	**71.30**
3.00–4.00 m	–	3.00	106.96	–	–	m^2	**106.96**
Spraying of vegetation; Glyphosate at 5l/ha by tractor drawn spray equipment	–		–	5.48	2.43	100 m^2	**7.91**
Excavated material; on site							
In spoil heaps							
average 25 m distance	–		–	3.57	–	m^3	**3.57**
average 50 m distance	–		–	4.12	–	m^3	**4.12**
average 100 m distance (1 dumpor)	–		–	5.16	–	m^3	**5.16**
average 100 m distance (2 dumpers)	–		–	6.31	–	m^3	**6.31**
average 200 m distance	–		–	6.22	–	m^3	**6.22**
average 200 m distance (2 dumpers)	–		–	7.59	–	m^3	**7.59**

5 EXCAVATING AND FILLING

Item	PC £	Labour hours	Labour £	Plant £	Material £	Unit	Total rate £
CLARIFICATION NOTES ON LABOUR COSTS IN THIS SECTION							
General groundworks team							
Generally a three man team is used in this section; The column 'Labour hours' reports team hours. The column 'Labour £' reports the total cost of the team for the unit of work shown							
3 man team	–	1.00	106.95	–	–	hr	**106.95**
banksman	–	1.00	35.65	–	–	hr	**35.65**
MACHINE SELECTION							
Machine volumes for excavating/filling only and placing excavated material alongside or to a dumper; no bulkages are allowed for in the material volumes; these rates should be increased by user-preferred percentages to suit prevailing site conditions; the figures in the next section for 'Excavation mechanical' and filling allow for the use of banksmen within the rates shown below							
1.5 tonne excavators: digging volume							
1 cycle/minute; 0.04 m^3	–	0.42	14.86	3.63	–	m^3	**18.49**
2 cycles/minute; 0.08 m^3	–	0.21	7.43	2.19	–	m^3	**9.62**
3 cycles/minute; 0.12 m^3	–	0.14	4.95	1.73	–	m^3	**6.68**
3 tonne excavators; digging volume							
1 cycle/minute; 0.13 m^3	–	0.13	4.57	1.88	–	m^3	**6.45**
2 cycles/minute; 0.26 m^3	–	0.06	2.29	2.68	–	m^3	**4.97**
3 cycles/minute; 0.39 m^3	–	0.04	1.52	1.20	–	m^3	**2.72**
5 tonne excavators; digging volume							
1 cycle/minute; 0.28 m^3	–		–	3.26	–	m^3	**3.26**
2 cycles/minute; 0.56 m^3	–		–	1.63	–	m^3	**1.63**
3 cycles/minute; 0.84 m^3	–		–	1.09	–	m^3	**1.09**
8 tonne excavators; supplied with operator; digging volume							
1 cycle/minute; 0.28 m^3	–		–	3.63	–	m^3	**3.63**
2 cycles/minute; 0.56 m^3	–		–	1.82	–	m^3	**1.82**
3 cycles/minute; 0.84 m^3	–		–	1.21	–	m^3	**1.21**
21 tonne excavators; supplied with operator; digging volume							
1 cycle/minute; 1.21 m^3	–		–	1.37	–	m^3	**1.37**
2 cycles/minute; 2.42 m^3	–		–	0.57	–	m^3	**0.57**
3 cycles/minute; 3.63 m^3	–		–	0.38	–	m^3	**0.38**

5 EXCAVATING AND FILLING

Item	PC £	Labour hours	Labour £	Plant £	Material £	Unit	**Total rate £**
Note: All volumes below are based on excavated 'earth' moist at 1,997 kg/m³ solid or 1,598 kg/m³ loose; a 25% bulkage factor has been used; the weight capacities below exceed the volume capacities of the machine in most cases; see the memorandum section at the back of this book for further weights of materials.							
Dumpers							
1 tonne high tip skip loader; volume 0.485 m³ (775 kg)							
5 loads per hour	–	0.41	14.70	1.91	–	m³	**16.61**
7 loads per hour	–	0.29	10.50	1.42	–	m³	**11.92**
10 loads per hour	–	0.21	7.35	1.06	–	m³	**8.41**
3 tonne dumper; excavated material volume 1.3 m³							
4 loads per hour			–	8.70	–	m³	**8.70**
5 loads per hour	–		–	6.96	–	m³	**6.96**
7 loads per hour	–		–	6.19	–	m³	**6.19**
10 loads per hour	–		–	3.39	–	m³	**3.39**
6 tonne dumper; maximum volume 3.40 m³ (5.4 t); available volume 3.77 m³							
4 loads per hour	–	0.07	2.36	0.47	–	m³	**2.84**
5 loads per hour	–	0.05	1.89	0.39	–	m³	**2.28**
7 loads per hour	–	0.04	1.35	0.29	–	m³	**1.64**
10 loads per hour	–	0.03	0.94	0.23	–	m³	**1.17**
EXCAVATING							
Market prices of topsoil; prices shown include for 20% settlement							
Multiple source screened topsoil	–		–	–	35.51	m³	**35.51**
Single source topsoil; British Sugar PLC	–		–	–	35.51	m³	**35.51**
High grade topsoil for planting	–		–	–	84.47	m³	**84.47**
Note: The figures in this section relate to the machine capacities shown earlier in this section. The figures below however allow for dig efficiency based on depth. A banksman is allowed for in all excavation build and disposal costs shown. Bulkages of 25% allowed; adjustments should be made for different soil types.							

5 EXCAVATING AND FILLING

Item	PC £	Labour hours	Labour £	Plant £	Material £	Unit	Total rate £
EXCAVATING – cont							
Excavating; mechanical; topsoil for preservation							
3 tonne tracked excavator (bucket volume 0.13 m³)							
average depth 100 mm	–		–	1.47	–	m²	**1.47**
average depth 150 mm	–		–	1.66	–	m²	**1.66**
average depth 200 mm	–		–	2.21	–	m²	**2.21**
average depth 250 mm	–		–	2.45	–	m²	**2.45**
average depth 300 mm	–		–	2.94	–	m²	**2.94**
8 tonne excavator							
average depth 100 mm	–	1.00	35.65	67.06	–	100 m²	**102.71**
average depth 150 mm	–	1.25	44.56	83.83	–	100 m²	**128.39**
average depth 200 mm	–	1.78	63.46	119.37	–	100 m²	**182.83**
average depth 250 mm	–	1.90	67.74	127.42	–	100 m²	**195.15**
average depth 300 mm	–	2.00	71.30	134.12	–	100 m²	**205.42**
Excavating; mechanical; to reduce levels							
5 tonne excavator (bucket volume 0.28 m³)							
maximum depth not exceeding 0.25 m	–	0.07	2.50	3.84	–	m³	**6.34**
maximum depth not exceeding 1.00 m	–	0.05	1.70	2.61	–	m³	**4.31**
maximum depth not exceeding 2.00 m	–	0.06	2.12	3.26	–	m³	**5.38**
8 tonne tracked excavator (bucket volume 0.28 m³)							
maximum depth not exceeding 1.00 m	–	0.06	2.23	3.81	–	m³	**6.04**
maximum depth not exceeding 2.00 m	–	0.07	2.55	4.35	–	m³	**6.90**
maximum depth not exceeding 3.00 m	–	0.09	3.24	5.54	–	m³	**8.78**
21 tonne 360 tracked excavator (bucket volume 1.21 m³)							
maximum depth not exceeding 1.00 m	–	0.01	0.39	0.91	–	m³	**1.30**
maximum depth not exceeding 2.00 m	–	0.02	0.59	1.36	–	m³	**1.95**
maximum depth not exceeding 3.00 m	–	0.03	1.19	2.75	–	m³	**3.94**
Pits; 3 tonne tracked excavator							
maximum depth not exceeding 0.25 m	–	0.33	11.88	17.18	–	m³	**29.06**
maximum depth not exceeding 1.00 m	–	0.25	8.91	12.89	–	m³	**21.80**
maximum depth not exceeding 2.00 m	–	0.40	14.26	20.62	–	m³	**34.88**

5 EXCAVATING AND FILLING

Item	PC £	Labour hours	Labour £	Plant £	Material £	Unit	**Total rate £**
Pits; 8 tonne tracked excavator							
maximum depth not exceeding 2.00 m	–	0.40	14.26	24.39	–	m^3	**38.65**
maximum depth not exceeding 3.00 m	–	0.50	17.83	30.48	–	m^3	**48.31**
Trenches; width not exceeding 0.30 m; 3 tonne excavator							
maximum depth not exceeding 0.25 m	–	1.00	35.65	6.31	–	m^3	**41.96**
maximum depth not exceeding 1.00 m	–	0.69	24.42	4.32	–	m^3	**28.74**
maximum depth not exceeding 2.00 m	–	0.60	21.39	3.78	–	m^3	**25.17**
Trenches; width exceeding 0.30 m; 3 tonne excavator							
maximum depth not exceeding 0.25 m	–	0.60	21.39	3.78	–	m^3	**25.17**
maximum depth not exceeding 1.00 m	–	0.50	17.83	3.15	–	m^3	**20.98**
maximum depth not exceeding 2.00 m	–	0.38	13.71	2.42	–	m^3	**16.14**
Extra over any types of excavating irrespective of depth for breaking out existing materials; Excavator with breaker attachment							
hard rock	–	0.50	17.83	149.90	–	m^3	**167.73**
concrete	–	0.50	17.83	56.21	–	m^3	**74.04**
reinforced concrete	–	1.00	35.65	86.91	–	m^3	**122.56**
brickwork, blockwork or stonework	–	0.25	8.91	56.21	–	m^3	**65.13**
Extra over any types of excavating irrespective of depth for breaking out existing hard pavings; Excavator with breaker attachment							
concrete; 100 mm thick	–		–	4.12	–	m^2	**4.12**
concrete; 150 mm thick	–		–	6.87	–	m^2	**6.87**
concrete; 200 mm thick	–		–	8.24	–	m^2	**8.24**
concrete; 300 mm thick	–		–	12.37	–	m^2	**12.37**
reinforced concrete; 100 mm thick	–	0.08	2.97	4.49	–	m^2	**7.46**
reinforced concrete; 150 mm thick	–	0.08	2.67	6.29	–	m^2	**8.97**
reinforced concrete; 200 mm thick	–	0.10	3.57	8.39	–	m^2	**11.96**
reinforced concrete; 300 mm thick	–	0.15	5.35	12.59	–	m^2	**17.94**
tarmacadam; 75 mm thick	–		–	4.12	–	m^2	**4.12**
tarmacadam and hardcore; 150 mm thick	–		–	6.60	–	m^2	**6.60**
Extra over any types of excavating irrespective of depth for taking up							
precast concrete paving slabs	–	0.07	2.38	0.97	–	m^2	**3.34**
natural stone paving	–	0.10	3.57	1.45	–	m^2	**5.01**
cobbles	–	0.13	4.46	1.81	–	m^2	**6.27**
brick paviors	–	0.13	4.46	1.81	–	m^2	**6.27**

5 EXCAVATING AND FILLING

Item	PC £	Labour hours	Labour £	Plant £	Material £	Unit	Total rate £
EXCAVATING – cont							
Excavating; hand							
Topsoil for preservation; loading to barrows							
average depth 100 mm	–	0.08	8.47	–	–	m^2	**8.47**
average depth 150 mm	–	0.12	12.71	–	–	m^2	**12.71**
average depth 200 mm	–	0.16	20.33	–	–	m^2	**20.33**
average depth 250 mm	–	0.20	25.41	–	–	m^2	**25.41**
average depth 300 mm	–	0.24	30.49	–	–	m^2	**30.49**
Topsoil to reduce levels							
maximum depth not exceeding 0.25 m	–	0.79	84.70	–	–	m^3	**84.70**
maximum depth not exceeding 1.00 m	–	1.03	110.12	–	–	m^3	**110.12**
Pits							
maximum depth not exceeding 0.25 m	–	0.88	94.12	–	–	m^3	**94.12**
maximum depth not exceeding 1.00 m	–	1.14	122.35	–	–	m^3	**122.35**
maximum depth not exceeding 2.00 m (includes earthwork support)	–	2.29	235.96	–	–	m^3	**327.66**
Trenches; width not exceeding 0.30 m							
maximum depth not exceeding 0.25 m	–	0.94	100.83	–	–	m^3	**100.83**
maximum depth not exceeding 1.00 m	–	1.23	131.32	–	–	m^3	**131.32**
maximum depth not exceeding 2.00 m (includes earthwork support)	–	1.23	126.87	–	–	m^3	**172.81**
Trenches; width exceeding 0.30 m wide							
maximum depth not exceeding 0.25 m	–	0.94	100.83	–	–	m^3	**100.83**
maximum depth not exceeding 1.00 m	–	1.32	141.17	–	–	m^3	**141.17**
maximum depth not exceeding 2.00 m (includes earthwork support)	–	1.98	203.36	–	–	m^3	**294.70**
Extra over any types of excavating irrespective of depth for breaking out existing materials; hand held pneumatic breaker							
rock	–	1.65	176.47	48.32	–	m^3	**224.79**
concrete	–	0.83	88.23	24.16	–	m^3	**112.40**
reinforced concrete	–	1.32	141.17	44.64	–	m^3	**185.81**
brickwork, blockwork or stonework	–	0.50	52.94	14.50	–	m^3	**67.44**

5 EXCAVATING AND FILLING

Item	PC £	Labour hours	Labour £	Plant £	Material £	Unit	**Total rate £**
Earthwork Retention							
Manhole boxes to excavations (not included); Timber plywood with timber braces to shore excavations to 1.00 m deep; opposing faces distance less than 1.00 m							
1.00 m deep	–	0.50	35.65	–	38.14	m^2	**73.79**
2.00 m deep	–	1.00	71.30	–	38.14	m^2	**109.44**
Trench boxes; A-Plant							
Trench box soil retention; mecahical placement of hydraulic trenchboxes	–	0.50	53.48	208.58	–	week	**262.06**
Manhole protection system to 3.0 m deep to enable earthwork retention; maximum distance between opposing faces 1.8 m	–	0.50	53.48	282.06	–	week	**335.54**
FILLING							
Filling to make up levels; mechanical							
Arising from the excavations							
average thickness not exceeding 0.25 m	–	0.06	2.62	4.48	–	m^3	**7.10**
average thickness less than 500 mm	–	0.05	2.23	3.81	–	m^3	**6.04**
average thickness 1.00 m	–	0.04	1.65	2.82	–	m^3	**4.47**
Obtained from on site spoil heaps; average 25 m distance; multiple handling							
average thickness less than 250 mm	–		–	12.30	–	m^3	**12.30**
average thickness less than 500 mm	–		–	10.13	–	m^3	**10.13**
average thickness 1.00 m	–		–	8.61	–	m^3	**8.61**
Obtained off site; planting quality topsoil PC £29.59/m^3							
average thickness less than 250 mm	–		–	12.30	40.83	m^3	**53.13**
average thickness less than 500 mm	–		–	10.13	40.83	m^3	**50.96**
average thickness 1.00 m				8.61	40.83	m^3	**49.44**
Obtained off site; hardcore; PC £24.37/m^3							
average thickness less than 250 mm	–		–	14.35	34.78	m^3	**49.12**
average thickness less than 500 mm	–		–	10.76	34.78	m^3	**45.54**
average thickness 1.00 m	–		–	9.06	34.78	m^3	**43.83**

5 EXCAVATING AND FILLING

Item	PC £	Labour hours	Labour £	Plant £	Material £	Unit	Total rate £
FILLING – cont							
Filling to make up levels; hand							
Arising from the excavations							
average thickness exceeding 0.25 m; depositing in layers 150 mm maximum thickness	–	0.60	21.39	–	–	m^3	**21.39**
Obtained from on site spoil heaps; average 25 m distance; multiple handling							
average thickness exceeding 0.25 m thick; depositing in layers 150 mm maximum thickness	–	1.00	35.65	–	–	m^3	**35.65**
DISPOSAL							
Note							
Most commercial site disposal is carried out by 20 tonne, 8 wheeled vehicles. It has been customary to calculate disposal from construction sites in terms of full 15 m^3 loads. Spon's research has found that based on weights of common materials such as clean hardcore and topsoil, vehicles could not load more than 12 m^3 at a time. Most hauliers do not make it apparent that their loads are calculated by weight and not by volume. The rates below reflect these lesser volumes which are limited by the 20 tonne limit. The volumes shown below are based on volumes 'in the solid'. Weight and bulking factors have been applied. For further information please see the weights of typical materials in the Earthworks section of the Memoranda at the back of this book.							
Disposal; mechanical (all rates include banksman); MCM-SE Ltd							
Excavated material; off site; to tip; mechanically loaded (360 Excavator)							
hardcore	–	0.02	0.71	1.42	–	m^3	**22.59**
inert (clean landfill, rubble concrete); loose	–	0.03	0.95	–	–	m^3	**38.42**
inert (clean excavated material); compacted	–	0.03	0.95	–	–	m^3	**47.78**

5 EXCAVATING AND FILLING

Item	PC £	Labour hours	Labour £	Plant £	Material £	Unit	Total rate £
slightly contaminated –							
non hazardous	–	0.03	0.95	1.42	–	m^3	**58.32**
clean concrete	–	0.03	0.95	1.43	–	m^3	**15.34**
macadam	–	0.02	0.71	1.42	–	m^3	**21.52**
broken out compacted materials							
such as road bases and the like	–	0.03	0.95	–	–	m^3	**57.14**
Light soils and loams (bulking factor of 1.25); 40 tonnes (2 loads per hour)							
soil (sandy and loam); dry	–	0.03	0.95	–	–	m^3	**38.42**
soil (sandy and loam); wet	–	0.03	0.95	–	–	m^3	**46.45**
soil (clay); dry	–	0.03	0.95	–	–	m^3	**53.39**
soil (clay); wet	–	0.03	0.95	–	–	m^3	**57.14**
Other materials							
rubbish (mixed loads)	–	0.03	1.14	2.27	–	m^3	**48.23**
green waste	–	0.03	1.14	2.28	–	m^3	**21.43**
Costs if allowing for 3 loads (36 m^3, 60 tonne) per hour loaded and removed							
inert material	–	0.03	0.95	1.90	–	m^3	**77.09**
Excavated material; off site to tip; mechanically loaded by grab; capacity of load 13 m^3 (18 tonne)							
inert material	–	0.50	17.27	–	–	m^3	**62.39**
hardcore	–	0.50	17.23	–	–	m^3	**77.92**
macadam	–	0.50	17.40	–	–	m^3	**39.28**
clean concrete	–	0.50	17.40	–	–	m^3	**39.28**
soil (sandy and loam); dry	–	0.50	17.30	–	–	m^3	**53.84**
soil (sandy and loam); wet	–	0.50	17.26	–	–	m^3	**65.09**
broken out compacted materials							
such as road bases and the like	–	0.50	17.25	–	–	m^3	**67.35**
soil (clay); dry	–	0.50	17.26	–	–	m^3	**65.09**
soil (clay); wet	–	0.50	17.23	–	–	m^3	**77.92**
rubbish (mixed loads)	–	0.50	17.21	–	–	m^3	**84.92**
green waste unchipped;							
compacted	–		–	–	–	m^3	**77.43**
Disposal by skip; 8 yd^3 (6.1 m^3)							
Excavated material loaded to skip							
by machine	–		–	67.64	–	m^3	**67.64**
by hand	–	3.00	106.95	63.80		m^3	**170.75**
Excavated material; on site							
In spoil heaps							
average 25 m distance	–		–	3.57	–	m^3	**3.57**
average 50 m distance	–		–	4.12	–	m^3	**4.12**
average 100 m distance							
(1 dumper)	–		–	5.16	–	m^3	**5.16**
average 100 m distance							
(2 dumpers)	–		–	6.31	–	m^3	**6.31**
average 200 m distance	–		–	6.22	–	m^3	**6.22**
average 200 m distance							
(2 dumpers)	–		–	7.59	–	m^3	**7.59**

5 EXCAVATING AND FILLING

Item	PC £	Labour hours	Labour £	Plant £	Material £	Unit	Total rate £
DISPOSAL – cont							
Excavated material – cont							
Spreading on site							
average 25 m distance	–	0.02	0.80	8.16	–	m^3	**8.96**
average 50 m distance	–	0.04	1.39	9.14	–	m^3	**10.53**
average 100 m distance	–	0.06	2.21	11.14	–	m^3	**13.35**
average 200 m distance	–	0.12	4.28	10.35	–	m^3	**14.63**
Disposal; hand							
Excavated material; onsite; in spoil heaps							
average 25 m distance	–	2.00	85.56	–	–	m^3	**85.56**
average 50 m distance	–	2.20	94.12	–	–	m^3	**94.12**
average 100 m distance	–	2.50	106.95	–	–	m^3	**106.95**
average 200 m distance	–	3.00	128.34	–	–	m^3	**128.34**
Excavated material; spreading on site							
average 25 m distance	–	2.20	94.12	–	–	m^3	**94.12**
average 50 m distance	–	2.50	106.95	–	–	m^3	**106.95**
average 100 m distance	–	3.00	128.34	–	–	m^3	**128.34**
average 200 m distance	–	3.50	149.73	–	–	m^3	**149.73**
Disposal of material from site clearance operations							
Shrubs and groundcovers less than 1.00 m height; disposal by 15 m^3 self loaded truck							
deciduous shrubs; not chipped; winter	–	0.05	1.72	1.15	–	m^2	**17.86**
deciduous shrubs; chipped; winter	–	0.05	1.75	3.48	–	m^2	**8.30**
evergreen or deciduous shrubs; chipped; summer	–	0.08	2.59	2.28	–	m^2	**16.94**
Shrubs 1.00–2.00 m height; disposal by 15 m^3 self loaded truck							
deciduous shrubs; not chipped; winter	–	0.17	5.75	1.15	–	m^2	**27.98**
deciduous shrubs; chipped; winter	–	0.25	8.71	2.30	–	m^2	**23.19**
evergreen or deciduous shrubs; chipped; summer	–	0.30	10.40	2.67	–	m^2	**37.28**
Shrubs or hedges 2.00–3.00 m height; disposal by 15 m^3 self loaded truck							
deciduous plants non-woody growth; not chipped; winter	–	0.25	8.59	1.15	–	m^2	**57.76**
deciduous shrubs; chipped; winter	–	0.50	17.36	2.29	–	m^2	**49.98**

5 EXCAVATING AND FILLING

Item	PC £	Labour hours	Labour £	Plant £	Material £	Unit	**Total rate £**
evergreen or deciduous shrubs non-woody growth; chipped; summer	–	0.25	8.71	2.30	–	m2	**23.19**
evergreen or deciduous shrubs woody growth; chipped; summer	–	0.67	23.18	3.44	–	m2	**63.05**
Shrubs and groundcovers less than 1.00 m height; disposal to spoil heap							
deciduous shrubs; not chipped; winter	–	0.05	1.78	–	–	m2	**1.78**
deciduous shrubs; chipped; winter	–	0.05	1.78	1.17	–	m2	**2.95**
evergreen or deciduous shrubs; chipped; summer	–	0.08	2.67	1.17	–	m2	**3.84**
Shrubs 1.00–2.00 m height; disposal to spoil heap							
deciduous shrubs; not chipped; winter	–	0.17	5.94	–	–	m2	**5.94**
deciduous shrubs; chipped; winter	–	0.25	8.91	1.17	–	m2	**10.08**
evergreen or deciduous shrubs; chipped; summer	–	0.30	10.70	1.56	–	m2	**12.25**
Shrubs or hedges 2.00–3.00 m height; disposal to spoil heaps							
deciduous plants non-woody growth; not chipped; winter	–	0.25	8.91	–	–	m2	**8.91**
deciduous shrubs; chipped; winter	–	0.50	17.83	1.17	–	m2	**18.99**
evergreen or deciduous shrubs non-woody growth; chipped; summer	–	0.67	23.77	2.34	–	m2	**26.10**
evergreen or deciduous shrubs woody growth; chipped; summer	–	1.50	53.47	4.67	–	m2	**58.14**

6 GROUND REMEDIATION AND SOIL STABILIZATION

Item	PC £	Labour hours	Labour £	Plant £	Material £	Unit	Total rate £
CLARIFICATION NOTES ON LABOUR COSTS IN THIS SECTION							
General groundworks team							
Generally a three man team is used in this section. The column Labour hours reports team hours. The column Labour £ reports the total cost of the team for the unit of work shown							
3 man team	–	1.00	106.95	–	–	hr	**106.95**
RETAINING WALLS AND REVETMENTS							
Preamble							
Earth retaining and stabilization can take numerous forms depending on the height, soil type and loadings. The main types include gravity walls, reinforced soil and soil nailing. Prices for these items depend on quantity, access for the installation and the strength of the soils. Estimates should be obtained from the manufacturer when site conditions and layout have been determined.							
Corten pre-fabricated retaining walls; Adezz limited							
200 mm high on prepared concrete base (not included); top edge of 50 × 25 mm; bolted together internally with 30 mm M8 bolts							
straight sections 1.00 long	35.51	0.20	14.26	–	40.84	m	**55.10**
straight sections 1.50 m long	35.61	0.17	11.89	–	40.95	m	**52.83**
straight sections 2.00 m long	34.86	0.13	8.91	–	40.08	m	**49.00**
inside corner or outside corner 500 mm × 500 mm	35.69	0.25	17.83	–	41.04	m	**58.87**
inside or outside corner 1.00 × 1.00	33.89	0.30	21.39	–	38.97	m	**60.36**
inside or outside curve 500 mm	124.04	0.20	14.26	–	142.65	m	**156.91**
inside or outside curve; 1.00 m lengths	100.83	0.25	17.83	–	115.95	m	**133.78**
inside or outside curve; 1.50 m lengths	94.79	0.17	11.89	–	109.00	m	**120.89**
inside or outside curve; 2.00 m lengths	96.81	0.14	10.19	–	111.33	m	**121.52**

6 GROUND REMEDIATION AND SOIL STABILIZATION

Item	PC £	Labour hours	Labour £	Plant £	Material £	Unit	**Total rate £**
400 mm high; on prepared concrete base (not included); top edge of 50 × 25 mm; bolted together internally with 30 mm M8 bolts							
straight sections 1.00 long	79.96	0.22	15.69	–	91.95	m	**107.64**
straight sections 1.50 m long	80.04	0.18	13.08	–	92.05	m	**105.12**
straight sections 2.00 m long	78.05	0.14	9.80	–	89.76	m	**99.56**
inside corner or outside corner 500 mm × 500 mm	96.70	0.28	19.61	–	111.21	m	**130.81**
inside or outside corner 1.00 × 1.00	90.71	0.33	23.53	–	104.31	m	**127.84**
inside or outside curve 500 mm	216.82	0.22	15.69	–	249.34	m	**265.03**
inside or outside curve; 1.00 m lengths	181.41	0.28	19.61	–	208.62	m	**228.23**
inside or outside curve; 1.50 m lengths	179.44	0.18	13.08	–	206.36	m	**219.43**
inside or outside curve; 2.00 m lengths	182.11	0.16	11.21	–	209.42	m	**220.63**
600 mm high; on prepared concrete base (not included); top edge of 50 × 25 mm; bolted together internally with 30 mm M8 bolts							
straight sections 1.00 long	120.94	0.24	17.25	–	139.08	m	**156.34**
straight sections 1.50 m long	121.02	0.20	14.38	–	139.17	m	**153.55**
straight sections 2.00 m long	116.87	0.15	10.79	–	134.40	m	**145.19**
inside corner or outside corner 500 mm × 500 mm	149.12	0.30	21.57	–	171.49	m	**193.06**
inside or outside corner 1.00 × 1.00	131.51	0.36	25.88	–	151.23	m	**177.11**
inside or outside curve 500 mm	298.24	0.24	17.25	–	342.98	m	**360.23**
inside or outside curve; 1.00 m lengths	271.08	0.30	21.57	–	311.74	m	**333.31**
inside or outside curve; 1.50 m lengths	249.87	0.20	14.38	–	287.35	m	**301.74**
inside or outside curve; 2.00 m lengths	244.11	0.17	12.33	–	280.72	m	**293.05**
Retaining walls; FP McCann Concrete							
Retaining walls of units with plain concrete finish; prices based on 24 tonne loads but other quantities available (excavation, temporary shoring, foundations and backfilling not included)							
1000 mm wide × 1000 mm high	176.37	0.50	35.65	33.53	262.73	m	**331.91**
1000 mm wide × 1500 mm high	224.58	0.42	30.16	33.53	331.94	m	**395.63**
1000 mm wide × 2000 mm high	374.31	0.42	30.20	33.53	533.60	m	**597.32**
1000 mm wide × 2500 mm high	482.15	0.42	30.19	33.53	696.77	m	**760.48**
1000 mm wide × 3000 mm high	522.76	0.42	30.16	71.16	761.85	m	**863.18**

6 GROUND REMEDIATION AND SOIL STABILIZATION

Item	PC £	Labour hours	Labour £	Plant £	Material £	Unit	Total rate £
RETAINING WALLS AND REVETMENTS – cont							
Retaining Walls; FP McCann Ltd; Easi-Bloc; (Lego Block); precast self-locking concrete gravity wall system; storage bays, partitions retaining walls, revetments; to firm and levelled base; concrete or granular; min. 800 mm wide (not included)							
Blocks 450 mm high laid stretcher bond							
1.20 m × 600 mm (750 kg per block)	206.91	0.46	49.52	28.23	237.95	m²	**315.70**
600 mm × 600 mm (375 kg per block)	248.31	0.50	53.48	30.48	285.56	m²	**369.52**
Retaining walls; Tensar International							
Tensartech TW1 retaining wall system; modular dry laid concrete blocks; 220 × 400 mm long × 150 mm high connected to Tensar RE geogrid with proprietary connectors; geogrid laid horizontally within the fill at 300 mm centres; on 150 × 450 mm concrete foundation; filling with imported granular material							
1.00 m high	153.75	1.00	71.30	16.77	194.57	m²	**282.63**
2.00 m high	153.75	1.33	95.06	16.77	194.57	m²	**306.40**
3.00 m high	153.75	1.50	106.95	16.77	194.57	m²	**318.28**
Retaining walls; Grass Concrete Ltd							
Betoflor precast concrete landscape retaining walls including soil filling to pockets (excavation, concrete foundations, backfilling stones to rear of walls and planting not included)							
Betoflor interlocking units; 250 mm long × 250 × 200 mm modular deep; in walls 250 mm wide	–		–	–	–	m²	**155.19**
extra over Betoflor interlocking units for colours	–		–	–	–	m²	**20.35**
Betoatlas earth retaining walls; 250 mm long × 500 mm wide × 200 mm modular deep; in walls 500 mm wide	–		–	–	–	m²	**238.77**
extra over Betoatlas interlocking units for colours	–		–	–	–	m²	**27.72**

6 GROUND REMEDIATION AND SOIL STABILIZATION

Item	PC £	Labour hours	Labour £	Plant £	Material £	Unit	Total rate £
Retaining walls; Forticrete Ltd							
Keystone precast concrete block retaining wall; geogrid included for walls over 1.00 m high; excavation, concrete foundation, stone backfill to rear of wall all measured separately; includes design							
1.0 m high	151.68	0.80	85.56	–	174.44	m^2	**260.00**
2.0 m high	151.68	0.80	85.56	5.60	197.44	m^2	**288.60**
3.0 m high	151.68	0.80	85.56	7.47	213.42	m^2	**306.45**
4.0 m high	151.68	0.80	85.56	8.97	225.79	m^2	**320.32**
Retaining walls; Anderton Concrete Ltd							
Stepoc Blocks; interlocking blocks; 10 mm reinforcing laid loose horizontally to preformed notches and vertical reinforcing nominal size 10 mm fixed to starter bars, infilling with concrete; foundations and starter bars measured separately							
type 325; 400 × 225 × 325 mm	73.85	0.47	49.91	–	124.86	m^2	**174.78**
type 256; 400 × 225 × 256 mm	72.84	0.40	42.78	–	118.56	m^2	**161.34**
type 200; 400 × 225 × 200 mm	63.49	0.33	35.65	–	102.68	m^2	**138.32**
Timber log retaining walls; timber posts are kiln dried redwood with 15 year guarantee; Machine rounded softwood logs to trenches priced separately; disposal of excavated material priced separately; inclusive of 75 mm hardcore blinding to trench and backfilling trench with site mixed concrete 1:3:6; geofabric pinned to rear of logs; heights of logs above ground							
100 mm dia.							
500 mm (constructed from 1.80 m lengths)	79.32	0.50	53.48	–	116.53	m	**170.01**
1.20 m (constructed from 1.80 m lengths)	158.64	0.43	46.34	–	235.47	m	**281.81**
1.60 m (constructed from 2.70 m lengths)	250.56	0.83	89.09	–	356.10	m	**445.19**
2.00 m (constructed from 3.00 m lengths)	265.20	1.17	124.77	–	387.85	m	**512.62**
150 mm dia.							
500 mm	98.70	0.83	89.09	–	136.51	m	**225.60**
1.20 m	322.00	0.58	62.38	–	423.33	m	**485.72**
1.60 m	322.32	1.00	106.95	–	438.63	m	**545.58**

6 GROUND REMEDIATION AND SOIL STABILIZATION

Item	PC £	Labour hours	Labour £	Plant £	Material £	Unit	Total rate £
RETAINING WALLS AND REVETMENTS – cont							
Recycled materials and Systems for External Works							
Planter retaining wall; Recycled plastic; Hahn Plastics Ltd; Hollow or Solid; inclusive of trenches, placement and surrounding with concrete; terram to rear of palisades; backfiling to height of palisade not included; brown; (grey also available)							
'Ogee' C Section interlocking profile palisades 100 mm dia.; hollow profile; retained height							
300 mm	67.77	0.25	26.67	–	107.83	m	**142.41**
600 mm	110.65	0.50	53.40	–	176.49	m	**237.80**
'Ogee' C Section interlocking profile palisades 100 mm dia.; hollow profile; retained height							
1.20 m	237.89	0.43	46.30	–	338.54	m	**393.94**
Round palisades; 100 mm dia.; hollow profile; retained height							
300 mm	49.71	0.25	26.69	–	87.16	m	**118.60**
800 mm	119.78	0.43	46.29	–	190.72	m	**244.12**
Round palisades; 100 mm dia.; solid profile; retained height							
1.00 m	119.78	0.43	46.29	–	190.72	m	**244.12**
1.20 m	168.51	0.50	53.41	–	263.57	m	**326.86**
1.50 m	219.47	1.00	106.72	–	442.66	m	**579.02**
Rectangular palisades; 120 × 160 mm ; hollow profile; retained height; Grey, Brown or Black							
300 mm	73.55	0.25	26.70	–	114.57	m	**146.01**
800 mm	219.47	0.43	46.30	–	305.33	m	**358.75**
1.00 m	261.13	0.50	53.42	–	370.07	m	**433.39**
Railway sleeper walls; retaining wall from railway sleepers; fixed with steel galvanized pins 12 mm driven into the ground; sleepers laid flat							
Grade 1 hardwood; 2590 × 250 × 150 mm							
150 mm; 1 sleeper high	2.68	0.17	17.82	–	4.27	m	**22.09**
300 mm; 2 sleepers high	5.35	0.33	35.65	–	8.54	m	**44.19**
450 mm; 3 sleepers high	7.96	0.50	53.48	–	12.29	m	**65.77**
600 mm; 4 sleepers high	10.73	0.67	71.30	–	15.47	m	**86.78**

6 GROUND REMEDIATION AND SOIL STABILIZATION

Item	PC £	Labour hours	Labour £	Plant £	Material £	Unit	**Total rate £**
Grade 1 hardwood; 2590 × 250 × 150 mm with 2 nr galvanized angle iron stakes set into concrete internally and screwed to the inside face of the sleepers							
750 mm; 5 sleepers high	13.27	0.83	89.12	–	78.52	m^2	**167.65**
900 mm; 6 sleepers high	15.88	0.92	98.04	–	88.70	m^2	**186.74**
New pine softwood; 2.40 m × 250 mm × 125 mm							
125 mm; 1 sleeper high	15.84	0.17	17.83	–	19.41	m	**37.24**
250 mm; 2 sleepers high	31.68	0.33	35.65	–	38.82	m	**74.47**
375 mm; 3 sleepers high	47.52	0.50	53.48	–	57.79	m	**111.26**
500 mm; 4 sleepers high	63.36	0.67	71.30	–	76.00	m	**147.31**
New pine softwood 2.40 m × 250 mm × 125 mm with 2 nr galvanized angle iron stakes set into concrete internally and screwed to the inside face of the sleepers							
625 mm; 5 sleepers high	79.20	0.83	89.12	–	154.35	m^2	**243.47**
775 mm; 6 sleepers high	95.04	0.92	98.04	–	179.73	m^2	**277.77**
New oak hardwood 2400 × 200 × 100 mm							
100 mm; 1 sleeper high	16.23	0.17	17.82	–	19.86	m	**37.68**
200 mm; 2 sleepers high	32.46	0.33	35.65	–	39.72	m	**75.37**
300 mm; 3 sleepers high	47.51	0.75	80.21	–	57.77	m	**137.98**
400 mm; 4 sleepers high	65.72	1.00	106.95	–	78.71	m	**185.66**
New oak hardwood 2400 × 200 × 100 mm but with 2 nr galvanized angle iron stakes set into concrete internally and screwed to the inside face of the sleepers							
500 mm; 5 sleepers high	82.34	0.83	89.12	–	161.55	m^2	**250.67**
600 mm; 6 sleepers high	98.97	1.10	117.65	–	184.25	m^2	**301.90**
Excavate foundation trench; set railway sleepers vertically on end in concrete 1:3:6 continuous foundation to 33.3% of their length to form retaining wall							
Grade 1 hardwood; finished height above ground level							
300 mm	21.48	1.00	106.90	2.63	36.09	m	**147.38**
500 mm	38.67	1.00	106.90	2.63	55.84	m	**167.13**
600 mm	46.41	1.00	106.91	2.63	76.13	m	**187.43**
750 mm	51.56	1.17	124.74	2.63	82.06	m	**211.18**
1.00 m	51.58	1.25	133.04	2.63	84.34	m	**222.37**

6 GROUND REMEDIATION AND SOIL STABILIZATION

Item	PC £	Labour hours	Labour £	Plant £	Material £	Unit	Total rate £
RETAINING WALLS AND REVETMENTS – cont							
Excavate and place vertical steel universal beams 165 mm wide in concrete base at 2.59 m centres; fix railway sleepers set horizontally between beams to form horizontal fence or retaining wall							
Grade 1 hardwood; bay length 2.59 m							
400 mm high; 2 sleepers	39.72	0.83	89.12	–	69.78	bay	**158.90**
600 mm high; 3 sleepers	58.38	0.87	92.69	–	102.36	bay	**195.06**
800 m high; 4 sleepers	78.24	1.00	106.95	–	127.81	bay	**234.76**
1.00 m high; 5 sleepers	99.90	1.17	124.78	–	170.53	bay	**295.31**
1.20 m high; 6 sleepers	117.36	1.00	106.95	–	200.81	bay	**307.76**
1.40 m high; 7 sleepers	136.61	1.00	106.95	–	230.87	bay	**337.82**
Waterside revetments; Willowbank Services Ltd; natural engineering solutions; ecological sustainable waterside stabilization systems							
Live willow spiling; 1 m high uprights at 0.50 m centres; woven geotextile backed; included filling with site based topsoil behind spiling to top of bank							
less than 50 m	–		–	–	–	m	**222.67**
50–100 m	–		–	–	–	m	**173.94**
Brushwood faggot and pre-planted coir revetments; 900 mm high revetment; 300 mm dia. faggot; 300 mm dia. pre-planted coir roll; staked in position with 1.8 m tanalized posts; 2 posts per m; faggots and coir fixed to posts with wire							
less than 50 m	–		–	–	–	m	**199.53**
50–100 m	–		–	–	–	m	**171.77**
Brushwood faggot revetments; 900 mm high revetment; 2 nr 300 mm dia. faggot staked in position with 1.8 m tanalized posts; 2 posts per m; faggots and coir fixed to posts with wire							
less than 50 m	–		–	–	–	m	**171.27**
50–100 m	–		–	–	–	m	**118.65**

6 GROUND REMEDIATION AND SOIL STABILIZATION

Item	PC £	Labour hours	Labour £	Plant £	Material £	Unit	Total rate £
GENERALLY							
Soil stabilization – General							
Preamble: Earth-retaining and stabilizing materials are often specified as part of the earth-forming work in landscape contracts, therefore this section lists a number of products specially designed for large-scale earth control. There are two types: rigid units for structural retention of earth on steep slopes; and flexible meshes and sheets for control of soil erosion where structural strength is not required. Prices for these items depend on quantity, difficulty of access to the site and availability of suitable filling material; estimates should be obtained from the manufacturer when the site conditions have been determined.							
GRADING AND SURFACE PREPARATION							
Grading operations; surface previously excavated to reduce levels to prepare to receive subsequent treatments; grading to accurate levels and falls 20 mm tolerances							
Clay or heavy soils or hardcore							
5 tonne excavator	–	0.04	1.43	0.24	–	m²	**1.67**
13 tonne excavator	–	0.01	0.36	0.71	–	m²	**1.07**
21 tonne 360 tracked excavator	–	0.01	0.24	0.55	–	m²	**0.79**
by hand	–	0.10	3.57	–	–	m²	**3.57**
Loamy topsoils							
5 tonne excavator	–	0.03	0.95	0.16	–	m²	**1.11**
13 tonne excavator	–	0.01	0.24	0.48	–	m²	**0.72**
21 tonne 360 tracked excavator	–	0.01	0.26	0.61	–	m²	**0.88**
by hand	–	0.05	1.78	–	–	m²	**1.78**
Sand or graded granular materials							
5 tonne excavator	–	0.02	0.71	0.12	–	m²	**0.83**
13 tonne excavator	–	0.01	0.24	0.48	–	m²	**0.72**
21 tonne 360 tracked excavator	–	0.01	0.20	0.47	–	m²	**0.67**
by hand	–	0.03	1.19	–	–	m²	**1.19**

6 GROUND REMEDIATION AND SOIL STABILIZATION

Item	PC £	Labour hours	Labour £	Plant £	Material £	Unit	Total rate £
GRADING AND SURFACE PREPARATION – cont							
Grading operations to bottoms of excavations or trenches to receive foundations or bases; surface previously excavated to reduce levels to prepare to receive subsequent treatments; grading to accurate levels and falls 20 mm tolerances							
Trenches							
5 tonne excavator	–	0.32	20.20	0.80	–	m^2	**21.01**
by hand	–	0.33	23.76	–	–	m^2	**23.76**
Excavated areas to receive bases or formwork							
5 tonne excavator	–	0.13	8.32	0.40	–	m^2	**8.72**
by hand	–	0.17	11.89	–	–	m^2	**11.89**
Grading operations; surface recently filled to raise levels to prepare to receive subsequent treatments; grading to accurate levels and falls 20 mm tolerances							
Clay or heavy soils or hardcore							
5 tonne excavator	–	0.03	1.07	0.18	–	m^2	**1.25**
21 tonne 360 tracked excavator	–	0.01	0.24	0.55	–	m^2	**0.79**
by hand	–	0.08	2.97	–	–	m^2	**2.97**
Loamy topsoils							
5 tonne excavator	–	0.02	0.73	0.12	–	m^2	**0.86**
21 tonne 360 tracked excavator	–	0.01	0.24	0.55	–	m^2	**0.79**
by hand	–	0.04	1.43	–	–	m^2	**1.43**
Sand or graded granular materials							
5 tonne excavator	–	0.02	0.53	0.09	–	m^2	**0.63**
13 tonne excavator	–	0.01	0.27	0.53	–	m^2	**0.80**
21 tonne 360 tracked excavator	–	0.01	0.18	0.41	–	m^2	**0.59**
by hand	–	0.03	1.02	–	–	m^2	**1.02**
Grading operations to bunds, earthbanks and landforms; Grading to required levels and falls of deposited, excavated or imported materials; cut and fill operations locally; hand grading to final tolerances of 50 mm; maximum distance 25 m by dumper							
max cut and fill depths							
up to 200 mm	–	0.11	11.88	13.50	–	m^3	**25.38**
300 mm	–	0.08	8.91	10.12	–	m^3	**19.03**
600 mm	–	0.07	7.13	8.10	–	m^3	**15.24**
up to 1.00 m deep	–	0.06	5.95	6.75	–	m^3	**12.70**

6 GROUND REMEDIATION AND SOIL STABILIZATION

Item	PC £	Labour hours	Labour £	Plant £	Material £	Unit	Total rate £
Surface grading; Agripower Ltd; using laser controlled equipment							
Maximum variation of existing ground level							
75 mm; laser box grader and tractor	–		–	–	–	ha	**2436.09**
100 mm; laser controlled low ground pressure bulldozer	–		–	–	–	ha	**3995.18**
Fraise mowing; scarification of existing turf layer to rootzone level; Agripower Ltd							
Cart arisings to tip on site							
average thickness 50–75 mm	–		–	–	–	ha	**4371.03**
Cultivating							
Ripping up subsoil; using approved subsoiling machine; minimum depth 250 mm below topsoil; at 1.20 m centres; in							
gravel or sandy clay	–		–	5.82	–	100 m²	**5.83**
soil compacted by machines	–		–	6.47	–	100 m²	**6.47**
clay	–		–	9.71	–	100 m²	**9.71**
chalk or other soft rock	–		–	9.71	–	100 m²	**9.71**
Extra for subsoiling at 1 m centres	–		–	3.75	–	100 m²	**3.75**
Breaking up existing ground; using pedestrian operated tine cultivator or rotavator; loam or sandy soil							
100 mm deep	–	0.20	7.84	1.62	–	100 m²	**9.46**
150 mm deep	–	0.25	9.80	2.03	–	100 m²	**11.83**
200 mm deep	–	0.33	13.07	2.70	–	100 m²	**15.77**
Breaking up existing ground; using pedestrian operated tine cultivator or rotavator; heavy clay or wet soils							
100 mm deep	–	0.40	15.69	3.24	–	100 m²	**18.93**
150 mm deep	–	0.60	23.53	4.86	–	100 m²	**28.39**
200 mm deep	–	0.75	29.41	6.08	–	100 m²	**35.49**
Breaking up existing ground; using tractor drawn tine cultivator or rotavator							
100 mm deep	–		–	1.47		100 m²	**1.47**
150 mm deep	–		–	1.84	–	100 m²	**1.84**
200 mm deep	–		–	2.45	–	100 m²	**2.45**
400 mm deep	–		–	7.35	–	100 m²	**7.35**
final levelling to ploughed and rotavated ploughed ground; using disc, drag, or chain harrow							
4 passes	–		–	1.34	–	100 m²	**1.34**
Rolling cultivated ground lightly; using self-propelled agricultural roller	–	0.06	1.98	0.76	–	100 m²	**2.75**
Importing and storing selected and approved topsoil; 20 tonne load = 11.88 m³ average							
20 tonne loads	35.51		–	–	40.83	m³	**40.83**

6 GROUND REMEDIATION AND SOIL STABILIZATION

Item	PC £	Labour hours	Labour £	Plant £	Material £	Unit	Total rate £
GRADING AND SURFACE PREPARATION – cont							
Surface treatments							
Compacting							
bottoms of excavations	–	0.01	0.18	0.04	–	m^2	**0.22**
Surface preparation							
Trimming surfaces of cultivated ground to final levels; removing roots, stones and debris exceeding 50 mm in any direction to tip off site; slopes less than 15°							
clean ground with minimal stone content	–	0.25	8.91	–	–	100 m^2	**8.91**
slightly stony; 0.5 kg stones per m^2	–	0.33	11.88	–	–	100 m^2	**11.89**
very stony; 1.0–3.0 kg stones per m^2	–	0.50	17.83	–	–	100 m^2	**17.87**
clearing mixed, slightly contaminated rubble; inclusive of roots and vegetation	–	0.50	17.82	–	–	100 m^2	**17.97**
clearing brick-bats, stones and clean rubble	–	0.60	21.39	–	–	100 m^2	**21.49**

10 CRIB WALLS, GABIONS AND REINFORCED EARTH

Item	PC £	Labour hours	Labour £	Plant £	Material £	Unit	Total rate £
CRIB WALLS, GABIONS AND REINFORCED EARTH							
Crib walls; Phi Group							
Permacrib (timber) or Andacrib (concrete) crib walling system; machine filled infill with crushed rock; inclusive of concrete footing and rear wall drain; excluding excavation and backfill material							
retaining walls up to 2.0 m	–		–	–	–	m²	**297.83**
retaining walls up to 4.0 m	–		–	–	–	m²	**321.65**
retaining walls up to 6.0 m	–		–	–	–	m²	**345.48**
Thie labour in this section is generally calculated on a 3 person team. Labour time below should be multiplied by 3 to give the cost as shown.							
Retaining walls							
Gabion Walls; Maccaferri Ltd							
Wire mesh gabions; galvanized mesh 80 × 100 mm; filling with broken stones 125–200 mm size; wire down securely to manufacturer's instructions; filling front face by hand							
2 × 1 × 0.50 m	49.14	0.67	71.30	22.35	201.21	nr	**294.87**
2 × 1 × 1.0 m	49.14	1.33	142.60	44.71	345.92	nr	**533.23**
PVC coated gabions; 120 year design life							
2 × 1 × 0.50 m	50.42	0.67	71.30	22.35	202.69	nr	**296.35**
2 × 1 × 1.0 m	68.98	1.33	142.60	44.71	368.74	nr	**556.04**
Mattress gabions							
6 × 2 × 0.17 m	58.03	1.00	106.95	33.53	361.93	nr	**502.41**
6 × 2 × 0.30 m	52.49	2.00	213.90	50.30	581.31	nr	**845.51**

10 CRIB WALLS, GABIONS AND REINFORCED EARTH

Item	PC £	Labour hours	Labour £	Plant £	Material £	Unit	Total rate £
CRIB WALLS, GABIONS AND REINFORCED EARTH – cont							
Embankments; Tensar International							
Embankments; reinforced with Tensar Geogrid RE520; geogrid laid horizontally with infill to 100% of vertical height of slope at 1.00 m centres; filling with excavated material							
slopes less than 45°; Tensartech Natural Green System	3.58	0.61	65.52	5.74	4.12	m^3	**75.37**
slopes exceeding 45°; Tensartech Greenslope System	7.70	0.06	6.61	10.00	8.85	m^3	**25.46**
extra over to slopes exceeding 45° for temporary shuttering; shuttering measured per m^2 of vertical elevation	–		–	–	32.20	m^2	**32.20**
Embankments; reinforced with Tensar Geogrid RE540; geogrid laid horizontally within fill to 100% of vertical height of slope at 1.00 m centres; filling with excavated material							
slopes less than 45°; Tensartech Natural Green System	3.58	0.61	65.52	69.75	4.12	m^3	**139.39**
slopes exceeding 45°; Tensartech Greenslope System	–	0.06	6.62	147.63	–	m^3	**154.25**
extra over to slopes exceeding 45° for temporary shuttering; shuttering measured per m^2 of vertical elevation	–		–	–	32.20	m^2	**32.20**
Extra for Tensar Mat 400; erosion control mats; to faces of slopes of 45° or less; filling with 20 mm fine topsoil; seeding	5.15	0.01	0.71	0.34	6.92	m^2	**7.97**

10 CRIB WALLS, GABIONS AND REINFORCED EARTH

Item	PC £	Labour hours	Labour £	Plant £	Material £	Unit	**Total rate £**
Embankments; reinforced with Tensar Geogrid RE540; geogrid laid horizontally within fill to 50% of horizontal length of slopes at specified centres; filling with imported fill 6F2 PC £14.00/m^3							
300 mm centres; slopes up to 45°; Tensartech Natural Green System	11.81	0.08	8.32	113.21	34.45	m^3	**155.99**
600 mm centres; slopes up to 45°; Tensartech Natural Green System	5.94	0.06	6.62	63.13	27.70	m^3	**97.45**
extra over to slopes exceeding 45° for temporary shuttering; shuttering measured per m^2 of vertical elevation	–		–	–	32.20	m^2	**32.20**
Embankments; reinforced with Tensar Geogrid RE540; geogrid laid horizontally within fill; filling with excavated material; wrapping around at faces							
300 mm centres; slopes exceeding 45°; Tensartech Natural Green System	13.10	0.06	6.61	218.90	15.07	m^3	**240.58**
600 mm centres; slopes exceeding 45°; Tensartech Greenslope System	7.12	0.06	6.61	112.00	8.19	m^3	**126.80**
extra over to slopes exceeding 45° for temporary shuttering; shuttering measured per m^2 of vertical elevation	–		–	–	32.20	m^2	**32.20**
Extra over embankments for bagwork face supports for slopes exceeding 45°	–	0.17	17.82	–	32.20	m^2	**32.20**
Extra over embankments for seeding of bags	–	0.03	3.56	–	0.43	m^2	**4.00**
Extra over embankments for Bodkin joints	–		–	–	1.79	m	**1.79**
Erosion control mats							
Extra for erosion control mats; to faces of slopes of 45° or less; filling with 20 mm fine topsoil; seeding	5.15	0.01	0.71	0.34	6.92	m^2	**7.97**

10 CRIB WALLS, GABIONS AND REINFORCED EARTH

Item	PC £	Labour hours	Labour £	Plant £	Material £	Unit	Total rate £
CRIB WALLS, GABIONS AND REINFORCED EARTH – cont							
Green faced reinforced soil; Phi Group							
Textomur reinforced soil system embankments; to reinforced soil slopes 55–70° to the horizontal; including facing and geogrid; excluding backfill material							
retaining walls up to 3.0 m	–		–	–	–	m²	**208.48**
retaining walls up to 6.0 m	–		–	–	–	m²	**214.43**
Split face concrete blocks reinforced soil; Phi Group							
Modular blocks reinforced soil system embankments; to reinforced soil slopes near vertical; including geogrid; excluding backfill material							
retaining walls up to 3.0 m	–		–	–	–	m²	**297.83**
retaining walls up to 6.0 m	–		–	–	–	m²	**339.52**
This section uses one person in a composite with a machine. The labour time is not shown.							
Embankments; Grass Concrete Ltd; In situ grass concrete; Fully installed by subcontractor to cleared and graded embankments							
Grasscrete; in situ reinforced concrete surfacing; to 20 mm thick sand blinding layer (not included); including soiling and seeding							
GC1; 100 mm thick	–		–	–	4.94	m²	**56.87**
GC2; 150 mm thick	–		–	–	7.31	m²	**76.34**
Grassblock 103; solid matrix precast concrete blocks; to 20 mm thick sand blinding layer; excluding edge restraint; including soiling and seeding							
406 × 406 × 103 mm; fully interlocking	–		–	–	4.94	m²	**66.41**

10 CRIB WALLS, GABIONS AND REINFORCED EARTH

Item	PC £	Labour hours	Labour £	Plant £	Material £	Unit	**Total rate £**
Thie labour in this section is generally calculated on a 3 person team. Labour time below should be multiplied by 3 to give the cost as shown.							
Flexible sheet materials; Tensar International							
Tensar Mat 400 erosion mats; 3.0–4.5 m wide; securing with Tensar pegs; lap rolls 100 mm; anchors at top and bottom of slopes; in trenches	5.57	0.01	0.59	–	6.41	m^2	**6.99**
Topsoil filling to Tensar Mat; including brushing and raking	0.92	0.01	0.59	0.11	1.06	m^2	**1.76**
Tensar Bi-axial Geogrid; to graded compacted base; filling with 200 mm granular fill; compacting (turf or paving to surfaces not included); 400 mm laps							
TriAx 150; 39 × 39 mm mesh	2.27	–	0.48	0.67	6.79	m^2	**7.94**
TriAx 160; 39 × 39 mm mesh	2.97	–	0.48	0.67	7.59	m^2	**8.75**
TriAx 170; 39 × 39 mm mesh	3.51	–	0.48	0.67	8.21	m^2	**9.36**
Flexible sheet materials; Terram Ltd							
Terram synthetic fibre filter fabric; to graded base (not included)							
Terram 1000; 0.70 mm thick; mean water flow 50 l/m²/s	0.66	0.01	0.59	–	0.76	m^2	**1.34**
Terram 2000; 1.00 m thick; mean water flow 33 l/m²/s	1.62	0.01	0.59	–	1.86	m^2	**2.45**
Terram Minipack	1.01	–	0.12	–	1.16	m^2	**1.28**
Flexible sheet materials; Greenfix Ltd							
Geojute lightweight biodegradeable mats for erosion control to pre-seeded slopes ; 10–15 mm thick; fixing with 4 nr crimped pins in accordance with manufacturer's instructions; to graded surface (not included)							
Geojute 500; 500 kg/ m²	159.53	0.67	71.30	–	183.46	100 m^2	**254.76**
Greenfix C−100-P Unseeded mat	206.17	0.67	71.30	–	237.10	100 m^2	**308.40**
Greenfix C−10-J Unseeded mat	244.54	0.67	71.30	–	281.22	100 m^2	**352.53**
Coir matting							
Geocoir 400	182.96	0.67	71.30	–	210.40	100 m^2	**281.70**
Geocoir 900	294.95	0.67	71.30	–	355.03	100 m^2	**426.34**
Bioroll; 300 mm dia.; to river banks and revetments	20.67	0.04	4.46	–	29.44	m	**33.90**

10 CRIB WALLS, GABIONS AND REINFORCED EARTH

Item	PC £	Labour hours	Labour £	Plant £	Material £	Unit	Total rate £
CRIB WALLS, GABIONS AND REINFORCED EARTH – cont							
Flexible sheet materials – cont							
Extra over Greenfix erosion control mats for fertilizer applied at 70 g/m^2	11.43	0.07	7.13	–	13.14	100 m^2	**20.28**
Extra over Greenfix erosion control mats for laying to slopes exceeding 30°	–		–	–	–	25%	**–**
Extra for the following operations							
Spreading 25 mm approved topsoil							
by machine	0.89	–	0.22	0.21	1.02	m^2	**1.45**
by hand	0.89	0.01	1.25	–	1.02	m^2	**2.27**
Grass seed; PC £6.44/kg; spreading in two operations; by hand							
35 g/m^2	20.83	0.06	5.95	–	23.95	100 m^2	**29.90**
50 g/m^2	29.75	0.06	5.95	–	34.21	100 m^2	**40.16**
70 g/m^2	41.65	0.06	5.95	–	47.90	100 m^2	**53.84**
100 g/m^2	59.50	0.07	7.13	–	68.43	100 m^2	**75.56**
125 g/m^2	74.38	0.67	71.29	–	85.53	100 m^2	**156.82**
Extra over seeding by hand for slopes over 30° (allowing for the actual area but measured in plan)							
35 g/m^2	3.09	–	0.06	–	3.56	100 m^2	**3.62**
50 g/m^2	4.46	–	0.06	–	5.13	100 m^2	**5.20**
70 g/m^2	6.25	–	0.06	–	7.19	100 m^2	**7.25**
100 g/m^2	8.93	–	0.06	–	10.26	100 m^2	**10.33**
125 g/m^2	11.13	–	0.06	–	12.80	100 m^2	**12.86**
Grass seed; PC £6.44/kg; spreading in two operations; by machine							
35 g/m^2	20.83		–	1.50	23.95	100 m^2	**25.45**
50 g/m^2	29.75		–	1.50	34.21	100 m^2	**35.71**
70 g/m^2	41.65		–	1.50	47.90	100 m^2	**49.40**
100 g/m^2	59.50		–	1.50	68.43	100 m^2	**69.93**
125 kg/ha	743.75		–	149.60	855.31	ha	**1004.91**
150 kg/ha	892.50		–	149.60	1026.38	ha	**1175.97**
200 kg/ha	1190.00		–	149.60	1368.50	ha	**1518.10**
250 kg/ha	1487.50		–	149.60	1710.63	ha	**1860.22**
300 kg/ha	1785.00		–	149.60	2052.75	ha	**2202.35**
350 kg/ha	2082.50		–	149.60	2394.88	ha	**2544.47**
400 kg/ha	2380.00		–	149.60	2737.00	ha	**2886.60**
500 kg/ha	2975.00		–	149.60	3421.25	ha	**3570.85**
700 kg/ha	4165.00		–	149.60	4789.75	ha	**4939.35**
1400 kg/ha	8330.00		–	149.60	9579.50	ha	**9729.10**

10 CRIB WALLS, GABIONS AND REINFORCED EARTH

Item	PC £	Labour hours	Labour £	Plant £	Material £	Unit	**Total rate £**
Extra over seeding by machine for slopes over 30° (allowing for the actual area but measured in plan)							
35 g/m^2	3.12		–	0.23	3.59	100 m^2	**3.82**
50 g/m^2	4.46		–	0.23	5.13	100 m^2	**5.36**
70 g/m^2	6.25		–	0.23	7.18	100 m^2	**7.41**
100 g/m^2	8.93		–	0.23	10.26	100 m^2	**10.49**
125 kg/ha	111.56		–	22.44	128.30	ha	**150.74**
150 kg/ha	133.88		–	22.44	153.96	ha	**176.40**
200 kg/ha	178.50		–	22.44	205.28	ha	**227.71**
250 kg/ha	223.13		–	22.44	256.59	ha	**279.03**
300 kg/ha	267.75		–	22.44	307.91	ha	**330.35**
350 kg/ha	312.38		–	22.44	359.23	ha	**381.67**
400 kg/ha	357.00		–	22.44	410.55	ha	**432.99**
500 kg/ha	446.25		–	22.44	513.19	ha	**535.63**
700 kg/ha	624.75		–	22.44	718.46	ha	**740.90**
1400 kg/ha	1249.50		–	22.44	1436.93	ha	**1459.36**

11 IN SITU CONCRETE WORKS

Item	PC £	Labour hours	Labour £	Plant £	Material £	Unit	Total rate £
CLARIFICATION NOTES ON LABOUR COSTS IN THIS SECTION							
In situ concrete team							
Generally a three man team is used in this section; The column 'Labour hours' reports team hours. The column 'Labour £' reports the total cost of the team for the unit of work shown							
3 man team	–	1.00	106.95	–	–	hr	**106.95**
2 man team	–	1.00	71.30	–	–	hr	**71.30**
CONCRETE MIX INFORMATION							
General							
The industry standard refences to concrete are Designed, Standard, Prescribed and Designated mixes. Most Items in this section refer to Standard mixes							
Designed mix							
User specifies the performance of the concrete. Producer is responsible for selecting the appropriate mix. Strength testing is essential.							
Prescribed mix							
User specifies the mix constituents and is responsible for ensuring that the concrete meets performance requirements. Accurate mix proportions are essential.							
Standard mix							
Made with a restricted range of materials. Specification to include the proposed use of the material as well as the standard mix reference, type of cement, type and size of aggregate, and slump (workability). Quality assurance is required.							
Designated mix							
Producer to hold current product conformity certification and quality approval. Quality assurance is essential. The mix must not be modified.							

11 IN SITU CONCRETE WORKS

Item	PC £	Labour hours	Labour £	Plant £	Material £	Unit	Total rate £
CONCRETE MIXES							
Note: The mixes below are all mixed in 113 litre mixers.							
Concrete mixes at cost; mixed on site; costs for producing concrete; prices for commonly used mixes for various types of work; based on bulk load 20 tonne rates for aggregates							
Roughest type mass concrete such as footings and road haunchings; 300 mm thick							
1:3:6	91.96	3.50	108.50	–	91.96	m^3	**200.46**
Concrete mixes; mixed on site; costs for producing concrete; prices for commonly used mixes for various types of work; materials delivered in 10 tonne loads							
Roughest type mass concrete such as footings and road haunchings; 300 mm thick							
1:3:6	113.89	3.50	108.50	–	113.89	m^3	**222.39**
Concrete mixes; mixed on site; costs for producing concrete; prices for commonly used mixes for various types of work; materials delivered in 850 kg bulk bags							
Roughest type mass concrete such as footings and road haunchings; 300 mm thick							
1:3:6	140.15	3.50	108.50	–	140.15	m^3	**248.65**
Most ordinary use of concrete such as mass walls above ground, road slabs, etc. and general reinforced concrete work							
1:2:4; aggregates delivered in 20 tonne ballast loads	109.58	3.50	108.50	–	109.58	m^3	**218.08**
1:2:4; aggregates delivered in 10 tonne ballast loads	132.99	3.50	108.50	–	132.99	m^3	**241.49**
1:2:4; aggregates delivered in 850 kg bulk bags	159.26	4.50	139.50	–	159.26	m^3	**298.76**

11 IN SITU CONCRETE WORKS

Item	PC £	Labour hours	Labour £	Plant £	Material £	Unit	Total rate £
CONCRETE MIXES – cont							
Concrete mixes – cont							
Watertight floors, pavements, walls, tanks, pits, steps, paths, surface of two course roads; reinforced concrete where extra strength is required							
1:1.5:3; aggregates delivered in 20 tonne ballast loads	126.41	3.50	108.50	–	126.41	m^3	**234.91**
1:1.5:3; aggregates delivered in 10 tonne ballast loads	149.82	3.50	108.50	–	149.82	m^3	**258.32**
1:1.5:3; aggregates delivered in 850 kg bulk bags	176.09	3.50	108.50	–	176.09	m^3	**284.59**
IN SITU CONCRETE							
The labour in this section is calculated on a 3 person team. The labour time below should be multiplied by 3 to calculate the cost as shown.							
Plain in situ concrete; site mixed; 10 N/mm²–40 mm aggregate (1:3:6) (aggregate delivery indicated)							
Foundations							
ordinary Portland cement; 20 tonne ballast loads	205.19	0.33	35.29	–	235.97	m^3	**271.27**
ordinary Portland cement; 10 tonne ballast loads	227.66	0.33	35.29	–	261.81	m^3	**297.11**
ordinary Portland cement; 850 kg bulk bags	254.59	0.33	35.29	–	292.78	m^3	**328.07**
Foundations; poured on or against earth or unblinded hardcore							
ordinary Portland cement; 20 tonne ballast loads	210.20	0.33	35.29	–	241.73	m^3	**277.02**
ordinary Portland cement; 10 tonne ballast loads	233.22	0.33	35.29	–	268.20	m^3	**303.49**
ordinary Portland cement; 850 kg bulk bags	260.80	0.33	35.29	–	299.92	m^3	**335.21**
Isolated foundations							
ordinary Portland cement; 20 tonne ballast loads	205.19	0.33	35.29	–	235.97	m^3	**271.27**
ordinary Portland cement; 850 kg bulk bags	254.59	0.33	35.29	–	292.78	m^3	**328.07**

11 IN SITU CONCRETE WORKS

Item	PC £	Labour hours	Labour £	Plant £	Material £	Unit	**Total rate £**
Plain in situ concrete; site mixed; 21 N/mm²–20 mm aggregate (1:2:4)							
Foundations							
ordinary Portland cement; 20 tonne ballast loads	223.26	0.33	35.29	–	256.74	m³	**292.04**
ordinary Portland cement; 850 kg bulk bags	305.87	0.33	35.29	–	351.75	m³	**387.04**
Foundations; poured on or against earth or unblinded hardcore							
ordinary Portland cement; 20 tonne ballast loads	228.70	0.33	35.29	–	263.01	m³	**298.30**
ordinary Portland cement; 850 kg bulk bags	313.33	0.33	35.29	–	360.33	m³	**395.62**
Isolated foundations							
ordinary Portland cement; 20 tonne ballast loads	223.26	0.33	35.29	–	256.74	m³	**292.04**
ordinary Portland cement; 850 kg bulk bags	305.87	0.33	35.29	–	351.75	m³	**387.04**
Ready mix concrete; concrete mixed on site; Easymix							
Ready mix concrete mixed on site; placed by barrow not more than 25 m distance (16 barrows/m³)							
concrete C16/20	158.00	0.25	26.74	–	181.70	m³	**208.44**
concrete 1:2:4	162.00	0.25	26.74	–	186.30	m³	**213.04**
concrete C30	168.00	0.25	26.74	–	193.20	m³	**219.94**
Reinforced in situ concrete; site mixed; 21 N/mm²–20 mm aggregate (1:2:4); aggregates delivered in 10 tonne loads							
Foundations							
ordinary Portland cement	223.26	0.67	71.23	–	256.74	m³	**327.97**
Foundations; poured on or against earth or unblinded hardcore							
ordinary Portland cement	223.26	0.67	71.23	–	256.74	m³	**327.97**
Isolated foundations							
ordinary Portland cement	223.26	0.83	89.09	–	256.74	m³	**345.83**
Plain in situ concrete; ready mixed; Easymix Ltd; 10 N/mm mixes; suitable for mass concrete fill and blinding							
Foundations							
C10, 10 N/mm²	94.61	0.20	21.39	–	108.80	m³	**130.19**
C25; 25 N/mm²	99.34	0.20	21.39	–	114.24	m³	**135.63**

11 IN SITU CONCRETE WORKS

Item	PC £	Labour hours	Labour £	Plant £	Material £	Unit	Total rate £
IN SITU CONCRETE – cont							
Plain in situ concrete – cont							
Foundations; poured on or against earth or unblinded hardcore							
C10; 10 N/mm²	94.61	0.20	21.39	–	108.80	m³	**130.19**
C20; 20 N/mm²	96.56	0.20	21.39	–	111.05	m³	**132.44**
Isolated foundations							
C10; 10 N/mm²	94.61	0.67	71.23	–	108.80	m³	**180.03**
C25; 25 N/mm²	99.94	0.67	71.23	–	114.94	m³	**186.17**
Plain in situ concrete; ready mixed; Easymix Ltd;15 N/mm mixes; suitable for oversite below suspended slabs and strip footings in non aggressive soils							
Foundations							
C15; 15 N/mm²	94.61	0.20	21.39	–	108.80	m³	**130.19**
C20; 20 N/mm²	96.56	0.20	21.39	–	111.05	m³	**132.44**
Foundations; poured on or against earth or unblinded hardcore							
C15; 15 N/mm²	94.61	0.20	21.39	–	108.80	m³	**130.19**
ST 3; 15 N/mm²	96.56	0.20	21.39	–	111.05	m³	**132.44**
Isolated foundations							
C15; 15 N/mm²	94.61	0.67	71.23	–	108.80	m³	**180.03**
C20; 20 N/mm²	–	0.67	71.23	–	111.05	m³	**182.28**
Plain in situ concrete; ready mixed; air entrained mixes suitable for paving							
Beds or slabs; house drives, parking and external paving							
PAV 1; 35 N/mm²; Designated mix 100 mm thick	109.77	0.25	26.74	–	126.23	m³	**152.97**
Beds or slabs; house drives, parking and external paving							
PAV 1; 35 N/mm²; Designated mix	109.77	0.25	26.74	–	126.23	m³	**152.97**
Reinforced in situ concrete; ready mixed; 35 N/mm² mix; suitable for foundations in class 2 sulphate conditions							
Foundations							
RC 35; Designated mix	111.65	0.20	21.39	–	128.40	m³	**149.79**
Foundations; poured on or against earth or unblinded hardcore							
RC 35; Designated mix	114.31	0.20	21.39	–	131.46	m³	**152.85**
Isolated foundations							
RC 35; Designated mix	106.34	0.67	71.23	–	122.29	m³	**193.52**

11 IN SITU CONCRETE WORKS

Item	PC £	Labour hours	Labour £	Plant £	Material £	Unit	Total rate £
Reinforced in situ concrete; 21 N/mm²–20 mm aggregate (1:2:4); aggregates delivered in 20 tonne loads							
Site mixed concrete; aggregates in 20 tonne loads ; Walls; thickness not exceeding 150 mm							
1:2:4: 21 N/mm²; 20 mm aggregate	223.26	0.67	71.23	4.38	256.74	m³	**332.36**
Ready mixed concrete							
C10 Concrete	94.61	0.33	35.65	4.38	108.80	m³	**148.83**
C15 Concrete	99.34	0.33	35.65	4.38	114.24	m³	**154.27**
C20 Concrete	101.39	0.33	35.65	4.38	116.60	m³	**156.63**
C25 Concrete	104.94	0.33	35.65	4.38	120.68	m³	**160.71**
RC 35; Designated mix	111.65	0.33	35.65	4.38	128.40	m³	**168.43**
Walls; thickness not exceeding 150–450 mm							
Site mixed concrete; aggregates in 20 tonne loads 1:2:4: 21 N/mm²; 20 mm aggregate	223.26	0.50	53.48	4.38	256.74	m³	**314.60**
Ready mixed concrete							
C10 Concrete	94.61	0.20	21.39	4.38	108.80	m³	**134.58**
C15 Concrete	94.61	0.20	21.39	4.38	108.80	m³	**134.58**
C20 Concrete	96.56	0.20	21.39	4.38	111.05	m³	**136.82**
RC 35; Designated mix	106.34	0.20	21.39	4.38	122.29	m³	**148.06**
Adjustments for site mixed concrete; add the following amounts for aggregates delivered in different load sizes							
850 kg bulk bags	–		–	–	92.69	m³	**92.69**
10 tonne loads	–		–	–	26.92	m³	**26.92**
PLAIN VERTICAL FORMWORK							
The labour in this section is calculated on a 2 person team. The labour time below should be multiplied by 2 to calculate the cost as shown.							
Plain vertical formwork; basic finish							
Sides of foundations							
height not exceeding 250 mm	–	0.25	17.83		3.29	m	**21.12**
height 250–500 mm	–	0.25	17.83	–	4.30	m	**22.13**
height 500 mm–1.00 m	–	0.33	23.76	–	14.53	m	**38.29**
height exceeding 1.00 m	–	0.50	35.65	–	20.45	m²	**56.10**
Sides of foundations; left in							
height not exceeding 250 mm	–	0.20	14.26	–	9.41	m	**23.67**
height 250–500 mm	–	0.20	14.26	–	16.55	m	**30.81**
height 500 mm–1.00 m	–	0.25	17.83	–	33.11	m	**50.93**
height over 1.00 m	–	0.40	28.52	–	12.92	m²	**41.44**

11 IN SITU CONCRETE WORKS

Item	PC £	Labour hours	Labour £	Plant £	Material £	Unit	Total rate £
PLAIN VERTICAL FORMWORK – cont							
Plain vertical formwork – cont							
Walls							
height not exceeding 250 mm	–	0.31	22.28	–	15.78	m	**38.06**
height 250–500 mm	–	0.33	23.76	–	24.05	m	**47.81**
height 500 mm–1.00 m	–	0.42	29.70	–	54.02	m	**83.72**
height exceeding 1.00 m	–	0.75	53.48	–	78.34	m^2	**131.81**
Walls curved to 6.00 radius							
height not exceeding 250 mm	–	0.38	26.74	–	12.10	m	**38.84**
height 250–500 mm	–	0.40	28.52	–	24.16	m	**52.68**
height 500 mm–1.00 m	–	0.50	35.64	–	54.25	m	**89.89**
height exceeding 1.00 m	–	0.90	64.17	–	78.57	m^2	**142.74**
Walls curved to 3.00 radius							
height not exceeding 250 mm	–	0.47	33.43	–	12.78	m	**46.21**
height 250–500 mm	–	0.50	35.65	–	24.05	m	**59.70**
height 500 mm–1.00 m	–	0.62	44.56	–	54.02	m	**98.57**
height exceeding 1.00 m	–	1.13	80.21	–	78.34	m^2	**158.55**
'Solid Wall'; Arbour 'Landscape Solutions' Recycled plastic blocks for in situ concrete formwork (formwork left in) 150 mm thick on footing (not included) inclusive of integral ties and external bracing; filling with reinforcement and concrete C30/40 in max. 2.14 m lifts; includes vertical reinforcement T12 at 200 ccs and 2 nr T12 bars horizontally per m²							
Blocks standard size; 600.5 mm × 530 mm high × 150 mm thick							
Block standard; to receive liner where wall does not need a finished surface	69.27	0.46	28.21	–	108.90	m^2	**137.11**
Block tile to receive subsequent tiled, fibre glass or rendered surface.	77.93	0.46	28.21	–	117.56	m^2	**145.77**
Block tile 2 sided; for feature walls or where a finish is required to two sides	93.51	0.46	28.21	–	133.15	m^2	**161.36**
Blocks small size; 600.5 mm × 440 mm high × 150 mm thick							
Block Liner small to receive liner where wall does not need a finished surface	83.26	0.38	23.25	–	122.89	m^2	**146.14**

11 IN SITU CONCRETE WORKS

Item	PC £	Labour hours	Labour £	Plant £	Material £	Unit	**Total rate £**
Blocks Flex; Flexible blocks to curves; 600.5 mm × 440 mm high × 140 mm thick							
Flex block liner	89.36	0.47	29.32	–	128.99	m²	**158.31**
Flex block tiled	98.71	0.47	29.32	–	138.34	m²	**167.66**
Extra over for platforms for walls over 1.50 m high	–		–	57.40	–	week	**57.40**
BAR REINFORCEMENT The rates for reinforcement shown for steel bars below are based on prices which would be supplied on a typical landscape contract. The steel prices shown have been priced on a selection of steel delivered to site where the total order quantity is in the region of 2 tonnes. The assumption is that should larger quantities be required, the work would fall outside the scope of the typical landscape contract defined in the front of this book. Keener rates can be obtained for larger orders.							
The labour in this section is calculated on a 2 person team. The labour time below should be multiplied by 2 to calculate the cost as shown.							
Reinforcement bars; hot rolled plain round mild steel; straight or bent							
Bars							
8 mm nominal size	1300.00	13.50	962.55	–	1495.00	tonne	**2457.55**
10 mm nominal size	1300.00	13.00	926.90	–	1495.00	tonne	**2421.90**
12 mm nominal size	1300.00	12.50	891.25	–	1495.00	tonne	**2386.25**
16 mm nominal size	1300.00	12.00	855.60	–	1495.00	tonne	**2350.60**
20 mm nominal size	1300.00	11.50	819.95	–	1495.00	tonne	**2314.95**
Weights of bar reinforcement (m/tonne) – kg/m							
6 mm (4505)	–		–	–	–	0.22 kg	–
8 mm (2534)	–		–	–	–	0.40 kg	–
10 mm (1622)	–		–	–	–	0.62 kg	–
12 mm (1126)	–		–	–	–	0.89 kg	–
16 mm (634)	–		–	–	–	1.58 kg	–
20 mm (405)	–		–	–	–	2.47 kg	–
25 mm (260)	–		–	–	–	3.85 kg	–
32 mm (158)	–		–	–	–	6.31 kg	–

11 IN SITU CONCRETE WORKS

Item	PC £	Labour hours	Labour £	Plant £	Material £	Unit	Total rate £
BAR REINFORCEMENT – cont							
Reinforcement bar to concrete formwork straight; Slabs and foundations; Straight bar							
8 mm bar							
100 ccs	5.07	0.17	11.76	–	5.83	m^2	**17.60**
200 ccs	2.47	0.13	8.91	–	2.84	m^2	**11.75**
300 ccs	1.69	0.08	5.35	–	1.94	m^2	**7.29**
tonne rate	1300.00	11.25	802.13	–	1495.00	tonne	**2297.13**
10 mm bar							
100 ccs	7.93	0.17	11.76	–	9.12	m^2	**20.88**
200 ccs	3.90	0.13	8.91	–	4.49	m^2	**13.40**
300 ccs	2.60	0.08	5.35	–	2.99	m^2	**8.34**
tonne rate	1300.00	11.25	802.13	–	1495.00	tonne	**2297.13**
12 mm bar							
100 ccs	11.44	0.17	11.76	–	13.16	m^2	**24.92**
200 ccs	5.72	0.13	8.91	–	6.58	m^2	**15.49**
300 ccs	3.77	0.08	5.35	–	4.34	m^2	**9.68**
tonne rate	1300.00	8.45	602.49	–	1495.00	tonne	**2097.49**
16 mm bar							
100 ccs	20.41	0.20	14.26	–	23.47	m^2	**37.73**
200 ccs	10.27	0.17	11.76	–	11.81	m^2	**23.58**
300 ccs	6.89	0.13	8.91	–	7.92	m^2	**16.84**
tonne rate	1300.00	6.00	427.80	–	1495.00	tonne	**1922.80**
25 mm bar							
100 ccs	50.05	0.20	14.26	–	57.56	m^2	**71.82**
200 ccs	24.96	0.17	11.76	–	28.70	m^2	**40.47**
300 ccs	16.64	0.17	11.76	–	19.14	m^2	**30.90**
tonne rate	1300.00	3.45	245.99	–	1495.00	tonne	**1740.99**
32 mm bar							
100 ccs	82.03	0.20	14.26	–	94.33	m^2	**108.59**
200 ccs	40.95	0.17	11.76	–	47.09	m^2	**58.86**
300 ccs	27.30	0.13	8.91	–	31.40	m^2	**40.31**
tonne rate	1300.00	2.85	203.21	–	1495.00	tonne	**1698.21**
The labour in this section is calculated on a 3 person team. The labour time below should be multiplied by 3 to calculate the cost as shown.							

11 IN SITU CONCRETE WORKS

Item	PC £	Labour hours	Labour £	Plant £	Material £	Unit	**Total rate £**
Alternative tonne rates for bar reinforcement fixing to retaining walls in external works environments; Vertical bar only; Horizontal tables shown in separate table set below							
8 mm Bar Rates per tonne based on fixing rates shown							
28.2 hrs/tonne	1300.00	28.20	1005.33	–	1495.00	tonne	**2500.33**
10 mm bar. Rates per tonne based on fixing rates shown							
22.5 hrs/tonne	1300.00	22.50	802.13	–	1495.00	tonne	**2297.13**
12 mm bar. Rates per tonne based on fixing rates shown							
16.9 hrs/tonne	1300.00	16.90	602.49	–	1495.00	tonne	**2097.49**
16 mm bar. Rates per tonne based on fixing rates shown							
12.0 hrs/tonne	1300.00	12.00	427.80	–	1495.00	tonne	**1922.80**
25 mm bar. Rates per tonne based on fixing rates shown							
6.9 hrs/tonne	1300.00	6.90	245.99	–	1495.00	tonne	**1740.99**
32 mm bar. Rates per tonne based on fixing rates shown							
5.7 hrs/tonne	1300.00	5.70	203.21	–	1495.00	tonne	**1698.21**
Alternative tonne rates for bar reinforcement fixing to retaining walls in external works environments; Horizontal bar only in conjuction with vertical bar in separate table							
8 mm bar. Rates per tonne based on fixing rates shown							
7 hrs/tonne	1300.00	7.00	249.55	–	1495.00	tonne	**1744.55**
10 mm bar. Rates per tonne based on fixing rates shown							
5.62 hrs/tonne	1300.00	5.62	200.35	–	1495.00	tonne	**1695.35**
12 mm bar. Rates per tonne based on fixing rates shown							
4.22 hrs/tonne	1300.00	4.22	150.44		1495.00	tonne	**1645.44**
16 mm bar. Rates per tonne based on fixing rates shown							
3.0 hrs/tonne	1300.00	3.00	106.95	–	1495.00	tonne	**1601.95**
25 mm bar. Rates per tonne based on fixing rates shown							
2.2 hrs/tonne	1300.00	2.20	78.43		1495.00	tonne	**1573.43**
32 mm bar. Rates per tonne based on fixing rates shown							
2 hrs/tonne	1300.00	2.00	71.30	–	1495.00	tonne	**1566.30**

11 IN SITU CONCRETE WORKS

Item	PC £	Labour hours	Labour £	Plant £	Material £	Unit	Total rate £
BAR REINFORCEMENT – cont							
The labour in this section is calculated on a 2 person team. The labour time below should be multiplied by 2 to calculate the cost as shown.							
Reinforcement bar to concrete formwork or footings; bent							
8 mm bar							
100 ccs	5.07	0.17	11.76	–	5.83	m²	**17.60**
200 ccs	2.47	0.13	8.91	–	2.84	m²	**11.75**
300 ccs	1.69	0.08	5.35	–	1.94	m²	**7.29**
10 mm bar							
100 ccs	7.93	0.17	11.76	–	9.12	m²	**20.88**
200 ccs	3.90	0.13	8.91	–	4.49	m²	**13.40**
300 ccs	2.60	0.08	5.35	–	2.99	m²	**8.34**
12 mm bar							
100 ccs	11.44	0.17	11.76	–	13.16	m²	**24.92**
200 ccs	5.72	0.13	8.91	–	6.58	m²	**15.49**
300 ccs	3.77	0.08	5.35	–	4.34	m²	**9.68**
16 mm bar							
100 ccs	20.41	0.20	14.26	–	23.47	m²	**37.73**
200 ccs	10.27	0.17	11.76	–	11.81	m²	**23.58**
300 ccs	6.89	0.13	8.91	–	7.92	m²	**16.84**
25 mm bar							
100 ccs	50.05	0.20	14.26	–	57.56	m²	**71.82**
200 ccs	24.96	0.13	8.91	–	28.70	m²	**37.62**
300 ccs	16.64	0.13	8.91	–	19.14	m²	**28.05**
32 mm bar							
100 ccs	82.03	0.20	14.26	–	94.33	m²	**108.59**
200 ccs	40.95	0.17	11.76	–	47.09	m²	**58.86**
300 ccs	27.30	0.13	8.91	–	31.40	m²	**40.31**
Reinforcement fabric; lapped; in beds or suspended slabs							
Fabric							
A98 (1.54 kg/m²)	2.09	0.02	1.55	–	2.41	m²	**3.96**
A142 (2.22 kg/m²)	4.58	0.02	1.55	–	5.27	m²	**6.82**
A193 (3.02 kg/m²)	5.90	0.02	2.08	–	6.78	m²	**8.86**
A252 (3.95 kg/m²)	7.10	0.03	2.07	–	8.16	m²	**10.23**
A393 (6.16 kg/m²)	11.09	0.03	2.37	–	12.76	m²	**15.13**

14 MASONRY

Item	PC £	Labour hours	Labour £	Plant £	Material £	Unit	Total rate £
MARKET PRICES OF MATERIALS							
Cement; Builder Centre							
Portland cement	–		–	–	4.95	25 kg	**4.95**
Sulphate-resistant cement	–		–	–	8.25	25 kg	**8.25**
White cement	–		–	–	16.91	25 kg	**16.91**
Sand; Builder Centre							
Building sand							
loose	55.44		–	–	55.44	m^3	**55.44**
850 kg bulk bags	–		–	–	46.95	nr	**46.95**
850 kg bulk bags	–		–	–	91.55	m^3	**91.55**
Sharp sand							
loose	–		–	–	45.36	m^3	**45.36**
850 kg bulk bags	–		–	–	46.95	nr	**46.95**
850 kg bulk bags	–		–	–	91.55	m^3	**91.55**
Sand; Yeoman Aggregates Ltd							
Sharp sand	–		–	–	22.55	tonne	**22.55**
Sharp sand	–		–	–	36.08	m^3	**36.08**
Soft sand	–		–	–	30.80	tonne	**30.80**
Soft sand	–		–	–	49.28	m^3	**49.28**
CLARIFICATION NOTES ON LABOUR COSTS IN THIS SECTION							
Brick/Block walling team							
Generally a three man team is used in this section; The column ‘Labour hours’ reports team hours. The column ‘Labour £’ reports the total cost of the team for the unit of work shown							
3 man team	–	1.00	106.95	–	–	hr	**106.95**
2 man team	–	1.00	71.30	–	–	hr	**71.30**

14 MASONRY

Item	PC £	Labour hours	Labour £	Plant £	Material £	Unit	Total rate £
BRICK/BLOCK WALLING DATA							
Batching quantities for these mortar mixes may be found in the Tables and Memoranda section of this book.							
Mortar mixes for brickwork and pavings; all mortars mixed on site and priced at delivery to waiting wheelbarrows at the cement mixer; cement and lime in 25 kg bags; aggregates as described							
Mortar grade 1:3; aggregates delivered in 20 tonne load; cement: sand volumes shown below (bags: litres)							
85 litre mixer (1.84:93.5)	–	0.75	26.74	–	185.30	m^3	**212.03**
150 litre mixer (3.24:165)	–	0.50	17.83	–	185.30	m^3	**203.12**
Mortar grade 1:3; aggregates delivered in 10 tonne load; cement: sand volumes shown below (bags: litres)							
85 litre mixer (1.84:93.5)	–	0.75	26.74	–	224.97	m^3	**251.71**
150 litre mixer (3.24:165)	–	0.50	17.83	–	224.97	m^3	**242.79**
Mortar grade 1:3; aggregates delivered in 850 kg bulk bags; cement: sand volumes shown below (bags: litres)							
85 litre mixer (1.84:93.5)	–	0.75	26.74	–	234.72	m^3	**261.46**
150 litre mixer (3.24:165)	–	0.50	17.83	–	234.72	m^3	**252.55**
Mortar grade 1:4; aggregates delivered in 20 tonne load; cement: sand volumes shown below (bags: litres)							
85 litre mixer (1.36:102)	–	0.75	26.74	–	159.09	m^3	**185.82**
150 litre mixer (2.4:180)	–	0.50	17.83	–	159.09	m^3	**176.91**
Mortar grade 1:4; aggregates delivered in 10 tonne load; cement: sand volumes shown below (bags: litres)							
85 litre mixer (1.36:102)	–	0.75	26.74	–	202.36	m^3	**229.10**
150 litre mixer (2.4:180)	–	0.50	17.83	–	202.36	m^3	**220.19**
Mortar grade 1:4; aggregates delivered in 850 kg bulk bags; cement: sand volumes shown below (bags: litres)							
85 litre mixer (1.36:102)	–	0.75	26.74	–	213.10	m^3	**239.84**
150 litre mixer (2.4:180)	–	0.50	17.83	–	213.10	m^3	**230.93**

14 MASONRY

Item	PC £	Labour hours	Labour £	Plant £	Material £	Unit	**Total rate £**
Mortar grade 1:1:6; aggregates delivered in 20 tonne load; cement: lime: sand volumes shown below (bags: bags: litres)							
85 litre mixer;(0.9:0.44:93.5)	–	0.75	26.74	–	214.72	m^3	**241.45**
150 litre mixer; (1.6: 0.8: 165)	–	0.50	17.83	–	214.72	m^3	**232.54**
Mortar grade 1:1:6; aggregates delivered in 10 tonne load; cement: lime: sand volumes shown below (bags: bags: litres)							
85 litre mixer (0.9:0.44:93.5)	–	0.75	26.74	–	263.66	m^3	**290.40**
150 litre mixer (1.6:0.8:165)	–	0.50	17.83	–	263.66	m^3	**281.49**
Mortar grade 1:1:6; aggregates delivered in 850 kg bulk bags; cement: lime: sand volumes shown below (bags: bags: litres)							
85 litre mixer (0.9:0.44:93.5)	–	0.75	26.74	–	264.14	m^3	**290.88**
150 litre mixer (1.6:0.8:165)	–	0.50	17.83	–	264.14	m^3	**281.97**
Mortar grade 1:2:9; aggregates delivered in 20 tonne load; cement: lime: sand volumes shown below (bags: bags: litres)							
85 litre mixer (0.7:0.5:102)	–	0.75	26.74	–	218.43	m^3	**245.17**
150 litre mixer (1.2:0.9:180)	–	0.50	17.83	–	212.76	m^3	**230.59**
Mortar grade 1:2:9; aggregates delivered in 10 tonne load; cement: lime: sand volumes shown below (bags: bags: litres)							
85 litre mixer (0.7:0.5:102)	–	0.75	26.74	–	261.71	m^3	**288.44**
150 litre mixer (1.2:0.9:180)	–	0.50	17.83	–	261.71	m^3	**279.53**
Mortar grade 1:2:9; aggregates delivered in 850 kg bulk bags; cement: lime: sand volumes shown below (bags: bags: litres)							
85 litre mixer (0.7:0.5:102)	–	0.75	26.74	–	272.45	m^3	**299.18**
150 litre mixer (1.2:0.9:180)	–	0.50	17.83	–	271.91	m^3	**289.73**
Variation in brick prices (area × price difference × value shown below)							
Add or subtract the following amounts for every £1.00/1000 difference in the PC price of the measured items below							
half brick thick	–		–	–	7.98	m^2	**7.98**
one brick thick	–		–	–	15.97	m^2	**15.97**
one and a half brick thick	–		–	–	23.95	m^2	**23.95**
two brick thick	–		–	–	31.94	m^2	**31.94**

14 MASONRY

Item	PC £	Labour hours	Labour £	Plant £	Material £	Unit	Total rate £
BRICK/BLOCK WALLING DATA – cont							
Mortar (1:3) required per m^2 of brickwork; brick size 215 × 102.5 × 65 mm							
Half brick wall (103 mm); 58 bricks/ m^2							
no frog	–		–	–	4.93	m^2	**4.93**
single frog	–		–	–	5.71	m^2	**5.71**
double frog	–		–	–	6.75	m^2	**6.75**
2 × half brick cavity wall (270 mm); 116 bricks/ m^2							
no frog	–		–	–	9.87	m^2	**9.87**
single frog	–		–	–	11.68	m^2	**11.68**
double frog	–		–	–	14.28	m^2	**14.28**
One brick wall (215 mm); 116 bricks/ m^2							
no frog	–		–	–	11.94	m^2	**11.94**
single frog	–		–	–	14.28	m^2	**14.28**
double frog	–		–	–	16.62	m^2	**16.62**
One and a half brick wall (328 mm) 174 bricks/m^2							
no frog	–		–	–	16.36	m^2	**16.36**
single frog	–		–	–	19.21	m^2	**19.21**
double frog	–		–	–	22.85	m^2	**22.85**
Mortar (1:3) required per m^2 of blockwork; blocks 440 × 215 mm							
Block thickness							
100 mm	–		–	–	1.82	m^2	**1.82**
140 mm	–		–	–	2.34	m^2	**2.34**
hollow blocks 440 × 215 mm	–		–	–	0.59	m^2	**0.59**
Movement of materials							
Loading to wheelbarrows and transporting to location; per 215 mm thick walls; maximum distance 25 m	–	0.42	14.86	–	–	m^2	**14.86**
Adjustments for mortars in the brick prices below							
Difference for cement: mortar (1:3) in lieu of gauged mortar							
half brick thick	–		–	–	-0.77	m^2	**0.77**
one brick thick	–		–	–	-1.93	m^2	**1.93**
one and a half brick thick	–		–	–	-2.59	m^2	**2.59**
two brick thick	–		–	–	-3.85	m^2	**3.85**

14 MASONRY

Item	PC £	Labour hours	Labour £	Plant £	Material £	Unit	**Total rate £**
BRICK WALLING AND BRICK PIERS							
Note: frog types and mortar quantities used in the calculations below are based on bricks with single frogs. Please make adjustments for mortar volumes used with varying brick frog formats. See data above for mortar quantities for various frog types.							
Delivery volumes of aggregates for mortars are based on 10 tonne deliveries. The figures below show the approximate areas of walling which would be constructed mixed with building sand from a 10 tonne (6.25 m³) load of building sand.							
half brick wall = 256 m²							
one brick wall = 124 m²							
The labour in this section is calculated on a 3 person team. The labour time below should be multiplied by 3 to calculate the cost as shown.							
Class B engineering bricks; PC £494.00/1000; double Flemish bond in cement: mortar (1:3)							
Offloading mechanically; loading to wheelbarrows; transporting to location maximum 25 m distance per 215 mm thick walls	–	0.42	14.86	–	–	m²	**14.86**
Construct walls							
half brick thick	37.17	0.60	70.46	–	48.19	m²	**118.64**
one brick thick	74.34	1.20	140.91	–	102.90	m²	**243.81**
one and a half brick thick	111.51	1.80	211.37	–	148.36	m²	**359.73**
two brick thick	148.68	2.40	281.71	–	200.90	m²	**482.61**
Construct walls; curved; mean radius 6 m							
half brick thick	37.10	0.90	105.68	–	48.10	m²	**153.78**
one brick thick	74.34	1.80	211.37	–	99.09	m²	**310.46**
Construct walls; curved; mean radius 1.50 m							
half brick thick	37.17	1.20	140.90	–	48.19	m²	**189.09**
one brick thick	74.34	3.60	293.94	–	99.09	m²	**393.03**

14 MASONRY

Item	PC £	Labour hours	Labour £	Plant £	Material £	Unit	Total rate £
BRICK WALLING AND BRICK PIERS – cont							
Class B engineering bricks – cont							
Construct walls; tapering; one face battering; average							
one and a half brick thick	111.51	2.22	260.25	–	146.53	m^2	**406.79**
two brick thick	148.68	2.96	347.00	–	198.18	m^2	**545.17**
Construct walls; battering (retaining)							
one and a half brick thick	111.51	2.22	260.25	–	146.53	m^2	**406.79**
two brick thick	148.68	2.96	347.00	–	198.18	m^2	**545.17**
Projections; vertical							
one brick × half brick	8.26	2.33	273.96	–	10.36	m	**284.32**
one brick × one brick	16.52	0.47	54.72	–	20.72	m	**75.44**
one and a half brick × one brick	24.78	0.70	82.20	–	31.09	m	**113.29**
two brick by one brick	33.04	0.77	89.95	–	41.46	m	**131.40**
Walls; half brick thick							
in honeycomb bond	24.19	0.60	70.46	–	32.76	m^2	**103.22**
in quarter bond	36.29	0.56	65.17	–	45.44	m^2	**110.61**
Facing bricks; PC £541.00/1000; English garden wall bond; in gauged mortar (1:1:6); facework one side							
Mechanically offloading; maximum 25 m distance; loading to wheelbarrows and transporting to location; walls	–	0.42	14.86	–	–	m^2	**14.86**
Construct walls							
half brick thick	–	0.60	70.34	–	52.34	m^2	**122.68**
half brick thick (using site cut snap headers to form bond)	–	0.80	94.41	–	52.34	m^2	**146.75**
one brick thick	–	1.20	140.91	–	104.68	m^2	**245.60**
one and a half brick thick	–	1.80	211.37	–	148.13	m^2	**359.50**
two brick thick	–	2.40	281.82	–	200.70	m^2	**482.52**
Walls; curved; mean radius 6 m							
half brick thick	–	0.90	105.68	–	49.66	m^2	**155.35**
one brick thick	–	1.80	211.37	–	241.73	m^2	**453.10**
Walls; curved; mean radius 1.50 m							
half brick thick	–	1.20	140.91	–	50.70	m^2	**191.61**
one brick thick	–	2.40	281.82	–	243.80	m^2	**525.63**
Walls; tapering; one face battering; average							
one and a half brick thick	–	2.20	258.34	–	157.45	m^2	**415.78**
two brick thick	–	3.00	352.28	–	213.12	m^2	**565.40**
Walls; battering (retaining)							
one and a half brick thick	–	2.00	234.85	–	157.45	m^2	**392.30**
two brick thick	–	2.70	317.05	–	199.05	m^2	**516.10**

14 MASONRY

Item	PC £	Labour hours	Labour £	Plant £	Material £	Unit	Total rate £
Projections; vertical							
one brick × half brick	–	0.23	27.01	–	10.64	m	**37.65**
one brick × one brick	–	0.50	58.71	–	21.29	m	**80.00**
one and a half brick × one brick	–	0.70	82.20	–	31.94	m	**114.13**
two brick × one brick	–	0.80	93.94	–	44.27	m	**138.21**
Brickwork fair faced both sides; facing bricks in gauged mortar (1:1:6)							
extra for fair face both sides; flush, struck, weathered, or bucket-handle pointing	–	0.67	23.76	–	–	m²	**23.76**
Bricks; PC £866.00/1000; English garden wall bond; in gauged mortar (1:1:6)							
Walls							
half brick thick (stretcher bond)	–	0.60	70.46	–	69.84	m²	**140.30**
half brick thick (using site cut snap headers to form bond)	–	0.80	93.94	–	69.84	m²	**163.79**
one brick thick	–	1.20	140.91	–	146.50	m²	**287.41**
one and a half brick thick	–	1.80	211.37	–	212.72	m²	**424.08**
two brick thick	–	2.40	281.82	–	289.62	m²	**571.45**
Walls; curved; mean radius 6 m							
half brick thick	–	0.90	105.68	–	71.40	m²	**177.08**
one brick thick	–	2.40	281.82	–	147.30	m²	**429.12**
Walls; curved; mean radius 1.50 m							
half brick thick	–	0.53	62.24	–	77.61	m²	**139.84**
one brick thick	–	1.07	125.29	–	159.72	m²	**285.01**
Walls; stretcher bond; wall ties at 450 mm centres vertically and horizontally							
one brick thick	0.84	1.09	127.99	–	145.77	m²	**273.77**
two brick thick	2.51	1.64	192.58	–	292.51	m²	**485.09**
Brickwork fair faced both sides; facing bricks in gauged mortar (1:1:6)							
extra for fair face both sides; flush, struck, weathered or bucket-handle pointing	–	0.67	23.76	–	–	m²	**23.76**
Brick copings							
Copings; all brick headers-on-edge; two angles rounded 53 mm radius; flush pointing top and both sides as work proceeds; one brick wide; horizontal							
machine-made specials	0.62	0.16	19.26	–	2.40	m	**21.66**
handmade specials	82.40	0.16	19.26	–	96.45	m	**115.71**
Extra over copings for two courses machine-made tile creasings; projecting 25 mm each side; 260 mm wide copings; horizontal	6.94	0.17	19.96	–	8.82	m	**28.79**

14 MASONRY

Item	PC £	Labour hours	Labour £	Plant £	Material £	Unit	Total rate £
BRICK WALLING AND BRICK PIERS – cont							
Brick copings – cont							
Copings; all brick headers-on-edge; flush pointing top and both sides as work proceeds; one brick wide; horizontal							
engineering bricks PC £456.00/1000	8.26	0.16	19.26	–	9.50	m	**28.75**
facing bricks PC £600.00/1000	8.60	0.16	19.26	–	9.89	m	**29.15**
facing bricks PC £900.00/1000	12.00	0.16	19.26	–	13.80	m	**33.05**
Isolated brick piers in English bond; in gauged mortar 1:1:6							
Engineering brick PC £494.00/1000							
one brick thick (225 mm)	–	2.33	273.96	–	102.38	m^2	**376.33**
one and a half brick thick (337.5 mm)	–	3.00	352.28	–	153.57	m^2	**505.85**
two brick thick (450 mm)	–	3.33	391.38	–	207.57	m^2	**598.95**
three brick thick (675 mm)	–	4.13	485.32	–	307.13	m^2	**792.46**
Engineering brick PC £494.00/1000 (not SMM)							
one brick (225 mm)	–	0.54	63.95	–	23.56	m	**87.51**
one and a half brick thick (337.5 mm)	–	1.00	117.43	–	66.50	m	**183.93**
two brick thick (450 mm)	–	1.50	176.14	–	107.90	m	**284.04**
three brick thick (675 mm)	–	2.79	327.62	–	223.78	m	**551.40**
Brick PC £541.00/1000							
one brick thick (225 mm)	–	2.33	273.96	–	103.83	m^2	**377.78**
one and a half brick thick (337.5 mm)	–	3.00	352.28	–	155.74	m^2	**508.02**
two brick thick (450 mm)	–	3.33	391.38	–	210.47	m^2	**601.85**
three brick thick (675 mm)	–	4.13	485.32	–	311.48	m^2	**796.80**
Brick PC £541.00/1000 (not SMM)							
one brick (225 mm)	–	0.54	63.88	–	23.90	m	**87.78**
one and a half brick thick (337.5 mm)	–	1.00	117.43	–	67.23	m	**184.66**
two brick thick (450 mm)	–	1.50	176.14	–	109.20	m	**285.34**
three brick thick (675 mm)	–	2.79	327.62	–	226.71	m	**554.33**
Brick PC £866.00/1000							
one brick thick (225 mm)	–	2.33	273.96	–	147.30	m^2	**421.25**
one and a half brick thick (337.5 mm)	–	3.00	352.28	–	220.94	m^2	**573.22**
two brick thick (450 mm)	–	3.33	391.38	–	297.41	m^2	**688.79**
three brick thick (675 mm)	–	4.13	485.32	–	441.89	m^2	**927.21**
Brick PC £866.00/1000 (not SMM)							
one brick (225 mm)	–	0.54	63.88	–	34.04	m	**97.92**
one and a half brick thick (337.5 mm)	–	1.00	117.43	–	89.24	m	**206.67**
two brick thick (450 mm)	–	1.50	176.14	–	148.33	m	**324.47**
three brick thick (675 mm)	–	2.79	327.62	–	314.74	m	**642.36**

14 MASONRY

Item	PC £	Labour hours	Labour £	Plant £	Material £	Unit	Total rate £
RESTORATION BRICKWORK							
Carefully dismantle existing coping to walls; Salvage copings (50%); Stack for reuse and dispose of balance off site							
Walls over 1.5 m high to 2.40 m high							
one brick thick	–	0.50	40.81	2.33	–	m	**43.56**
one and a half brick thick	–	0.75	61.21	2.33	–	m	**64.18**
two brick thick	–	1.00	81.61	2.33	–	m	**84.79**
Walls under 1.5 m high							
one brick thick	–	0.50	40.81	–	–	m	**41.23**
one and a half brick thick	–	0.75	61.21	–	–	m	**61.84**
two brick thick	–	1.00	81.61	–	–	m	**82.46**
Dismantle existing wall; salvage bricks and dispose damaged material off site (25%)							
half brick thick	–	0.40	32.60	0.37	–	m	**34.43**
one brick thick	–	0.70	57.04	0.37	–	m	**60.32**
one and a half brick thick	–	1.30	105.97	0.37	–	m	**110.71**
two brick thick	–	1.80	146.73	0.37	–	m	**152.92**
Cut out damaged brickwork in isolated areas							
half brick thick	–	0.17	5.94	–	–	brick	**5.94**
one brick thick to 2 bricks thick	–	0.14	5.09	–	–	brick	**5.09**
Isolated areas under 0.5 m^2							
half brick thick	–	0.50	17.83	–	–	m^2	**17.83**
one brick thick to 2 bricks thick per side	–	0.40	14.26	–	–	m^2	**14.26**
Isolated areas over 1.0 m^2							
half brick thick	–	1.00	35.65	–	–	m^2	**35.65**
one brick thick to 2 bricks thick	–	0.80	28.52	–	–	m^2	**28.52**
Cut out existing damaged pointing and replace with new mortar pointing							
half brick thick walls in strecher bond							
in standard mortar	0.37	0.42	14.86	–	0.43	m^2	**15.29**
in lime mortar	2.77	0.45	16.04	–	3.19	m^2	**19.23**
one brick thick walls in English garden wall bond							
in standard mortar	0.79	0.58	20.79	–	0.91	m^2	**21.71**
in lime mortar	5.55	0.58	20.79	–	6.38	m^2	**27.17**
flemish bond							
in standard mortar	0.84	0.69	24.45	–	0.96	m^2	**25.41**
in lime mortar	5.55	0.69	24.45	–	6.38	m^2	**30.82**

14 MASONRY

Item	PC £	Labour hours	Labour £	Plant £	Material £	Unit	Total rate £
RESTORATION BRICKWORK – cont							
Replace damaged brickwork in existing wall; Cut out damaged bricks and replace with new bricks to match; reclaimed bricks PC £650.00 / 1000							
One brick thick; one side fair faced							
per brick	0.60	0.27	9.50	–	0.82	nr	**10.32**
larger areas	72.00	2.50	157.95	–	96.88	m^2	**261.88**
One brick thick both sides fair faced							
per brick	0.60	0.28	10.10	–	0.82	nr	**10.92**
larger areas	72.00	3.50	239.57	–	96.91	m^2	**343.53**
One and a half brick thick; one side fair faced							
per brick	1.20	0.57	20.18	–	1.63	nr	**21.81**
larger areas	108.00	5.50	356.76	–	145.36	m^2	**512.70**
Construct new wall sections to match existing walls in lime mortar; Facing bricks; PC £1083.00/1000; English garden wall bond; in lime mortar							
Facing bricks; PC £1083.00/1000; English garden wall bond;							
half brick thick	36.16	0.64	75.34	–	119.39	m^2	**194.73**
half brick thick (using site cut snap headers to form bond)	36.16	0.84	99.13	–	119.39	m^2	**218.52**
one brick thick	66.78	1.26	147.96	–	232.39	m^2	**380.35**
one and a half brick thick	97.40	1.89	221.94	–	345.40	m^2	**567.34**
two brick thick	128.02	2.52	295.91	–	458.41	m^2	**754.32**
Walls; curved; mean radius 6 m							
half brick thick	36.16	0.95	110.97	–	121.28	m^2	**232.25**
one brick thick	66.78	1.89	221.94	–	236.19	m^2	**458.12**
Walls; curved; mean radius 1.50 m							
half brick thick	36.16	1.26	147.96	–	123.18	m^2	**271.14**
one brick thick	66.78	2.52	295.91	–	239.98	m^2	**535.90**
Walls; tapering; one face battering; average							
one and a half brick thick	97.40	2.31	271.26	–	362.48	m^2	**633.73**
two brick thick	128.02	3.15	369.89	–	481.18	m^2	**851.07**
Walls; battering (retaining)							
one and a half brick thick	97.40	2.10	246.60	–	362.48	m^2	**609.07**
two brick thick	97.40	2.84	332.90	–	445.97	m^2	**778.87**
Projections; vertical							
one brick × half brick	12.35	0.24	28.36	–	31.91	m	**60.27**
one brick × one brick	19.15	0.53	61.65	–	57.43	m	**119.08**
one and a half brick × one brick	25.96	0.74	86.31	–	82.98	m	**169.29**
two brick × one brick	32.76	0.84	98.64	–	108.51	m	**207.15**

14 MASONRY

Item	PC £	Labour hours	Labour £	Plant £	Material £	Unit	**Total rate £**
Brickwork fair faced both sides; facing bricks in gauged mortar (1:1:6)							
extra for fair face both sides; flush, struck, weathered, or bucket-handle pointing	5.55	0.70	24.95	–	6.38	m²	**31.33**
Brick copings; lime mortar bedding and jointing							
Copings; all brick headers-on-edge; two angles rounded 53 mm radius; flush pointing top and both sides as work proceeds; one brick wide; horizontal							
machine-made specials PC £3.80 each	8.04	0.17	20.22	–	9.25	m	**29.47**
handmade specials PC £4.90 each	94.75	0.17	20.22	–	108.96	m	**129.18**
Copings; all brick headers-on-edge; flush pointing top and both sides as work proceeds; one brick wide; horizontal							
reclaimed bricks PC £1000.00 /1000	7.22	0.17	20.22	–	25.58	m	**45.81**
Isolated brick piers in English bond; in gauged mortar 1:1:6							
Brick PC £1083.00/1000							
one brick thick (225 mm)	66.78	2.45	287.66	–	236.19	m²	**523.85**
one and a half brick thick (337.5 mm)	19.15	3.15	369.89	–	261.11	m²	**631.00**
two brick thick (450 mm)	128.02	3.50	410.96	–	466.00	m²	**876.96**
three brick thick (675 mm)	189.25	4.34	509.58	–	695.81	m²	**1205.39**
Brick PC £1083.00/1000 (not SMM)							
one brick (225 mm)	18.04	0.57	67.07	–	57.95	m	**125.02**
one and a half brick thick (337.5 mm)	106.77	1.05	123.30	–	122.78	m	**246.08**
two brick thick (450 mm)	62.60	1.58	184.95	–	215.44	m	**400.39**
three brick thick (675 mm)	135.23	2.93	344.00	–	478.28	m	**822.28**

14 MASONRY

Item	PC £	Labour hours	Labour £	Plant £	Material £	Unit	Total rate £
BLOCK WALLING AND BLOCK PIERS							
The labour in this section is calculated on a 3 person team. The labour time below should be multiplied by 3 to calculate the cost as shown.							
IN SITU CONCRETE FOOTINGS							
Excavate trench mechanically and dispose excavated material off site; Fix starter bars of T12 at 300 mm centres and horizontal bars of T25 in two lengths through the width of the trench; Pour ready mixed concrete C25 to 225 mm below ground level							
Size of footing							
600 × 350 mm deep	2696.56	1.45	68.63	2.28	53.76	m	**135.81**
800 × 450 mm deep	2696.56	1.78	82.38	3.30	68.25	m	**170.29**
800 × 600 mm deep	2696.56	2.04	92.05	4.18	79.84	m	**197.67**
1.00 × 500 mm deep	2696.56	2.20	98.73	4.65	81.77	m	**208.71**
1.20 × 600 mm deep	2696.56	2.71	118.89	6.27	103.01	m	**260.57**
Dense aggregate concrete blocks; Tarmac Topblock or other equal and approved; in gauged mortar (1:2:9); one course underground							
Walls							
Solid blocks 7 N/mm^2							
440 × 215 × 100 mm thick	15.30	0.40	46.97	–	19.57	m^2	**66.54**
440 × 215 × 140 mm thick	26.00	0.44	52.02	–	32.13	m^2	**84.15**
Solid blocks 7 N/mm^2 laid flat							
440 × 100 × 215 mm thick	31.21	1.07	125.65	–	38.12	m^2	**163.77**
Hollow concrete blocks							
440 × 215 × 215 mm thick	38.00	0.43	50.85	–	45.93	m^2	**96.77**
Filling of hollow concrete blocks with concrete as work proceeds; tamping and compacting							
440 × 215 × 215 mm thick	17.80	0.25	29.36	–	20.46	m^2	**49.82**
The labour in this section is calculated on a 2 person team. The labour time below should be multiplied by 2 to calculate the cost as shown.							

14 MASONRY

Item	PC £	Labour hours	Labour £	Plant £	Material £	Unit	**Total rate £**
Isolated blockwork piers; to receive facing treatments (not included)							
Solid blocks; 440 × 100 × 215 mm thick; laid on-flat; 7 N/mm²							
450 × 450 mm (not SMM)	33.66	2.00	163.30	–	39.50	m	**202.80**
450 × 450 mm	15.30	1.00	81.65	–	19.82	m²	**101.47**
690 × 690 with hollow core filled with concrete (not SMM)	42.23	3.00	244.95	–	67.38	m	**312.33**
690 × 690 mm with hollow core filled with concrete	15.30	1.00	81.65	–	24.90	m²	**106.55**
Extra over for RSJ	60.18	0.50	31.00	–	60.18	m	**91.18**
Retaining walls; Anderton Concrete Ltd							
Stepoc Blocks; interlocking blocks; 10 mm reinforcing laid loose horizontally to preformed notches and vertical reinforcing nominal size 10 mm fixed to starter bars; infilling with concrete; footings and starter bars measured separately							
type 325; 400 × 225 × 325 mm	73.85	0.47	49.91	–	124.86	m²	**174.78**
type 256; 400 × 225 × 256 mm	72.84	0.40	42.78	–	118.56	m²	**161.34**
type 200; 400 × 225 × 200 mm	63.49	0.33	35.65	–	102.68	m²	**138.32**
NATURAL STONE WALLING							
Expert stonework team							
Generally a two man team is used in this section; The column Labour hours reports team hours. The column Labour £ reports the total cost of the team for the unit of work shown							
2 man team	–	1.00	71.30	–	–	hr	**71.30**

14 MASONRY

Item	PC £	Labour hours	Labour £	Plant £	Material £	Unit	Total rate £
NATURAL STONE WALLING – cont							
Dry stone walling – General							
Preamble: In rural areas where natural stone is a traditional material, it may be possible to use dry stone walling or dyking as an alternative to fences or brick walls. Many local authorities are willing to meet the extra cost of stone walling in areas of high landscape value, and they may hold lists of available craftsmen. DSWA Office, Westmorland County Showground, Lane Farm, Crooklands, Milnthorpe, Cumbria, LA7 7NH; Tel: 01539 567953; E-mail: information@dswa.org.uk							
Note: Traditional walls are not built on concrete foundations.							
Dry stone wall; wall on concrete foundation (not included); dry stone coursed wall inclusive of locking stones and filling to wall with broken stone or rubble; walls up to 1.20 m high; battered; 2 sides fair faced							
Yorkstone	171.00	4.00	285.20	–	196.65	m^2	**481.85**
Cotswold stone	47.40	4.00	285.20	–	54.51	m^2	**339.71**
Purbeck	132.00	4.00	285.20	–	151.80	m^2	**437.00**
Drystone style walling; mortared facing stone to blockwork wall (not included); stainless steel ties between blockwork and stone; stone wall mortar raked subsequently raked out to provide drystone impression.							
Yorkstone	171.00	3.25	231.73	–	201.47	m^2	**433.19**
Cotswold stone	47.40	3.25	231.73	–	59.33	m^2	**291.05**
Purbeck	132.00	3.25	231.73	–	156.62	m^2	**388.34**
Portland stone walling							
Portland stone walling to external wall; Portland stone 300 × 100 mm × 250 mm thick; laid to white cement mortar; jointed with white cement mortar							
fair faced 1 side	400.71	1.00	106.95	–	473.10	m^2	**580.05**
fair faced 2 sides	400.71	1.33	142.24	–	473.10	m^2	**615.34**

14 MASONRY

Item	PC £	Labour hours	Labour £	Plant £	Material £	Unit	Total rate £
Portland stone walling to external wall; 440 × 300 mm × 225 mm thick; laid to white cement mortar; jointed with white cement mortar							
fair faced 1 side	924.14	1.25	133.69	–	1075.04	m^2	**1208.73**
fair faced 2 sides	924.14	1.54	164.54	–	1075.04	m^2	**1239.58**
Coping Portland stone							
Coping 380 mm wide × 50 mm thick; Bull nosed single side; double drip	199.81	0.50	35.65	–	241.16	m	**276.81**
RECONSTITUTED STONE WALLING							
The labour in this section is calculated on a 2 person team. The labour time below should be multiplied by 2 to calculate the cost as shown.							
Haddonstone Ltd; cast stone piers; ornamental masonry in Portland Bath or Terracotta finished cast stone							
Gate Pier S120; to foundations and underground work measured separately; concrete infill							
S120G base unit to pier; 699 × 699 × 172 mm	192.00	0.25	26.74	–	233.02	nr	**259.76**
S120F/F shaft base unit; 533 × 533 × 280 mm	156.00	0.33	35.61	–	191.96	nr	**227.58**
S120E/E main shaft unit; 533 × 533 × 280 mm; nr of units required dependent on height of pier	156.00	0.33	35.61	–	191.96	nr	**227.58**
S120D/D top shaft unit; 33 × 533 × 280 mm	156.00	0.33	35.61	–	191.96	nr	**227.58**
S120C pier cap unit; 737 × 737 × 114 mm	182.40	0.17	17.75	–	222.32	nr	**240.07**
S120B pier block unit; base for finial; 533 × 533 × 64 mm	80.40	0.11	11.76	–	92.80	nr	**104.56**
Pier blocks; flat to receive gate finial							
S100B; 440 × 440 × 63 mm	51.60	0.11	11.76	–	60.13	nr	**71.89**
S120B; 546 × 546 × 64 mm	80.40	0.11	11.76	–	93.25	nr	**105.01**
S150B; 330 × 330 × 51 mm	28.80	0.11	11.76	–	33.91	nr	**45.67**
Pier caps; part weathered							
S100C; 915 × 915 × 150 mm	432.00	0.25	26.74	–	498.38	nr	**525.11**
S120C; 737 × 737 × 114 mm	182.40	0.17	17.75	–	210.55	nr	**228.30**
S150C; 584 × 584 × 120 mm	135.60	0.17	17.75	–	156.73	nr	**174.48**

14 MASONRY

Item	PC £	Labour hours	Labour £	Plant £	Material £	Unit	**Total rate £**
RECONSTITUTED STONE WALLING – cont							
Haddonstone Ltd – cont							
Pier caps; weathered							
S230C; 1029 × 1029 × 175 mm	564.00	0.17	17.75	–	649.39	nr	**667.14**
S215C; 687 × 687 × 175 mm	249.60	0.17	17.75	–	287.83	nr	**305.58**
S210C; 584 × 584 × 175 mm	159.60	0.17	17.75	–	184.33	nr	**202.08**
Pier strings							
S100S; 800 × 800 × 55 mm	147.60	1.66	177.54	–	170.22	nr	**347.76**
S120S; 555 × 555 × 44 mm	68.40	0.17	17.75	–	79.14	nr	**96.89**
S150S; 457 × 457 × 48 mm	43.20	0.17	17.75	–	50.16	nr	**67.91**
Balls and bases							
E150 A Ball 535 mm and E150C collared base	444.00	0.17	17.75	–	511.08	nr	**528.83**
E120 A Ball 330 mm and E120C collared base	158.00	0.17	17.75	–	182.18	nr	**199.94**
E110 A Ball 230 mm and E110C collared base	105.00	0.17	17.75	–	121.23	nr	**138.98**
E100 A Ball 170 mm and E100B plain base	70.00	0.17	17.75	–	80.97	nr	**98.73**
Haddonstone Ltd; cast stone copings; ornamental masonry in Portland Bath or Terracotta finished cast stone							
Copings for walls; bedded, jointed and pointed in approved coloured cement-lime mortar 1:2:9							
T100 weathered coping; 102 mm high × 178 mm wide × 914 mm	47.09	0.11	11.87	–	54.83	m	**66.70**
T140 weathered coping; 102 mm high × 337 mm wide × 914 mm	77.17	0.11	11.87	–	89.43	m	**101.30**
T200 weathered coping; 127 mm high × 508 mm wide × 750 mm	116.41	0.11	11.87	–	134.55	m	**146.43**
T170 weathered coping; 108 mm high × 483 mm wide × 914 mm	124.26	0.11	11.87	–	143.58	m	**155.45**
T340 raked coping; 75–100 mm high × 290 mm wide × 900 mm	70.63	0.11	11.87	–	81.91	m	**93.78**
T310 raked coping; 76–89 mm high × 381 mm wide × 914 mm	82.40	0.11	11.87	–	95.45	m	**107.32**

14 MASONRY

Item	PC £	Labour hours	Labour £	Plant £	Material £	Unit	**Total rate £**
Slope-Loc – Anderton Concrete Ltd							
Dry stacked random sized units 150–400 mm long × 150 mm high cast stone wall mechanically interlocked with fibreglass pins; constructed to levelling pad of coarse compacted granular material (not included) back filled behind the elevation with 300 mm wide granular drainage material; walls to 5.00 m high (retaining walls over heights shown below require individual design)							
Slope-loc 300 × 245 × 125 mm; gravity wall up to 850 mm;	83.57	1.00	71.30	54.85	101.33	m²	**227.48**
Slope-loc 300 × 245 × 125 mm; reinforced wall; inclusive of design and reinforcement grids; retaining up to 1.70 m	136.49	1.00	71.30	54.85	162.18	m²	**288.33**
The labour in this section is calculated on a 3 person team. The labour time below should be multiplied by 3 to calculate the cost as shown.							
Retaining wall; but reinforced with Tensar Geogrid RE520 laid between every two courses horizontally into the face of the excavation (excavation not included)							
Near vertical; 250 mm thick; 1.50 m of geogrid length							
up to 1.20 m high; 2 layers of geogrid	83.57	0.50	53.48	–	107.87	m²	**161.35**
1.20–1.50 m high; 3 layers of geogrid	83.57	0.67	71.23	–	113.75	m²	**184.98**
1.50–1.80 m high; 4 layers of geogrid	83.57	0.75	80.21	–	119.63	m²	**199.85**
Battered; maximum 1:3 slope							
up to 1.20 m high; 3 layers of geogrid	83.57	0.83	89.09	–	113.75	m²	**202.84**
1.20–1.50 m high; 4 layers of geogrid	83.57	0.83	89.09		119.63	m²	**208.72**
1.50–1.80 m high; 5 layers of geogrid	83.57	1.00	106.95	–	124.53	m²	**231.48**

14 MASONRY

Item	PC £	Labour hours	Labour £	Plant £	Material £	Unit	Total rate £
COPINGS AND PIER CAPS							
Copings extra over to all copings for access over 1.50 m							
Bandstand	–		–	66.01	–	week	**66.01**
The labour in this section is calculated on a 2 person team. The labour time below should be multiplied by 2 to calculate the cost as shown.							
Portland stone copings; Albion stone; laid to 10 mm mortar bed; stainless steel dowels at 4 nr/m drilled and set to wall and coping							
Coping Portland stone							
Coping 380 mm wide × 50 mm thick; Bull nosed single side; double drip	199.81	0.50	35.65	–	244.96	m	**280.61**
Granite copings; CED Ltd; on 10 mm mortar bed with 4 nr. 12 mm stainless steel dowels/lm drilled to wall and coping							
Coping Granite; Bush hammered top and exposed sides with 30 mm single bullnose to long edge and exposed sides; straight copings							
110 × 50 mm	9.23	0.08	5.94	–	15.18	m	**21.12**
350 × 50 mm	25.85	0.17	11.89	–	34.80	m	**46.68**
500 × 50 mm	37.00	0.18	12.96	–	49.52	m	**62.48**
650 × 50 mm	44.33	0.20	14.26	–	57.95	m	**72.21**
400 × 100 mm	59.10	0.22	15.84	–	72.78	m	**88.62**
500 × 100 mm	73.87	0.25	17.83	–	90.02	m	**107.84**
600 × 100 mm	88.35	0.29	20.37	–	106.67	m	**127.04**
650 × 100 mm	96.03	0.36	25.92	–	115.76	m	**141.68**
750 × 100 mm	110.81	0.40	28.52	–	133.01	m	**161.53**
Extra over granite copings for bullnosing to additional side							
50 mm coping	23.00		–	–	23.00	m	**23.00**
100 mm coping	27.00		–	–	27.00	m	**27.00**
Coping Granite; Curved to external radius 3 m–12 m; Bush hammered top and exposed sides with 30 mm single bullnose							

14 MASONRY

Item	PC £	Labour hours	Labour £	Plant £	Material £	Unit	**Total rate £**
Copings to radius; Add to the coping prices Prime cost prices shown for the following for the following							
Radius 1.00–3.00 m	40.00		–	–	–	%	–
radius 3.00 m–12.00 m	20.00		–	–	–	%	–
Copings; London stone; on 10 mm mortar bed with 4 nr. 12 mm stainless steel dowels/lm drilled to wall and coping; 300 wide × 40 mm thick							
Sandstone; Rock faced and with double pencil edge and drip							
Autumn brown/Kandla grey/Mint/ Heath msandstones	26.78	0.13	8.91	–	35.35	m	**44.27**
Sandstone; Sawn and flamed with double pencil edge and drip							
Beige/Heath sandstone	40.60	0.13	8.91	–	51.26	m	**60.17**
Sandstone and Black basalt; Sawn with double pencil edge and drip							
Black basalt/Buff sandstone/ Harvest sandstone	–	0.13	8.91	–	71.77	m	**80.68**
Copings; Haddonstone; Cast stone in Portland, Bath or Terracotta; on 10 mm mortar bed with 4 nr. 12 mm stainless steel dowels/lm drilled to wall and coping; 300 wide × 40 mm thick							
Plain ended coping stones for walls up to 248 mm thick							
T130; flat topped 337 mm wide at top to 248 mm at base overall 89 mm thick	86.42	0.10	7.13	–	99.39	m	**106.52**
T140; twice weathered 337 mm wide at top to 248 mm wide at base; overall 102 mm thick	89.29	0.10	7.13	–	102.68	m	**109.81**
T150; twice weathered 368 mm wide at top to 279 mm at base; 102 mm thick overall	92.20	0.10	7.13	–	106.03	m	**113.16**
Plain ended coping stones for walls up to 394 mm thick							
T440; twice weathered 394 mm wide at top to 356 mm at base overall 76 mm thick	93.59	0.14	10.19	–	107.63	m	**117.82**
T470; twice weathered 410 mm wide at top to 250 mm wide at base; overall 114 mm thick	127.00	0.17	11.89	–	151.37	m	**163.26**

14 MASONRY

Item	PC £	Labour hours	Labour £	Plant £	Material £	Unit	Total rate £
COPINGS AND PIER CAPS – cont							
The labour in this section is calculated on a 3 person team. The labour time below should be multiplied by 3 to calculate the cost as shown.							
Precast concrete coping							
Copings; twice weathered; twice grooved							
600 × 300 mm, 50–30 mm thick	17.62	0.40	14.26	–	20.26	m	**34.52**
600 × 300 mm bevelled edge	22.09	0.40	14.26	–	26.03	m	**40.29**
600 × 600 mm, 75–50 mm thick	22.09	0.50	17.83	–	26.40	m	**44.22**
Pier caps; four sides weathered							
295 × 295 mm	13.00	1.00	35.65	–	15.24	nr	**50.89**
380 × 380 mm	25.99	1.00	35.65	–	30.19	nr	**65.84**
520 × 520 mm	38.99	1.20	42.78	–	45.13	nr	**87.91**

19 WATERPROOFING

Item	PC £	Labour hours	Labour £	Plant £	Material £	Unit	**Total rate £**
COLD APPLIED BITUMINOUS EMULSION							
Tanking and damp-proofing; 'Blackjack DPM'; cold applied bituminous emulsion waterproof coating							
to smooth finished concrete or screeded slabs; flat; blinding with sand							
two coats	2.86	0.22	7.92	–	3.49	m^2	**11.42**
three coats	4.29	0.31	11.05	–	5.14	m^2	**16.19**
Blackjack DPM; to fair faced brickwork with flush joints, rendered brickwork or smooth finished concrete walls; vertical							
two coats	3.14	0.29	10.19	–	3.82	m^2	**14.01**
three coats	5.00	0.40	14.26	–	5.96	m^2	**20.22**
SELF ADHESIVE							
Tanking and damp-proofing; Grace Construction Products							
Bitu-thene 3000; 1.50 mm thick; overlapping and bonding; including sealing all edges							
to concrete slabs; flat	10.84	0.25	8.91	–	13.45	m^2	**22.36**
to brick/concrete walls; vertical	10.84	0.40	14.26	–	13.45	m^2	**27.71**

21 CLADDING AND COVERING

Item	PC £	Labour hours	Labour £	Plant £	Material £	Unit	Total rate £
NATURAL STONE SLAB CLADDING/FEATURES							
The labour in this section is calculated on a 2 person team. The labour time below should be multiplied by 2 to calculate the cost as shown.							
Sawn Yorkstone cladding; Johnsons Wellfield Quarries Ltd; to maximum height of 1.60 m inclusive of fixings and resin							
Six sides sawn stone; rubbed face; sawn and jointed edges; fixed to blockwork (not included) with stainless steel fixings; Ancon Ltd; grade 304 stainless steel frame cramp and dowel 7 mm; cladding units drilled 4 × to receive dowels							
440 × 200 × 50 mm thick	126.44	3.33	237.66	–	193.56	m^2	**431.23**
Stone cladding; to maximum height of 1.60 m inclusive of fixings and resin							
Portland stone cladding 40 mm thick fixed with 6 mm stainless steel dowell supports to blockwork; fixings at 200 mm centres ; load support of 30 mm angle every 4 courses;							
Single size 400 × 400 mm	160.28	3.13	222.81	–	230.46	m^2	**453.28**
Coursed – 3 sizes 400 mm × 400 mm 400 mm × 200 mm	160.28	4.00	285.20	–	248.47	m^2	**533.67**
Coping Portland stone							
Coping 380 mm wide × 50 mm thick; Bull nosed single side; double drip	129.96	0.50	35.65	–	160.83	m	**196.48**
Granite Cladding							
Silver grey 30 mm thick; single size	61.80	2.80	173.62	–	103.68	m^2	**277.29**

21 CLADDING AND COVERING

Item	PC £	Labour hours	Labour £	Plant £	Material £	Unit	**Total rate £**
Tier natural stone cladding; CED Ltd; Tier cladding system; real stone panelling system, natural stone finish; interlocking tiers, fitting seamlessly together. For both interior and exterior walls; multiple finishes							
To exterior blockwork brickwork or concrete walls; inclusive of mechanical fixings and adhesive; Girth exceeding 400 mm							
multiple colours and stone types	68.94	3.00	213.90	–	106.49	m^2	**320.39**
To exterior blockwork brickwork or concrete walls; inclusive of mechanical fixings and adhesive; Girth not exceeding 400 mm							
multiple colours and stone types	72.23	4.00	285.20	–	111.66	m^2	**396.86**
CAST STONE SLAB CLADDING/ FEATURES							
Cast stone cladding; Haddonstone Ltd							
Reconstituted stone in Portland Bath or Terracotta; fixed to blockwork or concrete (not included) with stainless steel fixings; stainless steel frame cramp and dowel M6 mm; cladding units drilled 4 × to receive dowels							
1000 × 300 × 50 mm thick	60.00	1.87	66.67	–	117.16	m^2	**183.82**

25 BALUSTRADES

Item	PC £	Labour hours	Labour £	Plant £	Material £	Unit	Total rate £
GLASS BALUSTRADES							
The labour in this section is calculated on a 2 person team. The labour time below should be multiplied by 2 to calculate the cost as shown.							
Glass balustrades; Elite Balustrades Ltd; Glass 17 mm laminated toughened with radius corners.							
Installed to stainless steel uprights with bolt fixings to concrete base; Cranage or porting of units not included							
1.10 m high	454.86	0.33	23.76	–	523.09	m	**546.85**
1.25 m high	410.71	0.33	23.76	–	472.31	m	**496.08**
1.81 m high	517.10	0.50	35.65	–	594.67	m	**630.32**
1.95 m high	549.91	0.50	35.65	–	632.40	m	**668.05**
2.15 m high	694.69	2.00	142.60	–	798.89	m	**941.49**
Gates to glass balustrades; toughened 17 mm; with secure locking mechanisms							
1.81 m high	1026.68	1.00	71.30	–	1180.69	m	**1251.99**
1.90 m high	1061.34	1.20	85.56	–	1220.54	m	**1306.10**
2.05 m high	1070.00	1.33	94.83	–	1230.50	m	**1325.33**
RECONSTITUTED STONE BALUSTRADES							
The labour in this section is calculated on a 2 person team. The labour time below should be multiplied by 2 to calculate the cost as shown.							
Bottle balustrades reconstituted stone; Haddonstone Ltd; to prepared brickwork walls or bases (not included).							
Top and bottom rails to support balustrades; laid to builders work of brick block or other stonework supports (not included)							
Plinths T110 A 900 × 30 × 86 mm	56.61		–	–	65.10	m	**65.10**
Plinth T900 A 900 × 230 × 102 mm	55.94		–	–	64.34	m	**64.34**

25 BALUSTRADES

Item	PC £	Labour hours	Labour £	Plant £	Material £	Unit	**Total rate £**
Under coping							
Under coping 900 × 377 × 76 mm	82.58		–	–	94.97	m	**94.97**
Pier bases; 337 × 102 mm high to receive shafts							
337 × 337 × 102 mm high	39.60	0.25	17.83	–	45.54	nr	**63.37**
Pier under coping 377 × 377x 76 mm thick	–	0.50	31.00	–	–	nr	**31.00**
Pier shafts; at specified centres along the balustrade fixed to upstand (not included); subsequently filled with concrete							
Piers shaft 279 × 279 × 610 mm high	84.00	1.00	71.30	–	115.86	nr	**187.16**
Pier shafts 279 × 279 × 710 mm high	100.80	1.25	89.13	–	137.02	nr	**226.14**
Pier caps to Pier shafts;							
337 × 337 × 86 mm	46.80	0.25	17.83	–	57.47	nr	**75.29**
Bottle balustrades comprising reconstituted bottle balusters ; fixed to plinths, top rail, bottom rail and under coping (all shown separately) with steel rods and mortar, ; balusters							
K457G 150 mm wide base x127 mm top × 457 high; at 300 mm centres	155.84	1.00	71.30	–	182.76	m	**254.06**
K610G 140 mm wide base x127 mm top × 610 high; at 300 mm centres	195.80	1.00	71.30	–	229.10	m	**300.40**
K729G 140 mm wide base x127 mm top × 610 high; at 300 mm centres	231.77	1.00	71.30	–	270.46	m	**341.76**

28 FLOOR, WALL AND CEILING FINISHINGS

Item	PC £	Labour hours	Labour £	Plant £	Material £	Unit	Total rate £
PLASTERED/RENDERED/ ROUGHCAST COATINGS							
The labour in this section is calculated on a 2 person team. The labour time below should be multiplied by 2 to calculate the cost as shown.							
Specialist renders; K-Rend – London Stone; 2 coats to blockwork or in situ concrete or brickwork walls							
Silicone FT; scraped texture render, incorporating silicone technology low-maintenance, water-repellent finish; six colours; single or double coat 8–12 mm thick each							
Primer	1.12	0.10	6.87	–	1.25	m^2	**52.95**
Single coat 2.2 kg/m^2	1.12	0.10	6.87	–	1.25	m^2	**52.95**
2 coats; 4.4 kg/m^2	11.24	0.67	47.54	–	12.92	m^2	**60.46**
Beads and stops for K-Rend							
K Bead Stop Bead White 2.5 m × 4 mm	1.04	0.07	4.76	–	1.19	m	**5.95**
K Bead Stop Bead Ivory 2.5 m × 4 mm	1.39	0.07	4.76	–	1.60	m	**6.36**
K Bead Feather Edge Angle Ivory 3 m × 4 mm	2.02	0.07	4.76	–	2.33	m	**7.08**
Cement: lime: sand (1:1:6); 19 mm thick; two coats; wood floated finish							
Walls							
width exceeding 300 mm; to brickwork or blockwork base	–		–	–	–	m^2	**16.98**
Extra over cement: sand: lime (1:1:6) coatings for decorative texture finish with water repellent cement							
combed or floated finish	–		–	–	–	m^2	**4.53**

28 ROOF FINISHINGS – GREEN ROOFS

Item	PC £	Labour hours	Labour £	Plant £	Material £	Unit	**Total rate £**
PREAMBLE							
Preamble A variety of systems are available which address all the varied requirements for a successful Green Roof. For installation by approved contractors only. The prices shown are for budgeting purposes only, as each installation is site specific and may incorporate some or all of the resources shown. Specifiers should verify that the systems specified include for design liability and inspections by the suppliers. The systems below assume commercial insulation levels are required to the space below the proposed Green Roof. Extensive Green Roofs are those of generally lightweight construction with low maintenance planting and shallow soil designed for aesthetics only; Intensive Green Roofs are designed to allow use for recreation and trafficking. They require more maintenance and allow a greater variety of surfaces and plant types.							
GREEN ROOFS The following systems which are compliant with European Technical Approval ETA 13/0668 (Kits for Green roofs) are indicated as such. The rates below assume that all elements are in position in the works location; no allowances for access, lifting or materials movement to the roof areas.							

28 ROOF FINISHINGS – GREEN ROOFS

Item	PC £	Labour hours	Labour £	Plant £	Material £	Unit	Total rate £
ROOT BARRIERS							
The labour in this section is calculated on a 3 person team. The labour time below should be multiplied by 3 to calculate the cost as shown.							
Root barriers; Zinco Green Roof Systems Ltd							
Where waterproofing does not contain a root barrier							
WSB 100-PO Weldable reinforced flexible polyolefin. 1.1 mm thick. Root resistant tested according to FLL	17.25	0.03	2.67	–	19.84	m²	**22.51**
WSF 40 High pressure polyethylene 0.34 mm thick	2.96	0.02	2.14	–	3.40	m²	**5.54**
VOID FORMERS							
The labour in this section is calculated on a 2 person team. The labour time below should be multiplied by 2 to calculate the cost as shown.							
Void former; Filcor; lightweight expanded polystyrene; to roof or planter voids to make up levels. Sheets 2400 × 1200 mm; exclusive of carriage or cranage to point of laying; thickness							
Filcor 20							
165 mm	7.40	0.02	1.25	–	8.50	m²	**9.75**
200 mm	8.96	0.02	1.43	–	10.31	m²	**11.73**
245 mm	10.98	0.02	1.48	–	12.63	m²	**14.11**
300 mm	12.22	0.03	1.96	–	14.05	m²	**16.02**
360 mm	16.13	0.03	1.93	–	18.55	m²	**20.47**
450 mm	26.02	0.03	2.04	–	29.92	m²	**31.96**
Filcor 45							
165 mm	9.86	0.02	1.25	–	11.34	m²	**12.59**
200 mm	11.95	0.02	1.43	–	13.74	m²	**15.17**
245 mm	14.64	0.02	1.48	–	16.83	m²	**18.32**
300 mm	17.92	0.03	1.78	–	20.61	m²	**22.40**
360 mm	21.51	0.03	1.93	–	24.74	m²	**26.66**
450 mm	26.89	0.03	2.04	–	30.92	m²	**32.96**

28 ROOF FINISHINGS – GREEN ROOFS

Item	PC £	Labour hours	Labour £	Plant £	Material £	Unit	Total rate £
EXTENSIVE GREEN ROOFS							
The labour in this section is calculated on a 3 person team. The labour time below should be multiplied by 3 to calculate the cost as shown.							
Extensive Sedum type Green Roof System; Zinco Green Roof Systems Ltd. System for planting with Sedum plugs or overlaying with pre grown cultivated Sedum Carpet. (ETA 13/0668) Flat roofs with no standing water up to 10° or for Inverted roofs and as an alternative to single layer systems. Drainage, water retention and protection mat with attached filter sheet. 20 mm thick.							
Zinco Drainage Element Fixodrain XD20; 3L/m² reservoir	6.53	0.04	4.28	–	7.51	m²	**11.79**
Zinco System Substrate Growing medium 'Sedum Carpet' – laid to Fixodrain XD20 at min. 70 mm thickness.	140.80	0.67	71.30	–	161.92	m³	**233.22**
Zinco Premium Sedum Carpet Mat – Laid over System substrate	26.88	0.07	7.13	–	30.91	m²	**38.05**
Extensive Rockery type Green Roof System; Zinco Green Roof Systems Ltd. For planting with Sedum plugs or rockery type plants; flat roofs with no standing water up to 10°. (Not suitable for inverted roofs); Suitable system depth 90 mm. System saturated weight 110 kg/m²; system water storage capacity: 36 L/M2; (ETA 13/0668) Non rotting synthetic fibre mat for protection of the waterproofing layer and provision of additional water reservoir of 5 L/m²							
Zinco Moisture mat/protection fleece SSM45	2.75	0.04	4.28	–	3.16	m²	**7.44**

28 ROOF FINISHINGS – GREEN ROOFS

Item	PC £	Labour hours	Labour £	Plant £	Material £	Unit	Total rate £
EXTENSIVE GREEN ROOFS – cont							
Extensive Rockery type Green Roof System – cont							
Pressure resistant drainage and water storage element; recycled polyolefin; laid to SSM45; 25 mm thick with diffusion openings; water storage capacity 3L/m^2.							
Zinco Drainage Element Fixodrain FD25-E	8.90	0.04	4.28	–	10.23	m^2	**14.51**
Zinco Filter Sheet SF – Rolled to drainage layer FD25-E	1.09	0.04	4.28	–	1.25	m^2	**5.53**
Zinco System Substrate Growing medium ‘Rockery Type Plants’ laid to Filter Fleece SF min. 70 mm thick	140.80	0.67	71.30	–	161.92	m^3	**233.22**
Planting to substrates							
Zinco Premium Sedum Carpet Mat	29.57	0.04	4.28	–	34.00	m^2	**38.28**
Zinco Seed Mix – Meadow Scents; 15 g/m^2	0.84	0.01	1.07	–	0.97	m^2	**2.04**
Extensive Rockery type Green Roof System for ZERO degree roofs; Zinco Green Roof Systems Ltd; for Zero degree roofs where isolated standing water may be present; planting with Sedum plugs or rockery type plants; system depth 150 mm; saturated weight: 105 kg/m^2; water storage capacity; 33 L/m^2							
Non rotting synthetic fibre mat for protection of the waterproofing layer and provision of additional water reservoir of 3 L/m^2							
Zinco Moisture mat/protection fleece TSM32	2.25	0.04	4.28	–	2.59	m^2	**6.87**

28 ROOF FINISHINGS – GREEN ROOFS

Item	PC £	Labour hours	Labour £	Plant £	Material £	Unit	**Total rate £**
Drainage Element; laid to TSM32; ensures that a gap is maintained between any standing water and the soil substrate and the necessary aeration of the root zone; expanded polystyrene; 50 mm thick; lightweight (0.6 kg/m^2); water storage capacity of 3L/m^2							
Zinco Drainage Element; Floraset FS50	10.14	0.04	4.28	–	11.66	m^2	**15.94**
Zinco Filter Sheet SF – Rolled to drainage layer FS50	1.09	0.01	0.71	–	1.25	m^2	**1.96**
Zinco System Substrate Growing medium 'Rockery Type Plants' – laid to filter fleece SF; min 70 mm thick	140.80	0.67	71.30	–	161.92	m^3	**233.22**
Planting to substrates							
Zinco Premium Sedum Carpet Mat	29.57	0.04	4.28	–	34.00	m^2	**38.28**
Zinco Seed Mix – Meadow Scents; 15 g/m^2	0.84	0.01	1.07	–	0.97	m^2	**2.04**
Specialist substrates and growing mediums; Boughton Loam Limited. Lightweight or horticulturally efficient growing mediums for green roof podiums (to G.R.O. guidelines), and horticultural environments. Prices for mechanical placement by lightweight machinery maximum 25 m from location of soils; Cranage priced separately. Delivery payload size shown							
Intensive or Extensive green roof substrate; IN1 IN2 or EX1 or EX2; sand and organic matter; depth 100–500 mm thick determined by plant variety and green roof loading capability; placed to planting areas.							
Full loose loads; 29 tonne articulated	92.32	0.13	3.88	11.43	92.32	m^3	**107.63**
Full loose loads; 20 tonne rigid	95.46	0.13	3.88	11.43	95.46	m^3	**110.77**
Bulk bags; 29 tonne articulated	163.72	0.25	7.75	22.86	163.72	m^3	**194.33**
Boughton Podium 1 substrate; landscaping projects where weight loading is not an issue; high sand content.							
Full loose loads; 29 tonne articulated	61.12	0.13	3.88	11.43	61.12	m^3	**76.42**
Full loose loads; 20 tonne rigid	88.06	0.13	3.88	11.43	88.06	m^3	**103.36**
Bulk bags; 29 tonne articulated	164.94	0.25	7.75	22.86	164.94	m^3	**195.55**

28 ROOF FINISHINGS – GREEN ROOFS

Item	PC £	Labour hours	Labour £	Plant £	Material £	Unit	Total rate £
EXTENSIVE GREEN ROOFS – cont							
Specialist substrates and growing mediums – cont							
Boughton Podium 1 substrate – cont							
Bulk bags; 20 tonne rigid	176.25	0.25	7.75	22.86	176.25	m^3	**206.86**
Extra over for cranage using main contractors tower crane; loose loaded material mechanically loaded to hopper; hopper loaded at soil stockpile location							
loose loaded material mechanically loaded to hopper.	–	0.50	31.00	11.92	–	m^3	**42.92**
single lift bulk bagged material; banksmen and slingers by main contractor (not included)	–	0.67	61.99	4.06	–	m^3	**66.05**
EXTENSIVE GREEN ROOF PITCH SYSTEMS							
The labour in this section is calculated on a 3 person team. The labour time below should be multiplied by 3 to calculate the cost as shown.							
Extensive Green Roof System for Pitched roofs; for pitched roofs with slopes from 10° up to 25°; Zinco Green Roof Systems Ltd; suitable for planting with Sedum plugs or overlaying with pre-grown cultivated Sedum Carpet. System depth – 130 mm; system saturated weight – 115 kg/m^2; system water storage capacity – 38 L/m^2; Waterproofing layer must contain a root barrier as a separate root barrier cannot be used on pitched roofs							
Non rotting synthetic fibre mat for protection of the waterproofing layer and provision of additional water reservoir of 7 L/m^2; laid to water proofing layer							
Protection Mat BSM64	3.26	0.10	10.70	–	3.75	m^2	**14.45**

28 ROOF FINISHINGS – GREEN ROOFS

Item	PC £	Labour hours	Labour £	Plant £	Material £	Unit	**Total rate £**
Lightweight (1 kg/m^2) drainage element of expanded polystyrene. 75 mm thick laid with large studs upwards; laid to BSM64							
Zinco Drainage Element Floraset FS75	13.44	0.10	10.70	–	15.46	m^2	**26.15**
Zinco System Substrate Growing medium infilled to drainage board and to a level of 50 mm above the large studs.							
'Rockery Type Plants' – laid to FS75	12.32	0.10	10.70	–	14.17	m^2	**24.86**
Erosion net; laid over substrate to aid prevention of wind erosion.							
Zinco Jute Anti Erosion Net JEG	1.15	0.07	7.13	–	1.32	m^2	**8.46**
Vegetation layer; laid over System substrate							
Zinco Premium Sedum Carpet Mat	26.88	0.07	7.13	–	30.91	m^2	**38.05**
Extensive Green Roof System for Steep Pitched roofs with slopes from 25° up to 35°; Zinco Green Roof Systems Ltd. Pitched roofs with slopes from 25° up to 35°; Suitable for overlay with pre grown cultivated Sedum Carpet. System height – 120 mm; system saturated weight – 155 kg/m^2; water storage capacity – 64 L/m^2 (depending on slope); Waterproofing layer must contain a root barrier as a separate root barrier cannot be used on pitched roofs							
Protection mat; non rotting needle fleeced synthetic mat for protection of the waterproofing layer and provision of additional water reservoir of 12 L/m^2							
WSM150 – laid to waterproofing layer	7.62	0.10	10.70	–	8.76	m^2	**19.45**
Drainage element; Stable Grid element 100 mm high made from recycled polyethylene. Grid dimension 625 mm with integrated connecting plug							
Zinco Drainage Element Georaster – Laid to WSM150	18.75	0.13	14.26	–	21.56	m^2	**35.82**

28 ROOF FINISHINGS – GREEN ROOFS

Item	PC £	Labour hours	Labour £	Plant £	Material £	Unit	Total rate £
EXTENSIVE GREEN ROOF PITCH SYSTEMS – cont							
Extensive Green Roof System for Steep Pitched roofs with slopes from 25° up to 35° – cont							
Zinco System Substrate Growing medium 'Heather with Lavender' filled to Georaster.							
Zinco Premium Sedum Carpet Mat – Laid over Georaster	26.88	0.20	21.39	–	30.91	m^2	**52.30**
Shear barriers and brackets for pitched roofs; Zinco Green Roof Systems Ltd. Elements in pitched roof constructions transfer shear forces into the roof construction. Structural calculations required to establish the exact location and fixing of the shear barriers and brackets.							
Zinco Eaves Profile and support brackets							
Eaves Profile TRP140 E; Stainless steel angle 140 mm high × 3.00 m long	19.80	0.08	8.91	–	22.77	m	**31.68**
Support bracket TSH100	30.72	0.50	53.48	–	35.33	nr	**88.80**
Shear fix bracket LF300	108.80	0.50	53.48	–	125.12	nr	**178.60**
INTENSIVE GREEN ROOFS							
The labour in this section is calculated on a 3 person team. The labour time below should be multiplied by 3 to calculate the cost as shown.							
Flat roofs up to 10°; Intensive Simple and Wildflower Green Roof System; Zinco Green Roof Systems Ltd; (not suitable for Inverted roofs); for planting with small shrubs, blooming perennials and wildflowers; can be combined with patios and walkways if required. System depth – 145 mm–200 mm; saturated weight – 145 kg/m^2–245 Kg/m^2; water storage capacity – 62 L/m^2 to 80 L/m^2. (ETA 13/0668)							

28 ROOF FINISHINGS – GREEN ROOFS

Item	PC £	Labour hours	Labour £	Plant £	Material £	Unit	Total rate £
Non rotting synthetic fibre mat for protection of the waterproofing layer and provision of additional water reservoir of 5 L/m²							
Moisture mat/protection fleece SSM45.	2.75	0.04	4.28	–	3.16	m²	**7.44**
Pressure resistant drainage and water storage element made from recycled polyolefin laid to SSM45; 40 mm thick with diffusion openings and water storage capacity of 6 L/m².							
Fixodrain FD40-E	9.79	0.04	4.28	–	11.25	m²	**15.53**
Zinco Filter Sheet SF – Rolled to drainage layer FD40-E	1.20	0.04	4.28	–	1.38	m²	**5.65**
System Substrate; growing medium 'Heather with Lavender'; laid to Filter Fleece SF at 100 mm thickness up to 200 mm	176.00	0.20	21.39	–	202.40	m³	**223.79**
Flat roofs up to 10°; Intensive Simple and Wildflower Green Roof System; Zinco Green Roof Systems Ltd; (suitable for Inverted roofs); for planting with small shrubs, blooming perennials and wildflowers; can be combined with patios and walkways if required. System depth – 145 mm–200 mm; saturated weight – 145 kg/m²–245 Kg/m²; water storage capacity – 62 L/m² to 80 L/m².							
Water repellent, air permeable membrane used as a diffusionable separation layer; laid above waterproofing and separation membrane; rolled over the insulation.							
Separation Layer; TGV21	1.20	0.04	4.28	–	1.38	m²	**5.65**
Pressure resistant drainage and water storage element made from recycled polyolefin laid to SSM45; 40 mm thick with diffusion openings and water storage capacity of 6 L/m².							
Fixodrain FD40-E; laid to TGV21	9.79	0.04	4.28	–	11.25	m²	**15.53**
Zinco Filter Sheet SF – Rolled to drainage layer FD40-E	1.20	0.04	4.28	–	1.38	m²	**5.65**
System Substrate; growing medium 'Heather with Lavender'; laid to Filter Fleece SF at 100 mm thickness up to 200 mm	176.00	0.67	71.30	–	202.40	m³	**273.70**

28 ROOF FINISHINGS – GREEN ROOFS

Item	PC £	Labour hours	Labour £	Plant £	Material £	Unit	Total rate £
INTENSIVE GREEN ROOFS – cont							
Flat roofs up to 10°; Intensive Simple and Wildflower Green Roof System; Zinco Green Roof Systems Ltd; (suitable for Inverted roofs); for planting with lawn, shrubs, small trees, and perennials; can be combined with patios and walkways if required. System depth – min 160 mm; saturated weight – 342 kg/m^2 ; water storage capacity – 110 L/m^2.							
Water repellent, air permeable membrane used as a diffusionable separation layer; laid above waterproofing and separation membrane; rolled over the insulation.							
Separation Layer; TGV21	1.20	0.04	4.28	–	1.38	m^2	**5.65**
Pressure resistant drainage and water storage element made from thermoformed hard plastic; laid to TGV21; 60 mm thick filled with Zincolit Plus mineral aggregate							
Zincolit Plus; mineral aggregate infill; 27 L/m^2	4.01	0.03	2.67	–	4.61	m^2	**7.29**
Filter Sheet SF – Rolled to drainage layer FD 60	1.20	0.04	4.28	–	1.38	m^2	**5.65**
System Substrate Growing medium; 'Roof Garden' laid to Filter Fleece SF min.200 mm thick dependent on planting required and depth capacities	176.00	1.00	106.95	–	202.40	m^3	**309.35**

28 ROOF FINISHINGS – GREEN ROOFS

Item	PC £	Labour hours	Labour £	Plant £	Material £	Unit	Total rate £
Flat roofs up to 10°; Intensive Simple and Wildflower Green Roof System; Zinco Green Roof Systems Ltd; (suitable for Inverted roofs); for planting with lawn, shrubs, small trees, and perennials; System depth – min 270 mm; saturated weight – 365 kg/m^2 ; water storage capacity – 135 L/m^2. (ETA 13/0668)							
Moisture and protection mat; Synthetic fibre resistant to mechanical stress resistant for protection of waterproofing layer and provision of additional extra water reservoir of 4 L/m^2							
Zinco Moisture mat/protection fleece ISM50.	6.97	0.04	4.28	–	8.02	m^2	**12.29**
Zinco Drainage Element FD 60 – Laid to ISM50	22.39	0.04	4.28	–	25.75	m^2	**30.02**
Zincolit Plus mineral aggregate infill 27 Litres per m^2	4.01	0.03	2.67	–	4.61	m^2	**7.29**
Filter Sheet SF – Rolled to drainage layer FD 60	1.20	0.04	4.28	–	1.38	m^2	**5.65**
System Substrate Growing medium; ‘Roof Garden’ laid to Filter Fleece SF min.200 mm thick dependent on planting required and depth capacities	176.00	0.67	71.30	–	202.40	m^3	**273.70**
‘LAWN’ Green Roof System; Zinco Green Roof Systems Ltd. Lawns on roofs where build up depth is restricted; Flat roofs – zero degrees up to 5°; Lawn surface treatment; System depth 150 mm system saturated weight min 165 Kg/m^2. System water storage capacity; from 65 L/m^2							
Protection Layer – polypropylene mat, mechanical stress resistant for protection of the waterproofing layer							
Zinco Filter sheet PV	3.45	0.04	4.28	–	3.97	m^2	**8.25**
Drainage and distribution element; Water storage, distribution and drainage element made from thermoformed recycled hard plastic; 45 mm thick and preformed with clipping points for Zinco Irrigation Dripperline 100-L1							
Zinco Aquatec 45 – laid to Filter Sheet PV	21.76	0.07	7.13	–	25.02	m^2	**32.16**

28 ROOF FINISHINGS – GREEN ROOFS

Item	PC £	Labour hours	Labour £	Plant £	Material £	Unit	Total rate £
INTENSIVE GREEN ROOFS – cont							
'LAWN' Green Roof System – cont							
Drippers; clipped into Aquatec 45 and linked to Zinco irrigation manager BM–4							
Zinco Dripperline 100-L1	3.59	0.07	7.13	–	4.13	m²	**11.27**
Programmable Irrigation management unit in lockable steel box for automated irrigation of lawn and green roofs; linked to dripperlines 100-L1							
Zinco Irrigation Manager BM–4	1950.08	2.00	212.91	–	2232.26	nr	**2693.11**
Water distributing polyester fleece with capillary effective fibres; laid over Aquatec 45 containing dripperlines 100-L1							
Zinco Wicking Mat DV 40	11.97	0.03	3.06	–	13.76	m²	**16.82**
System Substrate Growing medium; Minimum thickness 100 mm up to 150 mm laid over Wicking mat DV 40							
Zinco 'LAWN'	17.60	0.07	7.13	–	20.24	m²	**27.37**
Top dressing; 15 mm layer laid over 'LAWN' substrate							
'ZINCOHUM'	1.83	0.01	1.42	–	2.10	m²	**3.53**
LANDSCAPE OPTIONS TO GREEN ROOF SYSTEMS							
The labour in this section is calculated on a 2 person team. The labour time below should be multiplied by 2 to calculate the cost as shown.							
Metal Edgings to green roofs; Kinley systems Ltd. RoofEdge with vertical perforations ensuring lateral drainage across roof surfaces for preventing excessive water accumulation. Aluminium; flexible; lightweight; fixed by drainage medium counterweight over angled edging profile							
Rigid edgings							
50 mm	49.77	0.06	3.96	–	57.23	m	**61.20**
100 mm	58.16	0.06	4.46	–	66.89	m	**71.34**
150 mm	84.81	0.07	5.28	–	97.53	m	**102.82**
200 mm	106.09	0.08	5.48	–	122.00	m	**127.48**
300 mm	151.88	0.08	5.94	–	174.67	m	**180.61**

28 ROOF FINISHINGS – GREEN ROOFS

Item	PC £	Labour hours	Labour £	Plant £	Material £	Unit	**Total rate £**
Flexible edgings; laid to mixed curves and staight sections; inclusive of laying of concrete hauch;							
50 mm	68.58	0.13	8.91	–	79.58	m	**88.50**
100 mm	81.10	0.13	9.50	–	94.12	m	**103.63**
150 mm	114.12	0.14	10.19	–	135.53	m	**145.72**
200 mm	128.56	0.14	10.19	–	152.13	m	**162.32**
300 mm	192.74	0.15	10.97	–	225.94	m	**236.90**
Green roof drainage							
Outlet inspection chambers							
150 mm deep	–		–	–	–	nr	**83.39**
250 mm deep	–		–	–	–	nr	**101.26**
350 mm deep	–		–	–	–	nr	**113.17**
Green roof drainage; Hydrotrench; Garden drainage UK; Plastic composite solution sheets with high permeability laid to green roof surface to drain water to drainage outlets (not included) on roof surface;							
laid in conjuction with reservoir boards (not included)							
45 mm thick; units 1.00 m × 400 mm wide × 45 mm thick (2.5 planks /m^2)	–		–	–	–	m^2	**–**
Adjustable pedestals to rooftops for decking and pavings							
Decking; 300 mm joists at 1 pedestal per linear metre (3.3 / m^2) along the length of the joists;							
17 mm high	3.15	0.08	5.88	–	3.62	m^2	**9.50**
28 mm high	4.29	0.08	5.88	–	4.93	m^2	**10.81**
175–285 mm high	25.95	0.08	6.00	–	29.84	m^2	**35.84**
285–400 mm high	44.89	0.08	6.00	–	51.62	m^2	**57.62**
355–515 mm high	55.18	0.08	6.00	–	63.46	m^2	**69.46**
465–625 mm high	63.87	0.09	6.30	–	73.45	m^2	**79.75**
545–740 mm high	74.16	0.09	6.48	–	85.28	m^2	**91.76**
645–850 mm high	82.81	0.09	6.48	–	95.23	m^2	**101.71**
720–960 mm high	93.10	0.09	6.48	–	107.07	m^2	**113.55**
830–1070 mm high	101.78	0.09	6.48	–	117.05	m^2	**123.53**
Decking; 400 mm joists at 1 pedestal per linear metre (2.5 / m^2) along the length of the joists;							
17 mm high	2.38	0.07	4.76	–	2.74	m^2	**7.50**
28 mm high	3.25	0.07	4.76	–	3.74	m^2	**8.49**
175–285 mm high	19.66	0.07	5.09	–	22.60	m^2	**27.70**

28 ROOF FINISHINGS – GREEN ROOFS

Item	PC £	Labour hours	Labour £	Plant £	Material £	Unit	Total rate £
LANDSCAPE OPTIONS TO GREEN ROOF SYSTEMS – cont							
Adjustable pedestals to rooftops for decking and pavings – cont							
Decking – cont							
285–400 mm high	34.01	0.07	5.09	–	39.11	m^2	**44.20**
355–515 mm high	41.80	0.07	5.09	–	48.07	m^2	**53.17**
465–625 mm high	48.38	0.08	5.48	–	55.64	m^2	**61.12**
545–740 mm high	56.18	0.08	5.48	–	64.61	m^2	**70.09**
645–850 mm high	62.73	0.08	5.48	–	72.14	m^2	**77.63**
720–960 mm high	70.53	0.08	5.48	–	81.11	m^2	**86.59**
830–1070 mm high	77.11	0.08	5.48	–	88.68	m^2	**94.16**
Adjustable pedestals to rooftops for decking and pavings; Note other heights available up to 1.07 m							
Paving slabs; laid at uniform sizes with one pedestal per corner of each paving slab; paving slab sizes; laid on adjustable pedestal 28 mm high above the slab							
300 × 300 mm	22.09	0.50	35.65	–	25.41	m^2	**61.06**
400 × 400 mm	13.00	0.50	35.65	–	14.95	m^2	**50.60**
450 × 450 mm	9.75	0.50	35.65	–	11.21	m^2	**46.86**
600 × 450 mm with additional central support per slab	8.58	0.33	23.76	–	9.86	m^2	**33.63**
900 × 600 mm with additional central support per slab	4.94	0.25	17.83	–	5.68	m^2	**23.50**
Paving slabs; laid at uniform sizes with one pedestal per corner of each paving slab; paving slab sizes; laid on adjustable pedestal 285–400 mm high above the slab							
300 × 300 mm	231.24	0.63	44.56	–	265.93	m^2	**310.49**
400 × 400 mm	136.02	0.63	44.56	–	156.43	m^2	**200.99**
450 × 450 mm	102.02	0.63	44.56	–	117.32	m^2	**161.88**
600 × 450 mm with additional central support per slab	89.78	0.42	29.70	–	103.24	m^2	**132.95**
900 × 600 mm with additional central support per slab	51.69	0.31	22.28	–	59.44	m^2	**81.72**
Paving slabs; laid at uniform sizes with one pedestal per corner of each paving slab; paving slab sizes; laid on adjustable pedestal 645–850 mm high above the slab							
300 × 300 mm	426.58	0.63	44.56	–	490.57	m^2	**535.13**
400 × 400 mm	250.93	0.63	44.56	–	288.57	m^2	**333.13**
450 × 450 mm	188.20	0.63	44.56	–	216.43	m^2	**260.99**

28 ROOF FINISHINGS – GREEN ROOFS

Item	PC £	Labour hours	Labour £	Plant £	Material £	Unit	Total rate £
600 × 450 mm with additional central support per slab	165.61	0.42	29.70	–	190.46	m²	**220.16**
900 × 600 mm with additional central support per slab	95.35	0.31	22.28	–	109.66	m²	**131.94**
BLUE ROOFS							
Blue roof systems; ZinCo Green roof systems Ltd (Stormwater management roof); For managment of stormwater under a green roof system; buildup of Blue roof system over roof surface with water proofing (not included) and root resistant protection							
Extensive Blue roof; Build-up height (inclusive of green roof components, priced separately: 200 mm; Saturated weight including plants 220 kg/m²; Water storage capacity 114 l/m²; water retention height 60 mm							
Root barrier WSF 40	2.96	0.02	1.19	–	3.40	m²	**4.59**
Filter sheet PV; above and below the water retention unit (2 m²/m²)	6.90	0.03	2.37	–	7.93	m²	**10.31**
Retention spacer RS 60	23.04	0.25	17.83	–	26.50	m²	**44.32**
Zinco Filter Sheet SF – Rolled to drainage layer FD40-E	1.20	0.04	4.28	–	1.38	m²	**5.65**
Optional retention spacer units							
Retention spacer RSX 100; Water retention capacity 95 l/m²; Total water storage of system 154 l/m²	48.64	0.29	20.37	–	55.94	m²	**76.31**
Flow control; Adjustable flow controller to be installed on flat roofs over water outlets with a contact flange; including a suitable inspection chamber; suitable for installation on Stormwater Management Roofs (Blue roofs) with rainwater attenuation up to ca. 85 mm. Adjustable discharge rate							
RDS 28 and inspection chamber	382.72	1.00	35.65	–	440.13	nr	**475.78**
Intensive Green roof; instead of root barrier WSF 40							
Root barrier WSB-100-PO	17.25	0.03	1.78	–	19.84	m²	**21.62**

29 DECORATION

Item	PC £	Labour hours	Labour £	Plant £	Material £	Unit	Total rate £
PAINTING/CLEAR FINISHING							
Prepare; touch up primer; two undercoats and one finishing coat of gloss oil paint; on metal surfaces							
General surfaces							
girth exceeding 300 mm	5.23	0.67	23.77	–	6.01	m^2	**29.78**
isolated surfaces; girth not exceeding 300 mm	15.84	0.40	14.26	–	18.22	m	**32.48**
isolated areas not exceeding 0.50 m^2 irrespective of girth	10.46	0.67	23.77	–	12.03	nr	**35.79**
Ornamental railings; each side measured separately							
girth exceeding 300 mm	5.23	0.57	20.37	–	6.01	m^2	**26.38**
Prepare; one coat primer; two finishing coats of gloss paint; on wood surfaces							
New wood surfaces							
girth exceeding 300 mm	3.62	0.80	28.52	–	4.17	m^2	**32.69**
isolated surfaces; girth not exceeding 300 mm	1.21	1.00	35.65	–	1.39	m	**37.04**
isolated areas not exceeding 0.50 m^2 irrespective of girth	4.23	0.50	17.83	–	4.86	nr	**22.69**
Previously painted wood surfaces							
girth exceeding 300 mm	2.01	0.50	17.83	–	2.31	m^2	**20.14**
isolated surfaces; girth not exceeding 300 mm	0.67	0.67	23.77	–	0.77	m	**24.54**
isolated areas not exceeding 0.50 m^2 irrespective of girth	4.02	0.40	14.26	–	4.63	nr	**18.89**
Fences and sheds; prepare; two coats of Protek wood preserver on wood surfaces							
Planed surfaces							
girth exceeding 300 mm	0.54	0.07	2.38	–	0.63	m^2	**3.00**
isolated surfaces; girth not exceeding 300 mm	0.18	0.17	5.94	–	0.21	m	**6.15**
isolated areas not exceeding 0.50 m^2 irrespective of girth	0.54	0.25	8.91	–	0.63	nr	**9.54**
Sawn surfaces							
girth exceeding 300 mm	0.20	0.10	3.57	–	0.23	m^2	**3.80**
isolated surfaces; girth not exceeding 300 mm	2.72	0.17	5.94	–	3.13	m	**9.07**
isolated areas not exceeding 0.50 m^2 irrespective of girth	0.82	0.25	8.91	–	0.94	nr	**9.85**

29 DECORATION

Item	PC £	Labour hours	Labour £	Plant £	Material £	Unit	Total rate £
Fences and sheds; prepare; two coats of Sadolin wood preserver on wood surfaces							
Planed surfaces							
girth exceeding 300 mm	2.62	0.07	2.38	–	3.01	m^2	**5.39**
isolated surfaces; girth not exceeding 300 mm	0.87	0.17	5.94	–	1.00	m	**6.95**
isolated areas not exceeding 0.50 m^2 irrespective of girth	1.31	0.25	8.91	–	1.50	nr	**10.42**
Sawn surfaces							
girth exceeding 300 mm	1.31	0.10	3.57	–	1.50	m^2	**5.07**
isolated surfaces; girth not exceeding 300 mm	1.40	0.17	5.94	–	1.60	m	**7.55**
isolated areas not exceeding 0.50 m^2 irrespective of girth	2.09	0.25	8.91	–	2.41	nr	**11.32**
Prepare; proprietary solution primer; two coats of dark stain; on wood surfaces							
General surfaces							
girth exceeding 300 mm	3.00	0.17	5.94	–	3.45	m^2	**9.39**
isolated surfaces; girth not exceeding 300 mm	1.00	0.13	4.46	–	1.15	m	**5.61**
Two coats Weathershield; to clean, dry surfaces; in accordance with manufacturer's instructions							
Brick or block walls							
girth exceeding 300 mm	3.24	0.40	14.26	–	3.73	m^2	**17.99**
Cement render or concrete walls							
girth exceeding 300 mm	3.24	0.33	11.88	–	3.73	m^2	**15.61**

33 DRAINAGE ABOVE GROUND

Item	PC £	Labour hours	Labour £	Plant £	Material £	Unit	Total rate £
DRAINAGE TO ROOF DECKS AND PLANTERS							
The labour in this section is calculated on a 3 person team. The labour time below should be multiplied by 3 to calculate the cost as shown.							
Leca (light expanded clay aggregate); drainage aggregate to roofdecks and planters							
Placed mechanically to planters; average 100 mm thick; by mechanical plant tipped into planters							
Placed mechanically to planters; average 100 mm thick; by mechanical plant tipped into planters							
aggregate size 10–20 mm; delivered in 30 m^3 loads	77.98	0.40	14.26	7.06	89.67	m^3	**110.99**
aggregate size 10–20 mm; delivered in 70 m^3 loads	55.05	0.20	7.13	10.12	63.31	m^3	**80.55**
Placed by light aggregate blower (maximum 40 m)							
aggregate size 10–20 mm; delivered in 30 m^3 loads	77.98	0.14	5.09	–	95.55	m^3	**100.65**
aggregate size 10–20 mm; delivered in 55 m^3 loads on blower vehicle / 165 m^3 per day max	64.30	0.14	5.09	–	73.94	m^3	**79.04**
aggregate size 10–20 mm; delivered in 70 m^3 loads	55.05	0.14	5.09	–	69.19	m^3	**74.28**
By hand							
aggregate size 10–20 mm; delivered in 30 m^3 loads	77.98	1.33	47.53	–	89.67	m^3	**137.20**
aggregate size 10–20 mm; delivered in 70 m^3 loads	55.05	1.33	47.53	–	63.31	m^3	**110.84**
Drainage boards laid to insulated slabs on roof decks; boards laid below growing medium and granulated drainage layer and geofabric (all not included) to collect and channel water to drainage outlets (not included)							
Zinco Floradrain; polyethylene irrigation/drainage layer; inclusive of geofabric laid over the surface of the drainage board							
Floradrain FD40; 0.96 × 2.08 m panel	3.03	0.05	1.78	–	14.73	m^2	**16.52**
Floradrain FD60; 1.00 × 2.00 m panel	18.44	0.07	2.38	–	21.21	m^2	**23.59**

34 DRAINAGE BELOW GROUND

Item	PC £	Labour hours	Labour £	Plant £	Material £	Unit	Total rate £
CLARIFICATION ON LABOUR COSTS IN THIS SECTION							
General groundworks team							
Generally a three man team is used in this section. The column Labour hours reports team hours. The column Labour £ reports the total cost of the team for the unit of work shown							
3 man team	–	1.00	106.95	–	–	hr	**106.95**
TRENCHES/PIPEWAYS/PITS FOR BURIED ENGINEERING							
Excavating trenches; using 3 tonne tracked excavator; to receive pipes; grading bottoms; earthwork support; filling with excavated material to within 150 mm of finished surfaces and compacting; completing fill with topsoil; disposal of surplus soil							
Services not exceeding 200 mm nominal size							
average depth of run not exceeding 0.50 m	1.60	0.12	4.28	1.40	1.84	m	**7.51**
average depth of run not exceeding 0.75 m	1.60	0.16	5.81	1.94	1.84	m	**9.58**
average depth of run not exceeding 1.00 m	1.60	0.28	10.10	3.38	1.84	m	**15.32**
average depth of run not exceeding 1.25 m	1.60	0.38	13.66	4.55	1.84	m	**20.05**
Excavating trenches; using 3 tonne tracked excavator; to receive pipes; grading bottoms; earthwork support; filling with imported granular material (6F2) and compacting; disposal of surplus soil							
Services not exceeding 200 mm nominal size							
average depth of run not exceeding 0.50 m	2.72	0.09	3.03	0.96	3.07	m	**12.81**
average depth of run not exceeding 0.75 m	4.08	0.11	3.78	1.20	4.60	m	**18.20**
average depth of run not exceeding 1.00 m	5.44	0.14	4.88	1.56	6.13	m	**24.06**
average depth of run not exceeding 1.25 m	6.80	0.23	7.99	2.67	7.67	m	**32.73**

34 DRAINAGE BELOW GROUND

Item	PC £	Labour hours	Labour £	Plant £	Material £	Unit	Total rate £
TRENCHES/PIPEWAYS/PITS FOR BURIED ENGINEERING – cont							
The labour in this section is calculated on a 2 person team. The labour time below should be multiplied by 2 to calculate the cost as shown.							
Excavating trenches; using 3 tonne tracked excavator; to receive pipes; grading bottoms; earthwork support; filling with lean mix concrete; disposal of surplus soil							
Services not exceeding 200 mm nominal size							
average depth of run not exceeding 0.50 m	14.19	0.11	3.76	0.50	16.14	m	**27.37**
average depth of run not exceeding 0.75 m	21.29	0.13	4.58	0.62	24.21	m	**39.86**
average depth of run not exceeding 1.00 m	28.38	0.17	5.88	0.83	32.27	m	**52.91**
average depth of run not exceeding 1.25 m	35.48	0.23	7.93	1.25	40.35	m	**66.94**
Earthwork support; providing support to opposing faces of excavation; moving along as work proceeds							
Maximum depth not exceeding 2.00 m							
trenchbox; distance between opposing faces not exceeding 2.00 m	–	0.80	28.52	39.39	–	m	**67.91**
timber; distance between opposing faces not exceeding 500 mm	–	0.50	35.65	–	4.13	m	**39.78**
CHANNELS							
Best quality vitrified clay half section channels; Hepworth Plc; bedding and jointing in cement: mortar (1:2)							
Channels; straight							
100 mm	14.74	0.80	28.52	–	22.78	m	**51.30**
150 mm	24.51	1.00	35.65	–	34.02	m	**69.67**
225 mm	55.03	1.35	48.13	–	69.12	m	**117.25**
300 mm	112.98	1.80	64.17	–	135.76	m	**199.93**

34 DRAINAGE BELOW GROUND

Item	PC £	Labour hours	Labour £	Plant £	Material £	Unit	Total rate £
Bends; 15, 30, 45 or 90°							
100 mm bends	13.26	0.75	26.74	–	18.17	nr	**44.91**
150 mm bends	22.92	0.90	32.09	–	30.74	nr	**62.82**
225 mm bends	88.82	1.20	42.78	–	107.97	nr	**150.75**
300 mm bends	181.05	1.10	39.22	–	215.50	nr	**254.72**
Best quality vitrified clay channels; Hepworth Plc; bedding and jointing in cement: mortar (1:2)							
Branch bends; 15, 30, 45 or 90°; left or right hand							
100 mm	13.26	0.75	26.74	–	18.17	nr	**44.91**
150 mm	22.92	0.90	32.09	–	30.74	nr	**62.82**
EXCAVATION FOR DRAINAGE							
Machine excavation							
Excavating pits; starting from ground level; works exclude for earthwork retention							
maximum depth not exceeding 1.00 m	–	0.50	17.83	25.77	–	m^3	**43.60**
maximum depth not exceeding 2.00 m	–	0.50	17.83	41.14	–	m^3	**58.96**
maximum depth not exceeding 4.00 m	–	0.50	17.83	100.59	–	m^3	**118.41**
Disposal of excavated material; depositing on site in permanent spoil heaps; average 50 m	–	0.04	1.49	3.10	–	m^3	**4.59**
Filling to excavations; obtained from on site spoil heaps; average thickness not exceeding 0.25 m	–	0.13	4.75	8.94	–	m^3	**13.69**
The labour in this section is calculated on a 2 person team. The labour time below should be multiplied by 2 to calculate the cost as shown.							
Hand excavation							
Excavating pits; starting from ground level; works exclude for earthwork retention							
maximum depth not exceeding 1.00 m	–	1.20	85.56	–	–	m^3	**85.56**
maximum depth not exceeding 2.00 m	–	1.50	106.95	–	–	m^3	**106.95**

34 DRAINAGE BELOW GROUND

Item	PC £	Labour hours	Labour £	Plant £	Material £	Unit	Total rate £
EXCAVATION FOR DRAINAGE – cont							
Works to pits or drainage excavations							
Surface treatments; compacting; bottoms of excavations	–	0.05	1.78	–	–	m^2	**1.78**
Earthwork support							
Manhole boxes to excavations (not included); Timber plywood with timber braces to shore excavations to 1.00 m deep; opposing faces distance less than 1.00 m							
1.00 m deep	–	0.50	35.65	–	38.14	m^2	**73.79**
2.00 m deep	–	1.00	71.30	–	38.14	m^2	**109.44**
Trench boxes; A-Plant							
Manhole box soil retention	–	0.50	53.48	208.58	–	week	**262.06**
Placement of manhole protection system to 3.0 m deep to enable earthwork retention; maximum distance between opposing faces							
1.8 m	–	0.50	53.48	282.06	–	week	**335.54**
GULLIES AND INTERCEPTION TRAPS							
Intercepting traps; Hepworth Plc							
Vitrified clay; inspection arms; brass stoppers; iron levers; chains and staples; galvanized; staples cut and pinned to brickwork; cement: mortar (1:2) joints to vitrified clay pipes and channels; bedding and surrounding in concrete; 11.50 N/mm^2–40 mm aggregate; cutting and fitting brickwork; making good facings							
100 mm inlet; 100 mm outlet	159.42	3.00	106.95	–	207.70	nr	**314.65**
150 mm inlet; 150 mm outlet	229.86	2.00	71.30	–	303.58	nr	**374.88**
Gullies; concrete; FP McCann							
The labour in this section is calculated on a 2 person team. The labour time below should be multiplied by 2 to calculate the cost as shown.							

34 DRAINAGE BELOW GROUND

Item	PC £	Labour hours	Labour £	Plant £	Material £	Unit	Total rate £
Concrete road gullies;with Aquamax cast iron grating; trapped; cement: mortar (1:2) joints to concrete pipes; bedding and surrounding in concrete; 11.50 N/mm²–40 mm aggregate; rodding eye; stoppers							
375 mm dia. × 750 mm deep	74.26	3.00	213.90	–	330.98	nr	**544.88**
375 mm dia. × 900 mm deep	73.39	3.00	213.90	–	329.98	nr	**543.88**
450 mm dia. × 750 mm deep	73.39	3.00	213.90	–	329.98	nr	**543.88**
450 mm dia. × 900 mm deep	79.53	3.00	213.90	–	337.04	nr	**550.94**
450 mm dia. × 1.05 m deep	85.73	3.50	249.55	–	344.17	nr	**593.72**
450 mm dia. × 1.20 m deep	110.26	4.00	285.20	–	372.38	nr	**657.58**
Accessories for clay gullies							
coverslab	31.37	0.25	17.83	–	36.08	nr	**53.90**
seating rings	33.16		–	–	38.14	nr	**38.14**
Gullies; vitrified clay; Hepworth Plc; bedding in concrete; 11.50 N/mm²–40 mm aggregate							
Yard gullies (mud); trapped; domestic duty (up to 1 tonne)							
100 mm outlet; 100 mm dia.; 225 mm internal width; 585 mm internal depth	250.50	3.50	124.78	–	289.58	nr	**414.36**
150 mm outlet; 100 mm dia.; 225 mm internal width; 585 mm internal depth	250.50	3.50	124.78	–	289.58	nr	**414.36**
Yard gullies (mud); trapped; medium duty (up to 5 tonnes)							
100 mm outlet; 100 mm dia.; 225 mm internal width; 585 mm internal depth	354.13	3.50	124.78	–	408.75	nr	**533.53**
150 mm outlet; 100 mm dia.; 225 mm internal width; 585 mm internal depth	388.02	3.50	124.78	–	447.73	nr	**572.50**
Combined filter and silt bucket for yard gullies							
225 mm dia.	90.58		–	–	104.17	nr	**104.17**
Road gullies; trapped with rodding eye							
100 mm outlet; 300 mm internal dia.; 600 mm internal depth	240.93	3.50	124.78	–	278.57	nr	**403.35**
150 mm outlet; 300 mm internal dia.; 600 mm internal depth	246.73	3.50	124.78	–	285.23	nr	**410.01**
150 mm outlet; 400 mm internal dia.; 750 mm internal depth	286.15	3.50	124.78	–	330.58	nr	**455.35**
150 mm outlet; 450 mm internal dia.; 900 mm internal depth	387.14	3.50	124.78	–	446.71	nr	**571.49**

34 DRAINAGE BELOW GROUND

Item	PC £	Labour hours	Labour £	Plant £	Material £	Unit	Total rate £
GULLIES AND INTERCEPTION TRAPS – cont							
Gullies – cont							
Hinged gratings and frames for gullies; alloy							
193 mm for 150 mm dia. gully	–		–	–	69.04	nr	**69.04**
120 × 120 mm	–		–	–	24.57	nr	**24.57**
150 × 150 mm	–		–	–	44.42	nr	**44.42**
230 × 230 mm	–		–	–	81.21	nr	**81.21**
316 × 316 mm	–		–	–	215.28	nr	**215.28**
Hinged gratings and frames for gullies; cast iron							
265 mm for 225 mm dia. gully	–		–	–	137.95	nr	**137.95**
150 × 150 mm	–		–	–	44.42	nr	**44.42**
230 × 230 mm	–		–	–	81.21	nr	**81.21**
316 × 316 mm	–		–	–	215.28	nr	**215.28**
Universal gully trap Plastech Southern; bedding in concrete; 11.50 N/mm²–40 mm aggregate							
Universal gully fitting; comprising gully trap only							
110 mm outlet; 110 mm dia.; 205 mm internal depth	11.97	3.50	124.78	–	15.26	nr	**140.04**
Vertical inlet hopper; c\w plastic grate							
272 × 183 mm	22.24	0.25	8.91	–	25.58	nr	**34.49**
Sealed access hopper							
110 × 110 mm	52.29	0.25	8.91	–	60.14	nr	**69.05**
Universal gully PVC-u; accessories to universal gully trap							
Hoppers; backfilling with clean granular material; tamping; surrounding in lean mix concrete							
plain hopper; with 110 mm spigot; 150 mm long	18.18	0.40	14.26	–	21.86	nr	**36.12**
vertical inlet hopper; with 110 mm spigot; 150 mm long	22.24	0.40	14.26	–	25.58	nr	**39.84**
sealed access hopper; with 110 mm spigot; 150 mm long	52.29	0.40	14.26	–	60.14	nr	**74.40**
plain hopper; solvent weld to trap	8.59	0.40	14.26	–	9.88	nr	**24.14**
vertical inlet hopper; solvent wed to trap	21.34	0.40	14.26	–	24.54	nr	**38.80**
sealed access cover; PVC-u	88.77	0.10	3.57	–	102.09	nr	**105.65**

34 DRAINAGE BELOW GROUND

Item	PC £	Labour hours	Labour £	Plant £	Material £	Unit	**Total rate £**
Gullies PVC-u; bedding in concrete; 11.50 N/mm²–40 mm aggregate							
Bottle gully; providing access to the drainage system for cleaning							
bottle gully; 228 × 228 × 317 mm deep	10.56	0.50	17.83	–	13.52	nr	**31.35**
sealed access cover; PVC-u; 217 × 217 mm	34.89	0.10	3.57	–	40.12	nr	**43.69**
grating; ductile iron; 215 × 215 mm	88.20	0.10	3.57	–	101.43	nr	**105.00**
bottle gully riser; 325 mm	20.97	0.50	17.83	–	27.11	nr	**44.94**
Yard gully; trapped; 300 mm dia. × 600 mm deep; including catchment bucket and ductile iron cover and frame; medium duty loading B–125							
300 mm dia. × 600 mm deep	115.78	2.50	89.13	–	141.11	nr	**230.24**
Kerbs to gullies							
One course Class B engineering bricks to four sides; rendering in cement: mortar (1:3); dished to gully gratings							
150 × 150 mm	2.95	0.33	11.88	–	4.01	nr	**15.89**
ACCESS COVERS AND FRAMES							
Load Classes for Access Covers							
FACTA (Fabricated Access Cover Trade Association) class:							
A – 0.5 tonne maximum slow moving wheel load							
AA – 1.5 tonne maximum slow moving wheel load							
AAA – 2.5 tonne maximum slow moving wheel load							
B – 5 tonne maximum slow moving wheel load							
C – 6.5 tonne maximum slow moving wheel load							
D – 11 tonne maximum slow moving wheel load							

34 DRAINAGE BELOW GROUND

Item	PC £	Labour hours	Labour £	Plant £	Material £	Unit	Total rate £
ACCESS COVERS AND FRAMES – cont							
Access covers and frames; solid top; galvanized; Steelway Brickhouse; Bristeel; bedding frame in cement: mortar (1:3); cover in grease and sand; clear opening sizes; base size shown in brackets (50 mm depth)							
FACTA AA; single seal							
450 × 450 mm (520 × 520 mm)	180.00	1.50	53.48	–	224.51	nr	**277.99**
600 × 450 mm (670 × 520 mm)	193.00	1.80	64.17	–	241.41	nr	**305.58**
600 × 600 mm (670 × 670 mm)	205.00	2.00	71.30	–	259.10	nr	**330.40**
FACTA AA; double seal							
450 × 450 mm (560 × 560 mm)	187.00	1.50	53.48	–	232.56	nr	**286.04**
600 × 450 mm (710 × 560 mm)	231.00	1.80	64.17	–	285.11	nr	**349.28**
600 × 600 mm (710 × 710 mm)	236.00	2.00	71.30	–	294.75	nr	**366.05**
FACTA AAA; single seal							
450 × 450 mm (520 × 520 mm)	187.00	1.50	53.48	–	232.56	nr	**286.04**
600 × 450 mm (670 × 520 mm)	203.00	1.80	64.17	–	252.91	nr	**317.08**
600 × 600 mm (670 × 670 mm)	213.00	2.00	71.30	–	268.30	nr	**339.60**
FACTA AAA; double seal							
450 mm × 450 mm (560 × 560)	227.00	1.50	53.48	–	278.56	nr	**332.04**
600 mm × 450 mm (710 × 560)	247.00	1.80	64.17	–	303.51	nr	**367.68**
600 mm × 600 mm (710 × 710)	279.00	2.00	71.30	–	344.20	nr	**415.50**
FACTA B; single seal							
450 × 450 mm (520 × 520 mm)	187.00	1.50	53.48	–	232.56	nr	**286.04**
600 × 450 mm (670 × 520 mm)	205.00	1.80	64.17	–	255.21	nr	**319.38**
600 × 600 mm (670 × 670 mm)	218.00	2.00	71.30	–	274.05	nr	**345.35**
FACTA B; double seal							
450 × 450 mm (560 × 560 mm)	238.00	1.50	53.48	–	291.21	nr	**344.69**
600 × 450 mm (710 × 560 mm)	266.00	1.80	64.17	–	325.36	nr	**389.53**
600 × 600 mm (710 × 710 mm)	286.00	2.00	71.30	–	352.25	nr	**423.55**
Access covers and frames; recessed; galvanized; Steelway Brickhouse; Bripave; bedding frame in cement: mortar (1:3); cover in grease and sand; clear opening sizes; base size shown in brackets (for block depths 50 mm, 65 mm, 80 mm or 100 mm)							
FACTA AA							
450 × 450 mm (562 × 562 mm)	214.00	1.50	53.48	–	263.61	nr	**317.09**
600 × 450 mm (712 × 562 mm)	221.00	1.80	64.17	–	273.61	nr	**337.78**
600 × 600 mm (712 × 712 mm)	227.00	2.00	71.30	–	283.67	nr	**354.97**
750 × 600 mm (862 × 712 mm)	288.00	2.40	85.56	–	356.50	nr	**442.06**

34 DRAINAGE BELOW GROUND

Item	PC £	Labour hours	Labour £	Plant £	Material £	Unit	Total rate £
FACTA B							
450 × 450 mm (562 × 562 mm)	221.00	1.50	53.48	–	271.66	nr	**325.14**
600 × 450 mm (712 × 562 mm)	229.00	1.80	64.17	–	282.81	nr	**346.98**
600 × 600 mm (712 × 712 mm)	234.00	2.00	71.30	–	292.45	nr	**363.75**
750 × 600 mm (862 × 712 mm)	299.00	2.40	85.56	–	369.15	nr	**454.71**
FACTA D							
450 × 450 mm (562 × 562 mm)	226.00	1.50	53.48	–	277.41	nr	**330.89**
600 × 450 mm (712 × 562 mm)	236.00	1.80	64.17	–	290.86	nr	**355.03**
600 × 600 mm (712 × 712 mm)	247.00	2.00	71.30	–	307.40	nr	**378.70**
750 × 600 mm (862 × 712 mm)	319.00	2.40	85.56	–	392.15	nr	**477.71**
Extra over manhole frames and covers for							
filling recessed manhole covers with brick paviors; PC £600.00/1000	–	1.00	35.65	–	28.29	m^2	**63.94**
filling recessed manhole covers with vehicular paving blocks; PC £12.94/m^2	14.26	0.75	26.74	–	16.39	m^2	**43.13**
filling recessed manhole covers with concrete paving flags; PC £19.06/m^2	21.00	0.35	12.48	–	24.15	m^2	**36.63**
Access covers and frames; Stainless steel recessed; Kent Stainless; bedding frame in cement: mortar (1:3); cover in grease and sand; clear opening sizes; (for depths 80 mm or 100 mm)							
Thickness or loadings							
450 × 450 mm	638.97	1.50	93.00	–	639.45	nr	**732.45**
450 × 600 mm	660.63	1.75	108.50	–	661.29	nr	**769.79**
600 × 600 mm	676.88	1.50	93.00	–	677.54	nr	**770.54**
900 × 600 mm	844.74	1.90	117.80	–	845.45	nr	**963.25**

34 DRAINAGE BELOW GROUND

Item	PC £	Labour hours	Labour £	Plant £	Material £	Unit	Total rate £
INSPECTION CHAMBERS AND MANHOLES							
The labour in this section is calculated on a 2 person team. The labour time below should be multiplied by 2 to calculate the cost as shown.							
Inspection chambers; in situ concrete; excavations; earthwork reinforcement all not included							
Beds; plain in situ concrete; 11.50 N/mm^2–40 mm aggregate							
thickness not exceeding 150 mm	–	2.00	71.30	–	250.48	m^3	**321.78**
thickness 150–450 mm	–	1.75	62.39	–	250.48	m^3	**312.87**
Benchings in bottoms; plain in situ concrete; 25.50 N/mm^2–20 mm aggregate							
thickness 150–450 mm	–	2.00	71.30	–	250.48	m^3	**321.78**
Isolated cover slabs; reinforced In situ concrete; 21.00 N/mm^2–20 mm aggregate							
thickness not exceeding 150 mm	217.81	4.00	142.60	–	250.48	m^3	**393.08**
Fabric reinforcement; A193 (3.02 kg/m^2) in cover slabs	5.90	0.06	2.23	–	6.78	m^2	**9.01**
Formwork to reinforced in situ concrete; isolated cover slabs							
soffits; horizontal	–	3.28	116.93	–	7.71	m^2	**124.64**
height not exceeding 250 mm	–	0.97	34.58	–	3.21	m	**37.79**
Inspection chambers (House – HIC); precast concrete units; to excavations, earthwork reinforcement, in situ bases benchings all priced separately; includes cover slab							
Precast concrete inspection chamber units; FP McCann Ltd; bedding, jointing and pointing in cement: mortar (1:3); 600 × 450 mm internally							
600 mm deep	74.06	6.50	231.73	–	94.59	nr	**326.32**
900 mm deep	98.64	7.00	249.55	–	120.86	nr	**370.41**
Drainage chambers; FP McCann Ltd; 1200 × 750 mm; no base unit; depth of invert							
1.00 m deep	269.77	3.00	320.85	–	322.60	nr	**643.45**
1.50 m deep	345.70	3.67	392.15	–	417.34	nr	**809.49**
2.00 m deep	421.63	4.17	445.63	–	509.60	nr	**955.23**

34 DRAINAGE BELOW GROUND

Item	PC £	Labour hours	Labour £	Plant £	Material £	Unit	Total rate £
Bases to chamber ring manholes; in situ concrete; to excavated ground (not included); on 50 mm C10 concrete blinding on 225 mm mass concrete to receive bottom manhole ring							
Dia. of manhole							
900 mm	–	4.00	285.20	–	76.29	nr	**361.49**
1.05 m	–	4.00	285.20	–	92.32	nr	**377.52**
1.20 m	–	4.50	320.85	–	128.94	nr	**449.79**
1.50 m	–	5.50	392.15	–	195.29	nr	**587.44**
Bottom manhole chamber ring; FP McCann Ltd; including pipe entrances to bases (not included); forming pipe entrances to manhole ring; seating and sealing to base; connecting pipes (not included); depth 1.00 m; dia.							
900 mm	115.15	3.00	320.85	121.93	144.79	nr	**587.57**
1.05 m	121.38	3.25	347.59	121.93	151.94	nr	**621.46**
1.20 m	148.25	4.00	427.80	152.41	182.84	nr	**763.06**
1.50 m	258.83	4.50	481.28	182.89	310.01	nr	**974.18**
Benchings to manholes; 400 mm concrete; dia. of manhole							
900 mm	–	0.75	53.48	–	72.12	nr	**125.59**
1.05 m	–	0.75	53.48	–	100.97	nr	**154.44**
1.20 m	–	0.90	64.17	–	129.81	nr	**193.98**
1.50 m	–	1.10	78.43	–	204.82	nr	**283.25**
'Easi-Base' Polypropylene liner; FP McCann Ltd; complete with prefabricated benching and channels as one unit; Easi-Base dia.							
1200 mm	627.49	1.00	71.30	60.96	721.61	nr	**853.88**
1500 mm	1255.01	1.00	71.30	60.96	1443.27	nr	**1575.53**

34 DRAINAGE BELOW GROUND

Item	PC £	Labour hours	Labour £	Plant £	Material £	Unit	Total rate £
INSPECTION CHAMBERS AND MANHOLES – cont							
The labour in this section is calculated on a 3 person team. The labour time below should be multiplied by 3 to calculate the cost as shown.							
Concrete manhole chamber rings; FP McCann Ltd; standard ring units to excavated ground (not included); sealed with bitumen tape; tongue and groove units installed to 1.00 m deep base units (not included); surrounding in concrete min 150 thick and backfilling; Note: The final depths shown below show operations onto the 1.00 m deep base rings above							
internal dia. 900 mm; final manhole depth							
1.00 m deep	–	1.00	106.95	60.96	187.13	nr	**355.04**
internal dia. 1.05 mm; final manhole depth							
1.00 m deep	–	1.00	106.95	121.93	187.13	nr	**416.01**
2.00 m deep	121.38	2.50	267.38	213.38	501.47	nr	**982.22**
3.00 m deep	242.75	4.00	427.80	243.86	813.04	nr	**1484.70**
internal dia. 1200 mm; final manhole depth							
1.00 m deep	–	1.00	106.95	182.89	187.13	nr	**476.97**
2.00 m deep	148.25	3.00	320.85	243.86	532.37	nr	**1097.08**
3.00 m deep	296.49	4.00	427.80	365.79	874.85	nr	**1668.43**
internal dia. 1500 mm							
1.00 m deep	–	2.00	213.90	182.89	187.13	nr	**583.92**
2.00 m deep	258.83	4.00	427.80	243.86	659.54	nr	**1331.20**
3.00 m deep	517.65	5.00	534.75	365.79	1129.18	nr	**2029.72**
Concrete manhole chamber rings; Wide wall; FP McCann Ltd; 130 mm thick precast manholes; (concrete surround on backfill not required) to excavated ground (not included); Water-tight, complete with sealant; to prepared base (or Easi-Base priced elsewhere)							
internal dia. 1200 mm							
1.00 m deep	249.49	3.00	320.85	182.89	286.91	nr	**790.66**
2.00 m deep	498.98	4.00	427.80	243.86	573.82	nr	**1245.48**
3.00 m deep	748.47	6.00	641.70	304.82	860.74	nr	**1807.26**

34 DRAINAGE BELOW GROUND

Item	PC £	Labour hours	Labour £	Plant £	Material £	Unit	**Total rate £**
internal dia. 1500 mm							
1.00 m deep	393.13	3.00	320.85	243.86	464.46	nr	**1029.17**
2.00 m deep	786.26	5.00	534.75	365.79	916.56	nr	**1817.10**
3.00 m deep	1179.39	7.00	748.65	426.75	1368.66	nr	**2544.06**
The labour in this section is calculated on a 2 person team. The labour time below should be multiplied by 2 to calculate the cost as shown.							
Cover slabs for chambers or shaft sections; FP McCann Ltd; heavy duty							
900 mm dia. internally	117.00	0.67	23.74	–	134.55	nr	**158.29**
1050 mm dia. internally	125.58	2.00	71.30	22.35	144.42	nr	**238.07**
1200 mm dia. internally	153.05	1.00	35.65	22.35	176.00	nr	**234.00**
1500 mm dia. internally	277.18	1.00	35.65	22.35	318.76	nr	**376.76**
1800 mm dia. internally	455.09	2.00	71.30	50.30	523.35	nr	**644.95**
Brickwork							
Walls to manholes; engineering bricks; in cement: mortar (1:3)							
one brick thick	74.34	3.00	106.95	–	100.33	m²	**207.28**
one and a half brick thick	111.51	4.00	142.60	–	150.49	m²	**293.09**
two brick thick projection of footing or the like	148.68	4.80	171.12	–	200.65	m²	**371.77**
Extra over common or engineering bricks in any mortar for fair face; flush pointing as work proceeds; English bond walls or the like	–	0.13	4.75	–	–	m²	**4.75**
In situ finishings; cement: sand mortar (1:3); steel trowelled; 13 mm one coat work to manhole walls; to brickwork or blockwork base; over 300 mm wide	–	0.80	28.52	–	4.95	m²	**33.47**
Building into brickwork; ends of pipes; making good facings or renderings							
small	–	0.20	7.13	–	–	nr	**7.13**
large	–	0.30	10.70	–	–	nr	**10.70**
extra large	–	0.40	14.26	–	–	nr	**14.26**
extra large; including forming ring arch cover	–	0.50	17.83	–	–	nr	**17.83**
Inspection chambers; polypropylene							
Mini access chamber; including cover and frame							
300 mm dia. × 600 mm deep; three 100/110 mm inlets	79.80	3.00	106.95	–	98.68	nr	**205.63**

34 DRAINAGE BELOW GROUND

Item	PC £	Labour hours	Labour £	Plant £	Material £	Unit	Total rate £
INSPECTION CHAMBERS AND MANHOLES – cont							
Inspection chambers – cont							
Up to 1200 mm deep; including polymer cover and frame with screw down lid							
475 mm dia. × 940 mm deep; five 100/110 mm inlets; supplied with four stoppers in inlets	652.89	4.00	142.60	–	757.73	nr	**900.33**
Extra for square ductile iron cover and frame to 1 tonne load; screw down lid	321.79		–	–	127.32	nr	**127.32**
Step irons; drainage systems; malleable cast iron; galvanized; building into joints							
General purpose pattern; for one brick walls	5.78	0.17	6.06	–	6.65	nr	**12.71**
Accessories in PVC-u							
110 mm screwed access cover	7.88		–	–	9.07	nr	**9.07**
110 mm rodding eye	12.30	0.50	17.83	–	25.66	nr	**43.48**
Kerbs; to gullies; in one course Class B engineering bricks; to four sides; rendering in cement: mortar (1:3); dished to gully gratings	4.96	1.00	35.65	–	7.18	nr	**42.83**
CAST IRON DRAINAGE							
The labour teams in this section are calculated on a 2 person team. The labour time below should be multiplied by 2 to calculate the cost as shown.							
Cast iron pipes and fittings; PAM; Timesaver							
100 mm cast iron pipes double spiggot; in trenches (trenches not included)							
laid straight	48.81	0.11	7.92	–	56.13	m	**64.06**
short runs under 3.00 m	62.91	0.17	11.89	–	72.34	m	**84.23**
Extra over 100 mm cast iron pipes for							
bends; medium radius – 87.5°–60°	108.54	0.33	11.76	–	124.82	nr	**296.33**
bends; medium radius – 45°–10°	90.58	0.33	11.76	–	104.16	nr	**115.93**
bends; long radius	162.49	0.25	8.91	–	186.86	nr	**195.78**
plain branch	139.75	0.25	8.91	–	160.71	nr	**169.62**

34 DRAINAGE BELOW GROUND

Item	PC £	Labour hours	Labour £	Plant £	Material £	Unit	**Total rate £**
150 mm cast iron pipes; in trenches (trenches not included)							
laid straight	97.61	0.13	8.91	–	112.25	m	**121.16**
short runs under 3.00 m	126.40	0.25	17.83	–	145.36	m	**163.19**
Extra over 150 mm cast iron pipes for							
bends; medium radius – 87.5°–60°	220.34	0.33	11.76	–	253.39	nr	**265.15**
bends; medium radius – 45°–10°	206.56	0.33	11.76	–	237.55	nr	**249.31**
225 mm cast iron pipes; in trenches (trenches not included)							
laid straight	329.31	0.20	14.26	–	378.71	m	**392.97**
short runs under 3.00 m	329.31	0.50	35.65	–	378.71	m	**414.36**
Extra over 225 mm cast iron pipes for							
bends; medium radius – 87.5°–10°	541.78	0.50	17.83	–	623.05	nr	**640.87**
Cast iron inspection chambers; PAM; Timesaver; drainage systems bolted flat covers; bedding in cement: mortar (1:3); mechanical coupling joints							
100 × 100 mm							
one branch each side	327.76	1.50	106.95	–	427.99	nr	**534.94**
two branches each side	645.87	2.50	178.25	–	891.03	nr	**1069.28**
100 × 150 mm							
one branch each side	553.71	2.00	142.60	–	637.27	nr	**779.87**
two branch each side	1078.66	2.50	178.25	–	1240.96	nr	**1419.21**
150 × 150 mm							
one branch each side	539.83	2.50	178.25	–	819.87	nr	**998.12**
two branches each side	1077.74	3.00	213.90	–	1537.75	nr	**1751.65**
Accessories in cast iron							
Gullies; including concrete bed and surround							
Gully trap with surface access 100 mm × 87.5°	277.47	1.00	71.30	–	388.45	nr	**459.75**
Gully trap with 225 mm inlet	242.03	1.00	71.30	–	347.68	nr	**418.98**
Trapless gully	266.48	1.00	71.30	–	375.81	nr	**447.11**
Gully trap square 225 × 225 mm with sediment trap and grate	568.96	1.00	71.30	–	723.66	nr	**794.96**
Doans trap gully 600 mm deep inclusive of sediment pan	672.29	1.00	71.30	–	842.49	nr	**913.79**
Road Gully including hinged grate and frame	785.72	3.00	213.90	–	972.93	nr	**1186.83**

34 DRAINAGE BELOW GROUND

Item	PC £	Labour hours	Labour £	Plant £	Material £	Unit	Total rate £
LINEAR DRAINAGE							
The labour in this section is calculated on a 2 person team. The labour time below should be multiplied by 2 to calculate the cost as shown.							
Marshalls Plc; Mini Beany combined kerb and channel drainage system; to trenches (not included)							
Precast concrete drainage channel base; 185–385 mm deep; bedding, jointing and pointing in cement: mortar (1:3); on 150 mm deep concrete (ready mixed) foundation; including haunching with in situ concrete; 11.50 N/mm^2–40 mm aggregate one side; channels 250 mm wide × 1.00 m long							
straight; 1.00 m long	50.11	1.00	35.65	–	62.94	m	**98.59**
straight; 500 mm long	50.11	1.05	37.43	–	62.94	m	**100.37**
radial; 30/10 or 9/6 internal or external	53.43	1.33	47.53	–	66.76	m	**114.30**
angles 45 or 90 degree	136.96	1.00	35.65	–	162.82	nr	**198.47**
Mini Beany Top Block; perforated kerb unit to drainage channel above; natural grey							
straight	50.45	0.33	11.88	–	61.47	m	**73.35**
radial; 30/10 or 9/6 internal or external	52.55	0.50	17.83	–	63.89	m	**81.71**
angles 45 or 90 degree	136.80	0.50	17.83	–	160.77	nr	**178.60**
Mini Beany; outfalls; two section concrete trapped outfall with Mini Beany cast iron access cover and frame; to concrete foundation							
high capacity outfalls; silt box 150/225 mm outlet; two section trapped outfall silt box and cast iron access cover	422.55	1.00	35.65	–	999.86	nr	**1035.51**
inline side or end outlet; outfall 150 mm; 2 section concrete trapped outfall; cast iron Mini Beany access cover and frame	402.01	1.00	35.65	–	463.55	nr	**499.20**
Ancillaries to Mini Beany							
end cap	25.82	0.25	8.91	–	29.70	nr	**38.61**
end cap outlets	67.20	0.25	8.91	–	77.28	nr	**86.20**

34 DRAINAGE BELOW GROUND

Item	PC £	Labour hours	Labour £	Plant £	Material £	Unit	**Total rate £**
Slot and Channel drains							
Loadings for slot drains							
A15; 1.5 tonne; pedestrian							
B125; 12.5 tonne; domestic use							
C250; 25 tonne; car parks, supermarkets, industrial units							
D400; 40 tonne; highways							
E600; 60 tonne; forklifts							
Concrete bases for channel and slot drains; no formwork requirement; Plain concrete bases; C20/25 bases; Constant depth channels; Other channels in each range may vary;							
C250 loading; Paving surface will vary the volumes shown marginally							
M100D; Channel 135 mm wide × 155 deep; 0.0915 m^3 /m	–	0.07	4.76	–	10.52	m	**15.27**
M100D; Slot 135 mm wide × 155 deep; 0.1097 m^3 /m	–	0.08	5.70	–	12.63	m	**18.33**
M150D; Channel 135 mm wide × 155 deep; 0.1138 m^3 /m	–	0.10	6.80	–	13.08	m	**19.88**
M150D; Slot 135 mm wide × 155 deep; 0.1310 m^3 /m	–	0.10	6.97	–	15.06	m	**22.03**
M200D; Channel 135 mm wide × 155 deep; 0.1343 m^3 /m	–	0.11	7.84	–	15.44	m	**23.27**
M200D; Slot 135 mm wide × 155 deep; 0.1507 m^3 /m	–	0.10	7.32	–	17.32	m	**24.64**
D400 loading; Paving surface will vary the volumes shown marginally							
M100D; Channel 135 mm wide × 155 deep; 0.1407 m^3 /m	–	0.10	7.31	–	16.17	m	**23.48**
M100D; Slot 135 mm wide × 155 deep; 0.1687 m^3 /m	–	0.12	8.76	–	19.39	m	**28.15**
M150D; Channel 135 mm wide × 155 deep; 0.1751 m^3 /m	–	0.13	9.10	–	20.13	m	**29.22**
M150D; Slot 135 mm wide × 155 deep; 0.2015 m^3 /m	–	0.15	10.47	–	23.16	m	**33.63**
M200D; Channel 135 mm wide × 155 deep; 0.2066 m^3 /m	–	0.15	10.74	–	23.75	m	**34.48**
M200D; Slot 135 mm wide × 155 deep; 0.2318 m^3 /m	–	0.17	12.04	–	26.64	m	**38.68**

34 DRAINAGE BELOW GROUND

Item	PC £	Labour hours	Labour £	Plant £	Material £	Unit	Total rate £
LINEAR DRAINAGE – cont							
Concrete bases for channel and slot drains; with formwork requirement; Plain concrete bases; C20/25 bases; Constant depth channels; Other channels in each range may vary;							
D400 loading; Paving surface will vary the volumes shown marginally							
M100D; Channel 135 mm wide × 155 deep; 0.0915 m^3 /m	9.15	0.23	16.58	–	19.88	m	**36.45**
M100D; Slot 135 mm wide × 155 deep; 0.1097 m^3 /m	10.98	0.24	17.04	–	21.99	m	**39.03**
M150D; Channel 135 mm wide × 155 deep; 0.1138 m^3 /m	11.37	0.24	17.15	–	22.44	m	**39.59**
M150D; Slot 135 mm wide × 155 deep; 0.1310 m^3 /m	13.09	0.25	17.58	–	24.42	m	**42.00**
M200D; Channel 135 mm wide × 155 deep; 0.1343 m^3 /m	13.42	0.25	17.66	–	24.79	m	**42.46**
M200D; Slot 135 mm wide × 155 deep; 0.1507 m^3 /m	15.06	0.25	18.07	–	26.68	m	**44.75**
D400 loading; Paving surface will vary the volumes shown marginally							
M100D; Channel 135 mm wide × 155 deep; 0.1407 m^3 /m	14.06	0.25	17.83	–	25.53	m	**43.36**
M100D; Slot 135 mm wide × 155 deep; 0.1687 m^3 /m	16.86	0.26	18.54	–	28.75	m	**47.29**
M150D; Channel 135 mm wide × 155 deep; 0.1751 m^3 /m	17.50	0.26	18.69	–	29.48	m	**48.18**
M150D; Slot 135 mm wide × 155 deep; 0.2015 m^3 /m	20.14	0.27	19.37	–	32.52	m	**51.88**
M200D; Channel 135 mm wide × 155 deep; 0.2066 m^3 /m	20.65	0.27	19.49	–	33.10	m	**52.60**
M200D; Slot 135 mm wide × 155 deep; 0.2318 m^3 /m	23.17	0.28	20.14	–	36.00	m	**56.14**

34 DRAINAGE BELOW GROUND

Item	PC £	Labour hours	Labour £	Plant £	Material £	Unit	Total rate £
Channel drains; ACO Technologies; ACO MultiDrain MD polymer concrete channel drainage system; traditional channel and grate drainage solution; on concrete base; Constant depth channels laid to falls; design consultation available and preferred for all complex systems							
ACO MultiDrain; M100D Overall widths 135 mm; overall depths 75–300 mm; installation on concrete bases (shown separately); gratings and slots shown separately; selected sizes only shown; D400; product name denotes internal width							
M100D	40.14	0.15	10.97	–	47.20	m	**58.17**
M100DS; Stainless	58.30	0.17	11.89	–	68.08	m	**79.97**
Aco MultiDrain; M150D; overall width 185 mm; overall depths 100–310 mm; installation on concrete bases (shown separately); gratings and slots shown separately; selected sizes only shown.							
M150D D400	47.87	0.17	12.40	–	56.09	m	**68.49**
M150DS D400; Stainless	101.54	0.17	12.40	–	117.81	m	**130.21**
ACO MultiDrain; M200D overall width 235 mm; overall depths 100 mm–350 mm; installation on concrete bases (shown separately); gratings and slots shown separately; selected sizes only shown.							
M200D D400	90.41	0.20	14.26	–	105.01	m	**119.27**
M200DS D400; Stainless	126.53	0.20	14.26	–	146.55	m	**160.81**
Gratings for Channel and Slot drains; ACO Technologies; ACO MultiDrain							
MultiDrain M100D load class D400							
brickslot; galvanized steel; 1000 mm	110.58	0.08	2.97	–	127.17	m	**130.14**
brickslot; stainless steel; 1000 mm	225.77	0.08	2.97	–	259.64	m	**262.61**
Heelguard mesh galvanized steel grating 410DL	54.06	0.08	5.94	–	62.17	m	**68.11**
12644 Slotted stainless steel	60.99	0.08	5.94	–	70.14	m	**76.08**
Heelguard composite – black 522DL	40.66	0.11	3.96	–	46.76	m	**50.72**

34 DRAINAGE BELOW GROUND

Item	PC £	Labour hours	Labour £	Plant £	Material £	Unit	Total rate £
LINEAR DRAINAGE – cont							
Gratings for Channel and Slot drains – cont							
MultiDrain M100D; Load class C250							
brickslot; galvanized steel; 1000 mm	110.58	0.08	2.97	–	127.17	m	**130.14**
Brickslot single slot stainless steel 23475	225.77	0.08	5.94	–	259.64	m	**265.57**
Heelguard mesh stainless steel 132887DL	86.44	0.08	5.94	–	99.41	m	**105.35**
Heelguard mesh stainless steel 132888DL 500 mm	136.70	0.08	5.94	–	157.21	m	**163.14**
Brickslot Twinslot offset stainless steel 23490	275.89	0.08	5.94	–	317.27	m	**323.21**
MultiDrain M150D load class D400							
brickslot; galvanized steel; 1000 mm	110.58	0.08	2.97	–	127.17	m	**130.14**
brickslot; stainless steel; 1000 mm	225.77	0.08	2.97	–	259.64	m	**262.61**
Heelguard ductile grating	74.48	0.08	5.94	–	85.65	m	**91.59**
Slotted ductile iron grating	71.18	0.11	7.92	–	81.86	m	**89.78**
Heelguard mesh stainless steel	–	0.11	3.96	–	191.89	m	**195.85**
MultiDrain M200D; load class D400							
brickslot; galvanized steel; 1000 mm	110.58	0.08	2.97	–	127.17	m	**130.14**
brickslot; stainless steel; 1000 mm	225.77	0.08	2.97	–	259.64	m	**262.61**
Heelguard ductile grating	115.36	0.08	5.94	–	132.66	m	**138.60**
Slotted ductile iron grating	92.82	0.11	7.92	–	106.74	m	**114.66**
Heelguard mesh stainless steel	179.71	0.11	3.96	–	206.67	m	**210.63**
Accessories for MultiDrain MD							
Brickslot single slot access unit stainless steel 23477; 500 mm	290.80	0.50	17.83	–	334.42	m	**352.25**
sump unit; complete with sediment bucket and access unit	116.13	1.00	35.65	–	298.28	nr	**333.93**
Brickslot Twinslot offset access unit stainless steel 23492; 500 mm	207.12	1.00	71.30	–	238.19	m	**309.49**
end cap; closing piece	12.65		–	–	14.55	nr	**14.55**
end cap; inlet/outlet	9.41		–	–	10.82	nr	**10.82**
PIPE LAYING							
The labour in this section is calculated on a 3 person team. The labour time below should be multiplied by 3 to calculate the cost as shown.							

34 DRAINAGE BELOW GROUND

Item	PC £	Labour hours	Labour £	Plant £	Material £	Unit	Total rate £
Excavating trenches; to receive pipes; grading bottoms to falls; backfilling with excavated material and compacting; disposal of surplus material off site; volumes allow for bedding materials which are priced separately below							
Trenches 300 mm wide							
depth of pipe 750 mm	–		–	14.51	–	m^3	**21.09**
depth of pipe 900 mm	–		–	16.56	–	m^3	**24.87**
depth of pipe 1.20 m	–		–	17.34	–	m^3	**29.08**
depth of pipe 1.50 m	–		–	19.23	–	m^3	**34.41**
depth of pipe 2.00 m	–		–	11.85	–	m^3	**32.61**
Trenches 300 mm wide							
depth of pipe 750 mm	–		–	4.80	–	m	**9.12**
depth of pipe 900 mm	–		–	5.83	–	m	**11.01**
depth of pipe 1.20 m	–		–	7.78	–	m	**14.68**
depth of pipe 1.50 m	–		–	9.73	–	m	**18.36**
depth of pipe 2.00 m	–		–	7.59	–	m	**28.26**
Earthwork support; providing support to opposing faces of excavation; moving along as work proceeds							
Maximum depth not exceeding 2.00 m							
distance between opposing faces not exceeding 2.00 m	–	0.80	28.52	35.73	–	m	**64.25**
Excavating trenches; using 3 tonne tracked excavator; to receive pipes; grading bottoms; earthwork support; filling with excavated material to within 150 mm of finished surfaces and compacting; completing fill with topsoil; disposal of surplus soil							
Services not exceeding 200 mm nominal size							
average depth of run not exceeding 0.50 m	1.60	0.12	4.28	1.40	1.84	m	**7.51**
average depth of run not exceeding 0.75 m	1.60	0.16	5.81	1.94	1.84	m	**9.58**
average depth of run not exceeding 1.00 m	1.60	0.28	10.10	3.38	1.84	m	**15.32**
average depth of run not exceeding 1.25 m	1.33	0.38	13.66	4.55	1.53	m	**19.74**
Granular beds to trenches; lay granular material to trenches excavated separately; to receive pipes (not included)							

34 DRAINAGE BELOW GROUND

Item	PC £	Labour hours	Labour £	Plant £	Material £	Unit	Total rate £
PIPE LAYING – cont							
300 mm wide × 100 mm thick							
reject sand	–	0.05	1.78	0.32	4.93	m	**7.03**
reject gravel	–	0.05	1.78	0.32	3.55	m	**5.65**
shingle 40 mm aggregate	–	0.05	1.78	0.32	6.69	m	**8.79**
sharp sand	4.99	0.05	1.78	0.32	5.74	m	**7.84**
300 mm wide × 150 mm thick							
reject sand	–	0.08	2.67	0.47	7.40	m	**10.54**
reject gravel	–	0.08	2.67	0.47	5.33	m	**8.47**
shingle 40 mm aggregate	–	0.08	2.67	0.47	10.04	m	**13.19**
sharp sand	7.48	0.08	2.67	0.47	8.61	m	**11.75**
Excavating trenches; using 3 tonne tracked excavator; to receive pipes; grading bottoms; earthwork support; filling with imported granular material type 2 and compacting; disposal of surplus soil							
Services not exceeding 200 mm nominal size							
average depth of run not exceeding 0.50 m	2.72	0.09	3.02	0.96	3.06	m	**13.94**
average depth of run not exceeding 0.75 m	4.08	0.11	3.77	1.19	4.59	m	**19.88**
average depth of run not exceeding 1.00 m	5.44	0.14	4.87	1.56	6.12	m	**26.31**
average depth of run not exceeding 1.25 m	6.80	0.23	7.98	2.66	7.66	m	**35.54**
Excavating trenches; using 3 tonne tracked excavator; to receive pipes; grading bottoms; earthwork support; filling with concrete, ready mixed C10; disposal of surplus soil							
Services not exceeding 200 mm nominal size							
average depth of run not exceeding 0.50 m	14.55	0.11	3.76	0.50	16.55	m	**27.78**
average depth of run not exceeding 0.75 m	21.82	0.13	4.58	0.62	24.82	m	**40.47**
average depth of run not exceeding 1.00 m	29.09	0.17	5.88	0.83	33.09	m	**53.73**
average depth of run not exceeding 1.25 m	36.37	0.23	7.93	1.25	41.36	m	**67.96**

34 DRAINAGE BELOW GROUND

Item	PC £	Labour hours	Labour £	Plant £	Material £	Unit	**Total rate £**
Earthwork support; providing support to opposing faces of excavation; moving along as work proceeds							
Maximum depth not exceeding 2.00 m							
trenchbox; distance between opposing faces not exceeding 2.00 m	–	0.80	28.52	35.73	–	m	**64.25**
timber; distance between opposing faces not exceeding 500 mm	–	0.20	14.27	–	0.03	m	**14.29**
Clay pipes and fittings; Hepworth Plc; Supersleve							
100 mm clay pipes; polypropylene slip coupling; in trenches (trenches not included)							
laid straight	11.58	0.25	8.91	–	19.49	m	**28.41**
short runs under 3.00 m	11.58	0.31	11.14	–	19.49	m	**30.63**
Extra over 100 mm clay pipes for							
bends; 15–90°; single socket	15.72	0.25	8.91	–	18.08	nr	**26.99**
junction; 45 or 90°; double socket	33.97	0.25	8.91	–	39.06	nr	**47.98**
slip couplings; polypropylene	8.59	0.08	2.97	–	9.88	nr	**12.85**
gully with P trap; 100 mm; 154 × 154 mm plastic grating	60.42	1.00	35.65	–	113.15	nr	**148.80**
150 mm clay pipes; polypropylene slip coupling; in trenches (trenches not included)							
laid straight	36.48	0.30	10.70	–	41.95	m	**52.65**
short runs under 3.00 m	36.48	0.60	21.39	–	41.95	m	**63.34**
Extra over 150 mm clay pipes for							
bends; 15–90 degree	31.23	0.28	9.98	–	53.24	nr	**63.22**
junction; 45 or 90°; 100 × 150 mm	41.80	0.40	14.26	–	82.71	nr	**96.97**
junction; 45 or 90°; 150 × 150 mm	45.89	0.40	14.26	–	87.41	nr	**101.67**
slip couplings; polypropylene	15.06	0.05	1.78	–	17.32	nr	**19.10**
tapered pipe; 100–150 mm	47.25	0.50	17.83	–	54.34	nr	**72.17**
tapered pipe; 150–225 mm	120.62	0.50	17.83	–	138.71	nr	**156.53**
socket adaptor; connection to traditional pipes and fittings	31.68	0.33	11.76		53.75	nr	**65.52**
Accessories in clay							
access pipe; 150 mm	117.76		–	–	170.07	nr	**170.07**
rodding eye; 150 mm	112.64	0.50	17.83	–	141.05	nr	**158.88**
gully with P traps; 150 mm; 154 × 154 mm plastic grating	132.89	0.80	28.52	–	174.74	nr	**203.26**

34 DRAINAGE BELOW GROUND

Item	PC £	Labour hours	Labour £	Plant £	Material £	Unit	Total rate £
PIPE LAYING – cont							
PVC-u pipes and fittings; Wavin Plastics Ltd; OsmaDrain; laid to and covered over with imported granular material							
110 mm PVC-u pipes; in trenches (trenches not included)							
laid straight	3.82	0.08	2.85	2.58	6.60	m	**12.03**
short runs under 3.00 m	4.53	0.12	4.28	2.58	7.43	m	**14.28**
Extra over 110 mm PVC-u pipes for							
bends; short radius	6.82	0.25	8.91	–	7.84	nr	**16.76**
bends; long radius	22.95	0.25	8.91	–	26.39	nr	**35.30**
junctions; equal; double socket	6.23	0.25	8.91	–	7.17	nr	**16.08**
slip couplings	0.70	0.25	8.91	–	0.80	nr	**9.71**
adaptors to clay	25.30	0.50	17.83	–	29.10	nr	**46.92**
160 mm PVC-u pipes; in trenches (trenches not included)							
laid straight	6.44	0.08	2.85	2.58	9.62	m	**15.05**
short runs under 3.00 m	8.11	0.12	4.28	2.58	11.54	m	**18.39**
Extra over 160 mm PVC-u pipes for							
bends; short radius	1.27	0.25	8.91	–	9.30	nr	**18.21**
bends; long radius	29.96	0.25	8.91	–	34.46	nr	**43.37**
junctions; single	22.07	0.33	11.88	–	25.37	nr	**37.26**
pipe coupler	0.63	0.05	1.78	–	0.73	nr	**2.51**
slip couplings PVC-u	0.61	0.05	1.78	–	0.70	nr	**2.49**
adaptors to clay	10.36	0.50	17.83	–	11.92	nr	**29.74**
level invert reducer	22.50	0.50	17.83	–	25.87	nr	**43.70**
spiggot	7.19	0.20	7.13	–	8.27	nr	**15.40**
CATCHWATER OR FRENCH DRAINS							
The labour in this section is calculated on a 2 person team. The labour time below should be multiplied by 2 to calculate the cost as shown.							
Catchwater or French drains							
excavation; diposal to spoil heaps 100 m max; lining trench; wrapping trench with filter fabric; backfilling with shingle							
depth to 1.5 m	57.71	0.17	11.89	13.62	66.36	m^3	**91.86**
pipe perforated to trench excavated perepared and backfilled separately							
80 mm Ø	0.81	0.01	0.36	–	0.94	m	**1.30**
100 mm Ø	1.19	0.01	0.36	–	1.38	m	**1.74**
160 mm Ø	2.53	0.01	0.36	–	2.93	m	**3.29**

34 DRAINAGE BELOW GROUND

Item	PC £	Labour hours	Labour £	Plant £	Material £	Unit	**Total rate £**
DITCHING							
Ditching; clear silt and bottom ditch not exceeding 1.50 m deep; strim back vegetation; disposing to spoil heaps; by machine							
Up to 1.50 m wide at top	–	6.00	213.90	37.83	–	100 m	**251.73**
1.50–2.50 m wide at top	–	8.00	285.20	50.44	–	100 m	**335.64**
2.50–4.00 m wide at top	–	9.00	320.85	56.75	–	100 m	**377.60**
Ditching; clear only vegetation from ditch not exceeding 1.50 m deep; disposing to spoil heaps; by strimmer							
Up to 1.50 m wide at top	–	5.00	178.25	12.45	–	100 m	**190.70**
1.50–2.50 m wide at top	–	6.00	213.90	22.42	–	100 m	**236.32**
2.50–4.00 m wide at top	–	7.00	249.55	34.87	–	100 m	**284.42**
Ditching; clear silt from ditch not exceeding 1.50 m deep; trimming back vegetation; disposing to spoil heaps; by hand							
Up to 1.50 m wide at top	–	15.00	534.75	12.45	–	100 m	**547.20**
1.50–2.50 m wide at top	–	27.00	962.55	22.42	–	100 m	**984.97**
2.50–4.00 m wide at top	–	42.00	1497.30	34.87	–	100 m	**1532.17**
Ditching; excavating and forming ditch and bank to given profile (normally 45°); in loam or sandy loam; by machine							
Width 300 mm							
depth 600 mm	–	3.70	131.91	46.66	–	100 m	**178.56**
depth 900 mm	–	5.20	185.38	65.57	–	100 m	**250.95**
depth 1200 mm	–	7.20	256.68	90.79	–	100 m	**347.47**
depth 1500 mm	–	9.40	335.11	103.72	–	100 m	**438.83**
Width 600 mm							
depth 600 mm	–		–	415.78	–	100 m	**415.78**
depth 900 mm	–		–	616.96	–	100 m	**616.96**
depth 1200 mm	–		–	838.27	–	100 m	**838.27**
depth 1500 mm				1207.10	–	100 m	**1207.10**
Width 900 mm							
depth 600 mm	–		–	640.13	–	100 m	**640.13**
depth 900 mm	–		–	1127.85	–	100 m	**1127.85**
depth 1200 mm	–		–	1515.58	–	100 m	**1515.58**
depth 1500 mm	–		–	1877.72	–	100 m	**1877.72**
Width 1200 mm							
depth 600 mm	–		–	1005.92	–	100 m	**1005.92**
depth 900 mm	–		–	1403.64		100 m	**1403.64**
depth 1200 mm	–		–	1978.31	–	100 m	**1978.31**
depth 1500 mm	–		–	2534.92	–	100 m	**2534.92**

34 DRAINAGE BELOW GROUND

Item	PC £	Labour hours	Labour £	Plant £	Material £	Unit	Total rate £
DITCHING – cont							
Ditching – cont							
Width 1500 mm							
depth 600 mm	–		–	1240.63	–	100 m	**1240.63**
depth 900 mm	–		–	1891.13	–	100 m	**1891.13**
depth 1200 mm	–		–	2514.80	–	100 m	**2514.80**
depth 1500 mm	–		–	3151.88	–	100 m	**3151.88**
Extra for ditching in clay	–		–	–	–	20%	**–**
Disposal of excavated material							
Excavated material; off site; to tip; mechanically loaded (360 Excavator)							
hardcore	–	0.02	0.71	1.42	–	m^3	**22.59**
inert (clean landfill, rubble concrete); loose	–	0.03	0.95	–	–	m^3	**38.42**
In spoil heaps							
average 25 m distance	–		–	3.57	–	m^3	**3.57**
average 50 m distance	–		–	4.12	–	m^3	**4.12**
average 100 m distance (1 dumper)	–		–	5.16	–	m^3	**5.16**
average 100 m distance (2 dumpers)	–		–	6.31	–	m^3	**6.31**
The labour in this section is calculated on a 3 person team. The labour time below should be multiplied by 3 to calculate the cost as shown.							
Ditching; excavating and forming ditch and bank to given profile (normal 45°); in loam or sandy loam; by hand							
Width 300 mm							
depth 600 mm	–	36.00	1283.40	–	–	100 m	**1283.40**
depth 900 mm	–	42.00	1497.30	–	–	100 m	**1497.30**
depth 1200 mm	–	56.00	1996.40	–	–	100 m	**1996.40**
depth 1500 mm	–	70.00	2495.50	–	–	100 m	**2495.50**
Width 600 mm							
depth 600 mm	–	56.00	1996.40	–	–	100 m	**1996.40**
depth 900 mm	–	84.00	2994.60	–	–	100 m	**2994.60**
depth 1200 mm	–	112.00	3992.80	–	–	100 m	**3992.80**
depth 1500 mm	–	140.00	4991.00	–	–	100 m	**4991.00**
Width 900 mm							
depth 600 mm	–	84.00	2994.60	–	–	100 m	**2994.60**
depth 900 mm	–	126.00	4491.90	–	–	100 m	**4491.90**
depth 1200 mm	–	168.00	5989.20	–	–	100 m	**5989.20**
depth 1500 mm	–	210.00	7486.50	–	–	100 m	**7486.50**

34 DRAINAGE BELOW GROUND

Item	PC £	Labour hours	Labour £	Plant £	Material £	Unit	**Total rate £**
Width 1200 mm							
depth 600 mm	–	112.00	3992.80	–	–	100 m	**3992.80**
depth 900 mm	–	168.00	5989.20	–	–	100 m	**5989.20**
depth 1200 mm	–	224.00	7985.60	–	–	100 m	**7985.60**
depth 1500 mm	–	280.00	9982.00	–	–	100 m	**9982.00**
Extra for ditching in clay	–		–	–	–	50%	**–**
Extra for ditching in compacted soil	–		–	–	–	90%	**–**
Piped ditching							
Jointed concrete pipes; FP McCann Ltd; including bedding, haunching and topping with 150 mm concrete; 11.50 N/mm^2–40 mm aggregate; to existing ditch							
300 mm dia.	32.08	0.67	23.76	13.41	85.12	m	**122.29**
450 mm dia.	47.79	0.67	23.76	13.41	127.31	m	**164.48**
600 mm dia.	76.77	0.67	23.76	13.41	191.02	m	**228.19**
900 mm dia.	203.47	1.00	35.65	20.12	264.67	m	**320.44**
1200 mm dia.	349.20	1.00	35.65	20.12	441.73	m	**497.50**
extra over jointed concrete pipes for bends to 45°	320.63	0.67	23.76	16.77	380.96	nr	**421.49**
extra over jointed concrete pipes for single junctions 300 mm dia.	219.67	0.67	23.76	16.77	264.86	nr	**305.39**
extra over jointed concrete pipes for single junctions 450 mm dia.	325.62	0.67	23.76	15.24	386.70	nr	**425.71**
extra over jointed concrete pipes for single junctions 600 mm dia.	527.79	0.67	23.76	15.24	619.19	nr	**658.20**
extra over jointed concrete pipes for single junctions 900 mm dia.	892.18	0.67	23.76	15.24	1038.24	nr	**1077.25**
extra over jointed concrete pipes for single junctions 1200 mm dia.	1362.93	0.67	23.76	15.24	1579.61	nr	**1618.61**
Outfalls							
Reinforced concrete outfalls to water course; flank walls; for 150 mm drain outlets; overall dimensions							
Small headwall FP McCann Ltd	937.49	4.00	427.80	60.96	1183.44	m	**1672.20**

34 DRAINAGE BELOW GROUND

Item	PC £	Labour hours	Labour £	Plant £	Material £	Unit	Total rate £
LAND DRAINS							
Market prices of backfilling materials							
Sand	46.77		–	–	46.77	m^3	**46.77**
Gravel rejects	32.34		–	–	32.34	m^3	**32.34**
Imported topsoil (allowing for 20% settlement)	35.51		–	–	35.51	m^3	**35.51**
Land drainage; calculation table							
For calculation of drainage per hectare the following table can be used; rates show the lengths of drains per unit and not the value							
Lateral drains							
10.00 m centres	–		–	–	–	m/ha	**1000.00**
15.00 m centres	–		–	–	–	m/ha	**650.00**
25.00 m centres	–		–	–	–	m/ha	**400.00**
30.00 m centres	–		–	–	–	m/ha	**330.00**
Main drains							
1 nr (at 100 m centres)	–		–	–	–	m/ha	**100.00**
2 nr (at 50 m centres)	–		–	–	–	m/ha	**200.00**
3 nr (at 33.3 m centres)	–		–	–	–	m/ha	**300.00**
4 nr (at 25 m centres)	–		–	–	–	m/ha	**400.00**
The labour in this section is calculated on a 3 person team. The labour time below should be multiplied by 3 to calculate the cost as shown.							
Land drainage; excavating							
Removing 150 mm depth of topsoil; 300 mm wide; depositing beside trench; by machine	–	1.50	53.48	18.92	–	100 m	**72.39**
Removing 150 mm depth of topsoil; 300 mm wide; depositing beside trench; by hand	–	8.00	285.20	–	–	100 m	**285.20**
Excavated material; disposal on site							
In spoil heaps							
average 25 m distance	–		–	3.57	–	m^3	**3.57**
average 50 m distance	–		–	4.12	–	m^3	**4.12**
average 100 m distance (1 dumper)	–		–	5.16	–	m^3	**5.16**
average 100 m distance (2 dumpers)	–		–	6.31	–	m^3	**6.31**
average 200 m distance	–		–	6.22	–	m^3	**6.22**
average 200 m distance (2 dumpers)	–		–	7.59	–	m^3	**7.59**

34 DRAINAGE BELOW GROUND

Item	PC £	Labour hours	Labour £	Plant £	Material £	Unit	Total rate £
Removing excavated material from site to tip; mechanically loaded							
excavated material and clean hardcore rubble	–		–	2.44	–	m^3	**37.92**
Land drainage; Agripower Ltd; excavating trenches (with minimum project size of 1000 m) by trenching machine; spreading arisings on site; laying perforated pipe; as shown below; backfilling with shingle to 350 mm from surface							
Width 150 mm; 80 mm perforated pipe							
depth 750 mm	–		–	–	–	m	**11.32**
Width 175 mm; 100 mm perforated pipe							
depth 800 mm	–		–	–	–	m	**11.76**
Width 250 mm; 150 mm perforated pipe							
depth 900 mm	–		–	–	–	m	**16.01**
Land drainage; excavating for drains; by 5 tonne tonne mini-excavator; including disposing spoil to spoil heaps not exceeding 100 m; boning to levels by laser							
Width 150–225 mm							
depth 450 mm	–	9.67	344.62	455.48	–	100 m	**800.09**
depth 600 mm	–	13.00	463.45	683.21	–	100 m	**1146.66**
depth 700 mm	–	15.50	552.58	854.01	–	100 m	**1406.59**
depth 900 mm	–	23.00	819.95	1366.42	–	100 m	**2186.37**
Width 300 mm							
depth 450 mm	–	9.67	344.62	455.48	–	100 m	**800.09**
depth 700 mm	–	14.00	499.10	751.53	–	100 m	**1250.63**
depth 900 mm	–	23.00	819.95	1366.42	–	100 m	**2186.37**
depth 1000 mm	–	25.00	891.25	1503.06	–	100 m	**2394.31**
depth 1200 mm	–	25.00	891.25	1708.03	–	100 m	**2599.28**
depth 1500 mm	–	31.00	1105.15	1912.99	–	100 m	**3018.14**
Land drainage; excavating for drains; by 8 tonne tracked excavator; including disposing spoil to spoil heaps not exceeding 100 m							
Width 225 mm							
depth 450 mm	–	7.00	249.55	300.19	–	100 m	**549.74**
depth 600 mm	–	7.17	255.49	312.70	–	100 m	**568.19**
depth 700 mm	–	7.35	261.95	326.29	–	100 m	**588.24**

34 DRAINAGE BELOW GROUND

Item	PC £	Labour hours	Labour £	Plant £	Material £	Unit	Total rate £
LAND DRAINS – cont							
Land drainage – cont							
Width 225 mm – cont							
depth 900 mm	–	7.76	276.71	357.37	–	100 m	**634.08**
depth 1000 mm	–	8.00	285.20	375.24	–	100 m	**660.44**
depth 1200 mm	–	8.26	294.58	394.99	–	100 m	**689.57**
Width 300 mm							
depth 450 mm	–	7.35	261.95	326.29	–	100 m	**588.24**
depth 600 mm	–	7.76	276.71	357.37	–	100 m	**634.08**
depth 700 mm	–	8.26	294.58	394.99	–	100 m	**689.57**
depth 900 mm	–	9.25	329.76	469.05	–	100 m	**798.81**
depth 1000 mm	–	10.14	361.59	536.06	–	100 m	**897.65**
depth 1200 mm	–	11.00	392.15	600.38	–	100 m	**992.53**
Width 600 mm							
depth 450 mm	–	13.00	463.45	750.48	–	100 m	**1213.93**
depth 600 mm	–	14.11	503.06	833.86	–	100 m	**1336.92**
depth 700 mm	–	16.33	582.28	1000.63	–	100 m	**1582.92**
depth 900 mm	–	17.29	616.24	1072.11	–	100 m	**1688.34**
depth 1000 mm	–	18.38	655.41	1154.58	–	100 m	**1809.99**
depth 1200 mm	–	21.18	755.13	1364.50	–	100 m	**2119.63**
Land drainage; excavating for drains; by hand, including disposing spoil to spoil heaps not exceeding 100 m							
Width 150 mm							
depth 450 mm	–	0.44	15.71	–	–	m	**15.71**
depth 600 mm	–	0.59	20.95	–	–	m	**20.95**
depth 700 mm	–	0.69	24.42	–	–	m	**24.42**
depth 900 mm	–	0.88	31.41	–	–	m	**31.41**
Width 225 mm							
depth 450 mm	–	0.66	23.56	–	–	m	**23.56**
depth 600 mm	–	0.88	31.41	–	–	m	**31.41**
depth 700 mm	–	1.03	36.66	–	–	m	**36.66**
depth 900 mm	–	1.32	47.13	–	–	m	**47.13**
Width 300 mm							
depth 450 mm	–	0.88	31.41	–	–	m	**31.41**
depth 600 mm	–	1.18	41.89	–	–	m	**41.89**
depth 700 mm	–	1.37	48.87	–	–	m	**48.87**
Width 375 mm							
depth 450 mm	–	1.10	39.27	–	–	m	**39.27**
depth 600 mm	–	1.47	52.36	–	–	m	**52.36**
depth 700 mm	–	1.71	61.09	–	–	m	**61.09**

34 DRAINAGE BELOW GROUND

Item	PC £	Labour hours	Labour £	Plant £	Material £	Unit	Total rate £
Land drainage; pipe laying							
Wavin Plastics Ltd; flexible plastic perforated pipes in trenches (not included); to a minimum depth of 450 mm (couplings not included)							
OsmaDrain; flexible plastic perforated pipes in trenches (not included); to a minimum depth of 450 mm (couplings not included)							
80 mm dia.; available in 100 m coil	82.89	2.00	71.30	–	95.32	100 m	**166.62**
100 mm dia.; available in 100 m coil	121.96	2.00	71.30	–	140.25	100 m	**211.55**
160 mm dia.; available in 25 m coil	259.73	2.00	71.30	–	298.68	100 m	**369.98**
WavinCoil; plastic pipe junctions							
100 × 100 mm	5.51	0.05	1.78	–	6.33	nr	**8.12**
WavinCoil; couplings for flexible pipes							
100 mm dia.	1.30	0.03	1.19	–	1.50	nr	**2.68**
160 mm dia.	1.45	0.03	1.19	–	1.67	nr	**2.86**
Land drainage; backfilling trench after laying pipes with gravel rejects or similar; blind filling with ash or sand; topping with 150 mm imported topsoil from dumps not exceeding 100 m; by machine							
Width 150 mm							
depth 450 mm	–	3.30	117.65	108.10	240.90	100 m	**466.64**
depth 600 mm	–	4.30	153.30	141.63	325.04	100 m	**619.96**
depth 900 mm	–	6.30	224.60	208.69	469.12	100 m	**902.40**
Width 225 mm							
depth 600 mm	–	6.45	229.94	212.44	487.79	100 m	**930.17**
depth 900 mm	–	9.45	336.89	313.03	716.01	100 m	**1365.93**
Land drainage; backfilling trench after laying pipes with gravel rejects or similar, blind filling with ash or sand; topping with 150 mm topsoil from dumps not exceeding 100 m; by hand							
Width 150 mm							
depth 450 mm	–	0.18	6.42	–	2.24	m	**8.66**
depth 600 mm	–	24.84	885.55	–	325.04	100 m	**1210.59**
depth 750 mm	–	31.05	1106.93	–	291.13	100 m	**1398.06**
depth 900 mm		37.26	1328.32		469.12	100 m	**1797.44**
Width 225 mm							
depth 450 mm	–	27.94	996.06	–	373.85	100 m	**1369.91**
depth 600 mm	–	37.26	1328.32	–	487.79	100 m	**1816.11**
depth 750 mm	–	46.57	1660.22	–	602.07	100 m	**2262.29**
depth 900 mm	–	55.89	1992.48	–	716.01	100 m	**2708.48**

34 DRAINAGE BELOW GROUND

Item	PC £	Labour hours	Labour £	Plant £	Material £	Unit	Total rate £
LAND DRAINS – cont							
Land drainage – cont							
Width 375 mm							
depth 450 mm	–	46.57	1660.22	–	622.82	100 m	**2283.04**
depth 600 mm	–	61.10	2178.22	–	812.83	100 m	**2991.05**
depth 750 mm	–	77.63	2767.51	–	1003.18	100 m	**3770.69**
depth 900 mm	–	93.15	3320.80	–	1193.19	100 m	**4513.99**
MARKET PRICES OF BACKFILLING MATERIALS							
Market prices of commonly used materials used for backfilling							
The prices below show the market price with standard settlement factors allowed for							
Sharp sand	–		–	–	47.63	m^3	**47.63**
Gravel (washed river or pit)	–		–	–	46.60	m^3	**46.60**
Shingle	–		–	–	56.83	m^3	**56.83**
Topsoil	–		–	–	35.51	m^3	**35.51**
Gravel rejects	–		–	–	30.14	m^3	**30.14**
Selected granular material 6F2	–		–	–	18.14	m^3	**18.14**
SAND SLITTING/GROOVING							
Sand slitting; Agripower Ltd							
Drainage slits; at 1.00 m centres; using spinning disc trenching machine; backfilling to 100 mm of surface with pea gravel; finished surface with medium grade sand 100; arisings to be loaded, hauled and tipped onsite; minimum pitch size 6000 m^2							
250 mm depth	–		–	–	–	m	**3.14**
300 mm depth	–		–	–	–	m	**3.44**
400 mm depth	–		–	–	–	m	**5.19**
Sand grooving or banding; Agripower Ltd							
Drainage bands at 260 mm centres; using Blec Sandmaster; to existing grass at 260 mm centres; minimum pitch size 4000 m^2							
sand banding or sand grooving	–		–	–	–	m^2	**1.87**

34 DRAINAGE BELOW GROUND

Item	PC £	Labour hours	Labour £	Plant £	Material £	Unit	**Total rate £**
SOAKAWAYS							
Preamble: Flat rate hourly rainfall = 50 mm/hr and assumes 100 impermeability of the run-off area. A storage capacity of the soakaway should be 1/3 of the hourly rainfall. Formulae for calculating soakaway depths are provided in the publications mentioned below and in the Tables and Memoranda section of this publication. The design of soakaways is dependent on, amongst other factors, soil conditions, permeability, groundwater level and runoff. The definitive documents for design of soakaways are CIRIA 156 and BRE Digest 365 dated September 1991. The suppliers of the systems below will assist through their technical divisions. Excavation earthwork support of pits and disposal not included.							
Soakaway excavation							
Pits; 3 tonne tracked excavator							
maximum depth not exceeding 1.00 m	–	0.25	8.91	12.89	–	m^3	**21.80**
maximum depth not exceeding 2.00 m	–	0.40	14.26	20.62	–	m^3	**34.88**
Pits; 8 tonne tracked excavator							
maximum depth not exceeding 2.00 m	–	0.40	14.26	24.39	–	m^3	**38.65**
maximum depth not exceeding 3.00 m	–	0.50	17.83	30.48	–	m^3	**48.31**
Disposal							
Excavated material; off site; to tip; mechanically loaded (8 tonne excavator							
inert	–		–	–	–	m^3	**37.43**

34 DRAINAGE BELOW GROUND

Item	PC £	Labour hours	Labour £	Plant £	Material £	Unit	Total rate £
SOAKAWAYS – cont							
The labour in this section is calculated on a 2 person team. The labour time below should be multiplied by 2 to calculate the cost as shown.							
In situ concrete ring beam foundations to base of soakaway; 300 mm wide × 250 mm deep; poured on or against earth or unblinded hardcore							
Internal dia. of rings							
900 mm	46.71	2.00	142.60	–	53.71	nr	**196.31**
1200 mm	62.25	2.25	160.43	–	71.59	nr	**232.02**
1500 mm	77.82	2.50	178.25	–	89.50	nr	**267.75**
2400 mm	124.53	2.75	196.08	–	143.21	nr	**339.28**
Concrete manhole rings; Milton Pipes Ltd; perforations and step irons to concrete rings at manufacturers recommended centres; placing of concrete ring manholes to in situ concrete ring beams (1:3:6) (not included); filling and surrounding base with gravel 225 mm deep (not included)							
Ring dia. 900 mm							
1.00 m deep; volume 636 litres	135.82	0.63	44.56	26.67	361.60	nr	**432.83**
1.50 m deep; volume 954 litres	203.73	0.83	58.82	70.41	577.40	nr	**706.64**
2.00 m deep; volume 1272 litres	271.64	0.90	64.17	70.41	766.49	nr	**901.07**
Ring dia. 1200 mm							
1.00 m deep; volume 1131 litres	177.44	1.50	106.95	64.01	470.76	nr	**641.73**
1.50 m deep; volume 1696 litres	266.16	2.25	160.43	96.02	697.14	nr	**953.59**
2.00 m deep; volume 2261 litres	354.88	2.50	178.25	96.02	923.52	nr	**1197.79**
2.50 m deep; volume 2827 litres	443.60	3.00	213.90	128.03	1056.38	nr	**1398.31**
Ring dia. 1500 mm							
1.00 m deep; volume 1767 litres	286.19	2.25	160.43	64.01	659.41	nr	**883.85**
1.50 m deep; volume 2651 litres	429.28	3.00	213.90	89.62	975.04	nr	**1278.56**
2.00 m deep; volume 3534 litres	572.37	3.50	249.55	89.62	1290.66	nr	**1629.83**
2.50 m deep; volume 4418 litres	443.60	3.75	267.38	160.03	1293.64	nr	**1721.05**
Ring dia. 2400 mm							
1.00 m deep; volume 4524 litres	1099.26	3.00	213.90	64.01	1714.68	nr	**1992.59**
1.50 m deep; volume 6786 litres	1648.88	3.75	267.38	89.62	2551.25	nr	**2908.25**
2.00 m deep; volume 9048 litres	2198.51	4.00	285.20	89.62	3401.19	nr	**3776.01**
2.50 m deep; volume 11310 litres	2748.14	4.50	320.85	160.03	4237.76	nr	**4718.64**

34 DRAINAGE BELOW GROUND

Item	PC £	Labour hours	Labour £	Plant £	Material £	Unit	**Total rate £**
Extra over for							
250 mm depth chamber ring	–		–	–	–	100%	**–**
500 mm depth chamber ring	–		–	–	–	50%	**–**
The labour in this section is calculated on a 3 person team. The labour time below should be multiplied by 3 to calculate the cost as shown.							
Cover slabs to soakaways							
Heavy duty precast concrete							
900 mm dia.	224.66	0.50	53.48	32.01	258.36	nr	**343.84**
1200 mm dia.	251.84	0.50	53.48	32.01	289.61	nr	**375.09**
1500 mm dia.	408.02	0.50	53.48	32.01	469.22	nr	**554.70**
2400 mm dia.	1463.67	0.50	53.48	32.01	1683.22	nr	**1768.70**
Step irons to concrete chamber rings	53.63		–	–	61.67	m	**61.67**
Extra over soakaways for filter wrapping with a proprietary filter membrane							
900 mm dia. × 1.00 m deep	2.32	0.50	53.48	–	2.67	nr	**56.15**
900 mm dia. × 2.00 m deep	4.65	0.75	80.21	–	5.35	nr	**85.56**
1050 mm dia. × 1.00 m deep	2.71	0.75	80.21	–	3.12	nr	**83.33**
1050 mm dia. × 2.00 m deep	5.42	1.00	106.95	–	6.23	nr	**113.18**
1200 mm dia. × 1.00 m deep	3.10	1.00	106.95	–	3.56	nr	**110.51**
1200 mm dia. × 2.00 m deep	6.19	1.25	133.69	–	7.12	nr	**140.81**
1500 mm dia. × 1.00 m deep	3.87	1.25	133.69	–	4.45	nr	**138.14**
1500 mm dia. × 2.00 m deep	7.72	1.25	133.69	–	8.88	nr	**142.56**
1800 mm dia. × 1.00 m deep	4.64	1.50	160.43	–	5.34	nr	**165.76**
1800 mm dia. × 2.00 m deep	9.28	1.63	173.79	–	10.67	nr	**184.47**
Gravel surrounding to concrete ring soakaway							
40 mm aggregate backfilled to vertical face of soakaway wrapped with geofabric (not included)							
250 mm thick	55.44	0.10	3.57	21.35	63.76	m³	**88.67**
Backfilling to face of soakaway; carefully compacting as work proceeds							
Arising from the excavations							
average thickness exceeding 0.25 m; depositing in layers 150 mm maximum thickness	–	0.03	1.19	7.12	–	m³	**8.30**

34 DRAINAGE BELOW GROUND

Item	PC £	Labour hours	Labour £	Plant £	Material £	Unit	Total rate £
SOAKAWAYS – cont							
The labour in this section is calculated on a 2 person team. The labour time below should be multiplied by 2 to calculate the cost as shown.							
Geofabric surround to Aquacell units; Terram Ltd							
Terram synthetic fibre filter fabric; to face of concrete rings (not included); anchoring whilst backfilling (not included)							
Terram 1000; 0.70 mm thick; mean water flow 50 l/m²/s	0.79	0.03	1.78	–	0.91	m²	**2.69**
Stormwater infltration and attenuation crates; Plastech Southern; preformed polypropylene soakaway infiltration crate units; to trenches; surrounded by geotextile and 40 mm aggregate laid 100 mm thick in trenches (excavation, disposal and backfilling not included)							
1.00 m × 1.00 × 400 mm; internal volume 400 litres; 20 tonne load capacity							
2 crates; 2.00 m × 1.00 m × 400 mm; 800 litres	88.14	0.80	57.04	28.17	182.85	nr	**268.06**
4 crates; 2.00 m × 1.00 m × 800 mm; 1600 litres	176.27	1.60	114.08	69.74	280.72	nr	**464.54**
6 crates; 2.00 m × 1.00 m × 800 mm; 1600 litres	176.27	1.60	114.08	69.74	280.72	nr	**687.42**
8 crates; 4.00 m × 1.00 m × 800 mm; 3200 litres	352.54	3.20	228.16	67.06	576.41	nr	**871.64**
10 crates; 2.00 m × 1.00 m × 800 mm; 1600 litres	176.27	1.60	114.08	69.74	280.72	nr	**1078.94**
16 crates; 2.00 m × 1.00 m × 800 mm; 1600 litres	176.27	1.60	114.08	69.74	280.72	nr	**1741.88**
24 crates; 2.00 m × 1.00 m × 800 mm; 1600 litres	176.27	1.60	114.08	69.74	280.72	nr	**2655.06**
36 crates; 2.00 m × 1.00 m × 800 mm; 1600 litres	176.27	1.60	114.08	69.74	280.72	nr	**3380.20**

34 DRAINAGE BELOW GROUND

Item	PC £	Labour hours	Labour £	Plant £	Material £	Unit	Total rate £
Stormwater infltration and attenuation crates; Plastech Southern; preformed polypropylene soakaway infiltration crate units; to trenches; surrounded by geotextile and 40 mm aggregate laid 100 mm thick in trenches (excavation, disposal and backfilling not included)							
1.00 m × 1.00 × 400 mm; internal volume 400 litres; 45 tonne load capacity							
2 crates; 2.00 m × 1.00 m × 400 mm; 800 litres	99.15	0.80	57.04	28.17	195.52	nr	**280.73**
4 crates; 2.00 m × 1.00 m × 800 mm; 1600 litres	198.31	1.60	114.08	69.74	306.06	nr	**489.88**
6 crates; 3.00 m × 1.00 mm × 800 mm; 2400 litres	297.46	2.40	171.12	75.11	479.20	nr	**725.43**
8 crates; 4.00 m × 1.00 m × 800 mm; 3200 litres	396.61	3.20	228.16	67.06	627.09	nr	**922.31**
10 crates; 5.00 m × 1.00 m × 800 mm; 4000 litres	495.77	4.00	285.20	66.39	790.70	nr	**1142.29**
16 crates; 8.00 m × 1.00 m × 400 mm; 6400 litres	793.22	6.00	427.80	181.07	1234.37	nr	**1843.24**
24 crates; 12.00 m × 1.00 m × 800 mm; 9600 litres	1189.84	10.00	713.00	236.73	1857.36	nr	**2807.09**
36 crates; 12.00 m × 1.00 m × 1.20 mm; 14,400 litres	1784.75	11.00	784.30	236.73	2587.22	nr	**3608.25**
CULVERTS							
Culverts; precast box sections FP McCann Ltd; For directing watercourses and underpasses; to trenches or prepared bases (not included) on 200 mm prepared granular or concrete base blinded with 50 mm sand							

34 DRAINAGE BELOW GROUND

Item	PC £	Labour hours	Labour £	Plant £	Material £	Unit	Total rate £
CULVERTS – cont							
Plain box culvert; 800 mm high; width							
1.20 m	803.92	3.00	320.85	101.60	936.87	m	**1359.32**
1.50 m	845.86	3.50	374.33	127.00	985.10	m	**1486.43**
Plain box culvert; 1.00 m high; width							
1.20 m	838.87	4.00	427.80	152.41	977.06	m	**1557.27**
1.50 m	978.68	4.50	481.28	177.81	1137.84	m	**1796.92**
1.80 m	1083.55	6.00	641.70	254.01	1258.44	m	**2154.15**
Plain box culvert; 1.20 m high; width							
1.50 m	1034.62	6.00	641.70	254.01	1202.17	m	**2097.88**
1.80 m	1118.50	8.00	855.60	304.81	1298.64	m	**2459.05**
Extra over for cantilevered animal ledges	279.62		–	–	321.56	m	**321.56**
HEADWALLS							
The labour in this section is calculated on a 3 person team. The labour time below should be multiplied by 3 to calculate the cost as shown.							
Headwalls; FP McCann Ltd; Precast concrete headwalls to outflows; Placing to prepared graded and blinded 300 mm thick (not included); connection of outfall pipe; 150 mm thick wall with non standard opening; excludes standard grating							
size on plan 1.618 m tapering to 889 mm × 1.32 m × 663 mm high							
Small Headwall; pipe size up to 300 mm	937.49	4.00	427.80	60.96	1183.44	nr	**1672.20**
size on plan 1.878 m tapering to 1.026 m × 1.37 m × 998 mm high							
Medium Headwall; pipe size up to 450 mm	1527.18	7.00	748.65	182.89	1921.13	nr	**2852.67**
size on plan 2.351 m tapering to 1.656 m × 2.150 m × 1.258 m high							
Large Headwall; pipe size up to 900 mm	2252.98	8.00	855.60	365.79	2590.93	nr	**3812.32**

34 DRAINAGE BELOW GROUND

Item	PC £	Labour hours	Labour £	Plant £	Material £	Unit	**Total rate £**
SPORTS AND AMENITY DRAINAGE							
Sports or amenity drainage; Agripower Ltd Ltd; excavating trenches (with minimum project size of 1000 m) by trenching machine; arisings to spoil heap on site maximum 100 m; backfill with shingle to within 125 mm from surface and with rootzone to ground level							
Width 145 mm; 80 mm perforated pipe							
depth 550 mm	–		–	–	–	m	**13.22**
Width 145 mm; 100 mm perforated pipe							
depth 650 mm	–		–	–	–	m	**14.27**
Width 145 mm; 150 mm perforated plpe							
depth 700 mm	–		–	–	–	100 m	**17.92**

35 SITE WORKS – BASES

Item	PC £	Labour hours	Labour £	Plant £	Material £	Unit	Total rate £
HARDCORE BASES							
The labour in this section is calculated on a 3 person team. The labour time below should be multiplied by 3 to calculate the cost as shown.							
Hardcore bases; obtained off site; PC £24.00/m³							
By machine							
100 mm thick	2.60	0.05	1.78	1.68	2.99	m²	**6.45**
150 mm thick	3.90	0.07	2.38	2.18	4.48	m²	**9.04**
200 mm thick	5.20	0.08	2.85	2.28	5.98	m²	**11.11**
300 mm thick	7.80	0.08	2.97	2.43	8.96	m²	**14.37**
exceeding 300 mm thick	31.18	0.17	5.94	5.60	35.86	m³	**47.40**
By hand							
100 mm thick	2.60	0.20	7.13	0.17	2.99	m²	**10.29**
150 mm thick	4.68	0.30	10.71	0.17	5.38	m²	**16.26**
200 mm thick	6.24	0.40	14.26	0.28	7.17	m²	**21.71**
300 mm thick	9.35	0.60	21.39	0.17	10.76	m²	**32.32**
exceeding 300 mm thick	31.18	2.00	71.30	0.57	35.86	m³	**107.73**
Hardcore; difference for each £1.00 increase/decrease in PC price per m³; price will vary with type and source of hardcore							
average 75 mm thick	–		–	–	0.10	m²	**0.10**
average 100 mm thick	–		–	–	0.13	m²	**0.13**
average 150 mm thick	–		–	–	0.19	m²	**0.19**
average 200 mm thick	–		–	–	0.26	m²	**0.26**
average 250 mm thick	–		–	–	0.32	m²	**0.32**
average 300 mm thick	–		–	–	0.39	m²	**0.39**
exceeding 300 mm thick	–		–	–	1.42	m³	**1.42**
TYPE 1 BASES							
The labour in this section is calculated on a 3 person team. The labour time below should be multiplied by 3 to calculate the cost as shown.							
Type 1 granular fill base; PC £26.53 /tonne (£53.90 / m³ compacted)							
By machine							
100 mm thick	6.04	0.02	1.96	0.84	6.94	m²	**9.74**
150 mm thick	9.06	0.03	2.94	0.84	10.41	m²	**14.19**
200 mm thick	12.07	0.04	3.93	1.68	13.88	m²	**19.49**
250 mm thick	15.09	0.05	4.91	1.68	17.36	m²	**23.95**
over 250 mm thick	60.37	0.18	19.63	5.17	69.42	m³	**94.22**

35 SITE WORKS – BASES

Item	PC £	Labour hours	Labour £	Plant £	Material £	Unit	**Total rate £**
By hand (mechanical compaction)							
100 mm thick	4.03	0.07	7.14	0.06	4.63	m^2	**11.82**
150 mm thick	9.06	0.08	8.91	0.09	10.41	m^2	**19.41**
250 mm thick	15.09	0.13	13.37	0.14	17.36	m^2	**30.87**
over 250 mm thick	30.18	0.20	21.39	0.14	34.71	m^3	**56.24**
Type 1 (crushed concrete) granular fill base; PC £11.00/ tonne (£22.00/m^3 compacted)							
By machine							
100 mm thick	2.46	0.02	1.96	0.56	2.83	m^2	**5.35**
150 mm thick	3.70	0.03	2.94	0.84	4.25	m^2	**8.03**
200 mm thick	6.16	0.04	3.93	1.40	7.08	m^2	**12.41**
250 mm thick	6.16	0.05	4.91	1.40	7.08	m^2	**13.39**
over 250 mm thick	11.20	0.18	19.63	5.17	12.88	m^3	**37.67**
By hand (mechanical compaction)							
100 mm thick	2.46	0.07	7.14	0.06	2.83	m^2	**10.03**
150 mm thick	3.70	0.08	8.91	0.09	4.25	m^2	**13.25**
250 mm thick	6.16	0.13	13.37	0.14	7.08	m^2	**20.60**
over 250 mm thick	11.20	0.20	21.39	0.14	12.88	m^3	**34.41**
TYPE 3 PERMEABLE BASES							
The labour in this section is calculated on a 3 person team. The labour time below should be multiplied by 3 to calculate the cost as shown.							
Type 3 open aggregate granular fill base; PC £53.61 / m^3; incoprating optional triaxial geogrid and recommended geofabric							
By machine							
100 mm thick	5.82	0.02	2.49	0.59	10.61	m^2	**13.69**
150 mm thick	8.73	0.03	3.48	1.29	13.96	m^2	**18.73**
200 mm thick	11.64	0.04	4.46	1.90	17.31	m^2	**23.67**
250 mm thick	14.55	0.05	5.44	1.90	20.66	m^2	**28.00**
over 250 mm thick	64.03	0.19	20.34	7.68	77.55	m^3	**105.58**
By hand (mechanical compaction)							
100 mm thick	5.82	0.09	9.11	0.06	14.58	m^2	**24.08**
150 mm thick	8.73	0.11	11.23	0.09	17.93	m^2	**29.24**
250 mm thick	29.64	0.17	18.36	0.14	38.01	m^2	**56.52**
over 250 mm thick	58.21	0.51	54.01	0.14	70.86	m^3	**125.01**

35 SITE WORKS – BASES

Item	PC £	Labour hours	Labour £	Plant £	Material £	Unit	Total rate £
CONCRETE BASES							
The labour in this section is calculated on a 3 person team. The labour time below should be multiplied by 3 to calculate the cost as shown.							
Concrete bases; on 50 mm blinding							
Plain concrete; Site mixed concrete 1:3:6							
100 mm thick	20.02	0.03	3.35	0.42	26.49	m²	**30.26**
150 mm thick	30.03	0.03	3.73	0.42	38.00	m²	**42.16**
200 mm thick	40.04	0.04	4.24	0.42	49.51	m²	**54.17**
250 mm thick	50.05	0.05	4.95	0.42	61.03	m²	**66.40**
over 250 mm thick	200.19	0.21	22.06	0.42	233.69	m³	**256.18**
Mesh reinforced concrete A 142; Site mixed concrete 1:3:6							
100 mm thick	20.02	0.03	3.73	0.42	31.76	m²	**35.92**
150 mm thick	30.03	0.04	4.24	0.42	43.28	m²	**47.93**
200 mm thick	40.04	0.04	4.63	0.42	54.79	m²	**59.84**
250 mm thick	50.05	0.05	5.54	0.42	66.30	m²	**72.26**
Mesh reinforced concrete A 142; Site mixed concrete 1:2:4							
100 mm thick	21.78	0.03	3.73	0.42	33.79	m²	**37.95**
150 mm thick	32.67	0.04	4.24	0.42	46.32	m²	**50.97**
200 mm thick	43.56	0.04	4.63	0.42	58.84	m²	**63.89**
250 mm thick	54.45	0.05	5.54	0.42	71.36	m²	**77.33**
Mesh reinforced concrete A 142; Ready mixed concrete C20							
100 mm thick	9.66	0.03	3.73	0.42	19.85	m²	**24.00**
150 mm thick	14.48	0.04	4.24	0.42	25.40	m²	**30.06**
200 mm thick	19.31	0.04	4.63	0.42	30.95	m²	**36.01**
250 mm thick	24.14	0.05	5.54	0.42	36.51	m²	**42.47**

35 SITE WORKS – BASES

Item	PC £	Labour hours	Labour £	Plant £	Material £	Unit	Total rate £
CELLULAR CONFINED BASES							
The labour in this section is calculated on a 3 person team. The labour time below should be multiplied by 3 to calculate the cost as shown.							
Geoweb; Greenfix; cellular confinement system for load support or permeable bases below pavings							
To graded and compacted substrate (not included); on 50 mm sharp sand base filled with angular drainage aggregate 20–5 mm							
Geoweb 75 mm depth	8.99	0.05	4.87	2.22	19.88	m^2	**26.97**
Geoweb 100 mm depth	10.35	0.05	5.35	2.50	23.61	m^2	**31.46**
Geoweb 150 mm depth	12.94	0.06	5.95	2.63	30.51	m^2	**39.09**
Geoweb 200 mm depth	18.23	0.08	8.91	2.86	40.50	m^2	**52.28**
Geoweb 300 mm depth	20.90	0.10	10.70	3.33	51.41	m^2	**65.44**
SUBBASE REPLACEMENT SYSTEMS							
The labour in this section is calculated on a 3 person team. The labour time below should be multiplied by 3 to calculate the cost as shown.							
Subbase replacement system for SUDS paving; Permavoid Ltd; geocellular, interlocking, high strength; high void capacity (95%); typically a 4:1 ratio when compared with alternative granular systems; allows for water storage to be confined at shallow depth within the subbase layer; inclusive of connection, ties, pins; does not include for encapsulation in geotextile/ geomembranes							
708 × 354 mm							
85 mm thick	60.89	0.06	5.96	–	70.02	m^2	**75.98**
150 mm thick	104.83	0.06	5.96	–	120.55	m^2	**126.51**
300 mm thick	209.65	0.11	11.88	–	241.10	m^2	**252.98**

35 SITE WORKS – BASES

Item	PC £	Labour hours	Labour £	Plant £	Material £	Unit	Total rate £
SUBBASE REPLACEMENT SYSTEMS – cont							
Subbase replacement system for SUDS paving – cont							
Encapsulation of Permavoid; 150 mm thick systems; geomembrane geofabric or combination of both for filtration or attenuation of water; please see the manufacturers design guide							
geomembrane, heat welded	–	0.02	1.79	–	6.07	m^2	**7.85**
Permatex 300 geotextile	–	0.11	11.88	–	5.26	m^2	**17.14**
SURFACE TREATMENTS TO GRANULAR BASES							
Surface treatments							
Sand blinding; to hardcore base (not included); 25 mm thick	1.13	0.03	1.19	–	1.30	m^2	**2.49**
Sand blinding; to hardcore base (not included); 50 mm thick	2.27	0.05	1.78	–	2.61	m^2	**4.39**
Filter fabrics; to hardcore base (not included)	0.69	0.01	0.36	–	0.79	m^2	**1.15**

35 SITE WORKS – KERBS AND EDGINGS

Item	PC £	Labour hours	Labour £	Plant £	Material £	Unit	Total rate £
FOUNDATIONS TO KERBS AND EDGINGS							
The labour in this section is calculated on a 3 person team. The labour time below should be multiplied by 3 to calculate the cost as shown.							
Foundations to kerbs and edgings							
Excavating trenches; width 300 mm; 3 tonne excavator; disposal off site							
depth 200 mm	–		–	3.78	–	m	**6.09**
depth 300 mm	–		–	5.45	–	m	**8.91**
depth 400 mm	–		–	5.92	–	m	**10.53**
By hand; barrowing to spoil heap and disposal off site							
depth 300	–	0.06	6.18	–	–	m	**43.85**
depth 400	–	0.08	8.25	–	–	m	**46.00**
Excavating trenches; width 450 mm; 3 tonne excavator; disposal off site							
depth 300 mm	–		–	6.79	–	m	**11.97**
depth 400 mm	–		–	7.75	–	m	**14.65**
Foundations; to kerbs, edgings or channels; in situ concrete; 21 N/mm²–20 mm aggregate (1:2:4) site mixed); one side against earth face, other against formwork (not included); site mixed concrete							
Site mixed concrete							
150 mm wide × 100 mm deep	–	0.04	4.76	–	3.76	m	**8.52**
150 mm wide × 150 mm deep	–	0.06	5.94	–	5.64	m	**11.57**
200 mm wide × 150 mm deep	–	0.07	7.13	–	7.51	m	**14.65**
300 mm wide × 150 mm deep	–	0.06	6.25	–	11.27	m	**17.52**
600 mm wide × 200 mm deep	–	0.06	6.24	–	30.06	m	**36.29**
Ready mixed concrete							
150 mm wide × 100 mm deep	–	0.04	4.76	–	1.67	m	**6.43**
150 mm wide × 150 mm deep	–	0.06	5.94	–	2.50	m	**8.43**
200 mm wide × 150 mm deep	–	0.07	7.13	–	3.33	m	**10.47**
300 mm wide × 150 mm deep	–	0.08	8.32	–	5.00	m	**13.32**
600 mm wide × 200 mm deep	–	0.10	10.18	–	13.33	m	**23.51**
Formwork; sides of foundations (this will usually be required to one side of each kerb foundation adjacent to road subbases)							
100 mm deep	–	0.01	1.49	–	0.12	m	**1.60**
150 mm deep	–	0.01	1.49	–	0.18	m	**1.66**

35 SITE WORKS – KERBS AND EDGINGS

Item	PC £	Labour hours	Labour £	Plant £	Material £	Unit	Total rate £
PRECAST CONCRETE KERBS							
The labour in this section is calculated on a 3 person team. The labour time below should be multiplied by 3 to calculate the cost as shown.							
Precast concrete kerbs, channels, edgings etc.; Marshalls Mono; bedding, jointing and pointing in cement: mortar (1:3); including haunching with in situ concrete; 11.50 N/mm²–40 mm aggregate one side							
Kerbs; straight							
150 × 305 mm; HB1	14.51	0.17	17.83	–	23.59	m	**41.42**
125 × 255 mm; HB2; SP	7.13	0.15	15.84	–	15.11	m	**30.95**
125 × 150 mm; BN	4.92	0.13	14.26	–	12.57	m	**26.83**
Dropper kerbs; left and right handed							
125 × 255–150 mm; DL1 or DR1	11.86	0.17	17.83	–	20.55	m	**38.37**
125 × 255–150 mm; DL2 or DR2	11.86	0.17	17.83	–	20.55	m	**38.37**
Quadrant kerbs							
305 mm radius	20.81	0.17	17.83	–	28.54	nr	**46.37**
455 mm radius	23.51	0.17	17.83	–	33.94	nr	**51.77**
Straight kerbs or channels; to radius; 125 × 255 mm							
3.00 m radius (5 units per quarter circle)	13.28	0.22	23.76	–	22.32	m	**46.09**
4.50 m radius (2 units per quarter circle)	13.28	0.20	21.41	–	22.32	m	**43.73**
6.10 m radius (11 units per quarter circle)	13.28	0.19	20.66	–	22.32	m	**42.98**
7.60 m radius (14 units per quarter circle)	13.28	0.19	20.37	–	22.32	m	**42.70**
9.15 m radius (17 units per quarter circle)	13.28	0.19	19.81	–	22.32	m	**42.13**
10.70 m radius (20 units per quarter circle)	13.28	0.19	19.81	–	22.32	m	**42.13**
12.20 m radius (22 units per quarter circle)	13.28	0.18	18.77	–	22.32	m	**41.09**
Kerbs; Conservation X Kerb units; to simulate natural granite kerbs							
255 × 150 × 914 mm; laid flat	37.47	0.19	20.37	–	55.48	m	**75.85**
150 × 255 × 914 mm; laid vertical	54.93	0.19	20.37	–	73.53	m	**73.82**
145 × 255 mm; radius internal 3.25 m	48.68	0.28	29.71	–	66.34	m	**96.05**

35 SITE WORKS – KERBS AND EDGINGS

Item	PC £	Labour hours	Labour £	Plant £	Material £	Unit	Total rate £
150 × 255 mm; radius external 3.40 m	54.93	0.19	20.37	–	73.53	m	**93.53**
145 × 255 mm; radius internal 6.50 m	54.93	0.19	20.37	–	73.53	m	**84.14**
145 × 255 mm; radius external 6.70 m	54.93	0.19	20.37	–	73.53	m	**83.86**
150 × 255 mm; radius internal 9.80 m	54.93	0.19	20.37	–	73.53	m	**83.58**
150 × 255 mm; radius external 10.00 m	54.93	0.19	20.37	–	73.53	m	**81.01**
305 × 305 × 255 mm; solid quadrants	54.93	0.19	20.37	–	73.53	nr	**93.90**
Marshalls Mono; small element precast concrete kerb system; Keykerb Large (KL) upstand of 100–125 mm; on 150 mm concrete foundation including haunching with in situ concrete (1:3:6) 1 side							
Bullnosed or half battered; 100 × 127 × 200 mm							
laid straight	21.34	0.27	28.52	–	26.96	m	**55.48**
radial blocks laid to curve; 8 blocks/1/4 circle – 500 mm radius	16.54	0.33	35.65	–	21.44	m	**57.09**
radial blocks laid to curve; 8 radial blocks, alternating 8 standard blocks/1/4 circle – 1000 mm radius	27.21	0.42	44.57	–	33.71	m	**78.28**
radial blocks, alternating 16 standard blocks/1/4 circle – 1500 mm radius	25.00	0.50	53.48	–	31.17	m	**84.64**
internal angle 90 degree	9.11	0.07	7.13	–	10.48	nr	**17.61**
external angle	9.11	0.07	7.13	–	10.48	nr	**17.61**
Marshalls Mono; small element precast concrete kerb system; Keykerb Small (KS) upstand of 25–50 mm; on 150 mm concrete foundation including haunching with in situ concrete (1:3:6) 1 side							
Half battered							
laid straight	14.98	0.27	28.52	–	19.65	m	**48.17**
radial blocks laid to curve, 8 blocks/1/4 circle; 500 mm radius	12.43	0.33	35.65	–	16.71	m	**52.36**
radial blocks laid to curve; 8 radial blocks, alternating 8 standard blocks/1/4 circle; 1000 mm radius	19.92	0.42	44.57	–	25.33	m	**69.90**
radial blocks, alternating 16 standard blocks/1/4 circle; 1500 mm radius	21.36	0.50	53.48	–	26.98	m	**80.45**
internal angle 90 degree	9.11	0.07	7.13	–	10.48	nr	**17.61**
external angle	9.11	0.07	7.13	–	10.48	nr	**17.61**

35 SITE WORKS – KERBS AND EDGINGS

Item	PC £	Labour hours	Labour £	Plant £	Material £	Unit	Total rate £
PRECAST CONCRETE EDGINGS							
The labour in this section is calculated on a 3 person team. The labour time below should be multiplied by 3 to calculate the cost as shown.							
Precast concrete edging units; including haunching with in situ concrete; 11.50 N/mm²–40 mm aggregate both sides							
Edgings; rectangular, bullnosed or chamfered							
50 × 150 mm	2.59	0.11	11.88	–	14.49	m	**26.37**
125 × 150 mm bullnosed	5.37	0.11	11.88	–	17.68	m	**29.57**
50 × 200 mm	3.93	0.11	11.88	–	16.03	m	**27.92**
50 × 250 mm	4.35	0.11	11.88	–	16.52	m	**28.40**
50 × 250 mm flat top	4.83	0.11	11.88	–	17.07	m	**28.95**
Conservation edgings							
150 × 915 × 63 mm	16.81	0.14	15.28	–	26.38	ea	**41.67**
Precast concrete block edgings; PC £14.01/m²; 200 × 100 × 60 mm; on prepared base (not included); haunching one side							
Edgings; butt joints							
stretcher course	1.43	0.06	5.94	–	1.90	m	**7.84**
header course	2.85	0.09	9.51	–	3.81	m	**13.31**

35 SITE WORKS – KERBS AND EDGINGS

Item	PC £	Labour hours	Labour £	Plant £	Material £	Unit	Total rate £
STONE KERBS							
Dressed natural stone kerbs – General							
Preamble: The kerbs are to be good, sound and uniform in texture and free from defects; worked straight or to radius, square and out of wind, with the top front and back edges parallel or concentric to the dimensions specified. All drill and pick holes shall be removed from dressed faces. Standard dressings shall be designated as either fine picked, single axed or nidged or rough punched.							
The labour in this section is calculated on a 3 person team. The labour time below should be multiplied by 3 to calculate the cost as shown.							
Dressed natural stone kerbs; CED Ltd; on concrete foundations (not included); including haunching with in situ concrete; 11.50 N/mm²–40 mm aggregate one side							
Granite kerbs; 125 × 250 mm							
special quality; straight; random lengths	55.00	0.27	28.52	–	71.09	m	**99.61**
Granite kerbs; 125 × 250 mm; curved to mean radius 3 m							
special quality; random lengths	67.00	0.30	32.44	–	84.89	m	**117.33**

35 SITE WORKS – KERBS AND EDGINGS

Item	PC £	Labour hours	Labour £	Plant £	Material £	Unit	**Total rate £**
STONE EDGINGS							
The labour in this section is calculated on a 3 person team. The labour time below should be multiplied by 3 to calculate the cost as shown.							
New granite sett edgings; CED Ltd; 100 × 100 × 100 mm; bedding in cement: mortar (1:4); including haunching with in situ concrete; 11.50 N/mm²–40 mm aggregate one side; concrete foundations (not included)							
Edgings 100 × 100 × 100 mm; ref 1R 'better quality' with 20 mm pointing gaps							
100 mm wide	4.88	0.44	47.41	–	10.82	m	**58.23**
2 rows; 220 mm wide	9.77	0.74	79.14	–	17.96	m	**97.10**
3 rows; 340 mm wide	14.65	1.00	106.95	–	22.14	m	**129.09**
Edgings 100 × 100 × 200 mm; ref 1R 'better quality' with 20 mm pointing gaps							
100 mm wide	4.88	0.25	26.74	–	10.82	m	**37.56**
2 rows; 220 mm wide	9.77	0.50	53.48	–	17.96	m	**71.43**
3 rows; 340 mm wide	14.65	0.75	80.21	–	22.14	m	**102.35**
Edgings 100 × 100 × 100 mm; ref 8R 'standard quality' with 20 mm pointing gaps							
100 mm wide	4.88	0.44	47.41	–	10.82	m	**58.23**
2 rows; 220 mm wide	9.77	0.74	79.14	–	17.96	m	**97.10**
3 rows; 340 mm wide	14.65	1.00	106.95	–	22.14	m	**129.09**
Edgings 100 × 100 × 200 mm; ref 8R 'standard quality' with 20 mm pointing gaps							
100 mm wide	5.22	0.25	26.74	–	11.20	m	**37.94**
2 rows; 220 mm wide	10.44	0.50	53.48	–	18.73	m	**72.21**
3 rows; 340 mm wide	15.66	0.75	80.21	–	23.30	m	**103.51**
Reclaimed granite setts edgings; CED Ltd; 100 × 100 mm; bedding in cement: mortar (1:4); including haunching with in situ concrete; 11.50 N/mm²–40 mm aggregate one side; on concrete foundations (not included)							

35 SITE WORKS – KERBS AND EDGINGS

Item	PC £	Labour hours	Labour £	Plant £	Material £	Unit	**Total rate £**
Edgings; with 20 mm pointing gaps							
100 mm wide	5.20	0.44	47.41	–	11.18	m	**58.59**
2 rows; 220 mm wide	10.39	0.74	79.14	–	18.68	m	**97.82**
3 rows; 340 mm wide	15.59	1.00	106.95	–	23.22	m	**130.17**
Natural Yorkstone edgings; Johnsons Wellfield Quarries; sawn 6 sides; 50 mm thick; on prepared base measured separately; bedding on 25 mm cement: sand (1:3); cement: sand (1:3) joints							
Edgings							
100 mm wide × random lengths	11.94	0.17	17.83	–	14.35	m	**32.18**
100 × 100 mm	11.83	0.17	17.83	–	19.78	m	**37.61**
100 × 200 mm	23.65	0.17	17.83	–	33.38	m	**51.21**
250 mm wide × random lengths	24.73	0.13	14.26	–	34.62	m	**48.88**
500 mm wide × random lengths	83.86	0.11	11.88	–	102.63	m	**114.51**
Yorkstone edgings; 600 mm long × 250 mm wide; cut to radius							
1.00 m to 3.00	66.38	0.17	17.83	–	76.96	m	**94.79**
3.00 m to 5.00 m	66.38	0.15	15.84	–	76.96	m	**92.80**
exceeding 5.00 m	66.39	0.13	14.26	–	76.96	m	**91.22**
Natural stone, slate or granite flag edgings; CED Ltd; on prepared base (not included); bedding on 25 mm cement: sand (1:3); cement: sand (1:3) joints							
Edgings; silver grey							
100 mm wide × random lengths	6.16	0.33	35.65	–	7.70	m	**43.35**
100 × 100 mm	7.04	0.50	53.48	–	8.71	m	**62.19**
100 mm long × 200 mm wide	11.44	0.20	21.39	–	13.77	m	**35.16**
250 mm wide × random lengths	12.10	0.25	26.74	–	20.10	m	**46.83**
300 mm wide × random lengths	13.20	0.25	26.74	–	21.36	m	**48.10**
BRICK OR BLOCK EDGINGS							
The labour in this section is calculated on a 3 person team. The labour time below should be multiplied by 3 to calculate the cost as shown.							

35 SITE WORKS – KERBS AND EDGINGS

Item	PC £	Labour hours	Labour £	Plant £	Material £	Unit	Total rate £
BRICK OR BLOCK EDGINGS – cont							
Brick or block stretchers; bedding in cement: mortar (1:4); on 150 mm deep concrete foundations (not included);							
Single course							
concrete paving blocks; PC £13.82/m^2; 200 × 100 × 60 mm	3.05	0.10	10.96	–	4.76	m	**15.72**
engineering bricks; PC £456.00/1000; 215 × 102.5 × 65 mm	2.75	0.13	14.26	–	4.42	m	**18.68**
Two courses							
concrete paving blocks; PC £13.82/m^2; 200 × 100 × 60 mm	6.09	0.13	14.26	–	9.27	m	**23.53**
engineering bricks; PC £456.00/1000; 215 × 102.5 × 65 mm	5.51	0.19	20.37	–	8.60	m	**28.97**
Three courses							
concrete paving blocks; PC £13.82/m^2; 200 × 100 × 60 mm	9.14	0.15	15.84	–	12.77	m	**28.61**
engineering bricks; PC £456.00/1000; 215 × 102.5 × 65 mm	8.26	0.22	23.76	–	11.76	m	**35.53**
Bricks on edge; bedding in cement: mortar (1:4); on 150 mm deep concrete foundations							
One brick wide							
engineering bricks; 215 × 102.5 × 65 mm	8.26	0.19	20.37	–	12.01	m	**32.38**
paving bricks; 215 × 102.5 × 65 mm	8.00	0.19	20.37	–	11.71	m	**32.08**
Two courses; stretchers laid on edge; 225 mm wide							
engineering bricks; 215 × 102.5 × 65 mm	16.52	0.40	42.78	–	23.75	m	**66.53**
paving bricks; 215 × 102.5 × 65 mm	16.00	0.40	42.78	–	23.15	m	**65.93**
Edge restraints; to brick paving; on prepared base (not included); 65 mm thick bricks; PC £650.00/1000							
Header course							
200 × 100 mm; butt joints	6.00	0.09	9.51	–	7.43	m	**16.93**
200 × 100 mm; mortar joints	5.59	0.17	17.83	–	12.41	m	**30.24**
200 × 100 mm × 50; on edge; mortar joints	8.38	0.23	24.96	–	10.85	m	**35.81**
215 × 102.5 mm mortar joints	5.33	0.17	17.83	–	7.34	m	**25.17**
215 × 102.5 mm × 65; on edge; mortar joints	8.00	0.25	26.75	–	10.41	m	**37.16**
Stretcher course; on flat or on edge							
200 × 100 mm; butt joints	3.00	0.06	5.94	–	3.71	m	**9.65**
210 × 105 mm; mortar joints	2.72	0.11	11.88	–	3.72	m	**15.60**

35 SITE WORKS – KERBS AND EDGINGS

Item	PC £	Labour hours	Labour £	Plant £	Material £	Unit	Total rate £
The labour in this section is calculated on a 2 person team. The labour time below should be multiplied by 2 to calculate the cost as shown.							
Header courses; laid on edge; 10 mm flowable mortar jointed on 35 mm mortar bed							
Blue Baggeridge 200 × 100 × 62 mm	14.89	0.33	23.76	–	18.90	m	**42.66**
Vande Moortel Taupe pavers 215 × 52 × 70 mm	24.39	0.33	23.76	–	29.82	m	**53.58**
Wienerberger Dutch tumbled pavers 150 × 150 × 65 mm	6.20	0.29	20.37	–	8.90	m	**29.27**
Stretcher courses; laid flat or on edge; butt jointed on mortar bed							
Blue Baggeridge 200 × 100 × 62 mm	5.36	0.08	5.94	–	6.54	m	**12.48**
Haunching							
Haunching with in situ concrete; 11.50 N/mm^2–40 mm aggregate one side							
one side	6.46	0.13	8.91	–	7.42	m	**16.34**
Variation in brick prices; add or subtract the following amounts for every £100/1000 difference in the PC price							
Edgings; mortar jointed							
100 mm wide	0.55		–	–	0.63	m	**0.63**
200 mm wide	1.05		–	–	1.21	m	**1.21**
102.5 mm wide	0.51		–	–	0.59	m	**0.59**
215 mm wide	1.03		–	–	1.18	m	**1.18**

35 SITE WORKS – KERBS AND EDGINGS

Item	PC £	Labour hours	Labour £	Plant £	Material £	Unit	Total rate £
CHANNELS							
The labour in this section is calculated on a 3 person team. The labour time below should be multiplied by 3 to calculate the cost as shown.							
Channels; bedding in cement: mortar (1:3); joints pointed flush; on concrete foundations (not included)							
Three courses stretchers; 350 mm wide; quarter bond to form dished channels							
engineering bricks; PC £456.00/1000; 215 × 102.5 × 65 mm	8.26	0.33	35.65	–	11.76	m	**47.41**
paving bricks; PC £600.00/1000; 215 × 102.5 × 65 mm	8.00	0.33	35.65	–	11.46	m	**47.11**
Three courses granite setts; 340 mm wide; to form dished channels							
340 mm wide	17.45	0.67	71.30	–	25.27	m	**96.57**
Precast concrete channels etc.; Marshalls Mono; bedding, jointing and pointing in cement: mortar (1:3); including haunching with in situ concrete; 11.50 N/mm^2–40 mm aggregate one side							
Channels; square							
125 × 225 × 915 mm long; CS1	7.43	0.13	14.26	–	27.43	m	**41.69**
125 × 150 × 915 mm long; CS2	6.37	0.13	14.26	–	22.26	m	**36.51**
Channels; dished							
305 × 150 × 915 mm long; CD	21.25	0.13	14.26	–	45.21	m	**59.46**
150 × 100 × 915 mm	12.27	0.13	14.26	–	25.63	m	**39.89**

35 SITE WORKS – KERBS AND EDGINGS

Item	PC £	Labour hours	Labour £	Plant £	Material £	Unit	Total rate £
TIMBER EDGINGS							
The labour in this section is calculated on a 2 person team. The labour time below should be multiplied by 2 to calculate the cost as shown.							
Timber edging boards; fixed with 50 × 50 × 750 mm timber pegs at 1000 mm centres (excavations and hardcore under edgings not included)							
Straight							
22 × 150 mm treated softwood edge boards	5.09	0.07	4.76	–	5.85	m	**10.61**
50 × 150 mm treated softwood edge boards	9.45	0.10	7.13	–	10.87	m	**18.00**
Curved							
22 × 150 mm treated softwood edge boards	5.09	0.13	8.91	–	5.85	m	**14.77**
50 × 150 mm treated softwood edge boards	9.45	0.14	10.19	–	10.87	m	**21.06**
METAL EDGINGS							
The labour in this section is calculated on a 2 person team. The labour time below should be multiplied by 2 to calculate the cost as shown.							
Metal edging; Everedge Ltd; for Hard flush surface interfaces; Steel galvanized; plain or powder coated RAL; Rigid or Flexible; Edging abutting hard surfaces; Asphalt, block paviours, kerbs, imprinted and plain concrete resin bound and loose gravels; wetpour gravels and other safety surfacing.							
Halestem angle edging; 2 mm thickness with 10 mm × 6 mm rolled top. 2500 mm long, base varies with height. Supplied with fixing nails 150–195 mm. Flexible or rigid format. Powder coating optional; laid straight							
50 mm	12.92	0.07	4.76	–	14.85	m	**19.61**
75 mm	13.44	0.08	5.48	–	15.46	m	**20.94**
100 mm	17.65	0.08	5.94	–	20.30	m	**26.24**
150 mm	22.18	0.09	6.48	–	25.51	m	**31.99**

35 SITE WORKS – KERBS AND EDGINGS

Item	PC £	Labour hours	Labour £	Plant £	Material £	Unit	Total rate £
METAL EDGINGS – cont							
Metal edging – cont							
Halestem angle edging; 2 mm thickness with 10 mm × 6 mm rolled top. 2500 mm long, base varies with height. Supplied with fixing nails 150–195 mm. Flexible or rigid format. Powder coating optional; laid to curves radius not less than 5.00 m							
50 mm	12.92	0.09	6.34	–	14.85	m	**21.19**
75 mm	13.44	0.10	6.79	–	15.46	m	**22.24**
100 mm	17.65	0.11	7.92	–	20.30	m	**28.22**
150 mm	22.18	0.12	8.64	–	25.51	m	**34.15**
ProEdge; Everedge; Pathways, driveways and light traffic areas. Hard/Soft surfaces. 2500 mm × 2.5 mm. Multiple integral 100 mm spikes along base; Galvanized steel; brown or slate; (other colours available)							
laid straight							
75 mm deep	13.60	0.07	4.76	–	15.64	m	**20.39**
100 mm deep	15.88	0.08	5.48	–	18.27	m	**23.75**
125 mm deep	18.93	0.08	5.94	–	21.77	m	**27.71**
150 mm deep	21.27	0.09	6.48	–	24.47	m	**30.95**
laid to curves not less than 5.00 m radius							
75 mm deep	13.60	0.09	6.34	–	15.64	m	**21.97**
100 mm deep	15.88	0.10	6.79	–	18.27	m	**25.05**
125 mm deep	18.93	0.11	7.92	–	21.77	m	**29.69**
150 mm deep	21.27	0.12	8.64	–	24.47	m	**33.11**
extra over for Sleeve and Pin Accessory as alternative to concrete haunch for soft ground or to maintain curves.	1.52	0.07	4.14	–	1.52	m	**5.65**
Extra over for concrete haunch in madeup ground or shallow environments where Sleeve and Pin are not viable	11.80	0.13	7.75	–	11.80	m	**19.55**
Cor-Ten; laid straight							
75 mm deep	11.04	0.07	4.76	–	12.70	m	**17.46**
100 mm deep	13.07	0.07	5.09	–	15.03	m	**20.12**
125 mm deep	15.11	0.09	6.48	–	17.38	m	**23.86**
Cor-Ten; laid straight							
75 mm deep	11.04	0.09	6.34	–	12.70	m	**19.04**
100 mm deep	13.07	0.10	6.79	–	15.03	m	**21.82**
125 mm deep	15.11	0.11	7.92	–	17.38	m	**25.30**

35 SITE WORKS – KERBS AND EDGINGS

Item	PC £	Labour hours	Labour £	Plant £	Material £	Unit	**Total rate £**
TITAN for high strength commercial applications; Galvanized + Powder coated; brown, slate (others available)							
150 mm × 2.5 mm	22.83	0.12	7.29	–	22.83	m	**30.12**
300 mm × 2.5 mm	33.07	0.14	8.86	–	33.07	m	**41.93**
150 mm × 4.00 mm	27.02	0.14	8.86	–	27.02	m	**35.88**
300 mm × 4.00 mm	41.13	0.15	9.54	–	41.13	m	**50.66**
150 mm × 6.00 mm	54.30	0.17	10.34	–	54.30	m	**64.64**
300 mm × 6.00 mm	86.57	0.20	12.40	–	86.57	m	**98.97**
TITAN Corten							
Corten, 150 mm × 2.5 mm	24.14	0.13	7.75	–	24.14	m	**31.89**
Corten, 300 mm × 2.5 mm	37.32	0.14	8.86	–	37.32	m	**46.18**
Corten, 150 mm × 4.00 mm	31.69	0.17	10.34	–	31.69	m	**42.03**
Corten, 300 mm × 4.00 mm	51.94	0.20	12.40	–	51.94	m	**64.34**
Corten, 150 mm × 6.00 mm	46.95	0.20	12.40	–	46.95	m	**59.35**
Corten, 300 mm × 6.00 mm	77.07	0.25	15.50	–	77.07	m	**92.57**
Corten pre-fabricated retaining walls; Adezz limited							
200 mm high on prepared concrete base (not included); top edge of 50 × 25 mm; bolted together internally with 30 mm M8 bolts							
straight sections 1.00 long	35.51	0.20	14.26	–	40.84	m	**55.10**
straight sections 1.50 m long	35.61	0.17	11.89	–	40.95	m	**52.83**
straight sections 2.00 m long	34.86	0.13	8.91	–	40.08	m	**49.00**
inside corner or outside corner 500 mm × 500 mm	35.69	0.25	17.83	–	41.04	m	**58.87**
inside or outside corner 1.00 × 1.00	33.89	0.30	21.39	–	38.97	m	**60.36**
inside or outside curve 500 mm	124.04	0.20	14.26	–	142.65	m	**156.91**
inside or outside curve; 1.00 m lengths	100.83	0.25	17.83	–	115.95	m	**133.78**
inside or outside curve; 1.50 m lengths	94.79	0.17	11.89	–	109.00	m	**120.89**
inside or outside curve; 2.00 m lengths	96.81	0.14	10.19	–	111.33	m	**121.52**
400 mm high; on prepared concrete base (not included); top edge of 50 × 25 mm; bolted together internally with 30 mm M8 bolts							
straight sections 1.00 long	79.96	0.22	15.69		91.95	m	**107.64**
straight sections 1.50 m long	80.04	0.18	13.08	–	92.05	m	**105.12**
straight sections 2.00 m long	78.05	0.14	9.80	–	89.76	m	**99.56**
inside corner or outside corner 500 mm × 500 mm	96.70	0.28	19.61	–	111.21	m	**130.81**
inside or outside corner 1.00 × 1.00	90.71	0.33	23.53	–	104.31	m	**127.84**
inside or outside curve 500 mm	216.82	0.22	15.69	–	249.34	m	**265.03**
inside or outside curve; 1.00 m lengths	181.41	0.28	19.61	–	208.62	m	**228.23**

35 SITE WORKS – KERBS AND EDGINGS

Item	PC £	Labour hours	Labour £	Plant £	Material £	Unit	Total rate £
METAL EDGINGS – cont							
Corten pre-fabricated retaining walls – cont							
400 mm high – cont							
inside or outside curve; 1.50 m lengths	179.44	0.18	13.08	–	206.36	m	**219.43**
inside or outside curve; 2.00 m lengths	182.11	0.16	11.21	–	209.42	m	**220.63**
600 mm high; on prepared concrete base (not included); top edge of 50 × 25 mm; bolted together internally with 30 mm M8 bolts							
straight sections 1.00 long	120.94	0.24	17.25	–	139.08	m	**156.34**
straight sections 1.50 m long	121.02	0.20	14.38	–	139.17	m	**153.55**
straight sections 2.00 m long	116.87	0.15	10.79	–	134.40	m	**145.19**
inside corner or outside corner 500 mm × 500 mm	149.12	0.30	21.57	–	171.49	m	**193.06**
inside or outside corner 1.00 × 1.00	131.51	0.36	25.88	–	151.23	m	**177.11**
inside or outside curve 500 mm	298.24	0.24	17.25	–	342.98	m	**360.23**
inside or outside curve; 1.00 m lengths	271.08	0.30	21.57	–	311.74	m	**333.31**
inside or outside curve; 1.50 m lengths	249.87	0.20	14.38	–	287.35	m	**301.74**
inside or outside curve; 2.00 m lengths	244.11	0.17	12.33	–	280.72	m	**293.05**

35 SITE WORKS – PAVINGS

Item	PC £	Labour hours	Labour £	Plant £	Material £	Unit	Total rate £
SPECIALIZED MORTARS FOR PAVINGS							
SteinTec Ltd							
Primer for specialized mortar; SteinTec Tuffbond priming mortar immediately prior to paving material being placed on bedding mortar							
maximum thickness 1.5 mm (1.5 kg/m²)	3.38	0.02	0.71	–	3.88	m²	**4.59**
Tuffbed; 2 pack hydraulic mortar for paving applications							
30 mm thick	111.60	0.17	5.94	–	128.34	m²	**134.28**
40 mm thick	148.80	0.20	7.13	–	171.12	m²	**178.25**
50 mm thick	186.00	0.25	8.91	–	213.90	m²	**222.81**
The labour in this section is calculated on a 3 person team. The labour time below should be multiplied by 3 to calculate the cost as shown.							
SteinTec jointing mortar; Grey or Natural; joint size							
100 × 100 × 100 mm; 10 mm joints; 31.24 kg/m²	28.63	0.04	3.82	–	32.92	m²	**36.74**
100 × 100 × 100 mm; 20 mm joints; 31.24 kg/m²	50.40	0.05	5.35	–	57.96	m²	**63.31**
200 × 100 × 65 mm; 10 mm joints; 15.7 kg/m²	14.39	0.02	2.37	–	16.55	m²	**18.92**
215 × 112.5 × 65 mm; 10 mm joints; 14.3 kg/m²	13.10	0.02	2.37	–	15.07	m²	**17.44**
300 × 450 × 65 mm; 10 mm joints; 6.24 kg/m²	5.72	0.02	2.37	–	6.58	m²	**8.95**
300 × 450 × 65 mm; 20 mm joints; 11.98 kg/m²	10.98	0.04	3.82	–	12.63	m²	**16.44**
450 × 450 × 65 mm; 20 mm joints; 9.75 kg/m²	8.93	0.15	7.78	–	10.28	m²	**18.05**
450 × 600 × 65 mm; 20 mm joints; 8.6 kg/m²	7.88	0.02	2.37	–	9.06	m²	**11.44**
600 × 600 × 65 mm; 20 mm joints; 7.43 kg/m²	6.81	0.02	2.37	–	7.83	m²	**10.20**
SteinTec jointing mortar; Dark grey or beige; joint size							
100 × 100 × 100 mm; 10 mm joints; 31.24 kg/m²	28.63	0.04	3.82	–	32.92	m²	**36.74**
100 × 100 × 100 mm; 20 mm joints; 31.24 kg/m²	50.40	0.05	5.35	–	57.96	m²	**63.31**
200 × 100 × 65 mm; 10 mm joints; 15.7 kg/m²	14.39	0.02	2.37	–	16.55	m²	**18.92**

35 SITE WORKS – PAVINGS

Item	PC £	Labour hours	Labour £	Plant £	Material £	Unit	Total rate £
SPECIALIZED MORTARS FOR PAVINGS – cont							
The labour in this section is calculated on a 3 person team. – cont							
SteinTec jointing mortar – cont							
215 × 112.5 × 65 mm; 10 mm joints; 14.3 kg/m^2	13.10	0.02	2.37	–	15.07	m^2	**17.44**
300 × 450 × 65 mm; 10 mm joints; 6.24 kg/m^2	5.72	0.02	2.37	–	6.58	m^2	**8.95**
300 × 450 × 65 mm; 20 mm joints; 11.98 kg/m^2	10.98	0.04	3.82	–	12.63	m^2	**16.44**
450 × 450 × 65 mm; 20 mm joints; 9.75 kg/m^2	8.93	0.15	7.78	–	10.28	m^2	**18.05**
450 × 600 × 65 mm; 20 mm joints; 8.6 kg/m^2	7.88	0.02	2.37	–	9.06	m^2	**11.44**
600 × 600 × 65 mm; 20 mm joints; 7.43 kg/m^2	6.81	0.02	2.37	–	7.83	m^2	**10.20**
TREE ROOTZONE PROTECTION							
The labour in this section is calculated on a 3 person team. The labour time below should be multiplied by 3 to calculate the cost as shown.							
Tree Root zone protection; Greenfix Ltd Carefully hand dig around base of tree but away from trunk to required depth; Lay Treetex geofabric to subgrade; Lay Geoweb cellular root protection to required depth and backfill with angular material.							
To graded and compacted substrate (not included); on 50 mm sharp sand base filled with angular drainage aggregate 20–5 mm; not mechanical traffic on the area of the rootzone permitted; Excavated material loaded outside the root protection zone to spoil heaps							
Geoweb 75 mm depth	8.99	0.07	7.13	1.33	19.88	m^2	**28.34**
Geoweb 100 mm depth	10.35	0.08	8.02	1.54	23.61	m^2	**33.18**
Geoweb 150 mm depth	12.94	0.10	10.16	1.67	30.51	m^2	**42.34**
Geoweb 200 mm depth	18.23	0.10	11.08	2.50	40.50	m^2	**54.09**
Geoweb 300 mm depth	20.90	0.13	14.09	4.00	51.41	m^2	**69.50**

35 SITE WORKS – PAVINGS

Item	PC £	Labour hours	Labour £	Plant £	Material £	Unit	Total rate £
EXCAVATION OF PATHWAYS							
The labour in this section is calculated on a 3 person team. The labour time below should be multiplied by 3 to calculate the cost as shown.							
Excavation and path preparation							
Excavating; 300 mm deep; to width of path; depositing excavated material at sides of excavation							
width 1.00 m	–		–	4.47	–	m^2	**4.47**
width 1.50 m	–		–	3.73	–	m^2	**3.73**
width 2.00 m	–		–	3.19	–	m^2	**3.19**
width 3.00 m	–		–	2.68	–	m^2	**2.68**
Excavating trenches; in centre of pathways; 100 mm flexible drain pipes; filling with clean broken stone or gravel rejects							
300 × 450 mm deep	5.02	0.03	2.37	2.23	5.77	m	**10.38**
Hand trimming and compacting reduced surface of pathway; by machine							
width 1.00 m	–	0.02	1.79	0.17	–	m	**1.96**
width 1.50 m	–	0.01	1.58	0.15	–	m	**1.74**
width 2.00 m	–	0.01	1.42	0.14	–	m	**1.56**
width 3.00 m	–	0.01	1.42	0.14	–	m	**1.56**
Permeable membranes; to trimmed and compacted surface of pathway							
Terram 1000	0.69	0.01	0.72	–	0.79	m^2	**1.51**
FILLING TO MAKE UP LEVELS FOR PAVINGS AND SURFACES							
The labour in this section is calculated on a 3 person team. The labour time below should be multiplied by 3 to calculate the cost as shown.							
Filling to make up levels							
Obtained off site; hardcore; PC £24.26/m^3							
150 mm thick	4.16	0.01	1.42	0.81	4.78	m^2	**7.01**
Type 1 granular fill base; PC £26.53/tonne (£53.90/m^3 compacted)							
100 mm thick	6.04	0.01	0.99	0.61	6.94	m^2	**8.55**
150 mm thick	9.06	0.01	0.89	0.92	10.41	m^2	**12.23**

35 SITE WORKS – PAVINGS

Item	PC £	Labour hours	Labour £	Plant £	Material £	Unit	Total rate £
FILLING TO MAKE UP LEVELS FOR PAVINGS AND SURFACES – cont							
Surface treatments							
Sand blinding; to hardcore (not included)							
50 mm thick	2.49	0.01	1.42	–	2.87	m^2	**4.29**
Filter fabric; to hardcore (not included)	0.69	–	0.35	–	0.79	m^2	**1.15**
BRICK PAVINGS							
Bricks – General							
Preamble: Bricks shall be hard, well burnt, non-dusting, resistant to frost and sulphate attack and true to shape, size and sample.							
Movement of materials							
Mechanically offloading bricks; loading wheelbarrows; transporting maximum 25 m distance	–	0.07	7.13	–	–	m^2	**7.13**
Variation in brick prices; add or subtract the following amounts for every £100/1000 difference in the PC price							
Edgings; mortar jointed							
100 mm wide	0.55		–	–	0.63	m	**0.63**
200 mm wide	1.05		–	–	1.21	m	**1.21**
102.5 mm wide	0.51		–	–	0.59	m	**0.59**
215 mm wide	1.03		–	–	1.18	m	**1.18**
The labour in this section is calculated on a 2 person team. The labour time below should be multiplied by 2 to calculate the cost as shown.							
Edge restraints; to brick paving; on prepared base (not included); haunching one side; Paving bricks by Vande Moortel bricks; haunched							
header courses; laid on edge; butt jointed on sand bed; haunched							
Amarant; 215 × 52 × 70 mm	12.95	0.20	14.26	–	22.17	m	**36.43**
Vanilla; 215 × 52 x70 mm	14.59	0.22	15.84	–	24.06	m	**39.90**
Bordeaux; Handmade	17.21	0.22	15.84	–	27.07	m	**42.92**

35 SITE WORKS – PAVINGS

Item	PC £	Labour hours	Labour £	Plant £	Material £	Unit	**Total rate £**
header courses; laid on edge; 10 mm mortar jointed on 35 mm mortar bed; haunched							
Amarant; 215 × 52 × 70 mm	12.95	0.40	28.52	–	23.57	m	**52.09**
SeptimA; Vanilla; 215 × 52 × 70 mm	14.59	0.36	25.46	–	25.46	m	**50.92**
Taupe; 215 × 52 × 70 mm	24.39	0.33	23.76	–	36.73	m	**60.49**
Stretcher courses; laid flat or on edge; butt jointed on mortar bed; haunched							
Amarant; 215 × 52 × 70 mm	5.90	0.08	5.94	–	14.06	m	**20.00**
SeptimA; Vanilla; 215 × 52 × 70 mm	3.53	0.08	5.94	–	11.34	m	**17.28**
Taupe; 215 × 52 × 70 mm	3.13	0.08	5.94	–	10.88	m	**16.82**
header courses; laid flat; 10 mm flowable mortar jointed on 35 mm mortar bed; haunched							
Blue Baggeridge 200 × 100 × 62 mm	9.75	0.29	20.37	–	19.89	m	**40.27**
The labour in this section is calculated on a 3 person team. The labour time below should be multiplied by 3 to calculate the cost as shown.							
Clay brick pavings; laid to running stretcher or stack bond only; on prepared base (not included); bedding on sharp sand bed 50 mm thick with kiln dried sand swept in to joints.							
Pavings; Vande Moortel SeptimA Collection 215 × 52 × 70–72 mm							
Amarant	60.21	0.44	46.91	–	73.62	m²	**120.53**
Vanilla	67.87	0.44	46.91	–	82.45	m²	**129.36**
Taupe	113.43	0.44	46.91	–	134.85	m²	**181.76**
Pavings; Vande Moortel; Elegantia collection; sanded or tumbled 230–240 × 36 × 72 mm							
Carmin	90.88	0.57	60.77	–	108.89	m²	**169.66**
Carbon	109.16	0.57	60.77	–	129.92	m²	**190.69**
Salvia	142.66	0.57	60.77	–	168.44	m²	**229.21**
Pavings; Vande Moortel; Handmade pavers; Ancienne Belgique collection, sanded or tumbled 230–183 × 43 × 88 mm							
Bordeaux	94.05	0.60	64.04	–	112.54	m²	**176.59**
Lava	102.57	0.60	64.04	–	122.34	m²	**186.38**
Oyster	148.50	0.60	64.04	–	175.18	m²	**239.22**

35 SITE WORKS – PAVINGS

Item	PC £	Labour hours	Labour £	Plant £	Material £	Unit	Total rate £
BRICK PAVINGS – cont							
Clay brick pavings; Wienerberger; on prepared base (not included); bedding on 50 mm sharp sand; kiln dried sand joints brushed in; Herringbone bond							
Pavings; Blue Baggeridge range 200 × 100 × 62 mm wirecut chamfered paviors							
laid flat	53.64	0.44	46.91	–	65.99	m^2	**112.90**
Clay brick pavings; laid to running stretcher or stack bond only; on prepared base (not included); bedding on cement: sand (1:4) pointing with flowable mortar – Steintec							
Pavings; Blue Baggeridge range 200 × 100 × 62 mm wirecut chamfered paviors							
laid flat	46.44	0.44	47.11	–	84.14	m^2	**131.25**
Cutting							
curved cutting	–	0.50	17.83	8.37	–	m	**26.20**
raking cutting	–	0.25	8.91	6.43	–	m	**15.35**
Clay brick pavings; Chelmer valley; on prepared base (not included); bedding on 50 mm sharp sand; kiln dried sand joints brushed in; Herringbone bond; Commercial paving range							
Moulded pavers; Suitable for vehicular traffic laid on edge; 206 mm × 51 mm × 85 mm thick; 93 units/m^2							
Malaga	66.60	0.49	52.43	–	80.77	m^2	**133.20**
Aragon	56.61	0.49	52.43	–	69.28	m^2	**121.71**
Bergerac	53.28	0.49	52.43	–	65.45	m^2	**117.88**
Extra over for							
Tumbled finish	–		–	–	9.23	m^2	**9.23**
Wire cut pavers; Pedestrian or occasional vehicular traffic; 204 mm × 50 mm × 67 mm thick; 96 units/m^2 ; Tumbled							
Gromo Antica	63.36	0.50	53.48	–	76.97	m^2	**130.45**
Ancona, Lucca or Roma	54.95	0.50	53.48	–	67.30	m^2	**120.78**

35 SITE WORKS – PAVINGS

Item	PC £	Labour hours	Labour £	Plant £	Material £	Unit	**Total rate £**
Wire cut pavers; Vehicular traffic; 200 mm × 100 mm × 62 mm thick; 48 units/m^2 ; Chamfered; laid on-flat							
Meissen	44.40	0.33	35.65	–	55.12	m^2	**90.77**
CELLULAR CONFINED GRANULAR PAVINGS							
The labour in this section is calculated on a 3 person team. The labour time below should be multiplied by 3 to calculate the cost as shown.							
StablePAVE; Cedar Nursery; honeycomb shaped cellular polypropylene containment grids for retaining filled decorative aggregate surfaces (not included) for driveways, cycle paths and footpaths; to graded compacted substrate (not included) and 50 mm sharp sand base; integrated geofabric base							
Foot traffic, wheelchairs, bikes and pushchairs							
StablePAVE Eco; 1200 × 800 × 30 mm deep	12.00	0.08	4.46	–	16.41	m^2	**20.86**
Driveways: vehicles, wheelchairs, bikes and pushchairs (requires compacted granular base to support vehicle loadings)							
StablePAVE Trade 30; 1200 × 800 × 30 mm deep	13.25	0.08	4.84	–	17.85	m^2	**22.69**
StablePAVE Trade 40; 1200 × 800 × 40 mm deep	15.63	0.08	4.84	–	20.58	m^2	**25.43**
StablePAVE HD; Cedar Nursery; honeycomb shaped polypropylene cellular containment system for retaining filled decorative aggregate surfaces for driveways, cyclepaths and footpaths; on compacted subgrade and subbase (both not included)							
Driveways: vehicles, wheelchairs, bikes and pushchairs (requires compacted granular base to support vehicle loadings)							
StablePAVE HD; 1200 × 600 × 38 mm deep	15.75	0.08	5.13	–	20.72	m^2	**25.85**

35 SITE WORKS – PAVINGS

Item	PC £	Labour hours	Labour £	Plant £	Material £	Unit	Total rate £
CELLULAR CONFINED GRANULAR PAVINGS – cont							
StablePAVE HD – cont							
Aggregate filling to StablePAVE							
shingle 30 mm deep	2.99	0.03	3.56	2.90	3.44	m^2	**9.90**
shingle 40 mm deep	3.99	0.03	3.69	2.90	4.59	m^2	**11.18**
decorative aggregate; PC £127.05/tonne; 30 deep	8.10	0.03	3.56	2.90	9.32	m^2	**15.77**
decorative aggregate; PC £127.05/tonne; 40 deep	10.80	0.03	3.69	2.90	12.42	m^2	**19.00**
StableGRASS 38 for grass reinforcement							
Polyethylene soil stabilizing panels; to soil surfaces brought to grade (not included); filling with excavated material							
panels; 50 cm × 50 cm × 38 mm deep	10.42	0.03	3.56	6.02	11.98	m^2	**21.56**
Extra for filling StableGRASS with topsoil; seeding with rye grass at 50 g/m^2	1.78	0.07	7.17	0.25	2.07	m^2	**9.49**
Ground Grid; Recycled plastic; Hahn Plastics Ltd; Interlocking; filling with angular granular material (5–15 mm)							
HDGG 8 cm thick × 40 cm wide × 60 cm long grid for HGV use; Strength: 18,600 kN/m^2 (1,896 tonnes/m^2); laid to prepared base (not included)							
HDGG; Units 40 cm wide × 60 cm long × 80 mm thick; 100–500 m^2 Volume of angular gravel fill is 1.8 tonnes per 25 m^2 or 7.2 kg/m^2	25.75	0.03	3.10	2.90	29.25	m^2	**35.24**
Geocell; Greenfix; cellular confinement system for load support or permeable bases below pavings							
To graded and compacted substrate (not included); on 50 mm sharp sand base filled with angular drainage aggregate 20–5 mm							
Geocell 75 mm depth	8.99	0.05	4.87	2.22	19.88	m^2	**26.97**
Geocell 100 mm depth	10.35	0.05	5.35	2.50	23.61	m^2	**31.46**
Geocell 150 mm depth	12.94	0.06	5.95	2.63	30.51	m^2	**39.09**
Geocell 200 mm depth	18.23	0.08	8.91	2.86	40.50	m^2	**52.28**
Geocell 300 mm depth	20.90	0.10	10.70	3.33	51.41	m^2	**65.44**

35 SITE WORKS – PAVINGS

Item	PC £	Labour hours	Labour £	Plant £	Material £	Unit	**Total rate £**
Cellular pavings by Grass Concrete							
Honeycomb cellular polyproylene interconnecting paviors with integral downstead anti-shear cleats including topsoil but excluding edge restraints; to granular subbase (not included)							
635 × 330 × 42 mm overall laid to a module of 622 × 311 × 32 mm	–		–	–	–	m²	**45.25**
GRASS CONCRETE							
Grass concrete – General							
Preamble: Grass seed should be a perennial ryegrass mixture, with the proportion depending on expected traffic. Hardwearing winter sportsground mixtures are suitable for public areas. Loose gravel, shingle or sand is liable to be kicked out of the blocks; rammed hoggin or other stabilized material should be specified.							
Grass concrete; Grass Concrete Ltd; on blinded granular Type 1 subbase (not included)							
Grasscrete in situ concrete continuously reinforced surfacing; including expansion joints at 10 m centres; soiling; seeding							
GC2; 150 mm thick; traffic up to 40.00 tonnes	–		–	–	–	m²	**68.73**
GC1; 100 mm thick; traffic up to 13.30 tonnes	–		–	–	–	m²	**51.73**
GC3; 76 mm thick; traffic up to 4.30 tonnes	–		–	–	–	m²	**45.42**
Embankments; Grass Concrete Ltd							
Grasscrete; In situ reinforced concrete surfacing; to 20 mm thick sand blinding layer (not included), including soiling and seeding							
GC1; 100 mm thick	–		–	–	4.94	m²	**56.87**
GC2; 150 mm thick	–		–	–	7.31	m²	**76.34**
Grassblock 103; solid matrix precast concrete blocks; to 20 mm thick sand blinding layer; excluding edge restraint; including soiling and seeding							
406 × 406 × 103 mm; fully interlocking	–		–	–	4.94	m²	**66.41**

35 SITE WORKS – PAVINGS

Item	PC £	Labour hours	Labour £	Plant £	Material £	Unit	Total rate £
GRASS CONCRETE – cont							
Grass concrete; Grass Concrete Ltd; 406 × 406 mm blocks; on 20 mm sharp sand; on blinded MOT type 1 subbase (not included); level and to falls only; including filling with topsoil; seeding with dwarf rye grass at PC £6.44/kg							
Pavings							
GB103; 103 mm thick	27.84	0.40	14.26	0.25	35.51	m^2	**50.02**
GB83; 83 mm thick	26.41	0.36	12.96	0.25	33.87	m^2	**47.08**
Grass concrete; Marshalls Plc; on 25 mm sharp sand; including filling with topsoil; seeding with dwarf rye grass at PC £6.44/kg							
Concrete grass pavings							
Grassguard 130; for light duty applications (80 mm prepared base not included)	35.45	0.38	13.37	0.25	44.54	m^2	**58.16**
Grassguard 160; for medium duty applications (80–150 mm prepared base not included)	39.75	0.46	16.45	0.25	49.49	m^2	**66.19**
Grassguard 180; for heavy duty applications (150 mm prepared base not included)	44.50	0.60	21.39	0.25	54.96	m^2	**76.60**
Full mortar bedding							
Extra over pavings for bedding on 25 mm cement: sand (1:4); in lieu of spot bedding on sharp sand	–	0.03	0.89	–	3.22	m^2	**4.11**

35 SITE WORKS – PAVINGS

Item	PC £	Labour hours	Labour £	Plant £	Material £	Unit	Total rate £
COBBLE PAVINGS							
Cobble pavings – General Cobbles should be embedded by hand, tight-butted, endwise to a depth of 60% of their length. A dry grout of rapid-hardening cement: sand (1:2) shall be brushed over the cobbles until the interstices are filled to the level of the adjoining paving. Surplus grout shall then be brushed off and a light, fine spray of water applied over the area.							
The labour in this section is calculated on a 3 person team. The labour time below should be multiplied by 3 to calculate the cost as shown.							
Cobble pavings Cobbles; to present a uniform colour in panels; or varied in colour as required							
Scottish Beach Cobbles; 50–75 mm	32.00	1.11	118.83	–	51.80	m²	**170.63**
Scottish Beach Cobbles; 75–100 mm	46.75	0.83	89.12	–	68.76	m²	**157.88**
Scottish Beach Cobbles; 100–200 mm	46.75	0.67	71.30	–	68.76	m²	**140.06**
GRANULAR PAVINGS							
The labour in this section is calculated on a 3 person team. The labour time below should be multiplied by 3 to calculate the cost as shown.							
Footpath gravels; porous self binding gravel CED Ltd; Cedec gravel; self-binding; laid to inert (non-limestone) base measured separately; compacting CED Ltd; Cedec gravel; self-binding; laid to inert (non-limestone) base measured separately; compacting							
red, silver or gold; 50 mm thick	16.00	0.01	0.89	0.49	18.40	m²	**19.77**

35 SITE WORKS – PAVINGS

Item	PC £	Labour hours	Labour £	Plant £	Material £	Unit	Total rate £
GRANULAR PAVINGS – cont							
Footpath gravels – cont							
Grundon Ltd; Coxwell self-binding path gravels laid and compacted to excavation or base measured separately							
50 mm thick	7.78	0.01	0.89	0.49	8.95	m²	**10.33**
Breedon Special Aggregates; Golden Gravel or equivalent; rolling wet; on hardcore base (not included); for pavements; to falls and crossfalls and to slopes not exceeding 15° from horizontal; over 300 mm wide							
50 mm thick	6.12	0.01	0.89	0.39	7.03	m²	**8.31**
75 mm thick	9.17	0.03	3.56	0.50	10.55	m²	**14.61**
Breedon Special Aggregates; Wayfarer specially formulated fine gravel for use on golf course pathways							
50 mm thick	6.30	0.01	0.89	0.39	7.25	m²	**8.53**
75 mm thick	9.45	0.02	1.79	0.68	10.86	m²	**13.32**
Hoggin (stabilized); PC £53.61/ tonne on hardcore base (not included); to falls and crossfalls and to slopes not exceeding 15° from horizontal; over 300 mm wide							
100 mm thick	14.97	0.01	1.19	0.65	17.21	m²	**19.05**
150 mm thick	22.45	0.02	1.79	0.97	25.82	m²	**28.58**
Ballast; as dug; watering; rolling; on hardcore base (not included)							
100 mm thick	6.17	0.01	1.19	0.65	7.10	m²	**8.94**
150 mm thick	9.26	0.02	1.79	0.97	10.65	m²	**13.41**
Footpath gravels; porous loose gravels							
Breedon Special Aggregates; Breedon Buff decorative limestone chippings in Warwick Gold, Coastal cream							
50 mm thick	3.94	–	0.35	0.13	4.53	m²	**5.01**
75 mm thick	8.03	–	0.45	0.16	9.23	m²	**9.84**
Breedon Special Aggregates; Brindle, Moorland Black, Shierglas and Galloway chippings							
50 mm thick	3.94	–	0.35	0.13	4.53	m²	**5.01**
75 mm thick	5.91	–	0.45	0.16	6.79	m²	**7.40**

35 SITE WORKS – PAVINGS

Item	PC £	Labour hours	Labour £	Plant £	Material £	Unit	Total rate £
Breedon Special Aggregates; 40 mm slate chippings; plum/blue							
50 mm thick	3.94	–	0.35	0.13	4.53	m²	**5.01**
75 mm thick	5.91	–	0.45	0.16	6.79	m²	**7.40**
Washed shingle; on prepared base (not included)							
25–50 mm size; 25 mm thick	1.46	0.01	0.89	0.11	1.67	m²	**2.67**
25–50 mm size; 75 mm thick	4.37	0.03	2.67	0.33	5.03	m²	**8.03**
50–75 mm size; 25 mm thick	1.46	0.01	0.72	0.13	1.67	m²	**2.52**
50–75 mm size; 75 mm thick	4.37	0.02	2.37	0.42	5.03	m²	**7.82**
Pea shingle; on prepared base (not included)							
10–15 mm size; 25 mm thick	1.46	0.01	0.63	0.11	1.67	m²	**2.42**
5–10 mm size; 75 mm thick	4.37	0.02	1.87	0.33	5.03	m²	**7.23**
PRECAST CONCRETE BLOCK PAVINGS							
The labour in this section is calculated on a 3 person team. The labour time below should be multiplied by 3 to calculate the cost as shown.							
Precast concrete block edgings; PC £14.01/m²; 200 × 100 × 60 mm; on prepared base (not included); haunching one side							
Edgings; butt joints							
stretcher course	1.43	0.06	5.94	–	7.89	m	**13.82**
header course	2.85	0.09	9.51	–	9.79	m	**19.30**
Precast concrete paving blocks; Keyblok Marshalls Plc; on prepared base (not included); on 50 mm compacted sharp sand bed; blocks laid in 7 mm loose sand and vibrated; joints filled with sharp sand and vibrated; level and to falls only							
Herringbone bond							
200 × 100 × 60 mm; natural grey	14.97	0.28	29.96	0.18	20.59	m²	**50.73**
200 × 100 × 60 mm; colours	15.99	0.28	29.96	0.18	21.76	m²	**51.90**
200 × 100 × 80 mm; natural grey	17.73	0.28	29.96	0.18	23.77	m²	**53.91**
200 × 100 × 80 mm; colours	18.34	0.28	29.96	0.18	24.46	m²	**54.60**
Basketweave bond							
200 × 100 × 60 mm; natural grey	14.97	0.25	26.74	0.18	20.59	m²	**47.51**
200 × 100 × 60 mm; colours	15.99	0.25	26.74	0.18	21.76	m²	**48.68**
200 × 100 × 80 mm; natural grey	17.73	0.25	26.74	0.18	23.77	m²	**50.69**
200 × 100 × 80 mm; colours	18.34	0.25	26.74	0.18	24.46	m²	**51.38**

35 SITE WORKS – PAVINGS

Item	PC £	Labour hours	Labour £	Plant £	Material £	Unit	**Total rate £**
PRECAST CONCRETE BLOCK PAVINGS – cont							
Tegula traditional Concrete setts; on 50 mm sand; compacted; vibrated; joints filled with sand; natural or coloured; well rammed hardcore base (not included)							
Marshalls Plc; Tegula cobble paving; Mixed sizes 240 × 160 mm; 160 x160 mm; 120 × 160 mm							
60 mm thick; random sizes	32.78	0.19	20.37	0.18	41.34	m^2	**61.89**
60 mm thick; single size	32.78	0.15	16.20	0.18	41.34	m^2	**57.72**
80 mm thick; random sizes	36.33	0.19	20.37	0.18	45.42	m^2	**65.98**
80 mm thick; single size	36.33	0.15	16.20	0.18	45.42	m^2	**61.80**
cobbles; 80 × 80 × 60 mm thick; traditional	58.68	0.19	19.81	0.18	71.12	m^2	**91.10**
Marshalls Plc Textured granite aggregate sett paving; on primary aggregate 6–2 mm; compacted; vibrated; joints filled with grit; Silver grey; Graphite; Charcoal; well rammed hardcore base (not included)							
Conservation X Riven; Mixed sizes 240 × 160 mm; 160 x160 mm; 120 × 160 mm							
60 mm thick; mixed sizes	43.71	0.25	26.74	0.18	55.11	m^2	**82.02**
80 mm thick; mixed sizes	49.82	0.29	30.56	0.18	62.13	m^2	**92.87**

35 SITE WORKS – PAVINGS

Item	PC £	Labour hours	Labour £	Plant £	Material £	Unit	Total rate £
Modal textured blocks; mixed shades and sizes; 200 × 100 mm–400 × 200 mm; (larger sizes shown under slab paving)							
single size 80 mm thick	51.05	0.29	30.56	0.18	63.55	m^2	**94.28**
mixed sizes to pattern 80 mm thick	51.05	0.33	35.65	0.18	63.55	m^2	**99.37**
Modal smooth blocks; mixed shades and sizes; 200 × 100 mm–400 × 200 mm; (larger sizes shown under slab paving)							
single size 80 mm thick	68.29	0.29	30.56	0.18	83.37	m^2	**114.10**
mixed sizes to pattern 80 mm thick	68.29	0.33	35.65	0.18	83.37	m^2	**119.19**
PRECAST CONCRETE SUDS PERMEABLE PAVING							
SUDS paving; Concrete Block Permeable Paving (CBPP)							
Note: Sustainable Drainage Systems (SUDS) aim to reduce flood risk by controlling the rate and volume of surface water run off from developments. Permeable Paving is a SUDS technique. In addition to managing the quantity of surface water run-off more effectively, one of the primary benefits of CBPP is the potential enhancement of water quality. Permeable pavements improve water quality by mirroring nature in providing filtration and allowing for natural biodegradation of hydrocarbons and the dilution of other contaminants as water passes through the system.							
The labour in this section is calculated on a 3 person team. The labour time below should be multiplied by 3 to calculate the cost as shown.							

35 SITE WORKS – PAVINGS

Item	PC £	Labour hours	Labour £	Plant £	Material £	Unit	Total rate £
PRECAST CONCRETE SUDS PERMEABLE PAVING – cont							
Bases for Sustainable Urban Drainage System (SUDS) Type 3 open graded subbase for use below permeable pavings							
40 down open graded material laid loose on prepared sub-grade; within edge restraints to paving area (not included); graded to levels and falls as per paving manufacturer's instructions							
100 mm thick	5.82	0.02	2.49	0.59	10.61	m^2	**13.69**
150 mm thick	8.73	0.03	3.48	1.29	13.96	m^2	**18.73**
200 mm thick	11.64	0.04	4.46	1.90	17.31	m^2	**23.67**
250 mm thick	14.55	0.05	5.44	1.90	20.66	m^2	**28.00**
over 250 mm thick	64.03	0.19	20.34	7.68	77.55	m^3	**105.58**
Pervious surfacing materials; Marshalls Mono SuDs blocks; permeable block paving system; incorporating a 5 mm spacer design that provides a 5 mm void allowing ingress of water through to the subbase storage system (priced separately); prices shown are for the highest cost product in each range							
Priora; Rectangular various colours; bedded on 50 mm thick 2–6.3 mm clean, angular, free draining uncompacted aggregate; joints infilled with 3 mm clean grit; various colours available; prices shown below are for the highest cost product							
200 × 100 × 60 mm	17.96	0.20	21.39	2.92	24.07	m^2	**48.38**
200 × 100 × 80 mm	22.27	0.25	26.74	2.92	29.03	m^2	**58.69**
Tegula Priora; various colours; bedded on 50 mm thick 2–6.3 mm clean, angular, free draining uncompacted aggregate; joints infilled with 3 mm clean grit; prices shown below are for the highest cost product							
120 × 160 × 60 mm	33.69	0.13	13.89	2.92	42.16	m^2	**58.98**
240 × 160 × 80 mm	39.37	0.13	13.89	2.92	48.70	m^2	**65.51**

35 SITE WORKS – PAVINGS

Item	PC £	Labour hours	Labour £	Plant £	Material £	Unit	**Total rate £**
Modal Priora; Textured; various colours; bedded on 50 mm thick 2–6.3 mm clean, angular, free draining uncompacted aggregate; joints infilled with 3 mm clean grit; prices shown below are for the highest cost product; 80 mm thick							
200 × 100 mm	54.63	0.17	17.83	2.92	66.25	m^2	**87.00**
200 × 200 mm	54.63	0.13	14.26	2.92	66.25	m^2	**83.42**
200 × 300 mm	54.63	0.13	13.37	2.92	66.25	t	**82.54**
300 × 300 mm	54.63	0.14	15.28	2.92	66.25	m^2	**84.45**
300 × 600 mm	54.63	0.25	26.74	2.92	66.25	m^2	**95.90**
Mixed size formats	54.63	0.29	30.56	2.92	66.25	m^2	**99.72**
Conservation X priora; mix of natural granite aggregates with a choice of surface finish; bedded on 50 mm thick 2–6.3 mm clean, angular, free draining uncompacted aggregate; joints infilled with 3 mm clean grit; prices shown below are for the highest cost product; 65 mm thick							
200 × 400 mm	54.74	0.17	17.83	2.92	66.37	m^2	**87.12**
400 × 400 mm	54.74	0.14	15.28	2.92	66.37	m^2	**84.58**
600 × 400 mm	54.74	0.25	26.74	2.92	66.37	m^2	**96.03**
Mixed formats	54.74	0.29	30.56	2.92	66.37	m^2	**99.85**
PRECAST CONCRETE FLAG PAVINGS							
The labour in this section is calculated on a 3 person team. The labour time below should be multiplied by 3 to calculate the cost as shown.							
Note: Precast flag pavings. The prices shown below do not always reflect the full ranges of colours and sizes avaialble. Readers should consult the supplier listed via the directory at the front of this book							
Flag paving; precast concrete pavings; Marshalls Mono; on prepared subbase (not included); bedding on 25 mm thick cement: sand mortar (1:4); butt joints; straight both ways; on 50 mm thick sharp sand base							

35 SITE WORKS – PAVINGS

Item	PC £	Labour hours	Labour £	Plant £	Material £	Unit	Total rate £
PRECAST CONCRETE FLAG PAVINGS – cont							
Pavings; Standard; Grey							
450 × 450 × 50 mm	21.49	0.15	15.84	–	32.93	m²	**48.77**
600 × 300 × 50 mm	25.52	0.15	15.84	–	37.57	m²	**53.40**
400 × 400 × 65 mm	21.48	0.13	14.26	–	32.92	m²	**47.18**
450 × 600 × 50 mm	15.87	0.15	15.84	–	26.47	m²	**42.31**
600 × 600 × 50 mm	13.92	0.13	14.26	–	24.23	m²	**38.48**
750 × 600 × 50 mm	13.05	0.13	14.26	–	23.22	m²	**37.48**
900 × 600 × 50 mm	11.06	0.13	14.26	–	20.93	m²	**35.19**
Pavings; Buff							
400 × 400 × 65 mm	21.48	0.13	14.26	–	32.92	m²	**47.18**
600 × 600 × 50 mm	21.33	0.13	14.26	–	32.74	m²	**47.00**
900 × 600 × 50 mm	17.85	0.13	14.26	–	28.75	m²	**43.00**
Pavings; Pimple textured Buff or Natural; (Prices shown are for Buff) Natural slightly less							
450 × 450 × 50 mm	17.82	0.15	15.84	–	28.71	m²	**44.55**
600 × 300 × 50 mm	17.82	0.15	15.84	–	28.71	m²	**44.55**
400 × 400 × 65 mm	17.82	0.13	14.26	–	28.71	m²	**42.96**
450 × 600 × 50 mm	17.82	0.15	15.84	–	28.71	m²	**44.55**
600 × 600 × 50 mm	17.82	0.13	14.26	–	28.71	m²	**42.96**
750 × 600 × 50 mm	17.82	0.13	14.26	–	28.71	m²	**42.96**
900 × 600 × 50 mm	17.82	0.13	14.26	–	28.71	m²	**42.96**
Pavings; Marshalls Urbex; Pimple textured Buff or Natural; (Prices shown are for Buff) Natural slightly less							
450 × 450 × 38 mm	18.40	0.15	15.84	–	29.37	m²	**45.21**
600 × 300 × 38 mm	19.59	0.15	15.84	–	30.74	m²	**46.58**
450 × 450 × 38 mm	24.31	0.13	14.26	–	36.17	m²	**50.42**
450 × 450 × 50 mm	30.30	0.15	15.84	–	43.06	m²	**58.90**
600 × 600 × 38 mm	17.82	0.13	14.26	–	28.71	m²	**42.96**
750 × 600 × 38 mm	17.82	0.13	14.26	–	28.71	m²	**42.96**
900 × 600 × 38 mm	17.82	0.13	14.26	–	28.71	m²	**42.96**
Precast concrete flag pavings; Marshalls Plc; on prepared subbase measured separately; bedding on 25 mm thick cement: sand mortar (1:4); pointed straight both ways Steintec jointing mixture							
Urbex paving; Practical economical paving alternative to Pimple paving; Riven or Textured in Natural or Buff							
450 × 450 × 35 mm	18.95	0.32	34.50	–	33.40	m²	**67.90**
600 × 300 × 35 mm	18.95	0.33	35.65	–	33.59	m²	**69.23**
600 × 600 × 35 mm	18.95	0.27	28.52	–	31.28	m²	**59.80**

35 SITE WORKS – PAVINGS

Item	PC £	Labour hours	Labour £	Plant £	Material £	Unit	**Total rate £**
Precast concrete flag pavings; Marshalls Plc; on prepared subbase measured separately; bedding on 25 mm thick cement: sand mortar (1:4); pointed straight both ways Steintec jointing mixture							
Modal textured flags; mixed shades and sizes; (smaller sizes shown under slab paving); 80 mm thick							
450 × 450 mm	51.05	0.16	16.63	–	66.93	m^2	**83.56**
400 × 400 mm	51.05	0.14	14.97	–	66.93	m^2	**81.90**
600 × 200 mm	51.05	0.16	16.63	–	66.93	m^2	**83.56**
600 × 300 mm	51.05	0.16	17.53	–	66.93	m^2	**84.45**
600 × 600 mm	51.05	0.14	14.97	–	66.93	m^2	**81.90**
900 × 600 mm	51.05	0.14	14.97	–	66.93	m^2	**81.90**
mixed sizes to pattern	51.05	0.26	28.07	0.18	67.96	m^2	**96.21**
Precast concrete flag pavings; Marshalls Plc; Conservation X paving; Natural granite aggregate appearance on prepared subbase measured separately; bedding on 25 mm thick cement: sand mortar (1:4); pointed straight both ways Steintec jointing mixture							
Conservation X paving; Smooth skimmed or textured in various colour options; Silver grey, Charcoal, Heather grey, and Buff with minor cost variations between colours and textures. Highest material cost used in the calculations below; Smooth pavings are generally higher priced							
450 × 450 × 50 mm	85.44	0.32	34.50	–	108.55	m^2	**143.05**
450 × 600 × 63 mm	85.44	0.15	15.84	–	107.74	m^2	**123.58**

35 SITE WORKS – PAVINGS

Item	PC £	Labour hours	Labour £	Plant £	Material £	Unit	Total rate £
PRECAST CONCRETE FLAG PAVINGS – cont							
Precast concrete vehicular paving flags; Marshalls Plc; on prepared base (not included); on 50 mm compacted sharp sand bed; flags laid in 7 mm loose sand and vibrated; joints filled with sharp sand and vibrated; level and to falls only							
Trafica paving blocks; 450 × 450 × 70 mm							
Perfecta finish; colour natural	51.89	0.17	17.83	0.18	62.54	m^2	**80.54**
Perfecta finish; colour buff	56.95	0.17	17.83	0.18	68.36	m^2	**86.37**
Saxon finish; colour natural	51.05	0.17	17.83	0.18	61.58	m^2	**79.58**
Saxon finish; colour buff	56.45	0.17	17.83	0.18	67.79	m^2	**85.79**
900 × 600 mm with additional central support per slab	95.35	0.31	22.28	–	109.66	m^2	**131.94**
The labour in this section is calculated on a 2 person team. The labour time below should be multiplied by 2 to calculate the cost as shown.							
Paving flags to pedestals on roof decks or podiums; Rates for installing flags to pedestals. Prices do not include supply or movement of the slabs to the laying position							
Paving flags; laid at uniform sizes with one pedestal per corner of each paving flag; paving flag sizes; laid on adjustable pedestal 285–400 mm high above the flag							
300 × 300 mm	231.24	0.63	44.56	–	265.93	m^2	**310.49**
400 × 400 mm	136.02	0.63	44.56	–	156.43	m^2	**200.99**
450 × 450 mm	102.02	0.63	44.56	–	117.32	m^2	**161.88**
600 × 450 mm with additional central support per flag	89.78	0.42	29.70	–	103.24	m^2	**132.95**
900 × 600 mm with additional central support per flag	51.69	0.31	22.28	–	59.44	m^2	**81.72**
Paving flags; laid at uniform sizes with one pedestal per corner of each paving flag; paving flag sizes; laid on adjustable pedestal 645–850 mm high above the flag							
300 × 300 mm	426.58	0.63	44.56	–	490.57	m^2	**535.13**
400 × 400 mm	250.93	0.63	44.56	–	288.57	m^2	**333.13**
450 × 450 mm	188.20	0.63	44.56	–	216.43	m^2	**260.99**

35 SITE WORKS – PAVINGS

Item	PC £	Labour hours	Labour £	Plant £	Material £	Unit	**Total rate £**
600 × 450 mm with additional central support per flag	165.61	0.42	29.70	–	190.46	m^2	**220.16**
900 × 600 mm with additional central support per flag	95.35	0.31	22.28	–	109.66	m^2	**131.94**
PRECAST CONCRETE FLAG PAVINGS – FIBRE REINFORCED							
The labour in this section is calculated on a 3 person team. The labour time below should be multiplied by 3 to calculate the cost as shown.							
Precast concrete flag pavings; Marshalls Plc; Fibre reinforced pavings to match other flag paving families; on prepared subbase measured separately; bedding on 25 mm thick cement: sand mortar (1:4); pointed straight both ways Steintec jointing mixture							
Fibre Reinforced Paving							
Conservation X							
450 × 600 × 63 mm Textured	83.50	0.27	28.52	–	105.51	m^2	**134.03**
Standard paving							
450 × 600 × 63 mm	29.59	0.27	28.52	–	43.51	m^2	**72.03**

35 SITE WORKS – PAVINGS

Item	PC £	Labour hours	Labour £	Plant £	Material £	Unit	**Total rate £**
DETERRENT AND TACTILE PAVINGS							
The labour in this section is calculated on a 3 person team. The labour time below should be multiplied by 3 to calculate the cost as shown.							
Pedestrian deterrent or tactile pavings; Marshalls Plc; on prepared base (not included); bedding on 25 mm cement: sand (1:3); cement: sand (1:3) joints							
Blister paving; 450 × 450 × 50 mm							
450 × 450 × 50 mm	48.27	0.15	16.45	–	62.52	m²	**78.97**
450 × 450 × 70 mm	54.04	0.17	17.83	–	69.16	m²	**86.99**
400 × 400 × 50 mm	40.21	0.15	16.45	–	53.25	m²	**69.70**
400 × 400 × 65 mm	57.50	0.16	17.11	–	73.13	m²	**90.24**
Hazard warning flags; 400 × 400 × 50 mm							
400 × 400 × 50 mm	54.26	0.20	21.39	–	69.41	m²	**90.80**

35 SITE WORKS – PAVINGS

Item	PC £	Labour hours	Labour £	Plant £	Material £	Unit	Total rate £
Edge restraints; to block paving; on prepared base (not included); 200 × 100 × 80 mm; PC £16.60/m²; haunching one side							
Header course							
200 × 100 mm; butt joints	3.38	0.09	9.51	–	10.40	m	**19.90**
Stretcher course							
200 × 100 mm; butt joints	1.69	0.06	5.94	–	8.19	m	**14.12**
STONE PAVINGS							
Natural stone paving – General							
Preamble: Provide paving slabs of the specified thickness in random sizes but not less than 25 slabs per 10 m² of surface area, to be laid in parallel courses with joints alternately broken and laid to falls.							
The labour in this section is calculated on a 3 person team. The labour time below should be multiplied by 3 to calculate the cost as shown.							
Natural stone, slate or granite flag pavings;on prepared base (not included); bedding on 25 mm cement: sand (1:3); Steintec Tuff top joints							
Yorkstone; riven laid random rectangular							
new slabs; 40–60 mm thick	107.80	0.57	61.12	–	134.73	m²	**195.85**
reclaimed slabs; Cathedral grade; 50–75 mm thick	165.00	0.93	99.82	–	208.60	m²	**308.42**
Cornish Granite De Lank Quarries 50 mm thick							
200 mm × random lengths	114.26	1.33	142.60	–	140.58	m²	**283.18**
300 mm × random lengths	114.26	1.25	133.69	–	139.91	m²	**273.60**
350 mm × random lengths	114.26	1.25	133.69	–	139.68	m²	**273.37**
400 mm × random lengths	114.26	1.11	118.83	–	139.46	m²	**258.29**
450 mm × random lengths	114.26	1.00	106.95	–	139.40	m²	**246.41**

35 SITE WORKS – PAVINGS

Item	PC £	Labour hours	Labour £	Plant £	Material £	Unit	Total rate £
STONE PAVINGS – cont							
Natural Yorkstone pavings; Johnsons Wellfield Quarries; sawn 6 sides; 50 mm thick; on prepared base measured separately; bedding on 25 mm cement: sand (1:3); Steintec Tuff top joints							
Paving							
fixed widths random lengths	113.08	0.91	97.23	–	140.49	m^2	**237.72**
laid to coursed laying pattern; 3 sizes	105.19	0.80	85.56	–	131.19	m^2	**216.75**
Paving; single size							
600 × 600 mm	105.19	0.67	71.30	–	130.52	m^2	**201.82**
600 × 400 mm	105.19	0.67	71.30	–	131.19	m^2	**202.50**
300 × 200 mm	105.19	0.83	89.12	–	135.24	m^2	**224.36**
215 × 102.5 mm	117.70	0.83	89.12	–	155.47	m^2	**244.59**
Paving; cut to template off site; 600 × 600 mm; radius							
1.00 m	252.34	1.00	106.95	–	300.64	m^2	**407.59**
2.50 m	252.34	0.91	97.23	–	299.74	m^2	**396.97**
5.00 m	252.34	0.87	93.00	–	299.74	m^2	**392.75**
CED Ltd; Indian sandstone, riven pavings or edgings; calibrated +/−26 mm thick; on prepared base measured separately; bedding on 25 mm cement: sand (1:3); Steintec Tuff top joints							
Paving							
laid to random rectangular pattern	31.50	0.67	71.30	–	46.68	m^2	**117.98**
laid to coursed laying pattern; 3 sizes	31.50	0.63	66.84	–	46.68	m^2	**113.52**
Paving; single size							
600 × 600 mm	31.50	0.67	71.30	–	43.75	m^2	**115.06**
600 × 400 mm	31.50	0.67	71.30	–	44.43	m^2	**115.73**
Natural stone, slate or granite flag pavings; CED Ltd; on prepared base (not included); bedding on 25 mm cement: sand (1:3); Steintec Tuff top joints							
Granite paving; sawn 6 sides; textured top							
new slabs; silver grey; 50 mm thick	101.85	0.67	71.30	–	127.66	m^2	**198.97**
new slabs; blue grey; 50 mm thick	101.85	0.67	71.30	–	127.66	m^2	**198.97**
new slabs; yellow grey; 50 mm thick	101.85	0.67	71.30	–	127.66	m^2	**198.97**
new slabs; black; 50 mm thick	101.85	0.67	71.30	–	127.66	m^2	**198.97**

35 SITE WORKS – PAVINGS

Item	PC £	Labour hours	Labour £	Plant £	Material £	Unit	**Total rate £**
Edgings; silver grey; blue grey yellow or black basalt textured							
100 mm wide × 150 deep	45.10	0.33	35.65	–	52.55	m	**88.20**
Natural Portland stone pavings; Albion Stone; sawn 6 sides; 40 mm thick; on prepared base measured separately; bedding on 25 mm cement: sand (1:3); white cement: sand (1:3) joints							
Paving; 40 mm thick							
laid to coursed laying pattern; 3 sizes	160.28	0.87	93.00	–	203.28	m^2	**296.29**
Paving 40 mm thick; single size							
1200 × 750 mm	160.28	1.11	118.83	–	203.28	m^2	**322.12**
300 × 200 mm	160.28	0.40	42.78	–	210.87	m^2	**253.65**
600 × 600 mm	160.28	0.32	33.87	–	203.28	m^2	**237.15**
600 × 400 mm	160.28	0.30	32.09	–	203.28	m^2	**235.37**
Paving; 50 mm thick							
laid to coursed laying pattern; 3 sizes	200.36	0.87	93.00	–	249.36	m^2	**342.37**
Paving 50 mm thick; single size							
1200 × 750 mm	200.36	1.25	133.69	–	249.36	m^2	**383.05**
300 × 200 mm	200.36	0.40	42.78	–	256.95	m^2	**299.73**
600 × 600 mm	200.36	0.32	33.87	–	249.36	m^2	**283.24**
600 × 400 mm	200.36	0.30	32.09	–	249.36	m^2	**281.45**
PORCELAIN PAVINGS							
The labour in this section is calculated on a 3 person team. The labour time below should be multiplied by 3 to calculate the cost as shown.							
Porcelain pavings; London Stone Ltd; 20 mm thick; on prepared base measured separately; bedding to 35 mm mortar bed primed with slurry primer; Grouted with Ardex flowable grout 3–5 mm joints							
Paving 20 mm thick in range of colours; Higher rate paving ranges; PC £63.36/m^2; less than 150 m^2							
596 mm × 596 mm	70.12	0.40	42.78	–	90.05	m^2	**132.83**
740 mm × 748 mm	70.12	0.36	38.89	–	89.73	m^2	**128.62**
895 mm × 895 mm	70.12	0.33	35.65	–	89.52	m^2	**125.17**
895 mm × 446 mm	70.12	0.40	42.78	–	90.07	m^2	**132.85**
1194 mm × 596 mm	70.12	0.33	35.65	–	89.66	m^2	**125.30**
1194 mm × 297 mm	70.12	0.33	35.65	–	90.47	m^2	**126.11**

35 SITE WORKS – PAVINGS

Item	PC £	Labour hours	Labour £	Plant £	Material £	Unit	**Total rate £**
PORCELAIN PAVINGS – cont							
Porcelain pavings – cont							
Paving 20 mm thick in range of colours; Lower rate paving ranges; PC £41.00/m^2							
596 mm × 596 mm	25.50	0.40	42.78	–	38.74	m^2	**81.52**
748 mm × 748 mm	25.50	0.36	38.89	–	38.42	m^2	**77.30**
895 mm × 895 mm	25.50	0.33	35.65	–	38.21	m^2	**73.85**
895 mm × 446 mm	25.50	0.40	42.78	–	38.75	m^2	**81.53**
1194 mm × 596 mm	25.50	0.33	35.65	–	38.34	m^2	**73.99**
1194 mm × 297 mm	25.50	0.33	35.65	–	39.15	m^2	**74.80**
Paving 20 mm thick in range of colours; Higher rate paving ranges; PC £64.00/m^2; above 150 m^2							
596 mm × 596 mm	70.12	0.40	42.78	–	90.05	m^2	**132.83**
748 mm × 748 mm	70.12	0.36	38.89	–	89.73	m^2	**128.62**
895 mm × 895 mm	70.12	0.33	35.65	–	89.52	m^2	**125.17**
895 mm × 446 mm	70.12	0.40	42.78	–	90.07	m^2	**132.85**
1194 mm × 596 mm	70.12	0.33	35.65	–	89.66	m^2	**125.30**
1194 mm × 297 mm	70.12	0.33	35.65	–	90.47	m^2	**126.11**
Paving 20 mm thick in range of colours; Lower rate paving ranges; PC £35.20/m^2							
596 mm × 596 mm	25.50	0.40	42.78	–	38.74	m^2	**81.52**
748 mm × 748 mm	25.50	0.36	38.89	–	38.42	m^2	**77.30**
895 mm × 895 mm	25.50	0.33	35.65	–	38.21	m^2	**73.85**
895 mm × 446 mm	25.50	0.40	42.78	–	38.75	m^2	**81.53**
1194 mm × 596 mm	25.50	0.33	35.65	–	38.34	m^2	**73.99**
1194 mm × 297 mm	25.50	0.33	35.65	–	39.15	m^2	**74.80**
Cutting Porcelain paving							
straight	–	0.50	17.83	1.56	–	m	**19.38**
curved	–	0.80	28.52	1.56	–	m	**30.08**

35 SITE WORKS – PAVINGS

Item	PC £	Labour hours	Labour £	Plant £	Material £	Unit	Total rate £
STONE SETTS							
The labour in this section is calculated on a 3 person team. The labour time below should be multiplied by 3 to calculate the cost as shown.							
Granite setts; CED Ltd bedding on 30 mm Steintec mortar; surfaces primed with Steintec primer; Jointing with Steintec Tufftop; 6 mm joints							
100 mm × 100 mm							
60 mm deep; silver grey; flamed top	106.38	0.35	36.36	–	285.29	m²	**321.65**
60 mm deep; silver grey; fine picked top	111.53	0.35	36.36	–	291.21	m²	**327.57**
100 mm deep; silver grey; flamed top	106.38	0.38	39.60	–	308.44	m²	**348.04**
100 mm deep; silver grey; fine picked top	106.38	0.38	39.60	–	308.44	m²	**348.04**
Granite setts; riven surface; silver grey; 10 mm joints; cropped setts; bedding on 30 mm Steintec mortar; jointed with Steintec Tufftop							
bedding on 30 mm Steintec mortar; jointed with Steintec Tufftop							
new; high grade 100 mm × 100 mm × 100 mm deep	49.65	0.50	53.48	–	260.53	m²	**314.00**
new; standard grade; 100 mm × 200 mm × 100 mm deep	53.07	0.50	53.48	–	248.33	m²	**301.80**
reclaimed; cleaned; 100 mm × 100 mm × 100 mm deep	52.84	0.50	53.48	–	264.20	m²	**317.67**
bedding on 30 mm mortar 1:3; jointed with Steintec Tufftop							
new; high grade 100 mm × 100 mm × 100 mm deep	49.65	0.50	53.48	–	193.04	m²	**246.52**
new; standard grade; 100 mm × 200 mm × 100 mm deep	53.07	0.50	53.48	–	123.02	m²	**176.50**
reclaimed; cleaned; 100 mm × 100 mm × 100 mm deep	52.58	0.50	53.48	–	138.20	m²	**191.67**

35 SITE WORKS – PAVINGS

Item	PC £	Labour hours	Labour £	Plant £	Material £	Unit	Total rate £
RECONSTITUTED STONE PAVINGS							
The labour in this section is calculated on a 3 person team. The labour time below should be multiplied by 3 to calculate the cost as shown.							
Reconstituted Yorkstone aggregate pavings; Marshalls Plc; Saxon on prepared subbase measured separately; bedding on 25 mm thick cement: sand mortar (1:4); on 50 mm thick sharp sand base							
Square and rectangular paving in buff; butt joints straight both ways							
300 × 300 × 35 mm	70.93	0.27	28.52	–	90.86	m^2	**119.39**
600 × 300 × 35 mm	48.86	0.22	23.18	–	65.49	m^2	**88.66**
450 × 450 × 50 mm	39.61	0.25	26.74	–	54.84	m^2	**81.58**
600 × 600 × 35 mm	35.04	0.17	17.83	–	49.60	m^2	**67.42**
600 × 600 × 50 mm	34.49	0.18	19.60	–	48.96	m^2	**68.56**
Square and rectangular paving in natural; butt joints straight both ways							
300 × 300 × 35 mm	59.31	0.27	28.52	–	77.50	m^2	**106.03**
450 × 450 × 50 mm	35.68	0.23	24.95	–	50.33	m^2	**75.28**
600 × 300 × 35 mm	42.94	0.25	26.74	–	58.68	m^2	**85.42**
600 × 600 × 35 mm	31.50	0.17	17.83	–	45.52	m^2	**63.35**
600 × 600 × 50 mm	31.62	0.20	21.39	–	45.66	m^2	**67.05**
RESIN BOUND SURFACING							
Resin bound paving; Sureset Ltd; fully mixed permeable decorative aggregate paving; laid to macadam binder and base course (not included); thickness dependent on loading application and aggregate size							
18 mm thick – 6 mm aggregate							
Natural gravel							
areas 100–300 m^2	–		–	–	–	m^2	**62.58**
areas 300–500 m^2	–		–	–	–	m^2	**57.67**
areas over 500 m^2	–		–	–	–	m^2	**49.08**

35 SITE WORKS – PAVINGS

Item	PC £	Labour hours	Labour £	Plant £	Material £	Unit	Total rate £
Crushed rock							
areas 100–300 m²	–		–	–	–	m²	**65.03**
areas 300–500 m²	–		–	–	–	m²	**56.44**
areas over 500 m²	–		–	–	–	m²	**52.76**
Marble							
areas 100–300 m²	–		–	–	–	m²	**92.03**
areas 300–500 m²	–		–	–	–	m²	**77.30**
areas over 500 m²	–		–	–	–	m²	**71.17**
Spectrum							
areas 100–300 m²	–		–	–	–	m²	**95.71**
areas 300–500 m²	–		–	–	–	m²	**80.98**
areas over 500 m²	–		–	–	–	m²	**73.62**
Bound aggregates; Addagrip Terraco Ltd; natural decorative resin bound surface dressing laid to concrete, macadam or to plywood panels priced separately							
Lucerne Silver; Pecan; Scandinavian Pearl; Blue Granite 3 mm aggregate							
6 mm depth	–		–	–	–	m²	**41.70**
Chocolate Terrabound; 6 mm							
16 mm depth	–		–	–	–	m²	**47.65**
Terracotta 6 mm							
16 mm or 18 mm depth	–		–	–	–	m²	**53.61**
Midnight Grey; 6 mm							
15 mm depth	–		–	–	–	m²	**53.61**
Chocolate; 6 mm							
15 mm depth	–		–	–	–	m²	**53.61**
18 mm depth	–		–	–	–	m²	**65.52**
Acorn 3 mm							
12 mm	–		–	–	–	m²	**53.61**
Terrabase; rustic bound porous paving							
16 mm deep	–		–	–	–	m²	**89.35**
The labour in this section is calculated on a 3 person team. The labour time below should be multiplied by 3 to calculate the cost as shown.							
Resin bound macadam pavings; Bituchem; to pedestrian or vehicular hard landscape areas; laid to base course (not included)							
Natratex wearing course; clear resin bound macadam							
25 mm thick to pedestrian areas	–		–	–	–	m²	**49.51**
30 mm thick to vehicular areas	–		–	–	–	m²	**57.10**

35 SITE WORKS – PAVINGS

Item	PC £	Labour hours	Labour £	Plant £	Material £	Unit	Total rate £
RESIN BOUND SURFACING – cont							
Resin bound macadam pavings – cont							
Colourtex coloured resin bound macadam							
25 mm thick	–		–	–	–	m²	**49.51**
Resin bound paving to individual treepits; SureSet Ltd; fully mixed permeable decorative aggregate paving; laid to well compacted stone Type 3 (not included)							
30–40 mm thick – average tree pit size 1.50 × 1.50 tree pit aggregate							
Natural gravel							
areas 100–300 m²	–		–	–	–	nr	**134.97**
SudsTech; Langford Direct Ltd; resin bound porous paving solution; fully sustainable urban drainage Systems (SuDS) compliant, blended aggregates mixed with tensile resins, in collaboration with proprietary flexible base							
Permeable wearing top course 20 mm (resin bound aggregate) to subbase of resin bound recycled tyres and aggregate 30 mm laid to Type 3 granular subbase (not included)							
SudsTech – 2 part system overall 50 mm thick; standard colours	–		–	–	–	m²	**94.11**
SudsTech – 2 part system; overall 50 mm thick; premium colours	–		–	–	–	m²	**106.03**
Trailflex; SudsTech Ltd; Single flexible surface of resin bound recycled tyres and aggregate; alternative to macadam or concretesurfacing; traffic of up to 18 tonnes							
Permeable flexible wearing course; 40 mm thick; resin bound recycled tyre and stone; laid to Type 3 base (not included);							
less than 300 m²	–		–	–	–	m²	**53.61**
300 m²–900 m²	–		–	–	–	m²	**50.03**
900 m²–2000 m²	–		–	–	–	m²	**47.65**
Over 2000 m²	–		–	–	–	m²	**46.46**

35 SITE WORKS – PAVINGS

Item	PC £	Labour hours	Labour £	Plant £	Material £	Unit	Total rate £
Permeable flexible wearing course; 35 mm thick; resin bound recycled tyre and stone; laid to Type 3 base (not included); footpaths; cycle tracks							
less than 300 m²	–		–	–	–	m²	**51.23**
300 m²–900 m²	–		–	–	–	m²	**46.46**
900 m²–2000 m²	–		–	–	–	m²	**45.27**
Over 2000 m²	–		–	–	–	m²	**44.67**
Surface to treepits; Langford Direct Ltd; resin bound porous recycled tyre and aggregate paving solution; fully sustainable urban drainage Systems (SuDS) compliant.							
laid to surface of treepits as permeable mulch layer; min 10 nr pits average size 1.20 m × 1.20 m; to Type 3 base (SudsTech only)							
SudsTech (2 part system) to Type 3 base (not included)	–		–	–	–	m²	**165.59**
Trailflex (1 part system)	–		–	–	–	m²	**116.75**
RESIN BONDED SURFACING							
Bonded aggregates; Addagrip Terraco Ltd; natural decorative resin bonded surface dressing laid to concrete, macadam or to plywood panels priced separately							
Primer coat to macadam or concrete base	–		–	–	–	m²	**7.15**
Brittany Bronze; 1–3 mm							
buff adhesive	–		–	–	–	m²	**34.55**
red adhesive	–		–	–	–	m²	**34.55**
green adhesive	–		–	–	–	m²	**34.55**
Brittany Bronze; 2–5 mm							
buff adhesive	–		–	–	–	m²	**38.12**
Chinese bauxite; 1–3 mm							
buff adhesive	–		–	–	–	m²	**39.31**

35 SITE WORKS – PAVINGS

Item	PC £	Labour hours	Labour £	Plant £	Material £	Unit	Total rate £
MACADAM SURFACING							
Coated macadam/asphalt roads/ pavings – General							
Preamble: The prices for all in situ finishings to roads and footpaths include for work to falls, crossfalls or slopes not exceeding 15° from horizontal; for laying on prepared bases (not included) and for rolling with an appropriate roller. Users should note the new terminology for the surfaces described below which is to European standard descriptions. The now redundant descriptions for each course are shown in brackets.							
Macadam surfacing; Centar Surfacing Ltd; Base courses (Roadbase); 28 mm dense bitumen macadam							
75 mm thick							
Machine lay; areas 1000 m² and over	–		–	–	–	m²	**18.17**
Hand lay; areas 400 m² and over	–		–	–	–	m²	**23.24**
100 mm thick							
Machine lay; areas 1000 m² and over	–		–	–	–	m²	**22.75**
Hand lay; areas 400 m² and over	–		–	–	–	m²	**25.68**
150 mm thick							
Machine lay; areas 1000 m² and over	–		–	–	–	m²	**22.75**
Hand lay; areas 400 m² and over	–		–	–	–	m²	**25.68**
Macadam surfacing; Centar Surfacing Ltd; Binder courses; AC20 mm dense bitumen macadam							
50 mm thick							
Machine lay; areas 1000 m² and over	–		–	–	–	m²	**13.78**
Hand lay; areas 400 m² and over	–		–	–	–	m²	**18.86**
60 mm thick							
Machine lay; areas 1000 m² and over	–		–	–	–	m²	**15.45**
Hand lay; areas 400 m² and over	–		–	–	–	m²	**20.70**

35 SITE WORKS – PAVINGS

Item	PC £	Labour hours	Labour £	Plant £	Material £	Unit	Total rate £
80 mm thick							
Machine lay; areas 1000 m² and over	–		–	–	–	m²	**19.28**
Hand lay; areas 400 m² and over	–		–	–	–	m²	**24.96**
Macadam surfacing; Centar Surfacing Ltd; surface (wearing) course; 20 mm of 6 mm dense bitumen macadam							
20 mm thick; surface (wearing) course; 6 mm dense bitumen macadam							
Machine lay; areas 1000 m² and over	–		–	–	–	m²	**10.16**
Hand lay; areas 400 m² and over	–		–	–	–	m²	**14.87**
25 mm thick; surface (wearing) course; 6 mm dense bitumen macadam							
Machine lay; areas 1000 m² and over	–		–	–	–	m²	**10.91**
Hand lay; areas 400 m² and over	–		–	–	–	m²	**15.61**
30 mm thick; surface (wearing) course; 10 mm dense bitumen macadam							
Machine lay; areas 1000 m² and over	–		–	–	–	m²	**11.78**
Hand lay; areas 400 m² and over	–		–	–	–	m²	**16.67**
40 mm thick; surface (wearing) course; 10 mm dense bitumen macadam							
Machine lay; areas 1000 m² and over	–		–	–	–	m²	**14.83**
Hand lay; areas 400 m² and over	–		–	–	–	m²	**20.05**
Macadam surfacing; binder (base) course; 50 mm of 20 mm dense bitumen macadam							
Machine lay; areas 1000 m² and over	–		–	–	–	m²	**13.78**
Hand lay; areas 400 m² and over	–		–	–	–	m²	**18.86**
Coloured macadam pavings; Bituchem by Centar Surfacing Ltd; to pedestrian or vehicular hard landscape areas; laid to base course (not included)							
Natratex wearing course; clear macadam							
25 mm thick to pedestrian areas	–		–	–	–	m²	**49.51**
30 mm thick to vehicular areas	–		–	–	–	m²	**57.10**

35 SITE WORKS – PAVINGS

Item	PC £	Labour hours	Labour £	Plant £	Material £	Unit	Total rate £
MACADAM SURFACING – cont							
Coloured macadam pavings – cont							
Colourtex coloured macadam (not green or blue)							
25 mm thick	–		–	–	–	m^2	**49.51**
30 mm thick	–		–	–	–	m^2	**52.01**
MARKING OF CAR PARKS							
Marking car parks							
Car parking space division strips; laid hot at 115° C; on bitumen macadam surfacing							
minimum daily rate	–		–	–	–	item	**774.35**
Stainless metal road studs							
100 × 100 mm	6.71	0.08	8.91	–	7.72	nr	**16.63**

35 SITE WORKS – PAVINGS

Item	PC £	Labour hours	Labour £	Plant £	Material £	Unit	Total rate £
SUSPENDED PAVING							
The labour in this section is calculated on a 2 person team. The labour time below should be multiplied by 2 to calculate the cost as shown.							
Beam and Block suspended base for external use; FP McCann Ltd; Pre-stressed precast concrete beam; profile 115 mm wide at base, 80 mm wide at top 152 mm deep; laid to perimeter or spanned support retaining walls (not included); block infill of 440 × 225 × 100 mm solid blocks; screed of 75 mm concrete reinforced with mesh to receive surface treatments.							
Block centres between beams							
520 mm; blocks laid 440 mm wide	14.68	0.30	21.55	8.31	57.43	m²	**87.29**
408 mm; alternate blocks 440 mm wide and 215 mm wide between beams	18.65	0.36	25.93	10.97	56.39	m²	**93.29**
295 mm; blocks laid 215 mm wide between beams	29.37	0.36	25.93	16.62	65.29	m²	**107.84**
FORMWORK FOR CONCRETE PAVINGS							
The labour in this section is calculated on a 3 person team. The labour time below should be multiplied by 3 to calculate the cost as shown.							
Formwork for in situ concrete							
Sides of foundations							
height not exceeding 250 mm	0.25	0.01	1.19	–	0.74	m	**1.93**
height 250–500 mm	0.34	0.01	1.12	–	1.44	m	**2.87**
height 500 mm–1.00 m	0.34	0.02	1.79	–	2.12	m	**3.91**
height exceeding 1.00 m	1.02	0.67	71.30	–	8.61	m²	**79.91**
Extra over formwork for curved work 6 m radius	–	0.25	17.83	–	–	m	**17.83**

35 SITE WORKS – PAVINGS

Item	PC £	Labour hours	Labour £	Plant £	Material £	Unit	Total rate £
REINFORCEMENT FOR CONCRETE PAVINGS							
The labour in this section is calculated on a 3 person team. The labour time below should be multiplied by 3 to calculate the cost as shown.							
Reinforcement mesh; side laps 150 mm; head laps 300 mm; mesh 200 × 200 mm; in roads, footpaths or pavings							
Fabric							
A142 (2.22 kg/m^2)	4.58	0.03	2.96	–	5.27	m^2	**8.23**
A193 (3.02 kg/m^2)	5.90	0.02	1.89	–	6.78	m^2	**8.68**
A252 (3.02 kg/m^2)	7.10	0.02	2.14	–	8.16	m^2	**10.30**
IN SITU CONCRETE PAVINGS							
The labour in this section is calculated on a 3 person team. The labour time below should be multiplied by 3 to calculate the cost as shown.							
Unreinforced concrete; on prepared subbase (not included)							
Roads; 21.00 N/mm^2–20 mm aggregate (1:2:4) mechanically mixed on site							
100 mm thick	22.87	0.04	4.46	–	26.30	m^2	**30.76**
150 mm thick	34.31	–	0.21	–	39.45	m^2	**39.66**
Reinforced in situ concrete; mechanically mixed on site; on hardcore base (not included); reinforcement (not included)							
Roads; 11.50 N/mm^2–40 mm aggregate (1:3:6)							
100 mm thick	20.52	0.03	3.56	–	23.60	m^2	**27.16**
150 mm thick	30.78	0.20	21.39	–	35.40	m^2	**56.79**
200 mm thick	42.04	0.27	28.52	–	48.35	m^2	**76.87**
250 mm thick	51.30	0.33	35.65	0.50	58.99	m^2	**95.14**
300 mm thick	61.56	0.40	42.78	0.50	70.79	m^2	**114.07**
Roads; 21.00 N/mm^2–20 mm aggregate (1:2:4)							
100 mm thick	22.33	0.13	14.26	–	25.67	m^2	**39.93**
150 mm thick	33.49	0.20	21.39	–	38.51	m^2	**59.90**
200 mm thick	44.65	0.27	28.52	–	51.35	m^2	**79.87**
250 mm thick	55.81	0.33	35.65	0.50	64.19	m^2	**100.33**
300 mm thick	66.98	0.40	42.78	0.50	77.02	m^2	**120.30**

35 SITE WORKS – PAVINGS

Item	PC £	Labour hours	Labour £	Plant £	Material £	Unit	**Total rate £**
Roads; 25.00 N/mm²–20 mm aggregate C 25/30 ready mixed							
100 mm thick	10.24	0.13	14.26	–	11.78	m²	**26.04**
150 mm thick	15.37	0.20	21.39	–	17.67	m²	**39.06**
200 mm thick	20.49	0.27	28.52	–	23.56	m²	**52.09**
250 mm thick	25.61	0.33	35.65	0.50	29.45	m²	**65.60**
300 mm thick	30.73	0.40	42.78	0.50	35.34	m²	**78.62**
Reinforced in situ concrete; ready mixed; discharged directly into location from supply vehicle; normal Portland cement; on hardcore base (not included); reinforcement (not included)							
Roads; 11.50 N/mm²–40 mm aggregate (1:3:6)							
100 mm thick	9.46	0.05	5.70	–	10.88	m²	**16.58**
150 mm thick	14.19	0.08	8.56	–	16.32	m²	**24.88**
200 mm thick	18.92	0.12	12.83	–	21.76	m²	**34.59**
250 mm thick	23.65	0.18	19.25	0.50	27.20	m²	**46.95**
300 mm thick	28.38	0.22	23.53	0.50	32.64	m²	**56.67**
Roads; 21.00 N/mm²–20 mm aggregate (1:2:4)							
100 mm thick	9.66	2.08	222.81	–	11.10	m²	**233.91**
150 mm thick	14.48	0.08	8.56	–	16.66	m²	**25.21**
200 mm thick	19.31	0.12	12.83	–	22.21	m²	**35.04**
250 mm thick	24.14	0.18	19.25	0.50	27.76	m²	**47.51**
300 mm thick	28.97	0.22	23.53	0.50	33.31	m²	**57.34**
Roads; 26.00 N/mm²–20 mm aggregate (1:1.5:3)							
100 mm thick	9.99	0.05	5.70	–	11.49	m²	**17.19**
150 mm thick	14.99	0.08	8.56	–	17.24	m²	**25.80**
200 mm thick	19.31	0.12	12.83	–	22.21	m²	**35.04**
250 mm thick	24.99	0.18	19.25	0.50	28.73	m²	**48.48**
300 mm thick	29.98	0.22	23.53	0.50	34.48	m²	**58.51**
Roads; PAV1 concrete – 35 N/mm²; Designated mix							
100 mm thick	10.72	0.05	5.70	–	12.32	m²	**18.02**
150 mm thick	16.07	0.08	8.56	–	18.48	m²	**27.04**
200 mm thick	21.43	0.12	12.83	–	24.65	m²	**37.48**
250 mm thick	26.79	0.18	19.25	0.50	30.81	m²	**50.56**
300 mm thick	32.15	0.22	23.53	0.50	36.97	m²	**60.99**

35 SITE WORKS – PAVINGS

Item	PC £	Labour hours	Labour £	Plant £	Material £	Unit	Total rate £
IN SITU CONCRETE PAVINGS – cont							
Concrete sundries							
The labour in this section is calculated on a 2 person team. The labour time below should be multiplied by 2 to calculate the cost as shown.							
Treating surfaces of unset concrete; grading to cambers; tamping with 75 mm thick steel shod tamper or similar	–	0.13	9.50	–	–	m²	**9.50**
Treating surfaces of unset concrete; grading to cambers; tamping with 75 mm thick steel shod tamper or similar	–	0.13	9.50	–	–	m²	**9.50**
EXPANSION JOINTS IN PAVINGS							
Expansion joints							
13 mm thick joint filler; formwork							
width or depth not exceeding 150 mm	1.79	0.04	3.19	–	3.76	m	**6.95**
width or depth 150–300 mm	3.58	0.06	3.98	–	7.53	m	**11.51**
width or depth 300–450 mm	5.37	0.07	4.78	–	11.29	m	**16.07**
25 mm thick joint filler; formwork							
width or depth not exceeding 150 mm	3.31	0.04	3.19	–	5.51	m	**8.69**
width or depth 150–300 mm	6.61	0.06	3.98	–	11.01	m	**14.99**
width or depth 300–450 mm	9.92	0.07	4.78	–	16.52	m	**21.30**
Sealants							
Mastic sealant; Geocell 945							
12 mm wide × 30 mm deep	4.93	0.10	3.10	–	4.93	m	**8.03**
15 mm wide × 30 mm deep	6.16	0.11	3.44	–	6.16	m	**9.60**
12 mm wide × 40 mm deep	6.57	0.11	3.44	–	6.57	m	**10.01**
20 mm wide × 30 mm deep	8.21	0.13	3.88	–	8.21	m	**12.08**
12 mm wide × 50 mm deep	8.21	0.13	3.88	–	8.21	m	**12.08**
sealing top 25 mm of joint with rubberized bituminous compound	1.38	0.06	3.98	–	1.58	m	**5.56**

35 SITE WORKS – PAVINGS

Item	PC £	Labour hours	Labour £	Plant £	Material £	Unit	Total rate £
WOODCHIP SURFACES							
The labour in this section is calculated on a 3 person team. The labour time below should be multiplied by 3 to calculate the cost as shown.							
Wood chip surfaces; Melcourt Industries Ltd (items labelled FSC are Forest Stewardship Council certified)							
Wood chips; to surface of pathways by machine; material delivered in 80 m³ loads; levelling and spreading by hand (excavation and preparation not included)							
Walk Chips; 100 mm thick; FSC (25 m³ loads)	5.42	0.01	1.19	0.56	6.24	m²	**7.98**
Walk Chips; 100 mm thick; FSC (80 m³ loads)	3.95	0.01	1.19	0.56	4.54	m²	**6.28**
Woodfibre; 100 mm thick; FSC (80 m³ loads)	4.09	0.01	1.19	0.56	4.70	m²	**6.45**
STEPS							
The labour in this section is calculated on a 3 person team. The labour time below should be multiplied by 3 to calculate the cost as shown.							
In situ concrete steps; ready mixed concrete; C15; on 100 mm hardcore base; base concrete thickness 100 mm treads and risers cast simultaneously; inclusive of A142 mesh within the concrete; step riser 170 mm high; inclusive of all shuttering and concrete labours; floated finish; to graded ground (not included)							
Treads 250 mm wide							
one riser	6.39	1.32	141.17	–	10.44	m	**151.61**
two risers	10.23	1.98	211.76	–	19.01	m	**230.77**
three risers	14.81	2.64	282.35	–	29.32	m	**311.67**
four risers	19.89	3.96	423.52	–	37.48	m	**461.00**
six risers	25.19	5.28	564.70	–	42.71	m	**607.40**

35 SITE WORKS – PAVINGS

Item	PC £	Labour hours	Labour £	Plant £	Material £	Unit	Total rate £
STEPS – cont							
In situ concrete steps – cont							
Lay surface to concrete in situ steps above							
granite treads; pre-cut off site; 305 mm wide × 50 mm thick bullnose finish inclusive of 20 mm overhang; granite riser 20 mm thick all laid on 15 mm mortar bed	95.00	0.66	70.59	–	111.10	m	**181.69**
riven Yorkstone tread and riser 50 mm thick; cut on site	45.08	0.83	88.23	–	53.70	m	**141.93**
riven Yorkstone treads 225 mm wide × 50 mm thick cut on site; two courses brick risers laid on flat stretcher bond; bricks PC £500.00/1000	22.05	0.99	105.88	–	33.79	m	**139.67**
Marshalls Saxon paving flag 450 × 450 mm cut to 225 mm wide × 38 mm thick; cut on site; two courses brick risers laid on flat stretcher bond; bricks PC £500.00/1000	8.08	0.99	105.88	–	17.73	m	**123.61**
Treads 450 mm wide							
one riser	11.50	1.65	176.47	–	18.24	m	**194.71**
two risers	23.48	2.31	247.05	–	37.45	m	**284.50**
three risers	33.19	2.97	317.64	–	51.54	m	**369.18**
four risers	57.41	4.29	458.82	–	86.86	m	**545.67**
six risers	72.96	5.94	635.28	–	103.24	m	**738.52**
Lay surface to concrete in situ steps							
granite treads; pre-cut off site; 470 mm wide × 50 mm thick bullnose finish inclusive of 20 mm overhang; granite riser 20 mm thick all laid on 15 mm mortar bed	95.00	1.32	141.17	–	111.57	m	**252.75**
riven Yorkstone tread and riser 50 mm thick; cut on site	61.25	1.49	158.82	–	72.76	m	**231.58**
riven Yorkstone treads 460 mm wide 50 mm thick cut on site; two courses brick risers laid on flat stretcher bond; bricks PC £500.00/1000	45.08	1.65	176.47	–	60.64	m	**237.11**
Marshalls Saxon pavingg flag 450 × 450 × 38 mm thick cut on site; two courses brick risers laid on flat stretcher bond; bricks PC £500.00/1000	16.17	0.91	97.06	–	27.02	m	**124.08**

35 SITE WORKS – PAVINGS

Item	PC £	Labour hours	Labour £	Plant £	Material £	Unit	Total rate £
Precast concrete steps; FP McCann Ltd; Precast units laid to graded blinded and levelled bases (not included); to receive surface treatments priced separately; waist 200 mm thick; finished sizes with paving surface treatments of 40 mm thick including bedding							
5 Risers; 280 mm deep × 175 mm high							
1.20 m wide	424.99	3.00	320.85	50.80	488.74	nr	**860.39**
2.00 m wide	849.98	4.00	427.80	101.60	977.48	nr	**1506.88**
2.50 m wide	1062.53	5.00	534.75	254.01	1221.91	nr	**2010.67**
3.00 m wide	1274.97	7.00	748.65	254.01	1466.22	nr	**2468.88**
4.00 m wide	1699.96	8.00	855.60	355.61	1954.96	nr	**3166.17**
10 Risers; 280 mm deep × 175 mm high							
1.20 m wide	1020.06	4.00	427.80	101.60	1173.07	nr	**1702.48**
2.00 m wide	1700.17	6.00	641.70	203.21	1955.20	nr	**2800.11**
2.50 m wide	2125.27	8.00	855.60	406.42	2444.06	nr	**3706.07**
3.00 m wide	2550.26	8.00	855.60	406.42	2932.80	nr	**4194.81**
4.00 m wide	3400.35	8.00	855.60	406.42	3910.40	nr	**5172.41**
5 Risers; 450 mm deep × 175 mm high							
1.20 m wide	782.45	3.00	320.85	50.80	899.82	nr	**1271.47**
2.00 m wide	1304.02	4.00	427.80	101.60	1499.62	nr	**2029.02**
2.50 m wide	1630.13	5.00	534.75	254.01	1874.65	nr	**2663.41**
3.00 m wide	1477.87	7.00	748.65	254.01	1699.55	nr	**2702.21**
4.00 m wide	2608.14	8.00	855.60	355.61	2999.36	nr	**4210.58**
10 Risers; 450 mm deep × 175 mm high							
1.20 m wide	1565.33	3.00	320.85	50.80	1800.12	nr	**2171.78**
2.00 m wide	2608.88	4.00	427.80	101.60	3000.21	nr	**3529.61**
2.50 m wide	3261.09	5.00	534.75	254.01	3750.26	nr	**4539.02**
3.00 m wide	3911.22	7.00	748.65	254.01	4497.90	nr	**5500.56**
4.00 m wide	5211.46	8.00	855.60	355.61	5993.18	nr	**7204.39**
Extra over for integrated landing in step flights; 1.00 m deep							
1.20 m wide	–		–	–	289.41	nr	**289.41**
2.00 m wide	–		–	–	482.35	nr	**482.35**
2.50 m wide	–				602.94	nr	**602.94**
3.00 m wide	–		–	–	723.52	nr	**723.52**
4.00 m wide	–		–	–	964.70	nr	**964.70**

35 SITE WORKS – PAVINGS

Item	PC £	Labour hours	Labour £	Plant £	Material £	Unit	Total rate £
STEPS – cont							
Step treads; Marshalls Plc; Precast step treads to match paving types							
Conservation X Textured paving slabs – Silver grey							
Step Tread 450 × 450 × 50 mm – Chamfered	180.45	1.00	106.95	–	210.31	m	**317.26**
Step Tread 450 × 450 × 50 mm	172.50	1.00	106.95	–	201.17	m	**308.12**
Step Tread 400 × 400 × 65 mm	192.49	1.05	112.30	–	224.16	m	**336.46**
Step Unit 350 × 1000 × 150 mm	292.27	1.00	106.95	–	338.90	m	**445.85**
Step Unit 400 × 1000 × 150 mm	312.53	1.00	106.95	–	362.20	m	**469.15**
Saxon Step Tread 400 × 400 × 50 mm	160.44	1.05	112.30	–	187.29	m	**299.59**
Saxon Step Tread 450 × 450 × 50 mm	150.54	1.00	106.95	–	175.91	m	**282.86**
Portland stone steps; Albion Stone							
Stone 40 mm thick; Laid to preformed step surface (not included) on 35 mm bedding of white cement mortar; jointed with white cement mortar							
Treads 350 mm deep	56.10	0.50	35.65	–	74.07	m	**109.72**
Treads 500 mm deep	80.14	0.75	53.48	–	145.24	m	**198.72**
Treads 600 mm deep	96.17	0.90	64.17	–	125.76	m	**189.93**
Landings; maximum stone size 1200 × 750 mm	160.28	1.25	133.69	–	210.87	m^2	**344.55**
Stone 50 mm thick; Laid to preformed step surface (not included) on 35 mm bedding of white cement mortar; jointed with white cement mortar							
Treads 350 mm deep	120.86	0.55	39.22	–	148.55	m	**187.76**
Treads 500 mm deep	148.91	0.75	53.48	–	224.33	m	**277.80**
Treads 600 mm deep	168.95	1.00	71.30	–	209.46	m	**280.76**
Landings	200.36	1.25	133.69	–	256.95	m^2	**390.64**
Risers to steps; fixed with white cement mortar							
140 mm high × 30 mm thick	22.48	0.50	35.65	–	25.85	m	**61.50**
Stone steps; Haddonstone Ltd to step base in concrete or brick substructure; Cast stone steps in Tecstone in colours Bath, Terracotta, Portland, Slate or Sienna; continuosly bedded on full mortar bed							

35 SITE WORKS – PAVINGS

Item	PC £	Labour hours	Labour £	Plant £	Material £	Unit	**Total rate £**
Steps							
Bullnose Step 1000 × 350 × 50	60.00	0.25	26.74	–	72.04	m	**98.78**
Bullnose Step 1000 × 375 × 50	70.80	0.25	26.74	–	84.66	m	**111.40**
Risers							
Step Riser 1000 × 175 × 50	34.80	0.13	13.37	–	40.12	m	**53.49**
Step Riser 1000 × 150 × 50	31.20	0.13	13.37	–	35.98	m	**49.35**

35 SITE WORKS – DECKING

Item	PC £	Labour hours	Labour £	Plant £	Material £	Unit	Total rate £
SUPPORTS FOR DECKING SYSTEMS							
Timber decking; Exterior Decking Ltd; timber decking substructure/ carcassing costs							
Roof decks exclusive of beams; timber craned (not included); joists at 400 mm centres							
47 × 47 mm	–		–	–	–	m²	**35.88**
47 × 100 mm	–		–	–	–	m²	**40.33**
47 × 150 mm	–		–	–	–	m²	**44.50**
Commercial applications; decks with double beams at 300 mm centres							
47 × 100 mm	–		–	–	–	m²	**31.99**
47 × 150 mm	–		–	–	–	m²	**37.19**
Domestic garden applications; decks with double beams at 300 mm centres							
47 × 150 mm	–		–	–	–	m²	**57.05**
Structural posts to decks; tanalized treated softwood; rebated including excavation and setting in concrete							
100 × 100 mm; single beam; single rebate	–		–	–	–	m	**13.75**
150 × 150 mm; double beam; double rebate	–		–	–	–	m	**19.33**
The labour in this section is calculated on a 2 person team. The labour time below should be multiplied by 2 to calculate the cost as shown.							
Adjustable pedestals to rooftops for decking and pavings;							
Decking; 300 mm joists at 1 pedestal per linear metre (3.3 / m²) along the length of the joists;							
17 mm high	3.15	0.08	5.88	–	3.62	m²	**9.50**
28 mm high	4.29	0.08	5.88	–	4.93	m²	**10.81**
175–285 mm high	25.95	0.08	6.00	–	29.84	m²	**35.84**
285–400 mm high	44.89	0.08	6.00	–	51.62	m²	**57.62**
355–515 mm high	55.18	0.08	6.00	–	63.46	m²	**69.46**
465–625 mm high	63.87	0.09	6.30	–	73.45	m²	**79.75**
545–740 mm high	74.16	0.09	6.48	–	85.28	m²	**91.76**
645–850 mm high	82.81	0.09	6.48	–	95.23	m²	**101.71**
720–960 mm high	93.10	0.09	6.48	–	107.07	m²	**113.55**
830–1070 mm high	101.78	0.09	6.48	–	117.05	m²	**123.53**

35 SITE WORKS – DECKING

Item	PC £	Labour hours	Labour £	Plant £	Material £	Unit	Total rate £
Decking; 400 mm joists at 1 pedestal per linear metre (2.5 / m²) along the length of the joists;							
17 mm high	2.38	0.07	4.76	–	2.74	m²	**7.50**
28 mm high	3.25	0.07	4.76	–	3.74	m²	**8.49**
175–285 mm high	19.66	0.07	5.09	–	22.60	m²	**27.70**
285–400 mm high	34.01	0.07	5.09	–	39.11	m²	**44.20**
355–515 mm high	41.80	0.07	5.09	–	48.07	m²	**53.17**
465–625 mm high	48.38	0.08	5.48	–	55.64	m²	**61.12**
545–740 mm high	56.18	0.08	5.48	–	64.61	m²	**70.09**
645–850 mm high	62.73	0.08	5.48	–	72.14	m²	**77.63**
720–960 mm high	70.53	0.08	5.48	–	81.11	m²	**86.59**
830–1070 mm high	77.11	0.08	5.48	–	88.68	m²	**94.16**
Adjustable pedestals to rooftops for decking and pavings; Note other heights available up to 1.07 m							
Paving slabs; laid at uniform sizes with one pedestal per corner of each paving slab; paving slab sizes; laid on adjustable pedestal 28 mm high above the slab							
300 × 300 mm	22.09	0.50	35.65	–	25.41	m²	**61.06**
400 × 400 mm	13.00	0.50	35.65	–	14.95	m²	**50.60**
450 × 450 mm	9.75	0.50	35.65	–	11.21	m²	**46.86**
600 × 450 mm with additional central support per slab	8.58	0.33	23.76	–	9.86	m²	**33.63**
900 × 600 mm with additional central support per slab	4.94	0.25	17.83	–	5.68	m²	**23.50**
TIMBER DECKING							
Market prices of timber decking materials; Exterior Decking Ltd							
Hardwood decking; stainless steel screw fixings included in rates shown							
Ipe; pre-drilled and countersunk	95.07		–	–	109.33	m²	**109.33**
Hardwood decking; 'Exterpark' invisible fixings included in rates shown							
Exterpark Magnet Accoya® 21 × 144 mm	70.95		–	–	81.59	m²	**81.59**
Exterpark Magnet Iroko 21 × 120 mm	77.94		–	–	89.64	m²	**89.64**
Teak Rustic FSC 21 mm thick	137.41		–	–	158.02	m²	**158.02**
Exterpark magnet Indonesian Teak 20 × 120 mm	204.28		–	–	234.92	m²	**234.92**

35 SITE WORKS – DECKING

Item	PC £	Labour hours	Labour £	Plant £	Material £	Unit	Total rate £
TIMBER DECKING – cont							
Market prices of timber decking materials – cont							
Softwood decking							
Pine smooth; 145 × 38 mm	48.20		–	–	55.43	m²	**55.44**
The labour in this section is calculated on a 2 person team. The labour time below should be multiplied by 2 to calculate the cost as shown.							
Timber decking; Exterior Decking Ltd; timber decking surfaces; fixed to substructure (not included)							
Softwood deck timbers; screw fixed							
pine; smooth; 145 × 38 mm fixed with 65 × 5 mm coated deck screws	52.89	0.25	17.83	–	60.83	m²	**78.65**
Hardwood deck timbers; screw fixed							
Ipe; S4S E4E; 145 × 21 mm; pre-drilled and countersunk; S4S e4e	95.07	0.75	53.48	–	114.72	m²	**168.19**
Hardwood deck timbers; invisibly fixed							
Ipe; 21 mm thick	92.21	0.50	17.75	–	105.61	m²	**135.12**
Teak Rustic FSC 21 mm	137.41	0.50	17.78	–	157.60	m²	**187.15**
Exterpark Ipe 21 mm thickness 21 × 120 mm	108.45	0.50	17.77	–	124.31	m²	**153.84**
Hardwood decking; Yellow Balau; grooved or smooth; 6 mm joints							
deck boards; 145 mm wide × 21 mm thick	61.51	1.00	35.65	–	73.53	m²	**109.18**
Western Red Cedar; Clear grade; 6 mm joints smooth or reeded							
90 mm wide × 25 mm thick	79.56	1.00	35.65	–	94.47	m²	**130.12**
145 mm wide × 20 mm thick	61.51	1.00	35.65	–	73.71	m²	**109.36**
Surface treatments to timber decks; 'Owatrol' decking oil							
Seasonite	3.68		–	–	4.23	m²	**4.23**
Textrol	5.91		–	–	6.80	m²	**6.80**
Handrails and base rail; fixed to posts at 2.00 m centres							
posts; 100 × 100 × 1370 mm high	30.06	1.00	35.65	–	35.31	m	**70.96**
posts; turned; 1220 mm high	29.98	1.00	35.65	–	35.23	m	**70.88**

35 SITE WORKS – DECKING

Item	PC £	Labour hours	Labour £	Plant £	Material £	Unit	Total rate £
Handrails; balusters							
square balusters at 100 mm centres	35.40	0.50	17.83	–	41.46	m	**59.28**
square balusters at 300 mm centres	11.79	0.33	11.87	–	14.16	m	**26.03**
turned balusters at 100 mm centres	69.00	0.50	17.83	–	80.10	m	**97.92**
turned balusters at 300 mm centres	22.98	0.33	11.76	–	27.02	m	**38.79**
COMPOSITE DECKING							
The labour in this section is calculated on a 2 person team. The labour time below should be multiplied by 2 to calculate the cost as shown.							
Composite decking; London stone Ltd; Millboard composite decking							
Substructure for composite or timber decking; Millboard; London Stone Ltd; Pro-Plas recyled substructural elements to support timber or composite decking boards; joists and double bearers 125 × 50 mm; joists laid at 300 mm centres							
$4\,m^2$ deck; 2.00 × 2.00	108.00	1.50	106.95	–	131.68	m^2	**238.63**
$10\,m^2$ deck; 3.33 × 3.33	112.32	1.20	85.56	–	136.64	m^2	**222.20**
$25\,m^2$ deck; 5.00 × 5.00	98.50	1.18	83.99	–	120.75	m^2	**204.74**
$10\,m^2$ deck; 10.00 × 5.00	97.63	1.00	71.30	–	119.75	m^2	**191.05**
Composite deck surface; Millboard; 176 mm wide; fixed with stainless screws 4.5 × 70 mm							
Enhanced grain range	94.50	0.33	23.76	–	113.02	m^2	**136.79**
Antique Oak range	110.80	0.33	23.76	–	131.77	m^2	**155.53**

35 SITE WORKS – STREET FURNITURE

Item	PC £	Labour hours	Labour £	Plant £	Material £	Unit	**Total rate £**
GENERALLY							
Note: The costs of delivery from the supplier are generally not included in these items. Readers should check with the supplier to verify the delivery costs.							
Furniture/equipment – General Preamble: The following items include fixing to manufacturer's instructions; holding down bolts or other fittings and making good (excavating, backfilling and tarmac, concrete or paving bases not included).							
BARRIERS							
Barriers – General Preamble: The provision of car and truck control barriers may form part of the landscape contract. Barriers range from simple manual counterweighted poles to fully automated remote-control security gates, and the exact degree of control required must be specified. Complex barriers may need special maintenance and repair.							
Barriers; Barriers Direct Manually operated pole barriers; counterbalance; to tubular steel supports; bolting to concrete foundation (foundation not included); aluminium boom; various finishes, including arm rest.							
clear opening up to 3.00 m	1241.77	6.00	213.90	–	1482.14	nr	**1696.04**
clear opening up to 4.00 m	1369.13	6.00	213.90	–	1628.60	nr	**1842.50**
clear opening 5.00 m	1369.13	6.00	213.90	–	1628.60	nr	**1842.50**
clear opening 6.00 m	1464.65	6.00	213.90	–	1738.45	nr	**1952.35**
clear opening 7.00 m	1703.45	6.00	213.90	–	2013.07	nr	**2226.97**

35 SITE WORKS – STREET FURNITURE

Item	PC £	Labour hours	Labour £	Plant £	Material £	Unit	**Total rate £**
Electrically operated pole barriers; enclosed fan-cooled motor; double worm reduction gear; overload clutch coupling with remote controls; aluminium boom; various finishes, including catch pole (exclusive of electrical connections by electrician)							
Barriers Direct 4.0 m boom with catchpost	2358.25	6.00	213.90	–	2766.10	nr	**2980.00**
Barriers Direct 5.0 m boom with catchpost	2358.25	6.00	213.90	–	2766.10	nr	**2980.00**
Barriers Direct 6.0 m boom with catchpost	2358.25	6.00	213.90	–	2766.10	nr	**2980.00**
BASES FOR STREET FURNITURE							
Bases for street furniture							
Excavating; filling with concrete 1:3:6; bases for street furniture							
300 × 450 × 500 mm deep	–	0.75	26.66	–	15.49	nr	**45.31**
300 × 600 × 500 mm deep	–	1.11	39.46	–	20.66	nr	**64.33**
300 × 900 × 500 mm deep	–	1.67	59.36	–	30.99	nr	**96.68**
1750 × 900 × 300 mm deep	–	2.33	82.68	–	108.28	nr	**213.04**
2000 × 900 × 300 mm deep	–	2.67	94.75	–	123.75	nr	**243.73**
2400 × 900 × 300 mm deep	–	3.20	113.55	–	148.49	nr	**292.34**
2400 × 1000 × 300 mm deep	–	3.56	126.33	–	164.99	nr	**324.98**
Precast concrete flags; to concrete bases (not included); bedding and jointing in cement: mortar (1:4)							
450 × 600 × 50 mm	15.87	1.17	41.59	–	29.47	m²	**71.06**
Precast concrete paving blocks; to concrete bases (not included); bedding in sharp sand; butt joints							
200 × 100 × 65 mm	14.61	0.50	17.83	–	19.91	m²	**37.74**
200 × 100 × 80 mm	16.89	0.50	17.83	–	22.77	m²	**40.60**
Engineering paving bricks; to concrete bases (not included); bedding and jointing in sulphate-resisting cement: lime: sand mortar (1:1:6)							
over 300 mm wide	24.43	0.56	20.03	–	52.16	m²	**72.19**
Edge restraints to pavings; haunching in concrete (1:3:6)							
200 × 300 mm	12.01	0.10	3.57	–	13.81	m	**17.38**

35 SITE WORKS – STREET FURNITURE

Item	PC £	Labour hours	Labour £	Plant £	Material £	Unit	Total rate £
BASES FOR STREET FURNITURE – cont							
Bases for street furniture – cont							
Bases; Earth Anchors Ltd							
Rootfast ancillary anchors; A1; 500 mm long × 25 mm dia. and top strap F2; including bolting to site furniture (site furniture not included)	12.89	0.50	17.83	–	14.82	set	**32.65**
installation tool for above	25.45		–	–	29.27	nr	**29.27**
Rootfast ancillary anchors; A4; heavy duty 40 mm square fixed head anchors; including bolting to site furniture (site furniture not included)	31.41	0.33	11.88	–	36.12	set	**48.00**
Rootfast ancillary anchors; F4; vertical socket; including bolting to site furniture (site furniture not included)	13.81	0.05	1.78	–	15.88	set	**17.66**
Rootfast ancillary anchors; F3; horizontal socket; including bolting to site furniture (site furniture not included)	15.16	0.05	1.78	–	17.44	set	**19.22**
installation tools for the above	76.89		–	–	88.43	nr	**88.43**
Rootfast ancillary anchors; A3; anchored bases; including bolting to site furniture (site furniture not included)	51.98	0.33	11.88	–	59.78	set	**71.66**
installation tools for the above	76.89		–	–	88.43	nr	**88.43**
BINS: LITTER/WASTE/ DOGWASTE/SALT & GRIT							
Dog waste bins							
Furnitubes International Ltd							
LUK745 P; Lucky; steel dog bin; post mounted; 47 l	276.98	1.50	53.48	–	334.55	nr	**388.03**
Broxap							
BG 2591-PM Sirius, post mounted, galvanized steel, top opening lid, handle and hook for dog lead. 40 l	155.00	1.00	71.30	–	195.63	nr	**266.93**
Furnitubes International Ltd							
LUK745 W; Lucky; steel dog bin; wall mounted; 47 l	247.79	0.75	26.74	–	300.99	nr	**327.73**

35 SITE WORKS – STREET FURNITURE

Item	PC £	Labour hours	Labour £	Plant £	Material £	Unit	Total rate £
Litter bins; precast concrete in textured white or exposed aggregate finish; with wire baskets and drainage holes							
Bins; Marshalls Plc							
Escofet Laurel and Hardy litter bin and ashtray 60 litre. Cast stone litter bin with steel or corten finish.	3573.52	0.50	17.83	–	4109.55	nr	**4127.37**
Litter bins; all-steel							
Bins; Broxap							
Derby BX45G 2554 round bin; 2 posting apertures; 130 l; bar hinge system	339.00	1.00	71.30	–	389.85	nr	**461.15**
Litter bins; cast iron							
Bins; Marshalls Plc							
MSF5501 Imperial Heritage; cast iron litter bin 85 l	944.95	1.00	35.65	3.50	1086.69	nr	**1125.85**
Litter bins; timber faced; hardwood slatted casings with removable metal litter containers; ground or wall fixing							
Bins; Woodscape Ltd							
square; 580 × 580 × 950 mm high; with lockable lid	822.00	0.50	17.83	–	945.30	nr	**963.13**
round; 580 mm dia. × 950 mm high; with lockable lid	1032.00	0.50	17.83	–	1186.80	nr	**1204.63**
Plastic litter and grit bins; glassfibre reinforced polyester grit bins; yellow body; hinged lids							
Bins; Wybone Ltd; Victoriana glass fibre; cast iron effect litter bins; including lockable liner							
LBV/7; 533 mm dia. × 838 mm high; semi closed top; oblong shape; 112 litre capacity; with lockable liner	379.04	1.00	35.65	–	444.79	nr	**480.44**
Plastic litter and grit bins;							
Broxap; vandal-resistant plastic, inner reinforced galvanized framework							
Maelor Trafflex round open top high security litter bin; 90 l;	188.00	1.00	71.30	–	216.20	nr	**287.50**
Maelor Trafflex round covered high security bin; 180 l;	269.00	1.00	71.30	–	309.35	nr	**380.65**
grit and salt bin, yellow UV stabilized polyethylene; 200 l	127.00	0.50	35.65	–	146.05	nr	**181.70**

35 SITE WORKS – STREET FURNITURE

Item	PC £	Labour hours	Labour £	Plant £	Material £	Unit	Total rate £
BINS: LITTER/WASTE/ DOGWASTE/SALT & GRIT – cont							
Recycled plastic street furniture; Hahn Plastics Ltd; inclusive of excavations concrete bases and anchors;							
Litter Bin							
Collec Litter Bin 48 cm × 66 cm; ground mounted with wire basket insert; circular; 70 litre	289.16	1.00	71.30	–	355.70	nr	**427.00**
Resco Litter Bin 45 cm × 45 cm × 66 cm high 50 litre; square; brown	283.75	1.00	71.30	–	349.48	nr	**420.78**
Cigarette bins							
Furnitubes International Ltd							
ZEN 275 Zenith cigarette bin; stainless steel; wall or post mounted; 1.7 L	42.24	0.75	26.74	–	64.60	nr	**91.34**
LVR250 PC Liverpool cigarette bin; steel; wall mounted; 1.9 L	35.73	0.50	17.83	–	41.09	nr	**58.91**
SMK500F Smoke King cigarette bin; cast aluminium; bolt down; 3.5 L	243.68	0.75	26.74	–	296.26	nr	**322.99**
Contemporary bins							
Marshalls							
Festival bin; round, powder coated slatted steel; flared or rolled open top 110 l; galvanized steel liner	460.24	0.50	35.65	–	529.27	nr	**564.92**
Rendezvous bin; circular tapered cast iron coated with vertical hardwood slats; 55 l	1232.08	0.50	35.65	–	1416.89	nr	**1452.54**
BOLLARDS							
Excavating; for bollards and barriers; by hand							
Holes for bollards							
400 × 400 × 400 mm; disposing off site	–	0.42	14.75	–	–	nr	**17.73**
600 × 600 × 600 mm; disposing off site	–	0.97	34.29	–	–	nr	**42.67**

35 SITE WORKS – STREET FURNITURE

Item	PC £	Labour hours	Labour £	Plant £	Material £	Unit	**Total rate £**
Concrete bollards – General							
Preamble: Precast concrete bollards are available in a very wide range of shapes and sizes. The bollards listed here are the most commonly used sizes and shapes; manufacturer's catalogues should be consulted for the full range. Most manufacturers produce bollards to match their suites of street furniture, which may include planters, benches, litter bins and cycle stands. Most parallel sided bollards can be supplied in removable form, with a reduced shank, precast concrete socket and lifting hole to permit removal with a bar.							
Concrete bollards							
Marshalls Plc, cylinder; straight or tapered; 200–400 mm dia.; plain grey concrete; setting into firm ground (excavating and backfilling not included)							
Bridgford; 915 mm high above ground	131.40	2.00	71.30	–	179.89	nr	**251.19**
Marshalls Plc; cylinder; straight or tapered; 200–400 mm dia.; Beadalite reflective finish; setting into firm ground (excavating and backfilling not included)							
Wexham concrete bollard; exposed silver grey	265.79	2.00	71.30	–	321.12	nr	**392.42**
Wexham Major concrete bollard; exposed silver grey	413.41	2.00	71.30	–	504.20	nr	**575.50**
Marshalls Plc; Precast concrete verge markers; various shapes;							
Spherical 500; exposed silver grey	560.67	1.00	35.65	–	655.82	nr	**691.47**
Cube 600; silver grey	542.41	1.00	35.65	–	634.82	nr	**670.47**
Monoscape Woodhouse; smooth grey	166.89	1.00	35.65	–	202.97	nr	**238.62**
Townscape Products Ltd; Linby; natural grey concrete							
seats; 1800 × 600 × 736 mm high	601.70	2.00	71.30	–	691.96	nr	**763.26**

35 SITE WORKS – STREET FURNITURE

Item	PC £	Labour hours	Labour £	Plant £	Material £	Unit	**Total rate £**
BOLLARDS – cont							
Bollards							
Service bollards; Furnitubes International Ltd; stainless steel; for housing services connection points to electricity, water etc. (services connections not included)							
Kenton KEN717; 900 × 250 mm dia.	715.86	2.00	71.30	–	838.71	nr	**910.01**
Zenith ZEN707; 900 × 250 mm dia.	671.46	2.00	71.30	–	787.65	nr	**858.95**
Other bollards							
Removable parking posts; Marshalls Plc							
RT\RD4; domestic telescopic bollard	243.62	2.00	71.30	–	295.63	nr	**366.93**
RT\R8; heavy duty telescopic bollard	346.97	2.00	71.30	–	414.49	nr	**485.79**
Plastic bollards; Marshalls Plc							
Lismore; three ring recycled plastic bollard	115.16	2.00	71.30	–	147.91	nr	**219.21**
Cast iron bollards – General							
Preamble: The following bollards are particularly suitable for conservation areas. Logos for civic crests can be incorporated to order.							
Cast iron bollards							
Bollards; Furnitubes International Ltd (excavating and backfilling not included)							
Doric Round; 920 mm high × 170 mm dia.	215.52	2.00	71.30	–	262.58	nr	**333.88**
Kenton; heavy duty galvanized steel; 900 mm high; 350 mm dia.	306.49	2.00	71.30	–	367.20	nr	**438.50**
Bollards; Marshalls Plc (excavating and backfilling not included)							
MSF103; cast iron bollard; Small Manchester	214.09	2.00	71.30	–	261.68	nr	**332.98**
MSF102; cast iron bollard; Manchester	214.09	2.00	71.30	–	261.68	nr	**332.98**

35 SITE WORKS – STREET FURNITURE

Item	PC £	Labour hours	Labour £	Plant £	Material £	Unit	Total rate £
Cast iron bollards with rails – General							
Preamble: The following cast iron bollards are suitable for conservation areas.							
Cast iron bollards with rails							
Cast iron posts with steel tubular rails; Broxap Street Furniture; setting into firm ground (excavating not included)							
Weaver Polyurethane single rail post	87.00	1.50	53.48	–	106.81	nr	**160.29**
Type C mild steel tubular rail; including connector	1.67	0.03	1.19	–	3.84	m	**5.02**
Kent Stainless; Stainless steel bollards; root or bolt fixed to paving including all excavation concrete and removal of site of excavated material							
Bollards in 316 stainless steel 3 mm thick; 900 high installed height; 101 mm dia.;							
dome or flat topped bollard	154.87	1.50	93.00	–	163.81	nr	**256.81**
dome or flat topped bollard; removable	333.56	1.50	93.00	–	342.51	nr	**435.51**
Steel bollards							
Steel bollards; Marshalls Plc (excavating and backfilling not included)							
SSB01; stainless steel bollard; 101 × 1250 mm	169.78	2.00	71.30	–	210.72	nr	**282.02**
RS001; stainless steel bollard; 114 × 1500 mm	177.16	2.00	71.30	–	219.21	nr	**290.51**
RB119 Brunel; steel bollard; 168 × 1500 mm	175.70	2.00	71.30	–	230.83	nr	**302.13**
Timber bollards							
Woodscape Ltd; durable hardwood							
RP 250/1500; 250 mm dia. × 1500 mm long	442.51	1.00	35.65	–	509.54	nr	**545.19**
SP 250/1500; 250 mm square × 1500 mm long	442.51	1.00	35.65	–	509.54	nr	**545.19**
SP 150/1200; 150 mm square × 1200 mm long	148.93	1.00	35.65	–	171.92	nr	**207.57**
SP 125/750; 125 mm square × 750 mm long	93.37	1.00	35.65	–	108.03	nr	**143.68**
RP 125/750; 125 mm dia. × 750 mm long	93.37	1.00	35.65	–	108.03	nr	**143.68**

35 SITE WORKS – STREET FURNITURE

Item	PC £	Labour hours	Labour £	Plant £	Material £	Unit	Total rate £
BOLLARDS – cont							
Deterrent bollards							
Semi-mountable vehicle deterrent and kerb protection bollards; Furnitubes International Ltd (excavating and backfilling not included)							
bell decorative	779.76	2.00	71.30	–	917.70	nr	**989.00**
half bell	825.25	2.00	71.30	–	970.01	nr	**1041.31**
full bell ®	1077.59	2.00	71.30	–	1260.20	nr	**1331.50**
three quarter bell	897.81	2.00	71.30	–	1053.45	nr	**1124.75**
CYCLE HOLDERS/CYCLE SHELTERS							
Cycle holders							
Cycle stands; Marshalls Plc							
Ollerton Sheffield; steel cycle stand; RCS1	66.43	0.50	17.83	–	76.40	nr	**94.22**
Ollerton Sheffield; stainless steel cycle stand; RSCS1	178.65	0.50	17.83	–	205.45	nr	**223.27**
Cycle holders; Townscape Products Ltd							
Guardian cycle holders; tubular steel frame; setting in concrete; 1250 × 550 × 775 mm high; making good	448.16	1.00	35.65	–	533.16	nr	**568.81**
Penny cycle stands; 600 mm dia.; exposed aggregate bollards with 8 nr cycle holders; in galvanized steel; setting in concrete; making good	988.74	1.00	35.65	–	1154.84	nr	**1190.49**
Cycle stands; Kent Stainless; root fixed including all excavation concrete and removal of excavated material;							
Birmingham cycle stand; heavy duty box section and tube;							
270 wide × 920 mm high	393.13	1.50	93.00	–	396.22	nr	**489.22**
Triangular cycle stand 1.00 m × 1.00 m; allowing 2 cycles per stand; pedestrianized and parkland usage							
1.00 m long × 1.00 m high	285.91	1.50	93.00	–	289.00	nr	**382.00**

35 SITE WORKS – STREET FURNITURE

Item	PC £	Labour hours	Labour £	Plant £	Material £	Unit	Total rate £
Cycle holders; Broxap Street Furniture							
Neath cycle rack; BX/MW/AG; for 6 nr cycles; semi-vertical; galvanized and polyester powder coated; 1.32 m wide × 2.542 m long × 1.80 m high	607.00	10.00	356.50	–	733.62	nr	**1090.12**
Harrogate double sided free-standing cycle rack; BX/MW/GH; for 10 nr cycles; galvanized and polyester coated; 3.25 m long	272.00	4.00	142.60	–	339.85	nr	**482.45**
Llanelli cycle rack; BXMW/LLA; for 10 nr cycles; alternate high/low level rack; galvanized and polyester powder coated; 350 mm centres	266.00	3.00	213.90	–	305.90	nr	**519.80**
Cycle shelters							
Furnitubes International Ltd; Academy free-standing bolt down shelter, galvanized steel frame and roof; 3150 mm overall width; 2150 mm maximum height; 2670 mm depth							
corrugated roof	2038.21	6.00	213.90	–	2376.00	nr	**2589.90**
clear polycarbonate roof	2015.46	6.00	213.90	–	2349.84	nr	**2563.74**
Broxap							
Wardale cycle shelter for 7 nr cycles; galvanized; 2.13 m wide × 3.6 m long × 2.15 m high	1861.00	10.00	356.50	–	2175.72	nr	**2532.22**
FLAGPOLES							
Flagpoles; ground mounted; in glass fibre; smooth white finish; including hinged baseplate, nylon halyard system and revolving gold onion finial; setting in concrete; to manufacturer's recommendations (excavating not included)							
6 m high; external halyard system	250.50	4.00	142.60	–	342.18	nr	**484.78**
6 m high; internal halyard system	405.50	4.00	142.60	–	520.43	nr	**663.03**
8 m high; external halyard system	348.50	6.00	213.90	–	454.88	nr	**668.78**
8 m high; internal halyard system	503.50	6.00	213.90	–	633.13	nr	**847.03**
10 m high; external halyard system	456.50	5.00	178.25	–	579.08	nr	**757.33**
10 m high; internal halyard system	611.50	5.00	178.25	–	757.33	nr	**935.58**
Flagpoles; wall mounted; vertical or angled poles; in glass fibre; smooth white finish; including base, top bracket, external nylon halyard rope, cleat and revolving gold onion finial							
3 m pole	260.00	2.00	71.30	–	299.00	nr	**370.30**
Banner flagpoles; 90 mm dia. aluminium pole; suitable for 1 × 2 m banner							
6 m high	498.81	4.00	142.60	–	627.73	nr	**770.33**

35 SITE WORKS – STREET FURNITURE

Item	PC £	Labour hours	Labour £	Plant £	Material £	Unit	Total rate £
HANDRAILS							
Handrails; stainless steel; Kent Stainless							
Handrail in 316L stainless steel with bright satin finish 48 mm dia. tube; with integral middle leg; 1.10 m high; root fixed.							
single rail	413.71	2.00	124.00	–	422.21	m	**546.21**
double rail	827.41	3.00	186.00	–	844.41	m	**1030.41**
Handrails with integrated LED lighting; stainless steel; Kent Stainless							
Handrail in 316L stainless steel with bright satin finish 48 dia. tube; with integral middle leg; 1.10 m high; incorporating; with high flux Rail LED luminaires (ducting cabling and connections to power supply not included); root fixed							
single rail	1353.75	3.00	186.00	–	1362.25	m	**1548.25**
double rail	2707.50	4.00	248.00	–	2709.20	m	**2957.20**
PERGOLAS							
Pergolas; construct timber pergola; posts 150 × 150 mm × 2.40 m finished height in 600 mm deep minimum concrete pits; beams of 200 × 50 mm × 2.00 m wide; fixed to posts with dowels; rafters 200 × 38 mm notched to beams; inclusive of all mechanical excavation and disposal off site							
Pergola 2.00 m wide in green oak; posts at 2.00 m centres; dowel fixed							
rafters at 600 mm centres	250.33	8.40	333.38	5.25	413.01	m	**768.45**
rafters at 400 mm centres	282.53	10.80	428.67	5.25	457.92	m	**908.65**
Pergola 3.00 m wide in green oak; posts at 1.50 m centres							
rafters at 600 mm centres	192.71	7.20	300.33	6.30	364.80	m	**689.80**
rafters at 400 mm centres	240.93	8.00	333.76	6.30	432.03	m	**790.46**
Pergola 2.00 m wide in prepared softwood; beams fixed with bolts; posts at 2.00 m centres							
rafters at 600 mm centres	90.11	5.00	208.49	5.25	229.81	m	**460.35**
rafters at 400 mm centres	112.52	6.00	250.26	5.25	263.48	m	**535.78**

35 SITE WORKS – STREET FURNITURE

Item	PC £	Labour hours	Labour £	Plant £	Material £	Unit	Total rate £
Pergola 3.00 m wide in prepared softwood; posts at 1.50 m centres							
rafters at 600 mm centres	70.14	7.20	300.28	6.30	223.97	m	**548.91**
rafters at 400 mm centres	86.05	8.00	333.70	6.30	254.06	m	**612.43**
PICNIC TABLES/SUNDRY FURNITURE							
Picnic benches – General							
Preamble: The following items include for fixing to ground to manufacturer's instructions or concreting in.							
Picnic tables and benches							
Picnic tables and benches; Broxap Street Furniture							
Springfield timber picnic unit 1800 mm long softwood	405.60	2.00	71.30	–	484.22	nr	**555.52**
Picnic tables and benches; Woodscape Ltd							
table and benches built in; 2 m long	3269.70	2.00	71.30	–	3777.94	nr	**3849.24**
The labour in this section is calculated on a 2 person team. The labour time below should be multiplied by 2 to calculate the cost as shown.							
Recycled plastic Street furniture; Hahn Plastics Ltd; inclusive of excavations concrete bases and anchors;							
Sets							
Calero Table & Bench set 150 cm × 180 cm × 75 cm	796.01	1.00	71.30	–	961.74	nr	**1033.04**
Alton Table & Bench set 200 cm × 205 cm × 80 cm	1700.31	1.00	71.30	–	2001.70	nr	**2073.00**
Serengeti picnic set 200 cm × 174 cm × 70 cm	826.33	1.00	71.30	–	996.62	nr	**1067.92**
Street furniture ranges							
Lifebuoy stations							
Lifebuoys Direct							
Designed in collaboration with RNLI; housing for 30 inch lifebuoy and encapsulated throw line	179.17	1.00	71.30	–	355.50	nr	**426.80**

35 SITE WORKS – STREET FURNITURE

Item	PC £	Labour hours	Labour £	Plant £	Material £	Unit	Total rate £
PLANT CONTAINERS							
Market prices of containers							
Plant containers; timber							
Plant containers; Broxap; hardwood woodstained treated or softwood pressure treated							
Hemsworth 600 mm square softwood planter; 3 × 200 mm panels; 80 l	366.00	0.33	20.65	–	366.00	nr	**386.65**
Bedlington timber with galvanized frame; 2000 mm square × 920 mm high; 3680 l	1936.80	0.33	20.65	–	1936.80	nr	**1957.45**
Plant containers; hardwood; Woodscape Ltd							
Versailles square; 900 × 900 × 420 mm high	624.46	0.25	17.83	–	718.12	nr	**735.95**
Modular Planter System; Hardwood Stakes and Peg joining system Slats140 mm (w) × 65 mm (thick), planter height in multiples of 140 mm; Planter 2.00 long × 840 mm high × 900 mm deep	1881.18	0.50	31.00	–	1881.18	nr	**1912.18**
Plant containers; precast concrete							
Plant containers; Marshalls Plc							
Boulevard 700; circular base and ring	921.33	2.00	71.30	33.53	1059.53	nr	**1164.36**
Boulevard 1200; circular base and ring	1163.48	1.00	35.65	33.53	1338.01	nr	**1407.19**
Broxap							
Circular grey washed fine gravel coating; 600 mm high × 1420 mm dia.	732.00	1.00	31.00	23.85	732.00	nr	**786.85**
Andromeda square; 1000 mm × 1000 mm × 1150 mm high	806.40	1.00	31.00	26.51	806.40	nr	**863.91**

35 SITE WORKS – STREET FURNITURE

Item	PC £	Labour hours	Labour £	Plant £	Material £	Unit	Total rate £
Bespoke Planters; Outdoor Design; planters to required size and shape; with seamless polished corners; preformed drainage holes and lip detail; inclusive of placement; drainage layer 50 mm thick and potting soil							
Galvanized 1.5 mm steel; powder coated							
500 mm × 500 mm × 500 mm	387.71	0.50	35.65	–	455.52	nr	**491.17**
1500 mm × 500 mm × 500 mm	479.77	0.80	57.04	–	578.53	nr	**635.57**
1800 mm × 750 mm × 600 mm	610.81	1.60	114.08	25.40	761.33	nr	**900.81**
2500 mm × 1000 mm × 600 mm	749.44	1.50	106.95	25.40	980.69	nr	**1113.04**
Corten 2.0 mm							
500 mm × 500 mm × 500 mm	400.71	0.50	35.65	–	470.46	nr	**506.11**
1500 mm × 500 mm × 500 mm	509.01	0.80	57.04	–	612.16	nr	**669.20**
1800 mm × 750 mm × 600 mm	656.30	1.60	114.08	25.40	813.64	nr	**953.12**
2500 mm × 1000 mm × 600 mm	817.67	1.50	106.95	25.40	1059.15	nr	**1191.50**
Stainless steel 1.5 mm; Grained finish							
500 mm × 500 mm × 500 mm	496.01	0.50	35.65	–	580.06	nr	**615.71**
1500 mm × 500 mm × 500 mm	716.95	0.80	57.04	–	851.29	nr	**908.33**
1800 mm × 750 mm × 600 mm	985.53	1.60	114.08	25.40	1192.26	nr	**1331.74**
2500 mm × 1000 mm × 600 mm	1318.01	1.50	106.95	25.40	1634.55	nr	**1766.90**
Planters; Kent Stainless; Grade 316L 10 mm thick polished finished suitable for continuous modular installation; installed as edging to form continuous planters; including drainage medium and planting medium.							
Shoreditch planter							
1300 long × 440 mm wide × 800 mm high	465.69	1.00	62.00	–	547.21	m	**609.21**
Ferrycarrig planter; tapered 400 × 400 mm at base to 950 × 950 mmtop							
848 mm high;	476.52	0.50	31.00	–	511.59	nr	**542.59**
Recycled plastic Street furniture; Hahn Plastics Ltd; inclusive of excavations concrete bases and anchors;							
Planters; inclusive of placement and filling with planting medium and drainage; excludes planting							
Vinca 2 Planter							
108 cm × 62 cm × 44 cm	528.50	1.00	71.30	–	668.00	nr	**739.30**
Iberis 2 Planter							
110 cm × 110 cm × 47 cm	472.19	1.00	71.30	–	608.95	nr	**680.25**

35 SITE WORKS – STREET FURNITURE

Item	PC £	Labour hours	Labour £	Plant £	Material £	Unit	Total rate £
SEATING							
Outdoor seats							
Outdoor seats; concrete framed – General Preamble: Prices for the following concrete framed seats and benches with hardwood slats include for fixing (where necessary) by bolting into existing paving or concrete bases (bases not included) or building into walls or concrete foundations (walls and foundations not included). Delivery generally not included.							
The labour in this section is calculated on a 2 person team. The labour time below should be multiplied by 2 to calculate the cost as shown.							
Outdoor seats; concrete Outdoor seats; Furnitubes International Ltd							
Amesbury concrete seat, 2.2 m long	836.08	2.00	142.60	–	997.81	nr	**1140.41**
Marlborough concrete seat, 1.8 m long	773.26	2.00	142.60	–	925.57	nr	**1068.17**
Outdoor seats; metal framed – General Preamble: Metal framed seats with hardwood backs to various designs can be bolted to ground anchors.							
Outdoor seats and benches; metal framed Outdoor seats; Furnitubes International Ltd							
NS6 Newstead; steel standards with iroko slats; 1.80 m long	352.79	2.00	142.60	–	442.03	nr	**584.63**

35 SITE WORKS – STREET FURNITURE

Item	PC £	Labour hours	Labour £	Plant £	Material £	Unit	**Total rate £**
Outdoor seats; Furnitubes International Ltd							
NEB6 New Forest Single Bench; cast iron standards with iroko slats; 1.83 m long	454.86	2.00	142.60	–	559.41	nr	**702.01**
EA6 Eastgate; cast iron standards with iroko slats; 1.86 m long	637.89	2.00	142.60	–	769.89	nr	**912.49**
NE6 New Forest Seat; cast iron standards with iroko slats; 1.83 m long	585.90	2.00	142.60	–	710.11	nr	**852.71**
Zenith long bench, satin stainless steel, iroko slats, 1.8 m long	1420.90	2.00	142.60	–	1670.35	nr	**1812.95**
Ashburton; steel with recycled plastic slats; 1.8 m long	635.72	2.00	142.60	–	767.40	nr	**910.00**
Outdoor seats; Broxap Street Furniture; metal and timber							
Eastgate 1800 mm l × 620 mm w × 830 mm h	502.80	2.00	142.60	–	596.00	nr	**738.60**
Outdoor seats; Marshalls plc; metal and timber							
MSF 502; cast iron Heritage seat	789.91	2.00	142.60	–	908.39	nr	**1050.99**
Outdoor seats; Woodscape Ltd							
Type 2 Backless Bench, 2 metres, Straight, Built-in leg length for 450 mm seat height	1152.90	2.00	142.60	–	1325.84	nr	**1468.44**
Tooting Seat, 2.5 m, straight back, built-in leg length for 450 mm seat height	1726.20	1.00	71.30	–	1985.13	nr	**2056.43**
Outdoor seats; all steel							
Outdoor seats; Marshalls Plc; stainless steel or powder coated steel							
Urban city, single seat 767 mm H × 500 mm W × 568 mm D	1190.97	1.50	106.95	–	1369.62	nr	**1476.57**
Urban city, triple seat 767 mm H × 1700 mm W × 568 mm D	2107.25	2.00	142.60	–	2423.34	nr	**2565.94**
Outdoor seats; all timber							
Outdoor seats; Woodcraft UK							
1.524 m long	880.80	1.00	71.30	–	1059.26	nr	**1130.56**
1.829 m long	940.80	1.00	71.30	–	1128.26	nr	**1199.56**
2.438 m long (including centre leg)	1238.40	1.00	71.30	–	1470.50	nr	**1541.80**
Hardwood backless bench; 1.50 m long	621.60	1.00	71.30	–	761.18	nr	**832.48**

35 SITE WORKS – STREET FURNITURE

Item	PC £	Labour hours	Labour £	Plant £	Material £	Unit	Total rate £
SEATING – cont							
Outdoor seats – cont							
Outdoor seats and benches; Woodscape Ltd; solid hardwood							
Clifton Seat with Single Backrest, 2 m length, 0.5 m width, 90x65 mm slats, Powder Coated Band legs	2577.01	2.00	142.60	–	2963.56	nr	**3106.16**
Type 2; backless bench double slatted on solid timber legs 2000 mm or 2500 mm length	1152.90	2.00	142.60	–	1325.84	nr	**1468.44**
Wall seats							
Woodscape ltd							
Woodberry wall seat, 0.4 m width, 40x40 mm slats, straight	660.84	0.50	35.65	–	759.97	m	**795.62**
Contemporary outdoor seating							
Marshalls Plc; Igneo cast concrete range, can be manufactured to any length. Arm rests in Ferrocast polyurethane ; any colour							
Single seat span 953 mm × 434 mm high	564.78	1.00	71.30	–	649.49	nr	**720.79**
Broxap							
Helston Curved Bench; Galvanized and powder coated mild steel baseplated frame with hardwood Iroko timber slatted top. 6.00 m long × 450 mm wide × 450 mm long	3320.40	6.00	427.80	–	3818.46	nr	**4246.26**
Recycled plastic street furniture; Hahn Plastics Ltd; inclusive of excavations concrete bases and anchors;							
Benches							
Wall Street 955 bench 200 × 56 cm × 45 cm	1002.86	0.75	53.48	–	1199.63	nr	**1253.10**
Hyde Park bench 165 cm × 46 cm × 48 cm	341.15	0.75	53.48	–	438.66	nr	**492.13**
High line bench 2.00 m × 480 mm × 450 mm	458.11	0.75	53.48	–	573.16	nr	**626.64**
Seats							
Caserata Wall mounted seat 150 cm × 70 cm × 65 cm	427.79	0.75	53.48	–	538.29	nr	**591.77**
Eclipse seat 150 cm × 60 cm × 89 cm	276.17	0.75	53.48	–	363.93	nr	**417.40**
Sapo Seniors' seat 200 cm × 61 cm × 97 cm	736.44	0.75	53.48	–	893.25	nr	**946.72**

35 SITE WORKS – STREET FURNITURE

Item	PC £	Labour hours	Labour £	Plant £	Material £	Unit	Total rate £
Turon Round tree seat – octagonal 330 cm × 305 cm × 80 cm	2419.42	3.00	213.90	–	2959.18	nr	**3173.08**
Outdoor seats; tree benches/ seats							
Broxap							
Wythenshawe Bench; curved galvanized polyester powder coated steel root fixed frame with armrests;hardwood iroko timber slats; external dia. 3000 mm × internal dia. 2200 mm × 400 mm high	2655.60	2.00	142.60	–	3053.94	nr	**3196.54**
Weybridge Bench, curved steel frame; Iroko timber slats; external dia. 2600 mm × internal dia. 1600 mm × 450 mm high	2994.00	2.00	142.60	–	3443.10	nr	**3585.70**
Hexagonal shaped bench ideal for a single tree or wooded area; Redwoodmanufactured in a variety of sizes to suit individual applications; with backrest	780.00	4.00	285.20	–	897.00	nr	**1182.20**
Anti skate board devices							
Stainless steel; Furnitubes International Ltd							
for fitting to seats	21.66	0.20	7.13	–	24.91	nr	**32.04**
for fitting to benches	21.66	0.20	7.13	–	24.91	nr	**32.04**
SIGNAGE							
Directional signage – Fingerposts; cast aluminium							
Signage; Furnitubes International Ltd							
FFL1 Lancer; cast aluminium finials	92.06	0.07	2.37	–	105.86	nr	**108.24**
FAAIS; arrow end type cast aluminium directional arms; single line; 90 mm wide	389.88	2.00	71.30	–	448.36	nr	**519.66**
FAAID; arrow end type cast aluminium directional arms; double line; 145 mm wide	476.52	0.13	4.76	–	548.00	nr	**552.75**
FAAIS; arrow end type cast aluminium directional arms; treble line; 200 mm wide	563.16	0.20	7.13	–	647.63	nr	**654.76**
FCK1211G Kingston; composite standard root columns	622.73	2.00	71.30	–	741.43	nr	**812.73**

35 SITE WORKS – STREET FURNITURE

Item	PC £	Labour hours	Labour £	Plant £	Material £	Unit	Total rate £
SIGNAGE – cont							
Park signage							
Entrance signs and map boards; vitreous enamel							
entrance map board; 1250 × 1000 mm high with two support posts; all associated works to post bases	–		–	–	–	nr	**115.32**
information board with two locking cabinets; 1250 × 1000 mm high with two support posts; all associated works to post bases	–		–	–	–	nr	**340.24**
Miscellaneous park signage; vitreous enamel							
'No dogs' sign; 100 × 100 mm; fixing to fencing or gates	–		–	–	–	nr	**21.47**
'Dog exercise area' sign; 300 × 400 mm; fixing to fencing or gates	–		–	–	–	nr	**36.31**
'Nature conservation area' sign; 900 × 400 mm high with two support posts; all associated works to post bases	–		–	–	–	nr	**56.86**
The labour in this section is calculated on a 2 person team. The labour time below should be multiplied by 2 to calculate the cost as shown.							
Fingerposts; Oak FSC; Landmark Timber Products							
Fresh-sawn oak post 100 × 100 mm ; Air-dried oak finger-arms; 550 × 150 × 25 mm; mortised and fixed with hardwood dowels; Engraved and in-filled with durable paint; in concrete footing; County range							
Post 2.80 m overall with single finger	187.20	1.00	71.23	25.75	322.60	nr	**428.70**
Post 3.00 m overall with 2 fingers	324.00	1.00	71.25	25.76	479.90	nr	**586.03**
Post 3.20 m overall with 3 fingers	432.00	1.20	85.51	25.76	604.06	nr	**724.45**
Post 3.20 m overall with 4 fingers	432.00	1.30	92.64	25.76	728.25	nr	**855.78**

35 SITE WORKS – STREET FURNITURE

Item	PC £	Labour hours	Labour £	Plant £	Material £	Unit	**Total rate £**
SMOKING SHELTERS							
Smoking and Vaping shelters							
Broxap; Harrowby Smoking & Vaping Shelter ; coated steel frame, cladded roof, ClearView PETg UV all round visibility; perch seating;							
2110 mm long × 1040 mm wide; maximum 3 people	1293.60	4.00	142.60	–	1519.70	nr	**1662.30**
3135 mm long × 1540 mm wide ; maximum 6 people	1851.60	5.00	178.25	–	2161.40	nr	**2339.65**
STONES & BOULDERS							
Standing stones; CED Ltd; erect standing stones; vertical height above ground; in concrete base; including excavation setting in concrete to 1/3 depth and crane offload into position							
Purple schist							
1.00 m high	272.00	1.50	53.44	25.32	350.57	nr	**435.79**
1.25 m high	520.00	1.50	53.44	25.32	617.71	nr	**707.12**
1.50 m high	520.00	2.00	71.26	30.39	626.45	nr	**735.93**
2.00 m high	520.00	3.00	106.82	50.62	680.05	nr	**860.55**
2.50 m high	520.00	1.50	53.36	101.14	757.57	nr	**955.81**
Rockery stone – General							
Preamble: Rockery stone prices vary considerably with source, carriage, distance and load. Typical PC prices are in the range of £ 60–80 per tonne collected.							
Rockery stone; CED Ltd							
Boulders; maximum distance 25 m; by machine							
750 mm dia.	272.00	0.60	21.39	36.58	312.80	nr	**370.77**
1 m dia.	272.00	2.00	71.30	30.48	312.80	nr	**414.58**
1.5 m dia.	272.00	2.00	71.30	80.95	312.80	nr	**465.05**
2 m dia.	272.00	2.00	71.30	80.95	312.80	nr	**465.05**
Boulders; maximum distance 25 m; by hand							
750 mm dia.	272.00	0.75	26.74	–	312.80	nr	**339.54**
1 m dia.	272.00	1.89	67.22	–	312.80	nr	**380.02**

35 SITE WORKS – STREET FURNITURE

Item	PC £	Labour hours	Labour £	Plant £	Material £	Unit	**Total rate £**
TREE GRILLES/TREE PROTECTION							
Tree grilles; cast iron							
Cast iron tree grilles; Marshalls Plc							
Heritage; cast iron grille plus frame; 1 × 1 m	583.58	3.00	106.95	–	695.78	nr	**802.73**
Cast iron tree grilles; Townscape Products Ltd							
Baltimore; 1200 mm square Ductile(SG) finished in Black RAL9004	603.43	2.00	71.30	–	693.94	nr	**765.24**
Cast iron tree grilles; Broxap							
Lea round Tree Grille 1200 mm dia. × 500 mm tree hole	433.20	2.00	62.00	–	433.20	nr	**495.20**
Kent Stainless; Tree grilles and Tree protection; 316 stainless steel; inclusive of frame							
Kent square tree pit angle; Grade 316L Stainless Steel; Satin 320 grit polished; 10 mm thick							
730 mm × 730 mm × 40 mm deep	671.46	3.00	186.00	–	680.33	nr	**866.33**
Broadgate round tree grille; Stainless steel tree surround; circular; grade 316L stainless steel plates; removable inside ring to facilitate future growth; various finishes; comprising of 1–5 concentric rings							
900 mm external diam; 300 mm internal diam; 50 mm thick	1597.43	3.00	186.00	–	1606.30	nr	**1792.30**
Heelmesh tree grille; Wedge Grating; Grade 316 Stainless Steel; Satin Finish 320 Grit Polished. Loading FACTA B.							
Inner Circle Radius:145 mm (or to order) W:595 mm, L:640 mm, D:50 mm;	1429.56	3.00	186.00	–	1438.43	nr	**1624.43**
Waterford; Recessed tree pit; grade 316L stainless steel; satin 320 grit polished; removable inner tray sections to facilitate growth; optional aeration, irrigation inlet and uplighters.							
1200 mmx 1200 mm × 106 mm deep	2382.60	5.00	310.00	–	2399.96	nr	**2709.96**

35 SITE WORKS – STREET FURNITURE

Item	PC £	Labour hours	Labour £	Plant £	Material £	Unit	Total rate £
Recessed load bearing tree grille; grade 316L stainless steel; for heavy traffic areas, that have no support underground; for heavy loads up to 44 ton; slow moving; infilled with block paviours (not included) central section can be made to suit nearby decorative tree grilles.							
3752 mmx 2630 mm × 10 mm; 216 mm deep	15378.60	8.00	496.00	–	15527.05	nr	**16023.05**
VENTILATION GRILLES							
Ventilation Grilles; Kent Solo vent grille; heavy duty heelproof grille c/w angle frame. 316L stainless steel; in accordance with Facta loading standards; 5 mm thick bars; laser cut lettering; load bearing bars stepped down by 2 mm–5 mm below the surface of the transverse bars.							
Load class 125; installed to stainless steel angle frame fixed to in situ concrete slab (priced separately)							
1.00 m × 1.00 m × 50 mm–100 mm deep;	1862.76	0.50	31.00	–	1862.76	nr	**1893.76**
1.00 m × 250 mm × 50 mm–100 mm deep;	–	0.33	20.46	–	–	nr	**20.46**
extra for stainless steel angle supports base to receive grilles; fixed with stainless steel anchors drilled to internal reveal of concrete floor or podium	41.88	0.17	10.34	–	53.56	m	**63.90**
WARNING STRIPS AND STUDS							
Tactile warning strips and studs; Kent Stainless; epoxy root fixed to paved surfaces							
Diamond tactile warning strip; 420 mm long 35 mm high 5 mm thick							
316 stainless steel	3.79	0.20	12.40	–	4.34	nr	**16.74**

35 SITE WORKS – STREET FURNITURE

Item	PC £	Labour hours	Labour £	Plant £	Material £	Unit	Total rate £
WARNING STRIPS AND STUDS – cont							
Stainless steel warning studs; Pedestrian; Kent Stainless. Epoxy fixed; drilled to paving surface							
25 mm dia. 5 mm top plate 3 rings for secure anchoring; 25 deep;							
fixed 25 mm deep	3.79	0.25	7.75	–	4.07	nr	**11.82**
Diamond tactile warning stud; 35 mm dia. 5 mm top plate 15 mm deep							
fixed 15 mm deep	3.79	0.20	6.20	–	4.07	m^2	**10.27**
Stainless steel warning studs; Road studs; Kent Stainless. Epoxy fixed to road surface							
Domed road studs 100 mm dia. 12 mm threaded bar 16.63 mm stud							
to macadammed road surface	43.32	0.25	7.75	–	43.60	nr	**51.35**
Pyramid road stud; 100 × 100 mm × 18 mm high; 3 mm thick 316 grade stainless steel							
100 × 100 mm × 18 mm high	21.66	0.25	7.75	–	21.94	nr	**29.69**
WINDBREAKS AND SCREENS							
Kent Stainless; 316 stainless steel street furniture							
Windbreaks; moveable; comprising dome topped bollards 101 mm with 3 mm thick stainless walls; surface integrated lockable and fully removable. 2.00 m–2.50 m long							
1287 mm high; including initial installation of buried or cast in flange;	536.09	1.00	31.00	–	536.09	nr	**567.09**

35 SITE WORKS – SPORTS, PLAY EQUIPMENT, SURFACES

Item	PC £	Labour hours	Labour £	Plant £	Material £	Unit	**Total rate £**
SPORTS EQUIPMENT							
Sports equipment							
Tennis posts; steel; suitable for hard or grass tennis courts; including winder, sockets and dust cap							
round	243.68	1.00	35.65	–	280.23	set	**315.88**
square	243.68	1.00	35.65	–	280.23	set	**315.88**
Tennis nets; not including posts or fixings							
Tournament	122.38	3.00	106.95	–	158.77	set	**265.72**
Match	99.64	3.00	106.95	–	132.62	set	**239.57**
Club	67.15	3.00	106.95	–	95.25	set	**202.20**
Football goals; full size; socketed; including international net supports and nets							
aluminium	996.36	4.00	142.60	–	1163.85	set	**1306.45**
steel; heavyweight	996.36	5.00	178.25	–	1646.78	set	**1825.03**
Football goals; full size; free-standing; including nets							
aluminium	1678.65	4.00	142.60	–	1930.45	set	**2073.05**
steel	2117.27	4.00	142.60	–	2434.85	set	**2577.45**
Mini-football goals; free-standing; including nets							
aluminium	1055.93	4.00	142.60	–	1214.31	set	**1356.91**
steel	888.06	4.00	142.60	–	1021.27	set	**1163.87**
Rugby posts; socketed							
aluminium; 10 m high	3037.82	6.00	213.90	–	3541.58	set	**3755.48**
aluminium; 12 m high	2236.40	6.00	213.90	–	2619.95	set	**2833.85**
steel; 12 m high	2534.22	6.00	213.90	–	2962.45	set	**3176.35**
Hockey goals; steel; including backboards and nets							
socketed	1277.94	1.00	35.65	–	1469.63	set	**1505.28**
free-standing	1191.30	1.00	35.65	–	1370.00	set	**1405.65**
Cricket cages; steel; including netting							
free-standing	1440.39	12.00	427.80	–	1656.45	set	**2084.25**
wheelaway	1716.56	12.00	427.80	–	1974.04	set	**2401.84**

35 SITE WORKS – SPORTS, PLAY EQUIPMENT, SURFACES

Item	PC £	Labour hours	Labour £	Plant £	Material £	Unit	Total rate £
PLAYGROUND EQUIPMENT							
Playground equipment – General Preamble: The range of equipment manufactured or available in the UK is so great that comprehensive coverage would be impossible, especially as designs, specifications and prices change fairly frequently. The following information should be sufficient to give guidance to anyone designing or equipping a playground. In comparing prices note that only outline specification details are given here and that other refinements which are not mentioned may be the reason for some difference in price between two apparently identical elements. The fact that a particular manufacturer does not appear under one item heading does not necessarily imply that they do not make it. Landscape designers are advised to check that equipment complies with ever more stringent safety standards before specifying.							
Playground equipment – Installation The rates below include for installation of the specified equipment by the manufacturers. Most manufacturers will offer an option to install the equipment they have supplied.							
Sports and social areas; Freegame multi-use games areas; Kompan Ltd; enclosed sports areas; complete with surfacing boundary and goals and targets; galvanized steel framework with high density polyethylene panels, galvanized steel goals and equipment; surfacing priced separately							
FRE1211 Classic Multigoal; 7 m	–		–	–	–	nr	**8558.00**

35 SITE WORKS – SPORTS, PLAY EQUIPMENT, SURFACES

Item	PC £	Labour hours	Labour £	Plant £	Material £	Unit	Total rate £
Pitch; complete; suitable for multiple ball sports; suitable for use with natural artificial or hard landscape surfaces; fully enclosed including two end sports walls							
FRE2110 Cosmos; 12 × 20 m	–		–	–	–	nr	**31614.00**
FRE2116 Cosmos; 19 × 36 m	–		–	–	–	nr	**45947.00**
FRE3000; Meeting Point; shelter or social area	–		–	–	–	nr	**4477.00**
Swings – General							
Preamble: Prices for the following vary considerably. Those given represent the middle of the range and include multiple swings and timber ,rubber or tyre seats; ground fixing and priming only.							
Swings; Lappset Ltd							
Swings							
Flat seat swing	–		–	–	–	nr	**2433.23**
Birds nest swing	–		–	–	–	nr	**3457.75**
Swing Q10144	–		–	–	–	nr	**4162.10**
Swing with 2 × Cradle seats 137414M	–		–	–	–	nr	**2817.42**
Swings; Wicksteed Ltd							
Flexi Swing; 1 bay 1 flat seat	–		–	–	–	nr	**1794.10**
Flexi Swing; 1 bay 2 flat seats	–		–	–	–	nr	**2034.74**
Flexi Swing; 1 bay 2 cradle seats	–		–	–	–	nr	**2599.42**
Flexi Swing; 1 bay 1 flat seat & 1 cradle seat	–		–	–	–	nr	**2430.25**
Hurricane Swing; 2 button seats	–		–	–	–	nr	**7239.53**
Pendulum Swing; 1 basket seat	–		–	–	–	nr	**6785.64**
1.8 m timber swing; 2 cradle seats	–		–	–	–	nr	**2041.89**
2.4 m timber swing; 2 cradle seats	–		–	–	–	nr	**2424.30**
2.4 m Timber Swing; 2 flat seats	–		–	–	–	nr	**3931.29**
2.4 m birds nest timber swing; 1 basket seat	–		–	–	–	nr	**1916.80**
Swings; Kompan Ltd							
M951P sunflower swing; 1–3 years	–		–	–	–	nr	**1683.00**
Slides							
Slides; Wicksteed Ltd							
Pedestal slides; 3.40 m	–		–	–	–	nr	**3940.82**
Pedestal slides; 4.40 m	–		–	–	–	nr	**5066.60**
Pedestal slides; 5.80 m	–		–	–	–	nr	**6192.38**
Slides; Kompan Ltd							
M351P slide	–		–	–	–	nr	**2915.00**
M326P Aladdin's Cave slide	–		–	–	–	nr	**3102.00**

35 SITE WORKS – SPORTS, PLAY EQUIPMENT, SURFACES

Item	PC £	Labour hours	Labour £	Plant £	Material £	Unit	Total rate £
PLAYGROUND EQUIPMENT – cont							
Moving equipment – General							
Preamble: The following standard items of playground equipment vary considerably in quality and price; the following prices are middle of the range.							
Moving equipment							
Lappset Ltd							
Spinning machine	–		–	–	–	nr	**12159.75**
Waltz	–		–	–	–	nr	**6185.53**
Polka	–		–	–	–	nr	**4783.22**
Inclusive roundabout	–		–	–	–	nr	**8324.21**
Roundabouts; Wicksteed Ltd							
Dual axis roundabout; Dizzy	–		–	–	–	nr	**3429.75**
Dual axis roundabout; Whizzy	–		–	–	–	nr	**2370.69**
Dual axis roundabout; Solar Spinner	–		–	–	–	nr	**3121.21**
Roundabouts; Kompan Ltd							
Supernova GXY 916; multifunctional spinning and balancing disc; capacity approximately 15 children	–		–	–	–	nr	**4796.00**
Seesaws							
Seesaws; Lappset UK Ltd							
137237M Finno seesaw	–		–	–	–	nr	**3195.22**
Octopus seesaw Q10279	–		–	–	–	nr	**1568.80**
Seesaws; Wicksteed Ltd							
Buddy Board	–		–	–	–	nr	**2495.77**
Glow Worm	–		–	–	–	nr	**1995.43**
Cobra	–		–	–	–	nr	**2495.77**
Sidewinder	–		–	–	–	nr	**3746.64**
Play sculptures – General							
Preamble: Many variants on the shapes of playground equipment are available, simulating spacecraft, trains, cars, houses etc., and these designs are frequently changed. The basic principles remain constant but manufacturer's catalogues should be checked for the latest styles.							

35 SITE WORKS – SPORTS, PLAY EQUIPMENT, SURFACES

Item	PC £	Labour hours	Labour £	Plant £	Material £	Unit	Total rate £
Climbing equipment and play structures – General							
Preamble: Climbing equipment generally consists of individually designed modules. Play structures generally consist of interlinked and modular pieces of equipment and sculptures. These may consist of climbing, play and skill based modules, nets and various other activities. Both are set into either safety surfacing or defined sand pit areas. The equipment below outlines a range from various manufacturers. Individual catalogues should be consulted in each instance. Safety areas should be allowed round all equipment.							
Climbing equipment and play structures							
Climbing equipment; Wicksteed Ltd							
Funrun Fitness Trail; Under Starter's Orders; set of 12 units	–		–	–	–	nr	**17929.07**
Climbing equipment; Lappset UK Ltd							
Santa Maria	–		–	–	–	nr	**141511.55**
Halo Cubic	–		–	–	–	nr	**114617.95**
Halo Climber	–		–	–	–	nr	**35537.97**
Finno Fishing Ship	–		–	–	–	nr	**36946.68**
Castle Finno	–		–	–	–	nr	**35153.77**
230015 Halo Activity Tower	–		–	–	–	nr	**16489.62**
Finno Activity Tower	–		–	–	–	nr	**16489.62**
Finno Hades	–		–	–	–	nr	**15175.67**
ABC Play tower Q10519	–		–	–	–	nr	**12934.54**
Flora Troll's shelter	–		–	–	–	nr	**7875.98**
Play tower Q10945	–		–	–	–	nr	**7997.65**
137072M Finno Climbing Frame	–		–	–	–	nr	**7274.08**
Finno 3D Cube Climber	–		–	–	–	nr	**7274.08**
Fire truck Q10259	–		–	–	–	nr	**4866.46**
Flora Goblin's track	–		–	–	–	nr	**2586.91**
Flora Goblin's climbing	–		–	–	–	nr	**1511.16**
Hags-SMP Playgrounds Ltd							
Nexus – The Core; multi-play structure	–		–	–	–	nr	**14074.02**
Spring equipment							
Spring based equipment for 1–8 year olds; Kompan Ltd							
M101P Crazy Hen	–		–	–	–	nr	**550.00**
M128P Crazy Daisy	–		–	–	–	nr	**1012.00**
M141P Spring Seesaw	–		–	–	–	nr	**1452.00**
M155P Quartet Seesaw	–		–	–	–	nr	**1331.00**

35 SITE WORKS – SPORTS, PLAY EQUIPMENT, SURFACES

Item	PC £	Labour hours	Labour £	Plant £	Material £	Unit	Total rate £
PLAYGROUND EQUIPMENT – cont							
Spring equipment – cont							
Spring based equipment for under 12's; Lappset UK Ltd							
010501 Horse Springer	–		–	–	–	nr	**1152.58**
Tractor rocker 010515	–		–	–	–	nr	**2279.55**
Pig rocker 01516	–		–	–	–	nr	**1152.58**
Froggy 22034	–		–	–	–	nr	**1536.78**
Sandpits							
Market prices							
play pit sand; Boughton Loam Ltd	235.88		–	–	271.26	m^3	**271.26**
Kompan Ltd							
Basic550; 276 × 154 × 31 cm deep	–		–	–	–	nr	**1386.00**
Play area Furniture							
Picnic table Accessibility adapted NF 2694	–		–	–	–	nr	**768.39**
NATURAL SPORTS PITCHES							
Natural sports pitches; Agripower Ltd; sports pitches; gravel raft construction. Prices shown are for sports pitches ordered as part of a landscape subcontract. If sports pitches are ordered directly from the sports pitch contractor these rates would reduce by approximately 10%.							
Professional standard							
cricket wicket; county standard	–		–	–	–	nr	**45380.76**
Playing fields; Sport England Compliant							
football pitch; 6500 m^2	–		–	–	–	nr	**95912.11**
cricket wicket; playing field standard	–		–	–	–	nr	**23247.20**
track and infield	–		–	–	–	nr	**135752.52**
Existing area to 1:100 existing grade; 6500 m^2	1638.44	16.00	568.22	1128.53	2188.84	nr	**4256.55**
SPORTS FIELD CONSTRUCTION							
Note: The playing fields in this section are constructed to a lower standard than the specialist sports fields elsewhere in this book and would comply with standards for general amenity in parks, recreation grounds or school playing fields.							

35 SITE WORKS – SPORTS, PLAY EQUIPMENT, SURFACES

Item	PC £	Labour hours	Labour £	Plant £	Material £	Unit	Total rate £
Surface preparations to existing playing area – Prices shown are per 1000 m²							
Spray turf; Glyphosate single application by tractor;	–		–	95.61	3.36	1000 m²	**98.97**
Strip existing turf 25 mm thick	–		–	278.36	–	1000 m²	**278.36**
Disposal green waste off site	–		–	68.24	–	1000 m²	**1262.46**
Cultivation and soil preparation – Prices shown are per 1000 m²							
Light to medium soils							
Decompaction operations by tractor; ripping and breaking up ground	–		–	402.09	–	1000 m²	**402.09**
Cultivation and stone burying harrow and power raking in single operation	–		–	298.91	–	1000 m²	**298.91**
Clay soils							
Decompaction operations by tractor; ripping and breaking up ground	–		–	584.86	–	1000 m²	**584.86**
Strip and remove 100 mm thick surface	–		–	284.66	–	1000 m²	**284.66**
Disposal on site	–		–	383.28	–	1000 m²	**383.28**
Disposal off site 100 mm thick arisings	–	2.78	94.91	–	–	1000 m²	**3842.40**
Cultivation, stone burying, harrow and power raking in single operation	–		–	520.60	–	1000 m²	**520.60**
Imported new topsoil rootzone 150 mm deep	–		–	1844.42	6124.62	1000 m²	**7969.04**
Surface operations to prepared ground							
fertilizing 35 g /m²	41.88	3.00	106.95	–	48.16	nr	**155.11**
seeding operations by mechanical seeder; sportsfield seed 35 g /m² by mechanical seeder	252.07		–	58.85	289.88	nr	**348.73**
PLAY SURFACES							
Playgrounds; Valley Provincial Ltd; Wetpour in situ playground surfacing, porous; on prepared stone/granular base Type 1 (not included) and macadam base course (not included)							

35 SITE WORKS – SPORTS, PLAY EQUIPMENT, SURFACES

Item	PC £	Labour hours	Labour £	Plant £	Material £	Unit	Total rate £
PLAY SURFACES – cont							
Black							
15 mm thick (critical fall height n/a)	–		–	–	–	m^2	**53.72**
35 mm thick (1.00 m critical fall height)	–		–	–	–	m^2	**81.33**
60 mm thick (1.50 m critical fall height)	–		–	–	–	m^2	**97.60**
Playgrounds; Valley Provincial Ltd; Wetpour safety surfacing in situ playground surfacing, porous; on prepared stone/ granular base Type 1 (not included) and macadam base course (not included)							
Coloured							
15 mm thick (0.50 m critical fall height)	–		–	–	–	m^2	**71.48**
35 mm thick (1.00 m critical fall height)	–		–	–	–	m^2	**101.90**
60 mm thick (1.50 m critical fall height)	–		–	–	–	m^2	**118.59**
Playgrounds; Valley Provincial Ltd; porous safety surfaces; on prepared Type 1 base (not included)							
Critical fall height 1.20 mm							
natural colours	–		–	–	–	m^2	**112.11**
bright colours	–		–	–	–	m^2	**117.09**
Critical fall height 1.70 m							
natural colours	–		–	–	–	m^2	**134.90**
bright colours	–		–	–	–	m^2	**140.10**
Playgrounds Valley Provincial Ltd; Rubber crumb resin surfacing							
Mulch resin surfacing; Black or colours; 50 mm thick							
10–50 m^2	–		–	–	–	m^2	**131.47**
over 50 m^2	–		–	–	–	m^2	**99.04**

35 SITE WORKS – SPORTS, PLAY EQUIPMENT, SURFACES

Item	PC £	Labour hours	Labour £	Plant £	Material £	Unit	Total rate £
Playgrounds; Melcourt Industries Ltd; specifiers and users should contact the supplier for performance specifications of the materials below							
Play surfaces; on drainage layer (not included); minimum 300 mm settled depth; prices for 80 m³ loads unless specified							
Playbark® 8/25; 8–25 mm particles; red/brown; 25 m³ loads	21.72	0.35	12.51	–	24.98	m²	**37.48**
Playbark® 8/25; 8–25 mm particles; red/brown	19.25	0.35	12.51	–	22.13	m²	**34.64**
Playbark® 10/50; 10–50 mm particles; red/brown	21.98	0.35	12.51	–	25.27	m²	**37.78**
Playchips®; FSC graded woodchips	13.27	0.35	12.51	–	15.26	m²	**27.76**
Kushyfall; fibreised woodchips	13.82	0.35	12.51	–	15.90	m²	**28.40**
TENNIS COURTS							
Tennis courts; En Tout Cas Ltd							
Hard playing surfaces to SAPCA Code of Practice minimum requirements; laid on 40 mm thick macadam base on 200 mm thick stone foundation; to include lines, nets, posts, brick edging, 2.75 m high fence and perimeter drainage; based on court size 573 m²							
Playdek; all weather porous macadam and acrylic colour coating hardcourt	–		–	–	–	nr	**61947.60**
TennisTurf Advantage Pro; artificial grass	–		–	–	–	nr	**61947.60**
Sportsturf; short pile sand dressed artificial grass	–		–	–	–	nr	**61947.60**
Matchplay; impervious acrylic hardcourt	–					nr	**60756.30**
Savanna; half sand filled synthetic turf	–		–			nr	**65521.50**
Matchplay; acrylic tournament surface	–		–	–	–	nr	**71478.00**
Omniclay; synthetic clay system	–		–	–	–	nr	**71478.00**

35 SITE WORKS – SPORTS, PLAY EQUIPMENT, SURFACES

Item	PC £	Labour hours	Labour £	Plant £	Material £	Unit	Total rate £
SPORTS SURFACES							
Artificial surfaces and finishes – General							
Preamble: Advice should also be sought from the Technical Unit for Sport, the appropriate regional office of the Sports Council or the National Playing Fields Association. Some of the following prices include base work whereas others are for a specialist surface only on to a base prepared and costed separately.							
Artificial sports pitches; Agripower Ltd; third generation rubber crumb							
Sports pitches to respective National governing body standards; inclusive of all excavation, drainage, lighting, fencing, etc.							
football pitch; 6500 m²	–		–	–	–	nr	**711336.52**
rugby union multi-use pitch; 120 × 75 m	–		–	–	–	nr	**988354.05**
hockey pitch; water-based; 101.4 × 63 m	–		–	–	–	nr	**829660.59**
athletic track; 8 lane; International amateur athletic federation	–		–	–	–	nr	**835228.78**
Multi-sport pitches; inclusive of all excavation, drainage, lighting, fencing, etc.							
sand dressed; 101.4 × 63 m	–		–	–	–	nr	**594404.48**
sand filled; 101.4 × 63 m	–		–	–	–	nr	**538722.56**
polymeric (synthetic bound rubber)	–		–	–	–	m²	**197.67**
macadam	–		–	–	–	m²	**124.59**
Sports areas; Agripower Ltd							
Sports tracks; polyurethane rubber surfacing; on bitumen-macadam (not included); prices for 5500 m² minimum							
International	–		–	–	–	m²	**66.71**
Club Grade	–		–	–	–	m²	**58.97**
Sports areas; multi-component polyurethane rubber surfacing; on bitumen-macadam; finished with polyurethane or acrylic coat; green or red							
Permaprene	–		–	–	–	m²	**69.10**

35 SITE WORKS – WATER AND WATER FEATURES

Item	PC £	Labour hours	Labour £	Plant £	Material £	Unit	Total rate £
CLARIFICATION NOTES ON LABOUR COSTS IN THIS SECTION							
General groundworks team							
Generally a three man team is used in this section; The column Labour hours reports team hours. The column Labour £ reports the total cost of the team for the unit of work shown							
3 man team	–	1.00	106.95	–	–	hr	**106.95**
2 man team	–	1.00	71.30	–	–	hr	**71.30**
BOREHOLES							
Borehole drilling; Agripower Ltd							
Drilling operations to average 100 m depth; lining with 150 mm dia. sieve							
drilling to 100 m	–		–	–	–	nr	**17706.85**
rate per m over 100 m	–		–	–	–	m	**177.07**
pump, rising main, cable and kiosk	–		–	–	–	nr	**7377.85**
COLD WATER							
Blue MDPE polythene pipes; type 50; for cold water services; with compression fittings; bedding on 100 mm DOT type 1 granular fill material							
Pipes							
20 mm dia.	0.49	0.08	2.85	–	6.81	m	**9.66**
25 mm dia.	0.58	0.08	2.85	–	6.91	m	**9.76**
32 mm dia.	0.97	0.08	2.85	–	7.36	m	**10.22**
50 mm dia.	2.45	0.10	3.57	–	9.06	m	**12.63**
60 mm dia.	3.96	0.10	3.57	–	10.80	m	**14.37**
Hose union bib taps; including fixing to wall; making good surfaces							
15 mm	18.50	0.75	26.74	–	22.80	nr	**49.54**
22 mm	48.37	0.75	26.74	–	57.90	nr	**84.64**
Stopcocks; including fixing to wall; making good surfaces							
15 mm	5.61	0.75	26.74	–	7.97	nr	**34.71**
22 mm	9.81	0.75	26.74	–	12.80	nr	**39.54**
Standpipes; to existing 25 mm water mains							
1.00 m high	–		–	–	–	nr	**208.48**
Hose junction bib taps; to standpipes							
19 mm	–		–	–	–	nr	**107.22**

35 SITE WORKS – WATER AND WATER FEATURES

Item	PC £	Labour hours	Labour £	Plant £	Material £	Unit	Total rate £
RAINWATER HARVESTING							
Rainwater harvesting tanks							
Notes: Rainwater harvesting tanks must be installed on granular or concrete bases. If installed below ground the tanks may require construction of drained underground chambers to support them. Please see the appropriate sections in this book for excavation, disposal, bases, retaining walls and drainage.							
Excavation for underground tanks; excavation inclusive of earthwork retention for self supporting tanks							
8 tonne tracked excavator (bucket volume 0.28 m^3)							
maximum depth not exceeding 1.00 m	–	0.06	2.23	3.81	–	m^3	**6.04**
maximum depth not exceeding 2.00 m	–	0.07	2.55	4.35	–	m^3	**6.90**
maximum depth not exceeding 3.00 m	–	0.09	3.24	5.54	–	m^3	**8.78**
Disposal							
Excavated material; off site; to tip; mechanically loaded (JCB)							
inert	–		–	–	–	m^3	**37.43**
Type 1 granular fill base; PC £26.53/tonne (£53.90/m^3 compacted)							
By machine							
100 mm thick	6.04	0.03	1.00	0.61	6.94	m^2	**8.55**
150 mm thick	9.06	0.03	0.89	0.92	10.41	m^2	**12.23**
Backfilling to surround of rainwater tank; carefully compacting as work proceeds							
Arising from the excavations							
average thickness exceeding 0.25 m; depositing in layers 150 mm maximum thickness	–	0.03	1.19	8.68	–	m^3	**9.87**
The labour in this section is calculated on a 2 person team. The labour time below should be multiplied by 2 to calculate the cost as shown.							

35 SITE WORKS – WATER AND WATER FEATURES

Item	PC £	Labour hours	Labour £	Plant £	Material £	Unit	Total rate £
Rainwater harvesting tanks; Rainwater Harvesting Ltd; self supporting underground or above ground tanks							
Rainwater tank with max 1.00 m cover in pedestrian areas; complete with tank dome and pedestrian lid, submersible automatic pump system, supra filtration system, 25 m black and green rainwater pipe, rainwater labelling kit, 125 mm fine filter (excavation, backfilling base, trenching and inlet pipework all not included)							
3600 litre; 1.78 m high × 1.8 m dia.	541.50	3.00	211.80	–	616.61	nr	**1075.05**
4500 litre; 1.96 m high × 2.07 m long × 2.01 m dia.	1128.60	3.50	248.03	–	1289.99	nr	**1785.59**
10000 litre – 2.3 m long × 2.3 m wide × 2.9 m high	2203.88	7.00	497.45	–	2526.09	nr	**3271.01**
Lighter weight tank; complete with tank dome and pedestrian lid, submersible automatic pump system, supra filtration system, 25 m black and green rainwater pipe, rainwater labelling kit, 125 mm fine filter (excavation, backfilling base, trenching and inlet pipework all not included)							
1650 litre; 2.10 × 1.05 × 1.22	1265.40	5.00	354.64	–	1447.62	nr	**2050.05**
2650 litre; 2.10 × 1.30 × 1.50	1772.70	6.00	426.09	–	2030.45	nr	**2704.63**
Low profile tanks for reduced excavation; complete with rainwater filters and pressure pump							
1500 litre; 2.10 × 1.25 × 1.02	1110.37	4.00	283.44	–	1269.04	nr	**1800.03**
3000 litre; 2.46 × 2.1 × 1.05 m	1600.56	6.00	425.94	–	1832.64	nr	**2506.59**
5000 litre; 2.89 × 2.30 × 1.26	2155.63	6.50	461.88	–	2470.56	nr	**3180.69**
10000 litre 2.89 × 4,60 × 1.26 m	2580.70	7.00	497.62	–	2959.03	nr	**3705.01**
Rainwater Harvesting Ltd; accessories for rainwater harvesting tanks							
Downpipe filter for connection to roof downpipes.							
for roof areas up to 50 m^2	107.03	0.25	8.91	–	123.09	nr	**132.00**
for roof areas up to 80 m^2	46.17	0.25	8.91	–	53.10	nr	**62.01**
for roof areas up to 100 m^2	105.55	0.25	8.91	–	121.39	nr	**130.30**

35 SITE WORKS – WATER AND WATER FEATURES

Item	PC £	Labour hours	Labour £	Plant £	Material £	Unit	Total rate £
RAINWATER HARVESTING – cont							
Rainwater Harvesting Ltd – cont							
Pumps; inclusive of electrical installation and connections							
HydroForce Series 3 pump	328.61	0.75	31.58	–	377.90	nr	**409.48**
Filters for rainwater tanks; complete with overflow siphons and rodent guard	409.97		–	–	471.46	nr	**471.46**
Extra over to cover rainwater tanks in geofabric prior to backfilling; 10 m^2 of terram per tank; average cost	–	0.25	17.83	–	7.55	nr	**25.38**
IRRIGATION							
Irrigation Infrastructure and pipework; main or ring main supply							
Excavate and lay mains supply pipe to supply irrigated area							
PE 80 20 mm	–		–	–	–	m	**7.34**
PE 80 25 mm	–		–	–	–	m	**7.64**
Pro-Flow PEHD 63 mm PE	–		–	–	–	m	**13.08**
Pro-Flow PEHD 50 mm PE	–		–	–	–	m	**10.69**
Pro-Flow PEHD 40 mm PE	–		–	–	–	m	**9.21**
Pro-Flow PEHD 32 mm PE	–		–	–	–	m	**8.28**
Irrigation systems; head control to sprinkler stations							
Valves installed at supply point on irrigation supply manifold; includes for pressure control and filtration							
2 valve unit	–		–	–	–	nr	**198.96**
4 valve unit	–		–	–	–	nr	**310.61**
6 valve unit	–		–	–	–	nr	**439.28**
12 valve unit	–		–	–	–	nr	**842.30**
Solenoid valve; 25 mm with chamber; extra over for each active station	–		–	–	–	nr	**159.99**
Cable to solenoid valves (alternative to valves on manifold)							
4 core	–		–	–	–	m	**2.37**
6 core	–		–	–	–	m	**11.50**
12 core	–		–	–	–	m	**4.81**

35 SITE WORKS – WATER AND WATER FEATURES

Item	PC £	Labour hours	Labour £	Plant £	Material £	Unit	Total rate £
Irrigation; water supply and control							
Header tank and submersible pump and pressure stat							
1000 litre (500 gallon 25 mm/10 days to 1500 m^2) 2 m^3/hr pump	–		–	–	–	nr	**737.94**
4540 litre (1000 gallon 25 mm/10 days to 3500 m^2)	–		–	–	–	nr	**1494.81**
9080 litre (2000 gallon 25 mm/10 days to 7000 m^2)	–		–	–	–	nr	**2474.95**
Electric multi-station controllers 240 V							
WeatherMatic Smartline; 4 station; standard	–		–	–	–	nr	**224.86**
WeatherMatic Smartline; 6 station; standard	–		–	–	–	nr	**252.60**
WeatherMatic Smartline; 12 station; standard	–		–	–	–	nr	**428.26**
WeatherMatic Smartline; 12 station; radio controlled	–		–	–	–	nr	**879.81**
WeatherMatic Smartline; 12 station, with moisture sensor	–		–	–		nr	**1479.03**
WeatherMatic Smartline; 12 station; radio controlled; with moisture sensor	–		–	–	–	nr	**1776.20**
Irrigation sprinklers; station consisting of multiple sprinklers; inclusive of all trenching, wiring and connections to ring main or main supply							
Sprayheads; WeatherMatic LX; including nozzle; 21 m^2 coverage							
100 mm (4 inch)	–		–	–	–	nr	**23.39**
150 mm (6 inch)	–		–	–	–	nr	**33.03**
300 mm (12 inch)	–		–	–	–	nr	**38.10**
Matched precipitation sprinklers; WeatherMatic LX							
MP 1000; 16 m^2 coverage	–		–	–	–	nr	**53.61**
MP 2000; 30 m^2 coverage	–		–	–	–	nr	**53.61**
MP 3000; 81 m^2 coverage	–		–	–	–	nr	**53.61**
Gear drive sprinklers; WeatherMatic; placed to provide head to head (100%) overlap, average							
T3; 121 m^2 coverage	–		–	–	–	nr	**70.41**
CT70; 255 m^2 coverage	–		–	–	–	nr	**124.12**
CT70SS; 361 m^2 coverage	–		–	–	–	nr	**137.83**

35 SITE WORKS – WATER AND WATER FEATURES

Item	PC £	Labour hours	Labour £	Plant £	Material £	Unit	Total rate £
IRRIGATION – cont							
Drip irrigation; driplines; inclusive of connections and draindown systems							
Metzerplas TechLand 30 cm drip emitter							
fixed on soil surface	–		–	–	–	m	**1.62**
subsurface	–		–	–	–	m	**2.37**
Metzerplas TechLand 50 cm drip emitter							
fixed on soil surface	–		–	–	–	m	**1.36**
subsurface	–		–	–	–	m	**2.12**
Commissioning and testing of irrigation system							
per station	–		–	–	–	nr	**59.62**
Annual maintenance costs of irrigation system							
Call out charge per visit	–		–	–	–	nr	**496.72**
Extra over per station	–		–	–	–	nr	**19.88**
Moisture Control; moisture sensing control system							
Large garden comprising 7000 m^2 and 24 stations with four moisture sensors							
turf only	–		–	–	–	nr	**20093.69**
turf/shrub beds; 70/30	–		–	–	–	nr	**11721.34**
Medium garden comprising 3500 m^2 and 12 stations with two moisture sensors							
turf only	–		–	–	–	nr	**13395.81**
turf/shrub beds; 70/30	–		–	–	–	nr	**17414.58**
Smaller gardens garden comprising 1000 m^2 and 6 stations with one moisture sensor							
turf only	–		–	–	–	nr	**8204.90**
turf/shrub beds; 70/30	–		–	–	–	nr	**9042.18**
turf/shrub beds; 50/50	–		–	–	–	nr	**9711.92**
Leaky Pipe Systems Ltd; Leaky Pipe; moisture leaking pipe irrigation system							
Main supply pipe inclusive of machine excavation; exclusive of connectors							
20 mm LDPE polytubing	0.83	0.05	1.78	0.58	0.96	m	**3.32**
16 mm LDPE polytubing	0.72	0.05	1.78	0.58	0.83	m	**3.20**

35 SITE WORKS – WATER AND WATER FEATURES

Item	PC £	Labour hours	Labour £	Plant £	Material £	Unit	**Total rate £**
Water filters and cartridges							
No 10; 20 mm	–		–	–	77.84	nr	**77.84**
Big Blue and RR30 cartridge; 25 mm	–		–	–	110.85	nr	**110.85**
Water filters and pressure regulator sets; complete assemblies							
No 10; flow rate 3.1–82 litres per minute	–		–	–	110.85	nr	**110.85**
Leaky pipe hose; placed 150 mm sub surface for turf irrigation; distance between laterals 350 mm; excavation and backfilling priced separately							
LP12L low leak	3.70	0.04	1.43	–	4.26	m^2	**5.69**
LP12H high leak	3.89	0.04	1.43	–	4.47	m^2	**5.90**
Leaky pipe hose; laid to surface for landscape irrigation; distance between laterals 600 mm							
LP12L low leak	2.21	0.03	0.89	–	2.54	m^2	**3.43**
LP12H high leak	2.32	0.03	0.89	–	2.67	m^2	**3.56**
Leaky pipe hose; laid to surface for landscape irrigation; distance between laterals 900 mm							
LP12L low leak	1.48	0.02	0.60	–	1.70	m^2	**2.30**
LP12H high leak	1.55	0.02	0.60	–	1.79	m^2	**2.38**
Leaky pipe hose; laid to surface for tree irrigation laid around circumference of tree pit							
LP12L low leak	2.19	0.13	4.46	–	2.52	nr	**6.97**
LP12H high leak	2.30	0.13	4.46	–	2.64	nr	**7.10**
Accessories							
automatic multi-station controller stations; inclusive of connections	324.90	2.00	90.68	–	373.64	nr	**464.32**
solenoid valves; inclusive of wiring and connections to a multi-station controller; nominal distance from controller 25 m	38.45	0.50	17.83	–	638.92	nr	**656.74**

35 SITE WORKS – WATER AND WATER FEATURES

Item	PC £	Labour hours	Labour £	Plant £	Material £	Unit	Total rate £
FOUNTAINS AND WATER FEATURES							
Prices shown in this section are assumed contracted directly with the water feature subcontractor subcontractor; Add 5%–10% to these prices if the water features are part of a construction project subcontracted through a main contractor							
Water features; Fairwater water gardens Ltd ; Water feature in public park; Formal pond 225 concrete filled blockwork on a reinforced concrete base; EPDM lined with shingle drainage; spider filters; Concrete surround to receive pavings (not included) to 25% of perimeter. Stainless steel knife edge; abutting approximately 75% of perimeter; graded shelves to support plant growth; maximum depth 1.00 m; balancing pipework and balancing tank							
Water feature 920 m^2; perimeter 190 m^2; excavation volume 180 m^3; including excavation; hollow blockwork retaining to perimeters; trenching and paving bases							
Constructional elements in the Approximate Estimates section of this book							
Specialist Water feature elements							
Stainless steel knife edge 2 mm × 125	–		–	–	–	m	**55.57**
Coir rolls tied back to perimeter block edge to form edges to marginal planting	–		–	–	–	m	**90.11**
Boulders; mixed sizes 500 mm dia. average	–		–	–	–	nr	**263.60**
Waterproofing to exterior of pump rooms	–		–	–	–	m^2	**39.19**
Pond liner: Firestone Geoguard EPDM 1.14 mm including protective fleece overlay and underlay	–		–	–	–	m^2	**21.38**
Spider filter 100 mm	–		–	–	–	m	**6.62**
Spider filter 200 mm distribution pot	–		–	–	–	nr	**178.11**

35 SITE WORKS – WATER AND WATER FEATURES

Item	PC £	Labour hours	Labour £	Plant £	Material £	Unit	Total rate £
Pressure feed pipe 50 mm	–		–	–	–	m	**4.50**
Balancing pipe and flanges 160 mm	–		–	–	–	m	**14.25**
Bends flanges connectors	–		–	–	–	nr	**1061.52**
M&E Including GRP reservoir, GRP lined concrete chamber, circulation pumps, UV clarification, all valves filters etc. as required. Water volume 150 m^3	–		–	–	–	nr	**10686.38**
WRAS Break tank	–		–	–	–	nr	**1816.69**
Connections	–		–	–	–	nr	**2137.28**
Water feature; formal; in ground ; Fairwater Water Gardens Ltd. Constructed water feature measuring 12.80 m × 2.80 m × 600 mm depth of water; 900 mm deep inclusive of filtration bed. External pump housing; stainless steel water spouts; automatic top-up and overflows to soakaway (not included) specialist elements only; full build-up in approximate estimates section.							
Detailed design	–		–	–	–	nr	**484.45**
Circulation pump for water spout for a chemically filtered pool	–		–	–	–	nr	**8911.02**
Alternative circulation pump for a biologically filtered pool	–		–	–	–	nr	**14978.03**
Structural waterproofing – GRP	–		–	–	–	m^2	**145.20**
Supply and planting of specialist planting commensurate with pool 30–40 m^2	–		–	–	–	nr	**897.66**
Commissioning	–		–	–	–	nr	**427.46**

35 SITE WORKS – WATER AND WATER FEATURES

Item	PC £	Labour hours	Labour £	Plant £	Material £	Unit	Total rate £
FOUNTAINS AND WATER FEATURES – cont							
Water feature; Chemically dosed; formal; in ground ; Fairwater Water Gardens Ltd; formal; in ground Specialist works only; all construction elements priced separately; full build-up in approximate estimates section; External pump housing; stainless steel water spouts; automatic top-up and overflows							
Subcontractor preliminaries for small pools							
Circulation pump for a biologically filtered pool; pump chamber and pump, including all pipe work, valve and fittings. Sum includes supply and installation of 2 PVC spider filters; shingle surounds							
Detailed design	–		–	–	–	nr	**484.45**
biologically filtered pool mechanisms.	–		–	–	–	nr	**14978.03**
Circulation pump for water spout (chemically filtered)	–		–	–	–	nr	**8911.02**
Structural waterproofing – GRP	–		–	–	–	m^2	**145.20**
Supply and planting of specialist planting	–		–	–	–	nr	**897.66**
Commissioning and maintenance	–		–	–	–	nr	**427.46**
Waterfall construction; Fairwater Water Garden Design Consultants Ltd							
Stone placed on top of butyl liner; securing with concrete and dressing to form natural rock pools and edgings							
Boulders; m^3 rate	–		–	–	–	m^3	**1577.64**
Boulders; tonne rate	–		–	–	–	tonne	**631.06**
Balancing tank; blockwork construction; inclusive of recirculation pump and pond level control; 110 mm balancing pipe to pond; waterproofed with preformed polypropylene membrane; pipework mains water top-up and overflow							
450 × 600 × 1000 mm	–		–	–	–	nr	**2379.23**
Extra for pump							
2000 gallons per hour; submersible	–		–	–	–	nr	**447.81**

35 SITE WORKS – WATER AND WATER FEATURES

Item	PC £	Labour hours	Labour £	Plant £	Material £	Unit	Total rate £
Water features; Fairwater Water Garden Design Consultants Ltd							
Natural stream; butyl lined level changes of 1.00 m; water pumped from lower pond (not included) via balancing tank; level changes via Purbeck stone water falls							
20 m long stream	–		–	–	–	nr	**33094.96**
Bespoke free-standing water wall; steel or glass panel; self contained recirculation system and reservoir							
water wall; 2.00 m high × 850 mm wide	–		–	–	–	nr	**15759.50**
INTERNAL WATER FEATURES							
Internal water features; Fairwater Water Garden Design Consultants Ltd;							
Stainless steel rill; powder coated black 3000 mm × 250 mm; 2 nr letterbox weirs to maintain water level in rill; WRAS compliant header tank; modified balancing tank with chemical dosing unit; circulation pump and pipework;							
3.00 m × 250 mm wide	–		–	–	–	nr	**7091.78**
Brimming bowl with catchment tray; Fairwater Water Garden Design Consultants Ltd;							
Marble brimming bowl. recessed Polypro catchment tray 1.50 × 1.50x 120 mm ; pipework; WRAS compliant header tank balancing tank with Chemical dosing unit; circulation pump							
1.00 m dia.	–		–	–	–	nr	**5909.84**
Design mockup and commissioning on the above	–		–	–	–	nr	**5515.83**
Maintenance 12 months; monthly	–		–	–	–	nr	**2363.92**

35 SITE WORKS – WATER AND WATER FEATURES

Item	PC £	Labour hours	Labour £	Plant £	Material £	Unit	Total rate £
WATER AND RIVER BANK REMEDIATION/ ECOLOGY							
River and lake bed dredging; Blackdown Environmental Limited; Specialist water reclamation engineers; Operations include for removal of excavated material away from the water body bank and spreading locally in layers not exceeding 300 mm deep							
By long reach excavator							
Mobilization for Desilting	–		–	–	–	nr	**4109.99**
Dewatering and mechanical excavation with long reach excavator and tracked excavator to stockpile on bank	–		–	–	–	m^3	**15.07**
Mobilization £ 12,000 craning in barges, pontoon and tug	–		–	–	–	nr	**14438.85**
Excavation by floating excavator, transport to bank by tug and barge a	–		–	–	–	m^3	**32.28**
Pumping of silt to geotubes for containment and drying	–		–	–	–	m^3	**27.08**
By pump dredger							
By long reach excavator using dewatering techniques							
Mobilization Truxor type pump dredger up to 1000 m^3	–		–	–	–	nr	**721.95**
Mobilization pontoon type pump dredger over 1000 m^3	–		–	–	–	nr	**6016.20**
Lake size 1.5 ha; 15000 m^3	–		–	–	–	item	**7219.43**
Lake size 3.0 ha; 30000 m^3	–		–	–	–	item	**16845.33**
The labour in this section is calculated on a 3 person team. The labour time below should be multiplied by 3 to calculate the cost as shown.							
Bioengineering techniques for reclamation or establishment of waterways; Blackdown Environmental Limited							
Erosion control							
Coir roll + Faggot	–		–	–	–	m^2	**57.84**
Coir roll and rock roll	–		–	–	–	m^2	**84.46**

35 SITE WORKS – WATER AND WATER FEATURES

Item	PC £	Labour hours	Labour £	Plant £	Material £	Unit	Total rate £
Brushwood mattress revetments to encourage deposition of sediments							
200 mm deep	37.45	0.07	7.13	–	43.07	m^2	**50.20**
500 mm deep	63.07	0.10	10.70	–	72.53	m^2	**83.23**
Habitat creation to waterways and water bodies; Blackdown Environmental Limited; Specialist works to SSSI's and SAC sites							
Grading and enablement of river bank to allow access; temporary trackways or haul road and removal of same	–	8.00	843.62	561.35	419.09	nr	**2697.31**
Creation of riffle bars and point bars using gravel beds	–		–	–	–	m^2	**93.64**
Wetland establishment; Blackdown Environmental limited;							
Restoration of riverbanks; to create habitats for wildlife							
greenfield sites	–		–	–	–	m^2	**18.05**
brownfield sites	–		–	–	–	m^2	**30.08**
The labour in this section is calculated on a 2 person team. The labour time below should be multiplied by 2 to calculate the cost as shown.							
Wetland and revetment establishment and erosion control; Blackdown Environmental Limited							
Preparation works to receive treatments below							
Grade out bank receive treatments	–	0.05	3.57	2.53	–	m^2	**6.10**
Remove silt and accumulated deposits to maximum 300 mm deep within water line to reinstate freeflow of water along shoreline, Spread on river bank	–	0.05	3.57	3.05	–	m^2	**6.61**
The labour in this section is calculated on a 3 person team. The labour time below should be multiplied by 3 to calculate the cost as shown.							

35 SITE WORKS – WATER AND WATER FEATURES

Item	PC £	Labour hours	Labour £	Plant £	Material £	Unit	Total rate £
WATER AND RIVER BANK REMEDIATION/ ECOLOGY – cont							
Placement of pre-grown coir rolls 200 mm dia. to base of revetments or within the water body to aid establishment of vegetation and habitat at the base of retained revetments (above)							
on tidal banks or wet river banks	30.32	0.11	11.88	–	34.87	m	**46.75**
on shorelines max 300 mm deep	48.71	0.11	12.15	–	56.02	m	**68.17**
on shorelines max 600 mm deep	74.34	0.13	13.37	–	85.49	m	**98.86**
on shorelines max 1.00 m deep	103.06	0.13	14.26	–	118.52	m	**132.77**
on shorelines max 1.50 m deep	109.82	0.17	17.83	–	126.29	m	**144.12**
Placement of pre grown coir rolls 300 mm dia. to base of revetments or within the water body to aid establishment of vegetation and habitat at the base of retained revetments (above)							
on tidal banks or wet river banks	33.15	0.11	11.88	–	38.12	m	**50.01**
on shorelines max 300 mm deep	51.54	0.11	12.15	–	59.27	m	**71.42**
on shorelines max 600 mm deep	74.34	0.13	13.37	–	85.49	m	**98.86**
on shorelines max 1.00 m deep	105.88	0.13	14.26	–	121.77	m	**136.02**
on shorelines max 1.50 m deep	112.64	0.17	17.83	–	129.54	m	**147.37**
Stronger tidal demands- placement of pre grown coir rolls 400 mm dia. to base of revetments or within the water body to aid establishment of vegetation and habitat at the base of retained revetments (above)							
on tidal banks or wet river banks	64.21	0.13	13.37	–	73.84	m	**87.21**
on shorelines max 300 mm deep	68.72	0.13	13.37	–	79.02	m	**92.39**
on shorelines max 600 mm deep	95.18	0.17	17.83	–	134.07	m	**151.90**
on shorelines max 1.00 m deep	123.91	0.20	21.39	–	142.49	m	**163.88**
on shorelines max 1.50 m deep	130.66	0.20	21.39	–	150.26	m	**171.65**
Placement of pre-planted coir pallets to establish reedbeds to wetlands; habitat creation projects where mature vegetation needs to be established rapidly							
on tidal banks or wet river banks	38.59	0.20	21.39	–	44.38	m	**65.77**
on shorelines max 300 mm deep	41.97	0.25	26.74	–	48.26	m	**75.00**
unplanted	25.91	0.25	26.74	–	84.19	m^2	**110.93**
Brushwood faggots fascine in Hazel or Willow; 300 mm dia. to form sediment traps or build up inter tidal sediment							
300 m dia.	–	0.10	10.70	–	30.44	m	**41.13**
400 mm dia.	–	0.10	10.70	–	33.03	m	**43.72**

35 SITE WORKS – WATER AND WATER FEATURES

Item	PC £	Labour hours	Labour £	Plant £	Material £	Unit	**Total rate £**
Rock rolls; permanent revetment for use around reservoirs, shorelines, lake edges, streams and river banks.							
300 mm dia.	–	0.07	7.13	4.75	129.44	m^2	**141.32**
400 mm dia.	–	0.10	10.70	7.12	129.56	m^2	**147.37**
rock mattresses to retain and in conjuction with rock rolls on riverbank	–	0.08	8.91	5.93	60.89	m^2	**75.73**

35 SITE WORKS – WATER AND WATER FEATURES

Item	PC £	Labour hours	Labour £	Plant £	Material £	Unit	Total rate £
LAKES AND PONDS							
Lakes and ponds – General Preamble: The pressure of water against a retaining wall or dam is considerable, and where water retaining structures form part of the design of water features, the landscape architect is advised to consult a civil engineer. Artificially contained areas of water in raised reservoirs over 25,000 m^3 have to be registered with the local authority and their dams will have to be covered by a civil engineer's certificate of safety.							
Typical linings – General Preamble: In addition to the traditional methods of forming the linings of lakes and ponds in puddled clay or concrete, there are a number of lining materials available. They are mainly used for reservoirs but can also help to form comparatively economic water features especially in soil which is not naturally water retentive. Information on the construction of traditional clay puddle ponds can be obtained from the British Trust for Conservation Volunteers, 36 St. Mary's Street, Wallingford, Oxfordshire, OX10 0EU. Tel: (01491) 39766. The cost of puddled clay ponds depends on the availability of suitable clay, the type of hand or machine labour that can be used and the use to which the pond is to be put.							

35 SITE WORKS – WATER AND WATER FEATURES

Item	PC £	Labour hours	Labour £	Plant £	Material £	Unit	Total rate £
Lake liners; Fairwater Water Garden Design Consultants Ltd; to evenly graded surface of excavations (excavating not included); all stones over 75 mm; removing debris; including all welding and jointing of liner sheets							
Geotextile underlay; inclusive of spot welding to prevent dragging							
to water features	–		–	–	–	m^2	**2.76**
to lakes or large features	–		–	–	–	$1000\ m^2$	**2773.66**
EPDM 0.75 mm liner; inclusive of site vulcanising laid to geotextile above							
0.75 mm thick	–	0.10	6.82	–	–	m^2	**17.00**
0.75 mm thick	–		–	–	–	$1000\ m^2$	**11383.21**
Operations over surfaces of lake liners							
Dug ballast; evenly spread over excavation already brought to grade							
150 mm thick	814.97	2.00	71.30	134.12	937.21	$100\ m^2$	**1142.64**
200 mm thick	1086.62	3.00	106.95	201.18	1249.62	$100\ m^2$	**1557.75**
300 mm thick	1629.94	3.50	124.78	234.71	1874.43	$100\ m^2$	**2233.92**
Imported topsoil; evenly spread over excavation							
100 mm thick	355.05	1.50	53.48	100.59	408.31	$100\ m^2$	**562.38**
150 mm thick	532.58	2.00	71.30	134.12	612.46	$100\ m^2$	**817.89**
200 mm thick	710.10	3.00	106.95	201.18	816.62	$100\ m^2$	**1124.75**
Blinding existing subsoil with 50 mm sand	249.48	1.00	35.65	134.12	286.90	$100\ m^2$	**456.67**
Topsoil from excavation; evenly spread over excavation							
100 mm thick	–		–	100.59	–	$100\ m^2$	**100.59**
200 mm thick	–		–	134.12	–	$100\ m^2$	**134.12**
300 mm thick	–		–	201.18	–	$100\ m^2$	**201.18**
Extra over for screening topsoil using a Powergrid screener; removing debris	–		–	9.74	–	m^3	**11.68**

35 SITE WORKS – WATER AND WATER FEATURES

Item	PC £	Labour hours	Labour £	Plant £	Material £	Unit	Total rate £
LAKES AND PONDS – cont							
Lake construction; Fairwater Water Garden Design Consultants Ltd; lake construction at ground level; excavation; forming of lake; commissioning; excavated material spread on site							
Lined with existing site clay; natural edging with vegetation meeting the water							
500 m^2	–		–	–	–	nr	**15759.50**
1000 m^2	–		–	–	–	nr	**26003.21**
1500 m^2	–		–	–	–	nr	**36246.87**
2000 m^2	–		–	–	–	nr	**44126.61**
Lined with 0.75 mm EPDM on geotextile underlay; natural edging with vegetation meeting the water							
500 m^2	–		–	–	–	nr	**24427.25**
1000 m^2	–		–	–	–	nr	**45702.59**
1500 m^2	–		–	–	–	nr	**59886.13**
2000 m^2	–		–	–	–	nr	**72805.79**
Extra over to the above for hard block edging to secure and protect liner; measured at lake perimeter	–		–	–	–	m	**74.80**
Lined with imported puddling clay; natural edging with vegetation meeting the water							
500 m^2	–		–	–	–	nr	**38610.80**
1000 m^2	–		–	–	–	nr	**66189.94**
1500 m^2	–		–	–	–	nr	**90617.21**
2000 m^2	–		–	–	–	nr	**113468.48**

35 SITE WORKS – WATER AND WATER FEATURES

Item	PC £	Labour hours	Labour £	Plant £	Material £	Unit	**Total rate £**
Ornamental pools – General							
Preamble: Small pools may be lined with one of the materials mentioned under lakes and ponds, or may be in rendered brickwork, puddled clay or, for the smaller sizes, fibreglass. Most of these tend to be cheaper than waterproof concrete. Basic prices for various sizes of concrete pools are given in the Approximate Estimates section. Prices for excavation, grading, mass concrete, and precast concrete retaining walls are given in the relevant sections. The manufacturers should be consulted before specifying the type and thickness of pool liner, as this depends on the size, shape and proposed use of the pool. The manufacturer's recommendation on foundations and construction should be followed.							
Ornamental pools; Fairwater Ltd							
Pool liners; to 50 mm sand blinding to excavation (excavating not included); all stones over 50 mm; removing debris from surfaces of excavation; including underlay, all welding and jointing of liner sheets							
EPDM 0.75 mm	–		–	–	–	m^2	**17.05**
Fine gravel; 100 mm; evenly spread over area of pool; by hand	4.55	0.13	4.46	–	5.23	m^2	**9.68**
Selected topsoil from excavation; 100 mm; evenly spread over area of pool; by hand	–	0.13	4.46	–	–	m^2	**4.46**
Extra over selected topsoil for spreading imported topsoil over area of pool; by hand	35.51		–	–	40.83	m^3	**40.83**

35 SITE WORKS – WATER AND WATER FEATURES

Item	PC £	Labour hours	Labour £	Plant £	Material £	Unit	Total rate £
LAKES AND PONDS – cont							
Pool surrounds and ornament; Haddonstone Ltd; Portland Bath or terracotta cast stone							
Pool surrounds; installed to pools or water feature construction priced separately; surrounds and copings to 112.5 mm internal brickwork							
C4HSKVP half small pool surround; internal dia. 1780 mm; kerb features continuous moulding enriched with ovolvo and palmette designs; inclusive of plinth and integral conch shell vases flanked by dolphins	1482.00	16.00	570.40	–	1736.15	nr	**2306.55**
C4SKVP small pool surround as above but with full circular construction; internal dia. 1780 mm	2956.00	48.00	1711.20	–	3438.02	nr	**5149.22**
C4MKVP medium pool surround; internal dia. 2705 mm; inclusive of plinth and integral vases	4434.00	48.00	1711.20	–	5179.35	nr	**6890.55**
C4XLKVP extra large pool surround; internal dia. 5450 mm	5912.00	140.00	4991.00	–	6891.58	nr	**11882.58**
Pool centre pieces and fountains; inclusive of plumbing and pumps							
C251 Gothic Fountain and Gothic Upper Base A350; free-standing fountain	1189.00	4.00	142.60	–	1372.96	nr	**1515.56**
HC521 Romanesque Fountain; free-standing bowl with self-circulating fountain; filled with cobbles; 815 mm dia. × 348 mm high	499.00	2.00	71.30	–	708.49	nr	**779.79**
C300 Lion Fountain; 610 mm high on fountain base C305; 280 mm high	264.00	2.00	71.30	–	309.21	nr	**380.51**

35 SITE WORKS – WATER AND WATER FEATURES

Item	PC £	Labour hours	Labour £	Plant £	Material £	Unit	**Total rate £**
Wall fountains, watertanks and fountains; inclusive of installation drainage, automatic top up, pump and balancing tank							
Fountain surrounds; Architectural Heritage Ltd; to pools constructe dseparately inclusive of installation, drainage, automatic top up, pump and balancing tank							
Decorative Circular Pool Surround; age patinated artificial stone; reproduction; Overall dia. 4.26 m, overall height 220 mm	4400.00	4.00	140.83	–	4997.08	nr	**7594.36**
Great Westwood pool surround; age-patinated artificial stone pool; reproduction; Overall dia. 6 m; height 350 mm	6600.00	4.00	141.27	–	7519.38	nr	**10124.89**
Fountain kits; typical prices of submersible units comprising fountain pumps, fountain nozzles, underwater spotlights, nozzle extension armatures, underwater terminal boxes and electrical control panels							
Single aerated white foamy water columns; ascending jet 70 mm dia.; descending water up to four times larger; jet height adjustable between 1.00–1.70 m	4479.23		–	–	5122.90	nr	**5864.84**
Single aerated white foamy water columns; ascending jet 110 mm dia.; descending water up to four times larger; jet height adjustable between 1.50–3.00 m	7465.39		–	–	8541.15	nr	**9687.03**

35 SITE WORKS – WATER AND WATER FEATURES

Item	PC £	Labour hours	Labour £	Plant £	Material £	Unit	Total rate £
SWIMMING POOLS							
Natural swimming pools; Fairwater Water Garden Design Consultants Ltd							
Natural swimming pool of 50 m² area; shingle regeneration zone 50 m² planted with marginal and aquatic plants	–		–	–	–	nr	**102436.82**
Swimming pools; Guncast Swimming Pools Ltd; Excavation and blinding to ground to receive swimming pool inclusive of disposal off site							
Regular shaped pools; Excavation and blinding to ground to receive swimming pool inclusive of disposal off site							
100 m² × 1.20 m deep	–		–	–	–	nr	**13151.64**
72 m² × 1.20 m deep	–		–	–	–	nr	**10274.72**
60 m² × 1.20 m deep	–		–	–	–	nr	**10274.72**
Irregular shaped pools							
100 m² × 1.20 m deep	–		–	–	–	nr	**13809.22**
72 m² × 1.20 m deep	–		–	–	–	nr	**10788.46**
60 m² × 1.20 m deep	–		–	–	–	nr	**10788.46**
Swimming pools; Guncast Swimming Pools Ltd; Pool construction to excavations priced separately							
Construction of a rectangular outdoor swimming pool in situ concrete corner steps, constant shell depths of 1.35 m, water depth of 1.2 m as standard skimmer pool, tiled corner steps;drainage layer, formwork reinforcement and all Gunite construction, rendered internally and bespoke mosaic surface finish; bullnosed Indian sandstone copings; excludes pump house construction							
10.0 × 5.0 × 1.20; 60,000 litre	–		–	–	–	nr	**114746.84**
15.0 × 5.0 × 1.20 m; 90,000 litre	–		–	–	–	nr	**149619.61**
20.0 × 5.0 × 1.20 m; 120,000 litre	–		–	–	–	nr	**190154.44**

35 SITE WORKS – WATER AND WATER FEATURES

Item	PC £	Labour hours	Labour £	Plant £	Material £	Unit	Total rate £
Swimming pools; Guncast Swimming Pools Ltd; Accessories to swimming pools							
Lighting for pools; inclusive of trenching cables and connections to distribution board							
12 nr underwater LED spot lights stainless steel shroud	–		–	–	–	nr	**10078.40**
6 nr underwater LED spot lights stainless steel shroud	–		–	–	–	nr	**5039.20**
Swimming pool accessories							
Automatic pool cleaner	–		–	–	–	nr	**1816.73**
Insulated automatic slatted pool cover, housed in side pit	–		–	–	–	nr	**22913.78**
Solar cover; 17 × 6 m Ocea	–		–	–	–	nr	**41202.19**
Ozone disinfection system	–		–	–	–	nr	**7028.67**
pool heater; propane boiler; 250,000 BTU	–		–	–	–	nr	**5328.36**
heat pump 100 × 1000 mm concrete base	–		–	–	–	nr	**6944.20**
Pool heating via heat pump	–		–	–	–	nr	**6944.20**
Diving Board	–		–	–	–	nr	**3276.08**
Pool maintenance kit	–		–	–	–	nr	**1194.01**

36 FENCING

Item	PC £	Labour hours	Labour £	Plant £	Material £	Unit	Total rate £
CLARIFICATION NOTES ON LABOUR COSTS IN THIS SECTION							
Fencing team							
Generally a two man fencing team is used in this section; The column 'Labour hours' reports team hours. The column 'Labour £' reports the total cost of the team for the unit of fencing shown							
2 man team	–	1.00	71.30	–	–	hr	**71.30**
TEMPORARY FENCING							
The labour in this section is calculated on a 2 person team. The labour time below should be multiplied by 2 to calculate the cost as shown.							
Temporary security fence; HSS Hire; mesh framed unclimbable fencing; including precast concrete supports and couplings							
Weekly hire; 2.85 × 2.00high							
weekly hire rate	–		–	1.96	–	m	**1.96**
erection of fencing; labour only	–	0.07	4.76	–	–	m	**4.76**
removal of fencing loading to collection vehicle	–	0.03	2.38	–	–	m	**2.38**
delivery charge	–		–	1.25	–	m	**1.25**
return haulage charge	–		–	0.82	–	m	**0.82**
Protective fencing; Chestnut pale fencing							
Cleft chestnut rolled fencing; fixing to 100 mm dia. chestnut posts; driving into firm ground at 3 m centres							
900 mm high	19.95	0.05	3.80	–	30.51	m	**34.31**
1200 mm high	7.85	0.08	5.70	–	16.59	m	**22.30**
Enclosures; Earth Anchors							
Rootfast anchored galvanized steel enclosures post ADP 20–1000; 1000 mm high × 20 mm dia. with AA25–750 socket and padlocking ring	38.85	0.05	3.57	–	50.89	nr	**54.46**
steel cable, orange plastic coated	0.48	–	0.07	–	0.55	m	**0.62**

36 FENCING

Item	PC £	Labour hours	Labour £	Plant £	Material £	Unit	Total rate £
TIMBER FENCING							
The labour in this section is calculated on a 2 person team. The labour time below should be multiplied by 2 to calculate the cost as shown.							
Timber fencing; tanalized softwood fencing; all timber posts are kiln dried redwood with 15 year guarantee							
Timber lap panels; pressure treated; fixed to timber posts 75 × 75 mm in 1:3:6 concrete; at 1.90 m centres							
900 mm high	11.84	0.33	23.76	–	32.79	m	**56.55**
1200 mm high	14.04	0.38	26.74	–	35.10	m	**61.84**
1500 mm high	14.04	0.40	28.52	–	34.95	m	**63.47**
1800 mm high	16.23	0.45	32.09	–	39.23	m	**71.32**
Timber lap panels; fixed to slotted concrete posts 100 × 100 mm 1:3:6 concrete at 1.88 m centres							
900 mm high	11.84	0.38	26.74	–	39.21	m	**65.94**
1200 mm high	13.30	0.40	28.52	–	40.88	m	**69.40**
1500 mm high	17.30	0.43	30.30	–	50.09	m	**80.39**
1800 mm high	19.51	0.50	35.65	–	53.55	m	**89.20**
extra for corner posts	23.95	0.45	32.09	–	45.96	nr	**78.05**
extra over panel fencing for 300 mm high trellis tops; slats at 100 mm centres; including additional length of posts	13.95	0.50	35.65	–	16.05	m	**51.70**
Closeboarded fencing; to concrete posts 100 × 100 mm; 2 nr softwood arris rails; 100 × 22 mm softwood pales lapped 13 mm; including excavating and backfilling into firm ground at 3.00 m centres; setting in concrete 1:3:6; concrete gravel board 150 × 150 mm							
1050 mm high	21.67	0.55	39.22	–	31.82	m	**71.04**
1500 mm high	21.67	0.58	41.00	–	31.82	m	**72.82**
Closeboarded fencing; to concrete posts 100 × 100 mm; 3 nr softwood arris rails; 100 × 22 mm softwood pales lapped 13 mm; including excavating and backfilling into firm ground at 3.00 m centres; setting in concrete 1:3:6							
1650 mm high	27.01	0.63	44.56	–	37.97	m	**82.54**
1800 mm high	27.13	0.65	46.35	–	38.11	m	**84.46**

36 FENCING

Item	PC £	Labour hours	Labour £	Plant £	Material £	Unit	Total rate £
TIMBER FENCING – cont							
Timber fencing; tanalized softwood fencing – cont							
Closeboarded fencing; to timber redwood posts 100 × 100 mm; 2 nr softwood arris rails; 100 × 22 mm softwood pales lapped 13 mm; including excavating and backfilling into firm ground at 3.00 m centres; setting in concrete 1:3:6							
1350 mm high	22.68	0.50	35.65	–	32.99	m	**68.64**
1650 mm high	24.30	0.63	44.56	–	34.86	m	**79.42**
1800 mm high	25.17	0.65	46.35	–	35.86	m	**82.20**
Closeboarded fencing; to timber posts 100 × 100 mm; 3 nr softwood arris rails; 100 × 22 mm softwood pales lapped 13 mm; including excavating and backfilling into firm ground at 3.00 m centres; setting in concrete 1:3:6							
1350 mm high	25.80	0.48	33.95	–	36.57	m	**70.53**
1650 mm high	27.42	0.65	46.35	–	38.44	m	**84.78**
1800 mm high	28.29	0.68	48.13	–	39.44	m	**87.56**
extra over for post 125 × 100 mm; for 1800 mm high fencing	29.14		–	–	33.51	m	**33.51**
extra over to the above for counter rail	1.83		–	–	2.10	m	**2.10**
extra over to the above for capping rail	1.83	0.05	3.57	–	2.10	m	**5.67**
Feather edge board fencing; to timber posts 100 × 100 mm; 2 nr softwood cant rails; 125 × 22 mm feather edge boards lapped 13 mm; including excavating and backfilling into firm ground at 3.00 m centres; setting in concrete 1:3:6							
1350 mm high	18.64	0.28	19.97	–	24.54	m	**44.51**
1650 mm high	21.04	0.28	19.97	–	27.31	m	**47.28**
1800 mm high	22.00	0.28	19.97	–	28.41	m	**48.38**
Feather edge board fencing; to timber posts 100 × 100 mm; 3 nr softwood cant rails; 125 × 22 mm feather edge boards lapped 13 mm; including excavating and backfilling into firm ground at 3.00 m centres; setting in concrete 1:3:6							
1350 mm high	20.93	0.29	20.37	–	27.18	m	**47.55**
1650 mm high	23.33	0.29	20.37	–	29.94	m	**50.31**
1800 mm high	21.57	0.29	20.37	–	31.04	m	**51.41**

36 FENCING

Item	PC £	Labour hours	Labour £	Plant £	Material £	Unit	**Total rate £**
extra over for post 125 × 100 mm; for 1800 mm high fencing	29.14		–	–	33.51	nr	**33.51**
extra over to the above for counter rail and capping	4.30	0.05	3.57	–	4.95	m	**8.51**
Palisade fencing; 22 × 75 mm softwood vertical palings with pointed tops; nailing to two 50 × 100 mm horizontal softwood arris rails; morticed to 100 × 100 mm softwood posts with weathered tops at 3.00 m centres; setting in concrete							
900 mm high	20.15	0.45	32.09	–	30.08	m	**62.17**
1200 mm high	28.06	0.48	33.87	–	39.18	m	**73.05**
extra over for rounded tops	0.86		–	–	0.99	m	**0.99**
Post-and-rail fencing; 90 × 38 mm softwood horizontal rails; fixing with galvanized nails to 150 × 75 mm softwood posts; including excavating and backfilling into firm ground at 1.80 m centres; all treated timber							
1200 mm high; 3 horizontal rails	31.38	0.18	12.48	–	36.34	m	**48.83**
1200 mm high; 4 horizontal rails	38.00	0.18	12.48	–	43.70	m	**56.18**
Cleft rail fencing; chestnut adze tapered rails 2.80 m long; morticed into joints; to 125 × 100 mm softwood posts 1.95 m long; including excavating and backfilling into firm ground at 2.80 m centres							
two rails	6.14	0.13	8.91	–	7.06	m	**15.98**
three rails	26.29	0.14	9.99	–	30.23	m	**40.22**
four rails	33.01	0.18	12.48	–	37.96	m	**50.44**
Hit and Miss horizontal rail fencing; 87 × 38 mm top and bottom rails; 100 × 22 mm vertical boards arranged alternately on opposite side of rails; to 100 × 100 mm posts; including excavating and backfilling into firm ground; setting in concrete at 1.8 m centres							
treated softwood; 1600 mm high	57.76	0.60	42.78	–	72.64	m	**115.42**
treated softwood; 1800 mm high	63.55	0.67	47.52	–	79.93	m	**127.45**
treated softwood; 2000 mm high	69.37	0.70	49.91	–	86.62	m	**136.53**
Trellis tops to fencing							
Extra over screen fencing for 300 mm high trellis tops; slats at 100 mm centres; including additional length of posts	13.95	0.05	3.57	–	16.05	m	**19.61**

36 FENCING

Item	PC £	Labour hours	Labour £	Plant £	Material £	Unit	Total rate £
TRELLIS							
Traditional trellis panels; The Garden Trellis Company; bespoke trellis panels for decorative, screening or security applications; timber planed all round; height of trellis 1800 mm							
Free-standing panels; posts 70 × 70 mm set in concrete; timber frames mitred and grooved 45 × 34 mm; heights and widths to suit; slats 32 × 10 mm joinery quality tanalized timber at 100 mm ccs; elements fixed by galvanized staples; capping rail 70 × 34 mm							
HV68 horizontal and vertical slats; softwood	188.96	1.00	35.65	–	233.65	m	**269.30**
D68 diagonal slats; softwood	209.95	1.00	35.65	–	257.79	m	**293.44**
HV68 horizontal and vertical slats; hardwood iroko	352.25	1.00	35.65	–	421.44	m	**457.09**
D68 diagonal slats; hardwood iroko or Western Red Cedar	326.59	1.00	35.65	–	391.93	m	**427.58**
Trellis panels fixed to face of existing wall or railings; timber frames mitred and grooved 45 × 34 mm; heights and widths to suit; slats 32 × 10 mm joinery quality tanalized timber at 100 mm ccs; elements fixed by galvanized staples; capping rail 70 × 34 mm							
HV68 horizontal and vertical slats; softwood	151.63	0.50	17.83	–	176.92	m	**194.74**
D68 diagonal slats; softwood	174.96	0.50	17.83	–	203.74	m	**221.57**
HV68 horizontal and vertical slats; iroko or Western Red Cedar	307.93	0.50	17.83	–	356.66	m	**374.48**
D68 diagonal slats; iroko or Western Red Cedar	331.26	0.50	17.83	–	383.49	m	**401.31**

36 FENCING

Item	PC £	Labour hours	Labour £	Plant £	Material £	Unit	**Total rate £**
Contemporary style trellis panels; The Garden Trellis Company; bespoke trellis panels for decorative, screening or security applications; timber planed all round							
Free-standing panels 30/15; posts 70 × 70 mm set in concrete with 90 × 30 mm top capping; slats 30 × 14 mm with 15 mm gaps; vertical support at 450 mm centres							
joinery treated softwood	244.94	1.00	35.65	–	298.03	m	**333.68**
hardwood iroko or Western Red Cedar	321.93	1.00	35.65	–	386.56	m	**422.21**
Panels 30/15 face fixed to existing wall or fence; posts 70 × 70 mm set in concrete with 90 × 30 mm top capping; slats 30 × 14 mm with 15 mm gaps; vertical support at 450 mm centres							
joinery treated softwood	221.62	0.50	17.83	–	257.40	m	**275.22**
hardwood iroko or Western Red Cedar	354.59	0.50	17.83	–	410.31	m	**428.14**
Integral arches to ornamental trellis panels; The Garden Trellis Company							
Arches to trellis panels in 45 × 34 mm grooved timbers to match framing; fixed to posts							
R450 1/4 circle; joinery treated softwood	79.06	1.50	53.48	–	107.26	nr	**160.74**
R450 1/4 circle; hardwood iroko or Western Red Cedar	89.42	1.50	53.48	–	119.18	nr	**172.66**
Arches to span 1800 mm wide							
joinery treated softwood	295.49	1.50	53.48	–	356.16	nr	**409.63**
hardwood iroko or Western Red Cedar	235.87	1.50	53.48	–	287.60	nr	**341.08**
Painting or staining of trellis panels; high quality coatings							
microporous opaque paint or spirit based stain	–		–	–	44.71	m²	**44.71**

36 FENCING

Item	PC £	Labour hours	Labour £	Plant £	Material £	Unit	Total rate £
STRAINED WIRE FENCING/FIELD FENCING							
The labour in this section is calculated on a 2 person team. The labour time below should be multiplied by 2 to calculate the cost as shown.							
Posts for strained wire and wire mesh							
Strained wire fencing; concrete inter posts only at 2750 mm centres; 610 mm below ground; excavating holes; filling with concrete; replacing topsoil; disposing surplus soil off site							
900 mm high	5.72	0.28	19.81	1.24	9.96	m	**31.01**
1200 mm high	6.54	0.28	19.81	1.24	10.91	m	**31.96**
1800 mm high	6.96	0.39	27.73	3.10	11.38	m	**42.21**
Extra over strained wire fencing for concrete straining posts with one strut; posts and struts 610 mm below ground; struts, cleats, stretchers, winders, bolts and eye bolts; excavating holes; filling to within 150 mm of ground level with concrete (1:12) – 40 mm aggregate; replacing topsoil; disposing surplus soil off site							
900 mm high	54.51	0.34	23.89	2.79	125.04	nr	**151.71**
1200 mm high	60.81	0.33	23.77	2.80	134.03	nr	**160.60**
1800 mm high	66.63	0.33	23.77	2.80	145.49	nr	**172.05**
Extra over strained wire fencing for concrete straining posts with two struts; posts and struts 610 mm below ground; excavating holes; filling to within 150 mm of ground level with concrete (1:12) – 40 mm aggregate; replacing topsoil; disposing surplus soil off site							
900 mm high	76.50	0.46	32.44	3.59	181.24	nr	**217.27**
1200 mm high	83.78	0.43	30.30	2.80	193.52	nr	**226.62**
1800 mm high	92.46	0.43	30.30	2.80	217.81	nr	**250.91**
Strained wire fencing; galvanized steel angle posts only at 2750 mm centres; 610 mm below ground; driving in							
900 mm high; 40 × 40 × 5 mm	5.27	0.02	1.08	–	6.06	m	**7.15**
1200 mm high; 40 × 40 × 5 mm	6.61	0.02	1.19	–	7.60	m	**8.79**
1500 mm high; 40 × 40 × 5 mm	5.03	0.03	2.16	–	5.78	m	**7.94**
1800 mm high; 40 × 40 × 5 mm	6.78	0.04	2.60	–	7.80	m	**10.40**

36 FENCING

Item	PC £	Labour hours	Labour £	Plant £	Material £	Unit	**Total rate £**
Galvanized steel straining posts with two struts for strained wire fencing; setting in concrete							
900 mm high; 50 × 50 × 6 mm	67.60	0.50	35.65	–	84.64	nr	**120.29**
1200 mm high; 50 × 50 × 6 mm	37.39	0.50	35.65	–	49.89	nr	**85.54**
1500 mm high; 50 × 50 × 6 mm	88.43	0.50	35.65	–	108.59	nr	**144.24**
1800 mm high; 50 × 50 × 6 mm	109.63	0.50	35.65	–	132.97	nr	**168.62**
Fixing of fencing to posts							
Strained wire; to posts (posts not included); 3 mm galvanized wire; fixing with galvanized stirrups							
900 mm high; 2 wire	0.31	0.02	1.18	–	0.48	m	**1.66**
1200 mm high; 3 wire	0.46	0.02	1.66	–	0.65	m	**2.31**
1400 mm high; 3 wire	0.46	0.02	1.66	–	0.65	m	**2.31**
1800 mm high; 3 wire	0.46	0.02	1.66	–	0.65	m	**2.31**
Barbed wire; to posts (posts not included); 3 mm galvanized wire; fixing with galvanized stirrups							
900 mm high; 2 wire	0.73	0.03	2.37	–	0.96	m	**3.33**
1200 mm high; 3 wire	1.09	0.05	3.32	–	1.38	m	**4.70**
1400 mm high; 3 wire	1.09	0.05	3.32	–	1.38	m	**4.70**
1800 mm high; 3 wire	1.09	0.05	3.32	–	1.38	m	**4.70**
Fencing only to posts and strained wire							
Chain link fencing; to strained wire and posts priced separately; 3 mm galvanized wire; 51 mm mesh; galvanized steel components; fixing to line wires threaded through posts and strained with eye-bolts; posts (not included)							
1800 mm high	9.21	0.05	3.57	–	10.71	m	**14.28**
Chain link fencing; to strained wire and posts priced separately; 3.15 mm plastic coated galvanized wire (wire only 2.5 mm); 51 mm mesh; galvanized steel components; fencing with line wires threaded through posts and strained with eye-bolts, posts (not included) (Note: plastic coated fencing can be cheaper than galvanized finish as wire of a smaller cross-sectional area can be used)							
900 mm high	3.34	0.03	2.38	–	4.09	m	**6.47**
1200 mm high	4.36	0.03	2.38	–	5.26	m	**7.64**
1800 mm high	5.96	0.06	4.46	–	7.85	m	**12.30**

36 FENCING

Item	PC £	Labour hours	Labour £	Plant £	Material £	Unit	Total rate £
STRAINED WIRE FENCING/FIELD FENCING – cont							
Fencing only to posts and strained wire – cont							
Extra over strained wire fencing for cranked arms and galvanized barbed wire							
1 row	2.75	0.01	0.60	–	3.17	m	**3.77**
2 row	3.12	0.03	1.78	–	3.59	m	**5.37**
3 row	3.48	0.03	1.78	–	4.00	m	**5.79**
Field fencing; woven wire mesh; fixed to posts and straining wires measured separately							
cattle fence; 1200 m high; 114 × 300 mm at bottom to 230 × 300 mm at top	1.23	0.05	3.57	–	1.87	m	**5.43**
sheep fence; 800 mm high; 140 × 300 mm at bottom to 230 × 300 mm at top	1.06	0.05	3.57	–	1.67	m	**5.23**
deer fence; 2.0 m high; 89 × 150 mm at bottom to 267 × 300 mm at top	1.19	0.06	4.46	–	1.82	m	**6.27**
Extra for concreting in posts	–	0.25	17.83	–	11.05	nr	**28.88**
Extra for straining post	34.65	0.38	26.74	–	39.84	nr	**66.58**
Boundary fencing; strained wire and wire mesh; Jacksons Fencing							
Tubular chain link fencing; galvanized; plastic coated; 60.3 mm dia. posts at 3.0 m centres; setting 700 mm into ground; choice of 10 mesh colours; including excavating holes, backfilling and removing surplus soil; with top rail only							
900 mm high	–		–	–	–	m	**59.99**
1200 mm high	–		–	–	–	m	**64.70**
1800 mm high	–		–	–	–	m	**71.02**
2000 mm high	–		–	–	–	m	**75.01**
Tubular chain link fencing; galvanized; plastic coated; 60.3 mm dia. posts at 3.0 m centres; cranked arms and three lines barbed wire; setting 700 mm into ground; including excavating holes; backfilling and removing surplus soil; with top rail only							
2000 mm high	–		–	–	–	m	**76.04**
1800 mm high	–		–	–	–	m	**73.89**

36 FENCING

Item	PC £	Labour hours	Labour £	Plant £	Material £	Unit	**Total rate £**
RABBIT NETTING							
The labour in this section is calculated on a 2 person team. The labour time below should be multiplied by 2 to calculate the cost as shown.							
Rabbit netting							
Timber stakes; peeled kiln dried pressure treated; pointed; 1.8 m posts driven 750 mm into ground at 3 m centres (line wires and netting priced separately)							
75–100 mm stakes	6.09	0.13	8.91	–	7.00	m	**15.91**
Corner posts or straining posts 150 mm dia. × 2.4 m high set in concrete; centres to suit local conditions or changes of direction							
1 strut	38.33	0.50	35.65	7.97	60.11	nr	**103.73**
2 strut	42.75	0.50	35.65	7.97	65.19	nr	**108.81**
Strained wire; to posts (posts not included); 3 mm galvanized wire; fixing with galvanized stirrups							
900 mm high; 2 wire	0.31	0.02	1.18	–	0.48	m	**1.66**
1200 mm high; 3 wire	0.46	0.02	1.66	–	0.65	m	**2.31**
Rabbit netting; 31 mm 19 gauge 1050 mm high netting fixed to posts (line wires and straining posts or corner posts all priced separately)							
900 mm high; turned in	0.75	0.02	1.43	–	0.86	m	**2.29**
900 mm high; buried 150 mm in trench	0.75	0.04	2.97	–	0.86	m	**3.83**
TUBULAR FENCING							
Boundary fencing; Steelway-Fensecure Ltd							
Classic two rail tubular fencing; top and bottom with stretcher bars and straining wires in between; comprising 60.3 mm tubular posts at 3.00 m centres; setting in concrete; 35 mm top rail tied with aluminium and steel fittings; 50 × 50 × 355/2.5 mm PVC coated chain link; all components galvanized and coated in green nylon							
964 mm high	30.08		–	–	50.90	m	**50.90**
1269 mm high	32.22		–	–	72.40	m	**72.40**
1574 mm high	33.20		–	–	66.98	m	**66.98**
11878 mm high	33.71		–	–	69.96	m	**69.96**
2188 mm high	34.36		–	–	76.95	m	**76.95**

36 FENCING

Item	PC £	Labour hours	Labour £	Plant £	Material £	Unit	Total rate £
TUBULAR FENCING – cont							
Boundary fencing – cont							
Classic two rail tubular fencing – cont							
2458 mm high	38.39		–	–	90.95	m	**90.95**
2948 mm high	48.28		–	–	97.12	m	**97.12**
3562 mm high	71.85		–	–	82.62	m	**82.62**
End posts; Classic range; 60.3 mm dia.; setting in concrete							
964 mm high	109.58	0.50	35.65	–	143.40	nr	**179.05**
1269 mm high	119.01	0.50	35.65	–	154.24	nr	**189.89**
1574 mm high	133.96	0.60	42.78	–	171.43	nr	**214.21**
1878 mm high	146.21	0.65	46.35	–	185.52	nr	**231.86**
2188 mm high	159.76	0.75	53.48	–	208.16	nr	**261.63**
2458 mm high	181.19	0.80	57.04	–	235.52	nr	**292.56**
2948 mm high	209.85	1.00	71.30	–	273.91	nr	**345.21**
3562 mm high	239.55	1.00	71.30	–	313.49	nr	**384.79**
Corner posts; 60.3 mm dia.; setting in concrete							
964 mm high	30.08	0.50	35.65	–	51.56	nr	**87.21**
1269 mm high	32.22	0.75	53.48	–	54.02	nr	**107.50**
1574 mm high	33.20	0.80	57.04	–	55.15	nr	**112.19**
1878 mm high	33.71	0.90	64.17	–	63.20	nr	**127.37**
2188 mm high	34.36	1.00	71.30	–	63.95	nr	**135.25**
2458 mm high	38.39	1.25	89.13	–	68.59	nr	**157.71**
2948 mm high	48.28	1.00	71.30	–	79.96	nr	**151.26**
3562 mm high	50.29	1.25	89.13	–	82.27	nr	**171.39**
WELDED MESH FENCING							
The labour in this section is calculated on a 2 person team. The labour time below should be multiplied by 2 to calculate the cost as shown.							
Boundary fencing							
Paladin welded mesh colour coated green fencing; fixing to metal posts at 2.975 m centres with manufacturer's fixings; setting 600 mm deep in firm ground; including excavating holes; backfilling and removing surplus excavated material; includes 10 year product guarantee							
1800 mm high	45.41	0.33	23.53	–	52.22	m	**75.75**
2000 mm high	29.24	0.33	23.53	–	45.71	m	**69.24**
2400 mm high	35.59	0.33	23.53	–	54.72	m	**78.25**

36 FENCING

Item	PC £	Labour hours	Labour £	Plant £	Material £	Unit	Total rate £
Extra over welded galvanized plastic coated mesh fencing for concreting in posts	19.22	0.06	3.96	–	22.10	m	**26.07**
SECURITY FENCING							
Security fencing; Jacksons Fencing							
Barbican galvanized steel paling fencing; on 60 × 60 mm posts at 3 m centres; setting in concrete							
1250 mm high	–		–	–	–	m	**77.86**
1500 mm high	–		–	–	–	m	**87.74**
2000 mm high	–		–	–	–	m	**105.64**
2500 mm high	–		–	–	–	m	**136.64**
Gates; to match Barbican galvanized steel paling fencing							
width 1 m	–		–	–	–	nr	**925.10**
width 2 m	–		–	–	–	nr	**988.90**
width 3 m	–		–	–	–	nr	**729.30**
width 4 m	–		–	–	–	nr	**800.80**
width 8 m	–		–	–	–	pair	**1856.80**
width 9 m	–		–	–	–	pair	**2473.90**
width 10 m	–		–	–	–	pair	**2764.30**
The labour in this section is calculated on a 2 person team. The labour time below should be multiplied by 2 to calculate the cost as shown.							
Security fencing; Jacksons Fencing; intruder guards							
Viper Spike Intruder Guards; to existing structures; including fixing bolts							
Viper 1; 40 × 5 mm × 1.1 m long; with base plate	58.80	1.00	71.30	–	67.62	nr	**138.92**
Viper 3; 160 × 190 mm wide; U shape; to prevent intruders climbing pipes	66.50	0.50	35.65	–	76.48	nr	**112.13**

36 FENCING

Item	PC £	Labour hours	Labour £	Plant £	Material £	Unit	Total rate £
SECURITY FENCING – cont							
Security fencing; Jacksons Fencing							
Razor Barb Concertina; spiral wire security barriers; fixing to 600 mm steel ground stakes							
Ref 3275; 450 mm dia. roll; medium barb; galvanized	4.24	0.05	3.57	–	5.16	m	**8.73**
Ref 3276; 730 mm dia. roll; medium barb; galvanized	6.40	0.05	3.57	–	7.65	m	**11.21**
Ref 3277; 950 mm dia. roll; medium barb; galvanized	7.89	0.05	3.57	–	9.36	m	**12.92**
Three lines Barbed Tape; medium barb; on 50 × 50 mm mild steel angle posts; setting in concrete							
Ref 3283; barbed tape; medium barb; galvanized	1.41	0.21	15.02	–	23.00	m	**38.02**
Five lines Barbed Tape; medium barb; on 50 × 50 mm mild steel angle posts; setting in concrete							
Ref 3283; barbed tape; medium barb; galvanized	2.36	0.17	11.89	–	24.08	m	**35.97**
TRIP RAILS							
The labour in this section is calculated on a 2 person team. The labour time below should be multiplied by 2 to calculate the cost as shown.							
Trip rails; Birdsmouth; Jacksons Fencing							
Diamond rail fencing; Birdsmouth; posts 100 × 100 mm softwood planed at 1.35 m centres set in 1:3:6 concrete; rail 75 × 75 mm nominal secured with galvanized straps nailed to posts							
posts 900 mm (600 mm above ground); at 1.35 m centres	15.87	0.25	17.83	0.53	22.85	m	**41.21**
posts 1.20 m (900 mm above ground); at 1.35 m centres	18.31	0.25	17.83	0.53	25.67	m	**44.02**
posts 900 mm (600 mm above ground); at 1.80 m centres	13.56	0.19	13.40	0.40	19.09	m	**32.89**
posts 1.20 m (900 mm above ground); at 1.80 m centres	15.39	0.19	13.40	0.40	21.15	m	**34.95**

36 FENCING

Item	PC £	Labour hours	Labour £	Plant £	Material £	Unit	**Total rate £**
WINDBREAK FENCING							
The labour in this section is calculated on a 2 person team. The labour time below should be multiplied by 2 to calculate the cost as shown.							
Windbreak fencing							
Fencing; English Woodlands; Shade and Shelter Netting windbreak fencing; green; to 100 mm dia. treated softwood posts; setting 450 mm into ground; fixing with 50 × 25 mm treated softwood battens nailed to posts; including excavating and backfilling into firm ground; setting in concrete at 3 m centres							
1200 mm high	2.04	0.08	5.55	–	11.86	m	**17.41**
1800 mm high	3.00	0.08	5.55	–	12.96	m	**18.51**
BALL STOP FENCING							
The labour in this section is calculated on a 2 person team. The labour time below should be multiplied by 2 to calculate the cost as shown.							
Ball stop fencing; Steelway-Fensecure Ltd							
Ball stop net; 30 × 30 mm netting fixed to 60.3 mm dia. 12 mm solid bar lattice; galvanized; dual posts; top, middle and bottom rails							
4.5 m high	123.34	0.33	23.76	–	221.73	m	**245.49**
5 .0 m high	133.16	0.36	25.46	–	233.02	m	**258.49**
6.0 m high	163.16	0.37	26.41	–	267.53	m	**293.94**
7.0 m high	180.66	0.37	26.41	–	297.88	m	**324.29**
8.0 m high	217.33	0.40	28.52	–	329.82	m	**358.34**
9.0 m high	246.46	0.42	29.71	–	372.19	m	**401.90**
10.0 m high	276.72	0.45	32.41	–	409.78	m	**442.18**
Ball stop net; corner posts							
4.5 m high	183.64	1.00	71.30	–	291.08	m	**362.38**
5.0 m high	197.25	1.00	71.30	–	306.73	m	**378.03**
6.0 m high	395.19	1.10	78.43	–	534.36	m	**612.79**
7.0 m high	273.40	1.15	82.00	–	394.30	m	**476.30**
8.0 m high	310.85	1.20	85.56	–	440.69	m	**526.25**
9.0 m high	348.69	1.35	96.26	–	480.89	m	**577.14**
10 m high	386.52	1.38	98.39	–	533.26	m	**631.66**

36 FENCING

Item	PC £	Labour hours	Labour £	Plant £	Material £	Unit	Total rate £
BALL STOP FENCING – cont							
Ball stop fencing – cont							
Ball stop net; end posts							
4.5 m high	149.95	0.50	35.65	–	252.33	m	**287.98**
5.0 m high	166.98	0.56	39.61	–	271.92	m	**311.53**
6.0 m high	204.85	0.57	40.74	–	315.47	m	**356.21**
7.0 m high	242.70	0.63	44.56	–	358.99	m	**403.56**
8.0 m high	280.52	0.67	47.54	–	405.82	m	**453.35**
9.0 m high	318.38	0.71	50.93	–	454.90	m	**505.83**
10 m high	360.01	0.80	57.04	–	505.55	m	**562.59**
RAILINGS							
The labour in this section is calculated on a 2 person team. The labour time below should be multiplied by 2 to calculate the cost as shown.							
Railings; Steelway-Fensecure Ltd							
Mild steel bar railings of balusters at 112 mm centres welded to flat rail top and bottom; bays 2.72 m long; bolting to 50 × 50 mm hollow square section posts; setting in concrete							
galvanized; 900 mm high	50.41	0.20	14.26	–	63.52	m	**77.78**
galvanized; 1200 mm high	61.05	0.21	15.01	–	97.95	m	**112.96**
galvanized; 1500 mm high	67.86	0.25	17.83	–	94.68	m	**112.51**
galvanized; 1800 mm high	81.41	0.27	19.02	–	121.36	m	**140.38**
galvanized & powder coated; 900 mm high	47.51	0.20	14.26	–	60.18	m	**74.44**
galvanized & powder coated; 1200 mm high	57.68	0.21	15.01	–	94.08	m	**109.09**
galvanized & powder coated; 1500 mm high	63.77	0.25	17.83	–	89.98	m	**107.80**
galvanized & powder coated; 1800 mm high	76.00	0.27	19.02	–	115.15	m	**134.16**
Mild steel blunt top railings of 20 mm balusters at 130 mm centres welded to bottom rail; passing through holes in top rail and welded; top and bottom rails 50 × 10 mm; bolting to 70 × 70 mm hollow square section posts; setting in concrete							
galvanized; 800 mm high	54.28	0.20	14.26	–	67.97	m	**82.23**
galvanized; 1000 mm high	57.00	0.20	14.26	–	71.10	m	**85.36**
galvanized; 1300 mm high	67.85	0.21	15.01	–	105.77	m	**120.78**

36 FENCING

Item	PC £	Labour hours	Labour £	Plant £	Material £	Unit	Total rate £
galvanized; 1500 mm high	73.28	0.25	17.83	–	100.92	m	**118.74**
galvanized & powder coated; 800 mm high	61.63	0.20	14.26	–	76.42	m	**90.68**
galvanized & powder coated; 1000 mm high	65.58	0.20	14.26	–	80.96	m	**95.22**
galvanized & powder coated; 1300 mm high	77.66	0.21	15.01	–	117.05	m	**132.06**
galvanized & powder coated; 1500 mm high	84.31	0.25	17.83	–	113.60	m	**131.42**
Railings; traditional pattern; 16 mm dia. verticals at 120 mm intervals with horizontal bars near top and bottom; balusters with spiked tops; 50 × 50 mm standards; including setting 520 mm into concrete at 2.72 m centres							
glavanized & powder coated; 1200 mm high	62.02	0.21	15.01	–	99.06	m	**114.07**
galvanized & powder coated; 1500 mm high	77.80	0.25	17.83	–	106.11	m	**123.94**
galvanized & powder coated; 1800 mm high	93.01	0.27	19.02	–	134.70	m	**153.72**
Interlaced bow-top mild steel railings; traditional park type; 16 mm dia. verticals at 80 mm intervals; welded at bottom to 40 × 10 mm flat and slotted through 38 × 8 mm top rail to form hooped top profile; 50 × 50 mm standards; setting 560 mm into concrete at 2.75 m centres							
galvanized; 900 mm high	44.70	0.20	14.26	–	56.96	m	**71.22**
galvanized; 1200 mm high	50.55	0.21	15.01	–	85.88	m	**100.89**
galvanized; 1500 mm high	69.31	0.25	17.83	–	96.35	m	**114.18**
galvanized; 1800 mm high	94.20	0.27	19.02	–	136.07	m	**155.09**
galvanized & powder coated; 900 mm high	34.03	0.20	14.26	–	44.68	m	**58.94**
galvanized & powder coated; 1200 mm high	45.12	0.21	15.01	–	79.63	m	**94.64**
galvanized & powder coated; 1500 mm high	56.40	0.25	17.83	–	81.50	m	**99.32**
galvanized & powder coated; 1800 mm high	67.68	0.27	19.02	–	105.57	m	**124.59**

36 FENCING

Item	PC £	Labour hours	Labour £	Plant £	Material £	Unit	Total rate £
ESTATE FENCING/HORIZONTAL BAR FENCING							
Metal estate and parkland fencing; Britiannia Metalwork Services Ltd; continously welded, mild steel, 40 × 10 mm uprights manually driven into normal soil to a minimum depth of 600 mm at 1.00 m centres; 40 × 10 mm flat steel supports welded to uprights beneath topsoil at typically every third upright; horizontal rails threaded through holes within uprights; 20 mm solid steel top rail; 4 nr 25 × 8 mm × 16 mm dia. solid steel lower rails; painted fencing shot-blasted, primed and finished with two coats of paint							
Bare metal; 1200 mm high							
10–50 m	–		–	–	–	m	**97.28**
51–100 m	–		–	–	–	m	**82.15**
101–500 m	–		–	–	–	m	**76.71**
above 500 m	–		–	–	–	m	**73.72**
Primed and painted							
10–50 m	–		–	–	–	m	**116.21**
51–100 m	–		–	–	–	m	**98.40**
101–500 m	–		–	–	–	m	**91.98**
above 500 m	–		–	–	–	m	**88.47**
Galvanized and painted							
10–50 m	–		–	–	–	m	**156.88**
51–100 m	–		–	–	–	m	**132.84**
101–500 m	–		–	–	–	m	**124.18**
above 500 m	–		–	–	–	m	**119.43**
End posts and corner posts; 40 × 40 mm; standard hollow section; end or corner posts with capped tops set into concrete							
bare metal	–		–	–	–	nr	**96.79**
primed and painted	–		–	–	–	nr	**126.58**
galvanized and painted	–		–	–	–	nr	**148.91**

36 FENCING

Item	PC £	Labour hours	Labour £	Plant £	Material £	Unit	Total rate £
Gates for metal estate fencing; Britannia Metalwork Services Ltd; fully welded gates to match fence; scroll at hinge end; turnover at latch end; posts topped with finials and set into concrete							
Pedestrian gates; 1.0–1.5 m wide							
bare metal	–		–	–	–	nr	**737.12**
galvanized and painted	–		–	–	–	nr	**1034.98**
Field gates; 1.50–4.0 m wide							
bare metal	–		–	–	–	nr	**1034.98**
galvanized and painted	–		–	–	–	nr	**1332.80**
Kissing gates							
bare metal	–		–	–	–	nr	**819.03**
galvanized and painted	–		–	–	–	nr	**1116.88**
Quadrant; 4 nr uprights with rails welded to fence							
bare metal	–		–	–	–	nr	**416.96**
primed and painted	–		–	–	–	nr	**521.19**
Step stile; 2 treads							
bare metal	–		–	–	–	nr	**223.37**
primed and painted	–		–	–	–	nr	**327.61**
RECYCLED PLASTIC FENCING							
The labour in this section is calculated on a 2 person team. The labour time below should be multiplied by 2 to calculate the cost as shown.							
Recycled plastic; Fencing and Barriers; Hahn Plastics Ltd; Recycled and recyclable; components							
Posts; inclusive of mechanical excavation; 400 × 400 × 400 mm deep; backfilling with site mixed concrete and disposal of arisings							
Hanit Ultra – square post 10 cm × 10 cm × 1.75 m	28.75	0.25	17.80	8.81	55.25	nr	**85.01**
Square post – reinforced 9 cm × 9 cm × 2 m	28.34	0.25	17.80	8.81	54.77	nr	**84.54**
Round post – reinforced 8 cm dia. × 2 m	20.56	0.25	17.79	8.81	45.83	nr	**75.59**
Square post 7 cm × 7 cm × 2 m	11.74	0.25	17.79	8.80	35.70	nr	**65.45**
Round post 8 cm dia. × 2 m; in cocrete inclusive of mechanical excavation and disposal	11.39	0.25	17.79	8.80	35.30	nr	**65.05**
Cross profile stake (wire fence) 7 cm × 7 cm × 2 m; driven in	5.61	0.07	4.76	–	6.45	nr	**11.21**

36 FENCING

Item	PC £	Labour hours	Labour £	Plant £	Material £	Unit	Total rate £
RECYCLED PLASTIC FENCING – cont							
Recycled Plastic fencing; Hahn Plastics Ltd; Hanit fencing							
Fencing							
Woven Privacy Screen – Ribbon 1.50 m wide panels × 1.80 m high inclusive of recycled 9 cm × 9 cm posts set in concrete at 1.545 m centres.	259.92	0.29	20.37	5.88	339.60	m	**367.95**
Picket fencing; Hahn Plastics Ltd; curved top profile; 100% recycled self-coloured hanit® material, non-rot post and panels with 20 year warranty; pre-assembled panels in grey, brown or green with either aluminium or plastic cross-bar with matching posts.hand excacation and disposal off site							
850 mm high with plastic cross bar	136.46	0.20	14.25	–	187.97	m	**204.48**
Pale & Post fencing (palisade): Hahn Plastics Ltd; 100% recycled self-coloured hanit® Ultra material, enhanced wood grain finish, non-rot post and pales with 20 year warranty							
In black or brown woodgrain for fences; two crossrails and recycled 10 cm × 10 cm posts included; Posts at 1.80 m centres							
1.80 m high	82.31	0.40	28.51	–	106.74	m	**136.38**
2.40 m high	125.63	0.40	28.51	–	156.56	m	**186.20**
Privacy fencing: Hahn Plastics Ltd; 100% recycled self-coloured hanit®, non rot post and tongue & grooved panels with 20 year warranty; dedicated brown plastic reinforced fence posts supplied with matching end caps with individual grey, brown or green tongue & groove panels inserted into built in recesses with a central support.							
In Grey Brown or Green; Prices shown for the average priced colour; Posts at 2.00 m centres							
1.95 m high;	186.28	0.20	14.26	–	226.06	m	**241.44**
Corner post or end post	131.04	0.50	35.64	–	174.41	nr	**211.18**

36 FENCING

Item	PC £	Labour hours	Labour £	Plant £	Material £	Unit	Total rate £
Pasture fencing: Hahn Plastics Ltd; 100% recycled self-coloured hanit®, non-rot post and tongue & grooved panels with 20 year warranty; equestrian and agricultural use.							
Pointed reinforced posts driven in.							
1.50 m high 3 rails (inc posts)	55.23	0.15	10.80	–	63.52	m	**74.32**
1.50 m high 5 rails (inc posts)	79.06	0.15	10.80	–	90.92	m	**101.72**
GATES, KISSING GATES AND STILES							
Gates – General							
Preamble: Gates in fences; see specification for fencing as gates in traditional or proprietary fencing systems are usually constructed of the same materials and finished as the fencing itself.							
The labour in this section is calculated on a 2 person team. The labour time below should be multiplied by 2 to calculate the cost as shown.							
Gates; hardwood							
Hardwood field gate; five bar diamond braced; planed iroko; fixed to 150 × 150 mm softwood posts; inclusive of hinges and furniture							
Ref 1100 100; 0.9 m wide	299.00	2.00	142.60	–	512.72	nr	**655.32**
Ref 1100 101; 1.2 m wide	324.00	2.00	142.60	–	541.47	nr	**684.07**
Ref 1100 102; 1.5 m wide	381.00	2.00	142.60	–	607.02	nr	**749.62**
Ref 1100 103; 1.8 m wide	431.00	2.00	142.60	–	664.52	nr	**807.12**
Ref 1100 104; 2.1 m wide	464.00	2.00	142.60	–	702.47	nr	**845.07**
Ref 1100 105; 2.4 m wide	512.00	2.00	142.60	–	757.67	nr	**900.27**
Ref 1100 106; 2.7 m wide	537.00	2.00	142.60	–	786.42	nr	**929.02**
Ref 1100 108; 3.3 m wide	603.50	2.00	142.60	–	862.89	nr	**1005.49**
Gates; treated softwood; Jacksons Fencing							
Timber field gates; including wrought iron ironmongery; five bar type; diamond braced; 1.80 m high; to 200 × 200 mm posts; setting 750 mm into firm ground							
width 2400 mm	135.55	3.00	213.90	–	435.74	nr	**649.64**
width 2700 mm	146.00	3.00	213.90	–	447.75	nr	**661.65**
width 3000 mm	156.55	3.50	249.55	–	459.89	nr	**709.44**
width 3300 mm	167.75	3.75	267.38	–	472.77	nr	**740.14**

36 FENCING

Item	PC £	Labour hours	Labour £	Plant £	Material £	Unit	Total rate £
GATES, KISSING GATES AND STILES – cont							
Gates – cont							
Featherboard garden gates; including ironmongery; to 100 × 120 mm posts; one diagonal brace							
1.0 × 1.2 m high	54.30	1.50	106.95	–	266.73	nr	**373.68**
1.0 × 1.5 m high	71.60	1.50	106.95	–	303.88	nr	**410.83**
1.0 × 1.8 m high	190.00	1.50	106.95	–	330.44	nr	**437.39**
Picket garden gates; including ironmongery; to match picket fence; width 1000 mm; to 100 × 120 mm posts; one diagonal brace							
950 mm high	108.30	1.50	106.95	–	207.28	nr	**314.23**
1200 mm high	113.85	1.50	106.95	–	213.66	nr	**320.61**
1800 mm high	136.85	1.50	106.95	–	269.32	nr	**376.27**
Gates; tubular mild steel; Jacksons Fencing							
Field gates; galvanized; including ironmongery; diamond braced; 1.80 m high; to tubular steel posts; setting in concrete							
width 3000 mm	187.61	2.50	178.25	–	428.24	nr	**606.49**
width 3300 mm	198.05	2.50	178.25	–	440.25	nr	**618.50**
width 3600 mm	208.71	2.50	178.25	–	452.51	nr	**630.76**
width 4200 mm	230.18	2.50	178.25	–	477.20	nr	**655.45**
Gates; sliding; Jacksons Fencing							
Sliding Gate; including all galvanized rails and vertical rail infill panels; special guide and shutting frame posts (Note: foundations installed by suppliers)							
access width 4.00 m; 1.5 m high gates	–		–	–	–	nr	**5625.40**
access width 4.00 m; 2.0 m high gates	–		–	–	–	nr	**5999.40**
access width 4.00 m; 2.5 m high gates	–		–	–	–	nr	**6373.40**
access width 6.00 m; 1.5 m high gates	–		–	–	–	nr	**6946.50**
access width 6.00 m; 2.0 m high gates	–		–	–	–	nr	**7401.90**
access width 6.00 m; 2.5 m high gates	–		–	–	–	nr	**7858.40**
access width 8.00 m; 1.5 m high gates	–		–	–	–	nr	**8203.80**

36 FENCING

Item	PC £	Labour hours	Labour £	Plant £	Material £	Unit	**Total rate £**
access width 8.00 m; 2.0 m high gates	–		–	–	–	nr	**8747.20**
access width 8.00 m; 2.5 m high gates	–		–	–	–	nr	**9288.40**
access width 10.00 m; 2.0 m high gates	–		–	–	–	nr	**10485.20**
access width 10.00 m; 2.5 m high gates	–		–	–	–	nr	**11133.10**
Kissing gates							
Kissing gates; Jacksons Fencing; in galvanized metal bar; fixing to fencing posts (posts not included); 1.65 × 1.30 × 1.00high	494.85	2.50	178.25	–	569.08	nr	**747.33**
Stiles							
Stiles; two posts; setting into firm ground; three rails; two treads	116.00	1.50	106.95	–	162.87	nr	**269.82**
GUARD RAILS							
The labour in this section is calculated on a 2 person team. The labour time below should be multiplied by 2 to calculate the cost as shown.							
Pedestrian guard rails and barriers							
Mild steel pedestrian guard rails; Broxap Street Furniture; 1.00 m high with 150 mm toe space; to posts at 2.00 m centres; galvanized finish							
vertical staggered in-line bar infill panel	62.27	1.00	71.30	–	87.64	m	**158.94**
vertical staggered bar infill with 230 mm visibility gap at top	49.82	1.00	71.30	–	73.32	m	**144.62**

37 SOFT LANDSCAPING

Item	PC £	Labour hours	Labour £	Plant £	Material £	Unit	Total rate £
GENERALLY							
Seeding/turfing – General							
Preamble: The following market prices generally reflect the manufacturer's recommended retail prices. Trade and bulk discounts are often available on the prices shown. The manufacturers of these products generally recommend application rates. Note: the following rates reflect the average rate for each product.							
CLARIFICATION NOTES ON LABOUR COSTS IN THIS SECTION							
General landscape team							
Generally a three man team is used in this section; The column Labour hours reports team hours. The column Labour £ reports the total cost of the team for the unit of work shown							
3 man team	–	1.00	106.95	–	–	hr	**106.95**
MARKET PRICES OF SEEDING MATERIALS							
Market prices of pre-seeding materials							
Rigby Taylor Ltd							
turf fertilizer; Mascot Outfield 16 +6+6	–		–	–	4.24	100 m^2	**4.24**
Boughton Loam							
screened topsoil; 100 mm	–		–	–	53.26	m^3	**53.26**
screened Kettering loam; 3 mm	–		–	–	205.66	m^3	**205.66**
screened Kettering loam; sterilized; 3 mm	–		–	–	209.56	m^3	**209.56**
top dressing; sand soil mixtures; 90/10 to 50/50	–		–	–	155.95	m^3	**155.95**

37 SOFT LANDSCAPING

Item	PC £	Labour hours	Labour £	Plant £	Material £	Unit	Total rate £
Market prices of turf fertilizers; recommended average application rates							
Everris; turf fertilizers							
grass fertilizer; slow release; Sierraform GT Preseeder 18–22–05	–		–	–	6.47	100 m²	**6.47**
grass fertilizer; slow release; Sierraform GT All Season 18–06–18	–		–	–	5.39	100 m²	**5.39**
grass fertilizer; slow release; Sierraform GT Anti-Stress 15-00–26	–		–	–	6.47	100 m²	**6.47**
grass fertilizer; slow release; Sierraform GT Momentum 22–05–11	–		–	–	4.31	100 m²	**4.31**
grass fertilizer; slow release; Sierraform GT Spring Start 16–00–16	–		–	–	5.39	100 m²	**5.39**
grass fertilizer; controlled release; Sierrablen 28–05–05;5–6 months	–		–	–	7.54	100 m²	**7.54**
grass fertilizer; Sierrablen Fine 38–00–00;4–5 months	–		–	–	2.83	100 m²	**2.83**
grass fertilizer; Sierrablen Mini 38–00–00;2–3 months	–		–	–	3.03	100 m²	**3.03**
grass fertilizer; Greenmaster Pro-Lite Turf Tonic 08–00–00	–		–	–	3.25	100 m²	**3.25**
grass fertilizer; Greenmaster Pro-Lite Spring & Summer 14–05–10	–		–	–	4.63	100 m²	**4.63**
grass fertilizer; Greenmaster Pro-Lite Mosskiller 14–00–00	–		–	–	5.27	100 m²	**5.27**
grass fertilizer; Greenmaster Pro-Lite Invigorator 04–00–08	–		–	–	3.51	100 m²	**3.51**
grass fertilizer; Greenmaster Pro-Lite Autumn 06–05–10	–		–	–	4.98	100 m²	**4.98**
grass fertilizer; Greenmaster Pro-Lite Double K 07–00–14	–		–	–	4.99	100 m²	**4.99**
grass fertilizer; Greenmaster Pro-Lite NK 12–00–12	–		–	–	4.98	100 m²	**4.98**
outfield turf fertilizer; Sportsmaster Standard Spring & Summer 09–07–07	–		–	–	4.03	100 m²	**4.03**
outfield turf fertilizer; Sportsmaster Standard Autumn 04–12–12	–		–	–	4.39	100 m²	**4.39**
outfield turf fertilizer; Sportsmaster Standard Zero Phosphate 12–00–09	–		–	–	4.59	100 m²	**4.59**
TPMC; tree planting and mulching compost	–		–	–	3.79	75 l	**3.79**

37 SOFT LANDSCAPING

Item	PC £	Labour hours	Labour £	Plant £	Material £	Unit	Total rate £
MARKET PRICES OF SEEDING MATERIALS – cont							
Market prices of turf fertilizers – cont							
Rigby Taylor Ltd; 35 g/m²							
grass fertilizer; Mascot Microfine 12+0+10 + 2% Mg + 2% Fe	–		–	–	7.09	100 m²	**7.09**
grass fertilizer; Mascot Microfine 8+0+6 + 2% Mg + 4% Fe	–		–	–	6.23	100 m²	**6.23**
grass fertilizer; ApexOC1 8+0+0 + 2% Fe	–		–	–	5.80	100 m²	**5.80**
grass fertilizer; Apex OC2 5+2 +10	–		–	–	5.94	100 m²	**5.94**
grass fertilizer; Mascot Delta Sport 12+4+8 + 0.5% Fe	–		–	–	5.85	100 m²	**5.85**
grass fertilizer; Mascot Fine Turf 12+0+9 + 1% Mg + 1% Fe + seaweed	–		–	–	7.09	100 m²	**7.09**
outfield fertilizer; Mascot Outfield 16+6+6	–		–	–	5.93	100 m²	**5.93**
outfield fertilizer; Mascot Outfield 4+10+10	–		–	–	4.38	100 m²	**4.38**
outfield fertilizer; Mascot Outfield 9+5+5	–		–	–	5.22	100 m²	**5.22**
outfield fertilizer; Mascot Outfield 12+4+4	–		–	–	4.97	100 m²	**4.97**
liquid fertilizer; Mascot Microflow-C 26+0+0 + trace elements; 200–1400 ml/100 m²	–		–	–	8.05	100 m²	**8.05**
liquid fertilizer; Mascot Microflow-CX 6+2+18 + trace elements; 200–1400 ml/100 m²	–		–	–	7.17	100 m²	**7.17**
liquid fertilizer; Mascot Microflow-CX 14+0+7 + trace elements; 200–1400 ml/100 m²	–		–	–	6.43	100 m²	**6.43**
liquid fertilizer; Mascot Microflow-CX 17+2+5 + trace elements; 200–1400 ml/100 m²	–		–	–	7.16	100 m²	**7.16**
Market prices of grass seed							
Preamble: The prices shown are for supply only at one number 20 kg or 25 kg bag purchase price unless otherwise stated. Rates shown are based on the manufacturer's maximum recommendation for each seed type. Trade and bulk discounts are often available on the prices shown for quantities of more than one bag.							

37 SOFT LANDSCAPING

Item	PC £	Labour hours	Labour £	Plant £	Material £	Unit	Total rate £
Bowling greens; fine lawns; ornamental turf; croquet lawns							
Germinal GB; A1 Green; 35 g/m²	–		–	–	33.77	100 m²	**33.77**
DLF Seeds Ltd; J Green; 34–50 g/m²	–		–	–	53.77	100 m²	**53.77**
DLF Seeds Ltd; J Premier Green; 34–50 g/m²	–		–	–	88.26	100 m²	**88.26**
DLF Seeds Ltd; Promaster 20; 35–50 g/m²	–		–	–	26.21	100 m²	**26.21**
DLF Seeds Ltd; Promaster 10; 35–50 g/m²	–		–	–	48.74	100 m²	**48.74**
Tennis courts; cricket squares							
Germinal GB; A2 Lawns; 35 g/m²	–		–	–	25.02	100 m²	**25.02**
Germinal GB; A5 Cricket; 35 g/m²	–		–	–	31.14	100 m²	**31.14**
DLF Seeds Ltd; J Premier Green; 34–50 g/m²	–		–	–	88.26	100 m²	**88.26**
DLF Seeds Ltd; J Intense; 18–25 g/m²	–		–	–	13.54	100 m²	**13.54**
DLF Seeds Ltd; Promaster 35; 35 g/m²	–		–	–	19.90	100 m²	**19.90**
Amenity grassed areas; general purpose lawns							
Germinal GB; A3 Banks; 25–50 g/m²	–		–	–	32.35	100 m²	**32.35**
DLF Seeds Ltd; J Rye Fairway; 18–25 g/m²	–		–	–	14.24	100 m²	**14.24**
DLF Seeds Ltd; J Intense; 18–25 g/m²	–		–	–	13.54	100 m²	**13.54**
DLF Seeds Ltd; Promaster 50; 25–35 g/m²	–		–	–	18.35	100 m²	**18.35**
DLF Seeds Ltd; Promaster 120; 25–35 g/m²	–		–	–	20.32	100 m²	**20.32**
Conservation; country parks; slopes and banks							
Germinal GB; A4 Parkland; 17–35 g/m²	–		–	–	22.46	100 m²	**22.46**
Germinal GB; A16 Country Park; 8–19 g/m²	–		–	–	15.32	100 m²	**15.32**
Germinal GB; A17 Legume; 2 g/m²	–		–	–	28.27	100 m²	**28.27**
Shaded areas							
Germinal GB; A6 Shade; 50 g/m²	–		–	–	45.76	100 m²	**45.76**
Germinal GB, J Green, 34–50 g /m²	–		–	–	53.77	100 m²	**53.77**
DLF Seeds Ltd; Promaster 60; 35–50 g/m²	–		–	–	29.57	100 m²	**29.57**
Sports pitches; rugby; football pitches							
Germinal GB; A7 Sportsground; 20 g/m²	–		–	–	14.40	100 m²	**14.40**
DLF Seeds Ltd; J Pitch; 18–30 g/m²	–		–	–	15.89	100 m²	**15.89**
DLF Seeds Ltd; Promaster 70; 15–35 g/m²	–		–	–	16.72	100 m²	**16.72**

37 SOFT LANDSCAPING

Item	PC £	Labour hours	Labour £	Plant £	Material £	Unit	Total rate £
MARKET PRICES OF SEEDING MATERIALS – cont							
Market prices of grass seed – cont							
Sports pitches – cont							
DLF Seeds Ltd; Promaster 75; 15–35 g/m²	–		–	–	20.47	100 m²	**20.47**
DLF Seeds Ltd; Promaster 80; 17–35 g/m²	–		–	–	17.47	100 m²	**17.47**
Rigby Taylor; Mascot R11 Football & Rugby; 35 g/m²	–		–	–	23.16	100 m²	**23.16**
Rigby Taylor; Mascot R13 General Playing Fields; 35 g/m²	–		–	–	19.84	100 m²	**19.84**
Outfields							
Germinal GB; A7 Sportsground; 20 g/m²	–		–	–	14.40	100 m²	**14.40**
Germinal GB; A9 Outfield; 17–35 g/m²	–		–	–	22.27	100 m²	**22.27**
DLF Seeds Ltd; J Fairway; 12–25 g/m²	–		–	–	14.08	100 m²	**14.08**
DLF Seeds Ltd; Promaster 40; 35 g/m²	–		–	–	19.29	100 m²	**19.29**
DLF Seeds Ltd; Promaster 70; 15–35 g/m²	–		–	–	16.72	100 m²	**16.72**
Rigby Taylor; R4 Species Rich Fairways & Outfields; 35 g/m²	–		–	–	26.24	100 m²	**26.24**
Hockey pitches							
DLF Seeds Ltd; J Fairway; 12–25 g/m²	–		–	–	14.08	100 m²	**14.08**
DLF Seeds Ltd; Promaster 70; 15–35 g/m²	–		–	–	16.72	100 m²	**16.72**
Parks							
Germinal GB; A7 Sportsground 20 g/m²	–		–	–	14.40	100 m²	**14.40**
Germinal GB; A9 Outfield; 17–35 g/m²	–		–	–	22.27	100 m²	**22.27**
DLF Seeds Ltd Promaster 120; 25–35 g/m²	–		–	–	20.32	100 m²	**20.32**
Informal playing fields							
DLF Seeds Ltd; J Rye Fairway; 18–25 g/m²	–		–	–	14.24	100 m²	**14.24**
DLF Seeds Ltd; Promaster 45; 35 g/m²	–		–	–	19.14	100 m²	**19.14**
Caravan sites							
Germinal GB; A9 Outfield; 17–35 g/m²	–		–	–	22.27	100 m²	**22.27**

37 SOFT LANDSCAPING

Item	PC £	Labour hours	Labour £	Plant £	Material £	Unit	**Total rate £**
Sports pitch re-seeding and repair							
Germinal GB; A8 Ultra Fine Greens; 20–35 g/m²	–		–	–	33.64	100 m²	**33.64**
Germinal GB; A20 Ryesport; 20–35 g/m²	–		–	–	22.93	100 m²	**22.93**
DLF Seeds Ltd; J Pitch; 18–30 g/m²	–		–	–	15.89	100 m²	**15.89**
DLF Seeds Ltd; Promaster 80; 17–35 g/m²	–		–	–	17.47	100 m²	**17.47**
DLF Seeds Ltd; Promaster 81; 17–35 g/m²	–		–	–	18.50	100 m²	**18.50**
Rigby Taylor; R14 Sports Pitches 100% premium perennial ryegrass; 35 g/m²	–		–	–	22.79	100 m²	**22.79**
Racecourses; gallops; polo grounds; horse rides							
Germinal GB; Racecourse; 25–30 g/m²	–		–	–	18.11	100 m²	**18.11**
DLF Seeds Ltd; J Intense; 18–25 g/m²	–		–	–	13.54	100 m²	**13.54**
DLF Seeds Ltd; Promaster 65; 17–35 g/m²	–		–	–	20.62	100 m²	**20.62**
Motorway and road verges							
Germinal GB; A18 Road Verge; 6–15 g/m²	–		–	–	11.01	100 m²	**11.01**
DLF Seeds Ltd; Promaster 85; 10 g/m²	–		–	–	5.52	100 m²	**5.52**
DLF Seeds Ltd; Promaster 120; 25–35 g/m²	–		–	–	20.32	100 m²	**20.32**
Golf courses; tees							
Germinal GB; A10 Tees, 35–50 g/m²	–		–	–	33.64	100 m²	**33.64**
DLF Seeds Ltd; J Premier Fairway; 18–30 g/m²	–		–	–	19.92	100 m²	**19.92**
DLF Seeds Ltd; J Intense; 18–25 g/m²	–		–	–	13.54	100 m²	**13.54**
DLF Seeds Ltd; Promaster 40; 35 g/m²	–		–	–	19.29	100 m²	**19.29**
DLF Seeds Ltd; Promaster 45; 35 g/m²			–	–	19.14	100 m²	**19.14**
Golf courses; greens							
Germinal GB; A11 Golf Greens; 35 g/m²	–		–	–	32.22	100 m²	**32.22**
Germinal GB; A13 Roughs; 8 g/m²	–		–	–	4.35	100 m²	**4.35**
DLF Seeds Ltd; J All Bent; 8 g/m²	–		–	–	19.93	100 m²	**19.93**
DLF Seeds Ltd; J Green; 34–50 g/m²	–		–	–	53.77	100 m²	**53.77**
DLF Seeds Ltd; Promaster 5; 35–50 g/m²	–		–	–	35.41	100 m²	**35.41**
Rigby Taylor; R1 Golf & Bowls Fescue/Bent; 35 g/m²	–		–	–	39.82	100 m²	**39.82**

37 SOFT LANDSCAPING

Item	PC £	Labour hours	Labour £	Plant £	Material £	Unit	Total rate £
MARKET PRICES OF SEEDING MATERIALS – cont							
Market prices of grass seed – cont							
Golf courses; fairways							
Germinal GB; A12 Fairways; 15–25 g/m²	–		–	–	14.55	100 m²	**14.55**
DLF Seeds Ltd; J Rye Fairway; 18–30 g/m²	–		–	–	17.09	100 m²	**17.09**
DLF Seeds Ltd; J Premier Fairway; 18–30 g/m²	–		–	–	19.92	100 m²	**19.92**
DLF Seeds Ltd; Promaster 40; 35 g/m²	–		–	–	19.29	100 m²	**19.29**
DLF Seeds Ltd; Promaster 45; 35 g/m²	–		–	–	19.14	100 m²	**19.14**
Rigby Taylor; Mascot R116 Fescue Rye Fairway; 35 g/m²	–		–	–	63.19	100 m²	**63.19**
Golf courses; roughs							
DLF Seeds Ltd; Promaster 25; 17–35 g/m²	–		–	–	22.17	100 m²	**22.17**
Rigby Taylor; Mascot R5 Golf Links & Rough; 35 g/m²	–		–	–	5.25	100 m²	**5.25**
Waste land; spoil heaps; quarries							
Germinal GB; A15 Reclamation; 15–20 g/m²	–		–	–	14.73	100 m²	**14.73**
DLF Seeds Ltd; Promaster 95; 12–35 g/m²	–		–	–	24.94	100 m²	**24.94**
Low maintenance; housing estates; amenity grassed areas							
Germinal GB; A19 Housing; 25–35 g/m²	–		–	–	21.09	100 m²	**21.09**
Germinal GB; A22 Low Maintenance; 25–35 g/m²	–		–	–	26.44	100 m²	**26.44**
DLF Seeds Ltd; Promaster 120; 25–35 g/m²	–		–	–	20.32	100 m²	**20.32**
Saline coastal; roadside areas							
Germinal GB; A21 Saline; 15–20 g/m²	–		–	–	16.79	100 m²	**16.79**
DLF Seeds Ltd; Promaster 90; 15–35 g/m²	–		–	–	21.19	100 m²	**21.19**
Turf production							
Germinal GB; A25 Meadow Ley; 160 kg/ha	–		–	–	1416.56	ha	**1416.56**
Germinal GB; A24 Wear & Tear; 185 kg/ha	–		–	–	1203.63	ha	**1203.63**

37 SOFT LANDSCAPING

Item	PC £	Labour hours	Labour £	Plant £	Material £	Unit	**Total rate £**
Market prices of wild flora seed mixtures							
Acid soils							
Germinal GB; WF1 (Annual Flowering); 1–2 g/m^2	–		–	–	26.89	100 m^2	**26.89**
Neutral soils							
Germinal GB; WF3 (Neutral Soils); 0.5–1 g/m^2	–		–	–	16.76	100 m^2	**16.76**
Market prices of wild flora and grass seed mixtures							
General purpose							
DLF Seeds Ltd; Pro Flora 8 Old English Country Meadow Mix; 5 g/m^2	–		–	–	18.41	100 m^2	**18.41**
DLF Seeds Ltd; Pro Flora 9 General purpose; 5 g/m^2	–		–	–	13.54	100 m^2	**13.54**
Acid soils							
Germinal GB; WFG2 (Annual Meadow); 5 g/m^2	–		–	–	29.78	100 m^2	**29.78**
DLF Seeds Ltd; Pro Flora 2 Acidic soils; 5 g/m^2	–		–	–	15.03	100 m^2	**15.03**
Neutral soils							
Germinal GB; WFG4 (Neutral Meadow); 5 g/m^2	–		–	–	32.17	100 m^2	**32.17**
DLF Seeds Ltd; Pro Flora 3 Damp loamy soils; 5 g/m^2	–		–	–	20.85	100 m^2	**20.85**
Calcareous soils							
Germinal GB; WFG5 (Calcareous Soils); 5 g/m^2	–		–	–	41.15	100 m^2	**41.15**
DLF Seeds Ltd; Pro Flora 4 Calcareous soils; 5 g/m^2	–		–	–	18.36	100 m^2	**18.36**
Heavy clay soils							
Germinal GB; WFG6 (Clay Soils); 5 g/m^2	–		–	–	40.13	100 m^2	**40.13**
DLF Seeds Ltd; Pro Flora 5 Wet loamy soils; 5 g/m^2	–		–	–	33.95	100 m^2	**33.95**
Sandy soils							
Germinal GB; WFG7 (Free Draining Soils); 5 g/m^2	–		–	–	42.78	100 m^2	**42.78**
DLF Seeds Ltd; Pro Flora 6 Dry free draining loamy soils; 5 g/m^2	–		–	–	30.90	100 m^2	**30.90**
Shaded areas							
Germinal GB; WFG8 (Woodland and Hedgerow); 5 g/m^2	–		–	–	45.22	100 m^2	**45.22**
DLF Seeds Ltd; Pro Flora 7 Hedgerow and light shade; 5 g/m^2	–		–	–	42.24	100 m^2	**42.24**

37 SOFT LANDSCAPING

Item	PC £	Labour hours	Labour £	Plant £	Material £	Unit	Total rate £
MARKET PRICES OF SEEDING MATERIALS – cont							
Market prices of wild flora and grass seed mixtures – cont							
Educational							
Wetlands							
Germinal GB; WFG9 (Wetlands and Ponds); 5 g/m²	–		–	–	46.57	100 m²	**46.57**
DLF Seeds Ltd; Pro Flora 5 Wet loamy soils; 5 g/m²	–		–	–	33.95	100 m²	**33.95**
Scrub and moorland							
Germinal GB; WFG10 (Cornfield Annuals); 5 g/m²	–		–	–	54.15	100 m²	**54.15**
Hedgerow							
DLF Seeds Ltd; Pro Flora 7 Hedgerow and light shade; 5 g/m²	–		–	–	42.24	100 m²	**42.24**
Vacant sites							
DLF Seeds Ltd; Pro Flora 1 Cornfield annuals; 5 g/m²	–		–	–	32.49	100 m²	**32.49**
Regional Environmental mixes							
Germinal GB; RE1 (Traditional Hay); 5 g/m²	–		–	–	43.03	100 m²	**43.03**
Germinal GB; RE2 (Lowland Meadow); 5 g/m²	–		–	–	43.74	100 m²	**43.74**
Germinal GB; RE3 (Riverflood Plain/Water Meadow); 5 g/m²	–		–	–	51.18	100 m²	**51.18**
Germinal GB; RE4 (Lowland Limestone); 5 g/m²	–		–	–	42.89	100 m²	**42.89**
Germinal GB; RE8 (Coastal Reclamation); 5 g/m²	–		–	–	57.40	100 m²	**57.40**
Germinal GB; RE9 (Farmland Mixture); 5 g/m²	–		–	–	41.31	100 m²	**41.31**
Germinal GB; RE10 (Marginal Land); 5 g/m²	–		–	–	52.53	100 m²	**52.53**
Germinal GB; RE11 (Heath Scrubland); 5 g/m²	–		–	–	27.70	100 m²	**27.70**
CULTIVATION AND SOIL PREPARATION							
Cultivation by tractor; Agripower Ltd							
Ripping up subsoil; using approved subsoiling machine; minimum depth 250 mm below topsoil; at 1.20 m centres; in							
gravel or sandy clay	–		–	10.39	–	100 m²	**10.39**
soil compacted by machines	–		–	11.54	–	100 m²	**11.54**
clay	–		–	14.85	–	100 m²	**14.85**

37 SOFT LANDSCAPING

Item	PC £	Labour hours	Labour £	Plant £	Material £	Unit	Total rate £
chalk or other soft rock	–		–	17.32	–	100 m²	**17.32**
Extra for subsoiling at 1 m centres	–		–	2.97	–	100 m²	**2.97**
Breaking up existing ground; using tractor drawn Blec ground preparation equipment in one operation							
Single pass							
100 mm deep	–		–	14.70	–	100 m²	**14.70**
150 mm deep	–		–	18.37	–	100 m²	**18.37**
200 mm deep	–		–	21.00	–	100 m²	**21.00**
Cultivating ploughed ground; using disc, drag or chain harrow							
4 passes	–		–	16.70	–	100 m²	**16.70**
Rolling cultivated ground lightly; using self-propelled agricultural roller	–	0.07	2.38	0.92	–	100 m²	**3.29**
Cultivation by pedestrian operated rotavator							
Breaking up existing ground; tine cultivator or rotavator							
100 mm deep	–	0.25	9.80	2.03	–	100 m²	**11.83**
150 mm deep	–	0.29	11.20	2.32	–	100 m²	**13.52**
200 mm deep	–	0.36	14.26	2.95	–	100 m²	**17.21**
Breaking up existing ground; tine cultivator or rotavator but in heavy clay or wet soils							
100 mm deep	–	0.40	15.69	3.24	–	100 m²	**18.93**
150 mm deep	–	0.60	23.53	4.86	–	100 m²	**28.39**
200 mm deep	–	0.75	29.41	6.08	–	100 m²	**35.49**
Importing and storing selected and approved topsoil; inclusive of settlement							
small quantities (less than 15 m³)	54.58		–	–	62.77	m³	**62.77**
over 15 m³	35.51		–	–	40.83	m³	**40.83**
Spreading and lightly consolidating approved topsoil (imported or from spoil heaps); in layers not exceeding 150 mm; travel distance from spoil heaps not exceeding 100 m; by machine (imported topsoil not included)							
minimum depth 100 mm	–	1.55	55.26	73.42	–	100 m²	**128.68**
minimum depth 150 mm	–	2.33	83.19	110.49	–	100 m²	**193.68**
minimum depth 300 mm	–	4.67	166.36	220.98	–	100 m²	**387.34**
minimum depth 450 mm	–	6.99	249.19	331.11	–	100 m²	**580.31**

37 SOFT LANDSCAPING

Item	PC £	Labour hours	Labour £	Plant £	Material £	Unit	Total rate £
CULTIVATION AND SOIL PREPARATION – cont							
Cultivation by pedestrian operated rotavator – cont							
Spreading and lightly consolidating approved topsoil (imported or from spoil heaps); in layers not exceeding 150 mm; travel distance from spoil heaps not exceeding 100 m; by hand (imported topsoil not included)							
minimum depth 100 mm	–	20.00	713.14	–	–	100 m²	**713.14**
minimum depth 150 mm	–	30.01	1069.71	–	–	100 m²	**1069.71**
minimum depth 300 mm	–	60.01	2139.43	–	–	100 m²	**2139.43**
minimum depth 450 mm	–	90.02	3209.14	–	–	100 m²	**3209.14**
Extra over for spreading topsoil to slopes 15–30°; by machine or hand	–		–	–	–	10%	**–**
Extra over for spreading topsoil to slopes over 30°; by machine or hand	–		–	–	–	25%	**–**
Extra over for spreading topsoil from spoil heaps; travel exceeding 100 m; by machine							
100–150 m	–	0.01	0.44	0.10	–	m³	**0.54**
150–200 m	–	0.02	0.66	0.15	–	m³	**0.81**
200–300 m	–	0.03	0.99	0.22	–	m³	**1.21**
Extra over spreading topsoil for travel exceeding 100 m; by hand							
100 m	–	1.20	42.78	–	–	m³	**42.78**
200 m	–	1.67	59.42	–	–	m³	**59.42**
300 m	–	2.50	89.13	–	–	m³	**89.13**
Evenly grading; to general surfaces to bring to finished levels							
by machine (tractor mounted rotavator)	–		–	0.07	–	m²	**0.07**
by pedestrian operated rotavator	–	–	0.15	0.03	–	m²	**0.18**
by hand	–	0.01	0.36	–	–	m²	**0.36**
Extra over grading for slopes 15–30°; by machine or hand	–		–	–	–	10%	**–**
Extra over grading for slopes over 30°; by machine or hand	–		–	–	–	25%	**–**
Apply screened topdressing to grass surfaces; spread using Tru-Lute							
sand soil mixes 90/10 to 50/50	–	–	0.07	0.11	0.27	m²	**0.45**
Spread only existing cultivated soil to final levels using Tru-Lute							
cultivated soil	–	–	0.07	0.11	–	m²	**0.18**

37 SOFT LANDSCAPING

Item	PC £	Labour hours	Labour £	Plant £	Material £	Unit	Total rate £
Clearing stones; disposing off site							
by hand; stones not exceeding 50 mm in any direction; loading to skip 4.6 m^3	–	0.01	0.36	0.05	–	m^2	**0.42**
by mechanical stone rake; stones not exceeding 50 mm in any direction; loading to 15 m^3 truck by mechanical loader	–	–	0.07	0.17	–	m^2	**0.25**
Lightly cultivating; weeding; to fallow areas; disposing debris off site							
by hand	–	0.01	0.50	–	–	m^2	**0.69**
SOIL ADDITIVES AND IMPROVEMENTS							
Surface applications; soil additives; pre-seeding; material delivered to a maximum of 25 m from area of application; applied; by machine							
Soil conditioners; to cultivated ground; mushroom compost; delivered in 25 m^3 loads; including turning in							
1 m^3 per 40 m^2 = 25 mm thick	0.80	0.02	0.59	0.24	0.92	m^2	**1.75**
1 m^3 per 20 m^2 = 50 mm thick	1.59	0.03	1.10	0.35	1.83	m^2	**3.28**
1 m^3 per 13.33 m^2 = 75 mm thick	2.39	0.04	1.43	0.46	2.75	m^2	**4.64**
1 m^3 per 10 m^2 = 100 mm thick	3.19	0.05	1.78	0.58	3.67	m^2	**6.03**
Soil conditioners; to cultivated ground; mushroom compost; delivered in 35 m^3 loads; including turning in							
1 m^3 per 40 m^2 = 25 mm thick	0.80	0.02	0.59	0.24	0.92	m^2	**1.75**
1 m^3 per 20 m^2 = 50 mm thick	1.59	0.03	1.10	0.35	1.83	m^2	**3.28**
1 m^3 per 13.33 m^2 = 75 mm thick	2.39	0.04	1.43	0.46	2.75	m^2	**4.64**
1 m^3 per 10 m^2 = 100 mm thick	3.19	0.05	1.78	0.58	3.67	m^2	**6.03**
Surface applications and soil additives; pre-seeding; material delivered to a maximum of 25 m from area of application; applied; by hand							
Soil conditioners; to cultivated ground; mushroom compost; delivered in 25 m^3 loads; including turning in							
1 m^3 per 40 m^2 = 25 mm thick	0.80	0.02	0.79	–	0.92	m^2	**1.71**
1 m^3 per 20 m^2 = 50 mm thick	1.59	0.04	1.58	–	1.83	m^2	**3.42**
1 m^3 per 13.33 m^2 = 75 mm thick	2.39	0.07	2.37	–	2.75	m^2	**5.13**
1 m^3 per 10 m^2 = 100 mm thick	3.19	0.08	2.85	–	3.67	m^2	**6.52**

37 SOFT LANDSCAPING

Item	PC £	Labour hours	Labour £	Plant £	Material £	Unit	Total rate £
SOIL ADDITIVES AND IMPROVEMENTS – cont							
Surface applications and soil additives – cont							
Soil conditioners; to cultivated ground; mushroom compost; delivered in 60 m^3 loads; including turning in							
1 m^3 per 40 m^2 = 25 mm thick	0.48	0.02	0.79	–	0.55	m^2	**1.34**
1 m^3 per 20 m^2 = 50 mm thick	0.85	0.04	1.58	–	0.98	m^2	**2.56**
1 m^3 per 13.33 m^2 = 75 mm thick	1.44	0.07	2.37	–	1.66	m^2	**4.03**
1 m^3 per 10 m^2 = 100 mm thick	1.92	0.08	2.85	–	2.21	m^2	**5.06**
PREPARATION OF SEED BEDS							
Preparation of seedbeds – General							
Preamble: For preliminary operations see ‘Cultivation’ section.							
Preparation of seedbeds; soil preparation							
Lifting selected and approved topsoil from spoil heaps; passing through 6 mm screen; removing debris	–	0.08	2.97	6.24	–	m^3	**9.26**
Topsoil; supply only; PC £29.59/m^3; Calculations allow for 20% settlement							
25 mm	–		–	–	1.02	m^2	**1.02**
50 mm	–		–	–	2.04	m^2	**2.04**
100 mm	–		–	–	4.08	m^2	**4.08**
150 mm	–		–	–	6.12	m^2	**6.12**
200 mm	–		–	–	8.17	m^2	**8.17**
250 mm	–		–	–	10.21	m^2	**10.21**
300 mm	–		–	–	12.25	m^2	**12.25**
400 mm	–		–	–	16.33	m^2	**16.33**
450 mm	–		–	–	18.37	m^2	**18.37**
Spreading topsoil to form seedbeds (topsoil not included); by machine							
25 mm deep	–	–	0.09	0.19	–	m^2	**0.28**
50 mm deep	–	–	0.12	0.25	–	m^2	**0.37**
75 mm deep	–	–	0.14	0.29	–	m^2	**0.42**
100 mm deep	–	0.01	0.18	0.38	–	m^2	**0.55**
150 mm deep	–	0.01	0.27	0.56	–	m^2	**0.83**

37 SOFT LANDSCAPING

Item	PC £	Labour hours	Labour £	Plant £	Material £	Unit	**Total rate £**
Spreading only topsoil to form seedbeds (topsoil not included); by hand							
25 mm deep	–	0.03	0.89	–	–	m^2	**0.89**
50 mm deep	–	0.03	1.19	–	–	m^2	**1.19**
75 mm deep	–	0.04	1.53	–	–	m^2	**1.53**
100 mm deep	–	0.05	1.78	–	–	m^2	**1.78**
150 mm deep	–	0.08	2.67	–	–	m^2	**2.67**
Bringing existing topsoil to a fine tilth for seeding; by raking or harrowing; stones not to exceed 6 mm; by machine	–		–	0.09	–	m^2	**0.09**
Bringing existing topsoil to a fine tilth for seeding; by raking; stones not to exceed 6 mm; by hand	–	0.01	0.51	–	–	m^2	**0.51**
Preparation of seedbeds; soil treatments							
For the following operations add or subtract the following amounts for every £0.10 difference in the material cost price							
35 g/m²	–		–	–	0.40	100 m^2	**0.40**
50 g/m²	–		–	–	0.58	100 m^2	**0.58**
70 g/m²	–		–	–	0.81	100 m^2	**0.81**
100 g/m²	–		–	–	1.15	100 m^2	**1.15**
125 kg/ha	–		–	–	14.38	ha	**14.38**
150 kg/ha	–		–	–	17.25	ha	**17.25**
175 kg/ha	–		–	–	20.13	ha	**20.13**
200 kg/ha	–		–	–	23.00	ha	**23.00**
225 kg/ha	–		–	–	25.88	ha	**25.88**
250 kg/ha	–		–	–	28.75	ha	**28.75**
300 kg/ha	–		–	–	34.50	ha	**34.50**
350 kg/ha	–		–	–	40.25	ha	**40.25**
400 kg/ha	–		–	–	46.00	ha	**46.00**
500 kg/ha	–		–	–	57.50	ha	**57.50**
700 kg/ha	–		–	–	80.50	ha	**80.50**
1000 kg/ha	–		–	–	115.00	ha	**115.00**
1250 kg/ha	–		–	–	143.75	ha	**143.75**
Pre-seeding fertilizers (12+00+09); PC £3.06/kg; to seedbeds; by machine							
35 g/m²	3.06		–	0.88	3.52	100 m^2	**4.40**
50 g/m²	4.37		–	0.88	5.03	100 m^2	**5.91**
70 g/m²	6.12		–	0.88	7.04	100 m^2	**7.92**
100 g/m²	8.75		–	0.88	10.06	100 m^2	**10.94**
Pre-seeding fertilizers (12+00+09); PC £3.06/kg; to seedbeds; by hand							
35 g/m²	3.06	0.17	5.94	–	3.52	100 m^2	**9.46**
50 g/m²	4.37	0.17	5.94	–	5.03	100 m^2	**10.97**
70 g/m²	6.12	0.17	5.94	–	7.04	100 m^2	**12.98**
100 g/m²	8.75	0.20	7.13	–	10.06	100 m^2	**17.19**

37 SOFT LANDSCAPING

Item	PC £	Labour hours	Labour £	Plant £	Material £	Unit	Total rate £
SEEDING							
Seeding labours only in two operations; by machine (for seed prices see above)							
35 g/m²	–		–	1.50	–	100 m²	**1.50**
Grass seed; spreading in two operations; PC £5.00,90/kg (for changes in material prices please refer to table above); by machine							
35 g/m²	–		–	1.50	23.95	100 m²	**25.45**
50 g/m²	–		–	1.50	34.21	100 m²	**35.71**
70 g/m²	–		–	1.50	47.90	100 m²	**49.40**
100 g/m²	–		–	1.50	68.43	100 m²	**69.93**
125 kg/ha	–		–	149.60	855.31	ha	**1004.91**
150 kg/ha	–		–	149.60	1026.38	ha	**1175.97**
200 kg/ha	–		–	149.60	1368.50	ha	**1518.10**
250 kg/ha	–		–	149.60	1710.63	ha	**1860.22**
300 kg/ha	–		–	149.60	2052.75	ha	**2202.35**
350 kg/ha	–		–	149.60	2394.88	ha	**2544.47**
400 kg/ha	–		–	149.60	2737.00	ha	**2886.60**
500 kg/ha	–		–	149.60	3421.25	ha	**3570.85**
700 kg/ha	–		–	149.60	4789.75	ha	**4939.35**
Extra over seeding by machine for slopes over 30° (allowing for the actual area but measured in plan)							
35 g/m²	–		–	0.23	3.59	100 m²	**3.82**
50 g/m²	–		–	0.23	5.13	100 m²	**5.36**
70 g/m²	–		–	0.23	7.18	100 m²	**7.41**
100 g/m²	–		–	0.23	10.26	100 m²	**10.49**
125 kg/ha	–		–	22.44	128.30	ha	**150.74**
150 kg/ha	–		–	22.44	153.96	ha	**176.40**
200 kg/ha	–		–	22.44	205.28	ha	**227.71**
250 kg/ha	–		–	22.44	256.59	ha	**279.03**
300 kg/ha	–		–	22.44	307.91	ha	**330.35**
350 kg/ha	–		–	22.44	359.23	ha	**381.67**
400 kg/ha	–		–	22.44	410.55	ha	**432.99**
500 kg/ha	–		–	22.44	513.19	ha	**535.63**
700 kg/ha	–		–	22.44	718.46	ha	**740.90**
Seeding labours only in two operations; by hand (for seed prices see above)							
35 g/m²	–	0.17	5.94	–	–	100 m²	**5.94**
Grass seed; spreading in two operations; PC £6.44/kg (for changes in material prices please refer to table above); by hand							
35 g/m²	–	0.17	5.94	–	23.95	100 m²	**29.89**
50 g/m²	–	0.17	5.94	–	34.21	100 m²	**40.16**
70 g/m²	–	0.17	5.94	–	47.90	100 m²	**53.84**
100 g/m²	–	0.20	7.13	–	68.43	100 m²	**75.56**
125 g/m²	–	0.20	7.13	–	85.53	100 m²	**92.66**

37 SOFT LANDSCAPING

Item	PC £	Labour hours	Labour £	Plant £	Material £	Unit	Total rate £
Extra over seeding by hand for slopes over 30° (allowing for the actual area but measured in plan)							
35 g/m²	3.09	–	0.13	–	3.56	100 m²	**3.69**
50 g/m²	4.46	–	0.13	–	5.13	100 m²	**5.26**
70 g/m²	6.25	–	0.13	–	7.18	100 m²	**7.31**
100 g/m²	8.93	–	0.15	–	10.26	100 m²	**10.42**
125 g/m²	11.13	–	0.15	–	12.80	100 m²	**12.95**
Harrowing seeded areas; light chain harrow	–		–	0.45	–	100 m²	**0.45**
Raking over seeded areas							
by mechanical stone rake	–		–	6.41	–	100 m²	**6.41**
by hand	–	0.80	28.52	–	–	100 m²	**28.52**
Rolling seeded areas; light roller							
by tractor drawn roller	–		–	2.74	–	100 m²	**2.74**
by pedestrian operated mechanical roller	–	0.08	2.97	0.57	–	100 m²	**3.54**
by hand drawn roller	–	0.17	5.94	–	–	100 m²	**5.94**
Extra over harrowing, raking or rolling seeded areas for slopes over 30°; by machine or hand	–		–	–	–	25%	**–**
Turf edging; to seeded areas; 300 mm wide	–	0.05	1.70	–	3.99	m²	**5.68**
PREPARATION OF TURF BEDS							
Preparation of turf beds							
Rolling turf to be lifted; lifting by hand or mechanical turf stripper; stacks to be not more than 1 m high							
cutting only preparing to lift; pedestrian turf cutter	–	0.75	26.74	17.44	–	100 m²	**44.17**
lifting and stacking; by hand	–	8.33	297.08	–	–	100 m²	**297.08**
Rolling up; moving to stacks							
distance not exceeding 100 m	–	2.50	89.13	–	–	100 m²	**89.13**
extra over rolling and moving turf to stacks to transport per additional 100 m	–	0.83	29.71	–	–	100 m²	**29.71**
Lifting selected and approved topsoil from spoil heaps							
passing through 6 mm screen; removing debris	–	0.17	5.94	12.48	–	m³	**18.43**
Extra over lifting topsoil and passing through screen for imported topsoil; plus 20% allowance for settlement	–		–	–	40.83	m³	**40.83**
Topsoil; PC £29.59/m³; plus 20% allowance for settlement							
25 mm deep	–		–	–	1.02	m²	**1.02**
50 mm deep	–		–	–	2.04	m²	**2.04**
100 mm deep	–		–	–	4.08	m²	**4.08**

37 SOFT LANDSCAPING

Item	PC £	Labour hours	Labour £	Plant £	Material £	Unit	Total rate £
PREPARATION OF TURF BEDS – cont							
Preparation of turf beds – cont							
Topsoil – cont							
150 mm deep	–		–	–	6.12	m²	**6.12**
200 mm deep	–		–	–	8.17	m²	**8.17**
250 mm deep	–		–	–	10.21	m²	**10.21**
300 mm deep	–		–	–	12.25	m²	**12.25**
400 mm deep	–		–	–	16.33	m²	**16.33**
450 mm deep	–		–	–	18.37	m²	**18.37**
Spreading topsoil to form turf beds (topsoil not included); by machine							
25 mm deep	–		–	0.32	–	m²	**0.32**
50 mm deep	–		–	0.42	–	m²	**0.42**
75 mm deep	–		–	0.48	–	m²	**0.48**
100 mm deep	–		–	0.64	–	m²	**0.64**
150 mm deep	–		–	0.96	–	m²	**0.96**
Spreading topsoil to form turf beds (topsoil not included); by hand							
25 mm deep	–	0.03	0.89	–	–	m²	**0.89**
50 mm deep	–	0.03	1.19	–	–	m²	**1.19**
75 mm deep	–	0.04	1.53	–	–	m²	**1.53**
100 mm deep	–	0.05	1.78	–	–	m²	**1.78**
150 mm deep	–	0.08	2.67	–	–	m²	**2.67**
Cultivation by pedestrian operated rotavator							
Breaking up existing ground; tine cultivator or rotavator							
100 mm deep	–	0.25	9.80	2.03	–	100 m²	**11.83**
150 mm deep	–	0.29	11.20	2.32	–	100 m²	**13.52**
200 mm deep	–	0.36	14.26	2.95	–	100 m²	**17.21**
Breaking up existing ground; tine cultivator or rotavator in heavy clay or wet soils							
100 mm deep	–	0.40	15.69	3.24	–	100 m²	**18.93**
150 mm deep	–	0.60	23.53	4.86	–	100 m²	**28.39**
200 mm deep	–	0.75	29.41	6.08	–	100 m²	**35.49**
Bringing existing topsoil to a fine tilth for turfing by raking or harrowing; stones not to exceed 6 mm; by machine	–	–	0.14	0.09	–	m²	**0.24**
Bringing existing topsoil to a fine tilth for turfing by raking; stones not to exceed 6 mm; by hand							
general commercial or amenity turf	–	0.01	0.45	–	–	m²	**0.45**
higher grade fine culivation	–	0.02	0.71	–	–	m²	**0.71**

37 SOFT LANDSCAPING

Item	PC £	Labour hours	Labour £	Plant £	Material £	Unit	Total rate £
TURFING							
Turfing (see also Preparation of Turf beds – above)							
Hand load and transport and unload turf from delivery area to laying area							
by small site dumper up to 50 m	–	0.01	0.72	0.05	–	m²	**0.77**
by hand 25 m	–	0.03	0.89	–	–	m²	**0.89**
by hand 50 m	–	0.03	1.02	–	–	m²	**1.02**
Turfing; laying only; to stretcher bond; butt joints; including providing and working from barrow plank runs where necessary to surfaces not exceeding 30° from horizontal							
specially selected lawn turves from previously lifted stockpile	–	0.08	2.67	–	–	m²	**2.67**
cultivated lawn turves; to large open areas	–	0.06	2.07	–	–	m²	**2.07**
cultivated lawn turves; to domestic or garden areas	–	0.08	2.77	–	–	m²	**2.77**
road verge quality turf	–	0.04	1.42	–	–	m²	**1.42**
Industrially grown turf; PC prices listed represent the general range of industrial turf prices for sportsfields and amenity purposes; prices will vary with quantity and site location							
Rolawn							
RB Medallion; sports fields, domestic lawns, general landscape; full loads 1720 m²	3.09	0.05	5.70	–	3.55	m²	**9.25**
RB Medallion; sports fields, domestic lawns, general landscape; part loads	3.47	0.05	5.70	–	3.99	m²	**9.69**
Tensar Ltd							
Tensar Mat 400 reinforced turf for embankments; laid to embankments 3.3 m² turves	5.03	0.10	3.92	–	5.78	m²	**9.70**
Inturf							
Inturf Masters; formal lawns, golf greens, bowling greens and low maintenance areas	7.75	0.05	5.70	–	8.91	m²	**14.61**
Inturf Classic; golf tees, surrounds, lawns, parks, general purpose landscaping areas and winter sports	7.00	0.05	5.70	–	8.05	m²	**13.75**
Turf laid in Big roll; Inturf							
Inturf Classic; golf tees, surrounds, lawns, parks, general purpose landscaping areas and winter sports	5.75	0.01	0.86	–	8.80	m²	**9.65**

37 SOFT LANDSCAPING

Item	PC £	Labour hours	Labour £	Plant £	Material £	Unit	Total rate £
TURFING – cont							
Specialized turf products; Lindum Turf Ltd; Laid to prepared and cultivated soil (not included); Wildflower Meadow Turf; British wildflowers and grasses 50:50; P							
Flowering height 40–60 cm	24.60	0.08	2.97	–	29.21	m²	**32.18**
Instant use turf; LT6 Lindum turf; grown in an imported sand/soil rootzone, harvested up to 40 mm thickness and installed using specialist machinery. Compatibility with existing rootzones for sports pitches or high wear areas where instant use is required inclusive of removal of existing rootzone;							
minimum area 150 m²	24.00		–	–	53.82	m²	**53.82**
Reinforced turf; ABG ATS Advanced Turf; blended mesh elements incorporated into root zone; root zone spread and levelled over cultivated, prepared and reduced and levelled ground (not included); compacted with vibratory roller							
ABG ATS 400/B with selected turf and fertilizer 100 mm thick							
100–500 m²	30.87	0.03	1.19	3.13	35.50	m²	**39.81**
over 500 m²	29.24	0.03	1.19	3.13	33.63	m²	**37.94**
ABG ATS Turf 400/B with selected turf and fertilizer 150 mm thick							
100–500 m²	36.82	0.04	1.43	3.58	42.35	m²	**47.35**
over 500 m²	34.66	0.04	1.43	3.58	39.85	m²	**44.86**
ABG ATS Turf 400/B with selected turf and fertilizer 200 mm thick							
100–500 m²	43.32	0.05	1.78	4.25	49.82	m²	**55.85**
over 500 m²	41.15	0.05	1.78	4.25	47.33	m²	**53.36**
Firming turves with wooden beater	–	0.01	0.36	–	–	m²	**0.36**
Dressing with finely sifted topsoil; brushing into joints	0.03	0.05	1.79	–	0.03	m²	**1.82**
Turfing; laying only							
to slopes over 30°; to diagonal bond (measured as plan area – add 15% to these rates for the incline area of 30 degree slopes)	–	0.12	4.28	–	–	m²	**4.28**
Extra over laying turfing for pegging down turves							
wooden or galvanized wire pegs; 200 mm long; 2 pegs per 0.50 m²	1.87	0.01	0.47	–	2.15	m²	**2.63**

37 SOFT LANDSCAPING

Item	PC £	Labour hours	Labour £	Plant £	Material £	Unit	**Total rate £**
ARTIFICIAL TURF							
The labour in this section is calculated on a 2 person team. The labour time below should be multiplied by 2 to calculate the cost as shown.							
Artificial grass; Easigrass Limited							
base for area to recive artificial lawn							
Reduce area by 80 mm; lay 50 mm Type 1 and 50 mm sharp sand; remove excavated material from site	–	0.05	3.51	–	3.25	m²	**10.84**
Artificial lawns laid to cleared and levelled area; laid to 50 mm Type 1 drainage layer on 25 mm sharp sand bed and geofabric							
Lawn replacement areas							
Easi-Mayfair; 50 mm; deep pile;	–		–	–	–	m²	**42.72**
Easi-Belgravia; 40 mm; deep pile; realistic lawn	–		–	–	–	m²	**40.28**
Easi-Chelsea 40 mm	–		–	–	–	m²	**36.62**
Multi-use							
Easi-Kensington 35 mm deep pile	–		–	–	–	m²	**32.94**
Schools and Leisure							
Easi-Holland Park; 30 mm deep pile	–		–	–	–	m²	**32.09**
Patios, balconies terraces and golf; lighter loading solutions							
Easi-Wentworth; 15 mm pile height; 3.15 kg/m²	–		–	–	–	m²	**39.05**
Easi-Knightsbridge; 30 mm deep pile; 2.94 kg /m²	–		–	–	–	m²	**31.72**
WILD FLOWER MEADOWS							
Note for this section; Prices exclude for delivery and collection of specialized machinery where used to the site							
Topsoil stripping for wild flower planting							
Stripping to subsoil layer 300 mm deep							
moving to stockpile	–		–	2.79	–	m²	**2.79**
spreading locally	–		–	3.80	–	m²	**3.80**

37 SOFT LANDSCAPING

Item	PC £	Labour hours	Labour £	Plant £	Material £	Unit	Total rate £
WILD FLOWER MEADOWS – cont							
Soil testing for wildflower meadows							
Collection of samples and submission to soil laboratory for analysis and recommendation for wildflower meadow establishment							
3 nr random samples	–		–	–	–	nr	**953.04**
Cultivation and preparation of seedbeds for wildflower meadows							
Cultivate stripped, filled or existing surface prior to application of herbicides and leave fallow for seedbank eradication period;							
by tractor drawn implements including rough grading to medium tilth	–		–	1.47	–	100 m²	**1.47**
by hand drawn rotavator including hand raking, and hand grading, to medium tilth	–	0.70	25.67	1.62	–	100 m²	**27.29**
Seedbank eradication for wildflower meadows							
Treat stripped intended meadow area to eradicate inherent seedbank; allow optimum 2 years of repeat eradication by herbicide returning regularly to reapply herbicides							
Existing or reclaimed brownfield area by tractor drawn application; 4 applications	–	0.04	1.43	2.19	0.39	100 m²	**4.01**
Existing or reclaimed brownfield area by tractor drawn application; 3 applications	–	0.03	1.07	1.64	0.29	100 m²	**3.00**
Existing paddocks or agricultural area; knapsack; 4 applications	–	2.12	75.58	–	0.39	100 m²	**75.97**
Existing or reclaimed brownfield area by backpack spray application; 3 applications	–	1.59	56.68	–	0.29	100 m²	**56.97**
Subsoil filling to wildflower meadows							
Average thickness 300 mm; filling of poor nutrient value soil to form seedbed for wildflower area; grading to level and preparing to medium tilth							
from stockpile on site	–		–	322.04	–	100 m²	**322.04**
imported subsoil fill; tipped directly	1065.15	2.00	71.30	134.12	1224.93	100 m²	**1430.35**

37 SOFT LANDSCAPING

Item	PC £	Labour hours	Labour £	Plant £	Material £	Unit	Total rate £
Seeding to wildflower meadows; Spreading of selected wildflower seed to manufacturers specifications; by tractor drawn seeder							
General areas							
DLF Seeds Ltd; Pro Flora 8 Old English Country Meadow Mix; 5 g/m²	–		–	4.38	21.17	100 m²	**25.56**
Acid soils							
Germinal GB; WFG2 (Annual Meadow); 5 g/m²	–		–	4.38	34.25	100 m²	**38.63**
Neutral soils							
Germinal GB; WFG4 (Neutral Meadow); 5 g/m²	–		–	4.38	36.99	100 m²	**41.37**
DLF Seeds Ltd; Pro Flora 3 Damp loamy soils; 5 g/m²	–		–	4.38	23.97	100 m²	**28.36**
Heavy clay soils							
Germinal GB; WFG6 (Clay Soils); 5 g/m²	–		–	4.38	46.14	100 m²	**50.53**
Seeding to wildflower meadows; Spreading of selected wildflower seed to manufacturers specifications; by pedestrian drawn seeder							
General areas							
DLF Seeds Ltd; Pro Flora 8 Old English Country Meadow Mix; 5 g/m²	–	0.25	8.91	–	21.17	100 m²	**30.09**
Acid soils							
Germinal GB; WFG2 (Annual Meadow); 5 g/m²	–	0.25	8.91	–	34.25	100 m²	**43.16**
Neutral soils							
Germinal GB; WFG4 (Neutral Meadow); 5 g/m²	–	0.25	8.91	–	36.99	100 m²	**45.90**
DLF Seeds Ltd; Pro Flora 3 Damp loamy soils; 5 g/m²	–	0.25	8.91	–	23.97	100 m²	**32.89**
Heavy clay soils							
Germinal GB; WFG6 (Clay Soils); 5 g/m²		0.25	8.91	–	46.14	100 m²	**55.06**
Wildflower Turf; Lindum Turf							
Wildflower Meadow Turf; British wildflowers and grasses 50:50; P							
Flowering height 40–60 cm	24.60	0.08	2.97	–	29.21	m²	**32.18**

37 SOFT LANDSCAPING

Item	PC £	Labour hours	Labour £	Plant £	Material £	Unit	Total rate £
MAINTENANCE OF SEEDED/ TURFED AREAS							
Maintenance operations (Note: the following rates apply to aftercare maintenance executed as part of a landscaping contract only)							
Initial cutting; to turfed areas							
20 mm high; using pedestrian guided power driven cylinder mower; including boxing off cuttings (stone picking and rolling not included)	–	0.18	6.42	0.46	–	100 m²	**6.88**
Repairing damaged grass areas							
scraping out; removing slurry; from ruts and holes; average 100 mm deep	–	0.13	4.75	–	–	m²	**4.75**
100 mm topsoil	–	0.13	4.75	–	4.08	m²	**8.84**
Repairing damaged grass areas; sowing grass seed to match existing or as specified; to individually prepared worn patches							
35 g/m²	0.24	0.01	0.36	–	0.28	m²	**0.63**
50 g/m²	0.34	0.01	0.36	–	0.39	m²	**0.75**
Sweeping leaves; disposing off site; motorized vacuum sweeper or rotary brush sweeper							
areas of maximum 2500 m² with occasional large tree and established boundary planting; 4.6 m³ (1 skip of material to be removed)	–	0.40	14.26	4.91	–	100 m²	**19.17**
Leaf clearance; clearing grassed area of leaves and other extraneous debris							
Using equipment towed by tractor							
large grassed areas with perimeters of mature trees such as sports fields and amenity areas	–	0.01	0.45	0.05	–	100 m²	**0.49**
large grassed areas containing ornamental trees and shrub beds	–	0.03	0.89	0.06	–	100 m²	**0.96**
Using pedestrian operated mechanical equipment and blowers							
grassed areas with perimeters of mature trees such as sports fields and amenity areas	–	0.04	1.43	0.03	–	100 m²	**1.46**
grassed areas containing ornamental trees and shrub beds	–	0.10	3.57	0.08	–	100 m²	**3.65**
verges	–	0.07	2.38	0.06	–	100 m²	**2.44**

37 SOFT LANDSCAPING

Item	PC £	Labour hours	Labour £	Plant £	Material £	Unit	Total rate £
By hand							
grassed areas with perimeters of mature trees such as sports fields and amenity areas	–	0.05	1.78	0.22	–	100 m²	**2.00**
grassed areas containing ornamental trees and shrub beds	–	0.08	2.97	0.36	–	100 m²	**3.33**
verges	–	1.00	35.65	4.36	–	100 m²	**40.01**
Removal of arisings							
areas with perimeters of mature trees	–	0.01	0.19	0.12	–	100 m²	**3.18**
areas containing ornamental trees and shrub beds	–	0.02	0.56	0.47	–	100 m²	**8.21**
Cutting grass to specified height; per cut							
multi unit gang mower	–	0.59	20.97	24.10	–	ha	**45.06**
ride-on triple cylinder mower	–	0.01	0.50	0.18	–	100 m²	**0.68**
ride-on triple rotary mower	–	0.01	0.50	–	–	100 m²	**0.50**
pedestrian mower	–	0.18	6.42	1.09	–	100 m²	**7.51**
Cutting grass to banks; per cut							
side arm cutter bar mower	–	0.02	0.83	0.45	–	100 m²	**1.28**
Cutting rough grass; per cut							
power flail or scythe cutter	–	0.04	1.25	–	–	100 m²	**1.25**
Extra over cutting grass for slopes not exceeding 30°	–		–	–	–	10%	**–**
Extra over cutting grass for slopes exceeding 30°	–		–	–	–	40%	**–**
Cutting fine sward							
pedestrian operated seven-blade cylinder lawn mower	–	0.14	4.99	0.35	–	100 m²	**5.34**
Extra over cutting fine sward for boxing off cuttings							
pedestrian mower	–	0.14	1.00	0.07	–	100 m²	**1.07**
Cutting areas of rough grass							
scythe	–	1.00	35.65	–	–	100 m²	**35.65**
sickle	–	2.00	71.30	–	–	100 m²	**71.30**
petrol operated strimmer	–	0.30	10.71	0.75	–	100 m²	**11.45**
Cutting areas of rough grass which contain trees or whips							
petrol operated strimmer	–	0.40	14.26	1.00	–	100 m²	**15.26**
Extra over cutting rough grass for on site raking up and dumping	–	0.33	11.88	–	–	100 m²	**11.88**
Trimming edge of grass areas; edging tool							
with petrol powered strimmer	–	0.13	4.75	0.33	–	100 m	**5.08**
by hand	–	0.67	23.77	–	–	100 m	**23.77**

37 SOFT LANDSCAPING

Item	PC £	Labour hours	Labour £	Plant £	Material £	Unit	Total rate £
MAINTENANCE OF SEEDED/ TURFED AREAS – cont							
Marking out pitches using approved line marking compound; including initial setting out and marking							
discus, hammer, javelin or shot putt area	5.67	2.00	71.30	–	6.52	nr	**77.82**
cricket square	3.78	2.00	71.30	–	4.35	nr	**75.65**
cricket boundary	13.24	8.00	285.20	–	15.22	nr	**300.42**
grass tennis court	5.67	4.00	142.60	–	6.52	nr	**149.12**
hockey pitch	18.91	8.00	285.20	–	21.75	nr	**306.95**
football pitch	18.91	8.00	285.20	–	21.75	nr	**306.95**
rugby pitch	18.91	8.00	285.20	–	21.75	nr	**306.95**
eight lane running track; 400 m	37.82	16.00	570.40	–	43.49	nr	**613.89**
Re-marking out pitches using approved line marking compound							
discus, hammer, javelin or shot putt area	3.78	0.50	17.83	–	4.35	nr	**22.17**
cricket square	3.78	0.50	17.83	–	4.35	nr	**22.17**
grass tennis court	13.24	1.00	35.65	–	15.22	nr	**50.87**
hockey pitch	13.24	1.00	35.65	–	15.22	nr	**50.87**
football pitch	13.24	1.00	35.65	–	15.22	nr	**50.87**
rugby pitch	13.24	1.00	35.65	–	15.22	nr	**50.87**
eight lane running track; 400 m	37.82	2.50	89.13	–	43.49	nr	**132.62**
Rolling grass areas; light roller							
by tractor drawn roller	–		–	0.69	–	100 m²	**0.69**
by pedestrian operated mechanical roller	–	0.08	2.97	0.57	–	100 m²	**3.54**
by hand drawn roller	–	0.17	5.94	–	–	100 m²	**5.94**
Aerating grass areas; to a depth of 100 mm							
using tractor-drawn aerator	–	0.06	2.08	3.62	–	100 m²	**5.70**
using pedestrian-guided motor powered solid or slitting tine turf aerator	–	0.18	6.24	2.80	–	100 m²	**9.04**
using hollow tine aerator; including sweeping up and dumping corings	–	0.50	17.83	5.60	–	100 m²	**23.43**
using hand aerator or fork	–	1.67	59.42	–	–	100 m²	**59.42**
Extra over aerating grass areas for on site sweeping up and dumping corings	–	0.17	5.94	–	–	100 m²	**5.94**
Switching off dew; from fine turf areas	–	0.20	7.13	–	–	100 m²	**7.13**
Scarifying grass areas to break up thatch; removing dead grass							
using tractor-drawn scarifier	–	0.07	2.50	2.37	–	100 m²	**4.86**
using self-propelled scarifier; including removing and disposing of grass on site	–	0.33	11.88	0.83	–	100 m²	**12.72**

37 SOFT LANDSCAPING

Item	PC £	Labour hours	Labour £	Plant £	Material £	Unit	Total rate £
Harrowing grass areas							
using drag mat	–	0.03	1.00	0.38	–	100 m²	**1.38**
using chain harrow	–	0.04	1.25	0.48	–	100 m²	**1.73**
using drag mat	–	2.80	99.83	38.44	–	ha	**138.27**
using chain harrow	–	3.50	124.78	48.05	–	ha	**172.83**
Extra for scarifying and harrowing grass areas for disposing excavated material off site; to tip; loading by machine							
slightly contaminated	–		–	3.76	–	m³	**48.59**
rubbish	–		–	3.76	–	m³	**48.59**
inert material	–		–	2.50	–	m³	**40.04**
For the following topsoil improvement and seeding operations add or subtract the following amounts for every £0.10 difference in the material cost price							
35 g/m²	–		–	–	0.40	100 m²	**0.40**
50 g/m²	–		–	–	0.58	100 m²	**0.58**
70 g/m²	–		–	–	0.81	100 m²	**0.81**
100 g/m²	–		–	–	1.15	100 m²	**1.15**
125 kg/ha	–		–	–	14.38	ha	**14.38**
150 kg/ha	–		–	–	17.25	ha	**17.25**
175 kg/ha	–		–	–	20.13	ha	**20.13**
200 kg/ha	–		–	–	23.00	ha	**23.00**
225 kg/ha	–		–	–	25.88	ha	**25.88**
250 kg/ha	–		–	–	28.75	ha	**28.75**
300 kg/ha	–		–	–	34.50	ha	**34.50**
350 kg/ ha	–		–	–	40.25	ha	**40.25**
400 kg/ha	–		–	–	46.00	ha	**46.00**
500 kg/ha	–		–	–	57.50	ha	**57.50**
700 kg/ha	–		–	–	80.50	ha	**80.50**
1000 kg/ha	–		–	–	115.00	ha	**115.00**
1250 kg/ha	–		–	–	143.75	ha	**143.75**
Top dressing fertilizers (7+7+7); PC £0.77/kg; to seedbeds; by machine							
35 g/m²	2.49		–	0.69	2.86	100 m²	**3.54**
50 g/m²	3.55		–	0.69	4.08	100 m²	**4.77**
300 kg/ha	213.12		–	68.50	245.09	ha	**313.59**
350 kg/ha	248.64		–	68.50	285.94	ha	**354.44**
400 kg/ha	284.16		–	68.50	326.78	ha	**395.28**
500 kg/ha	355.20		–	109.60	408.48	ha	**518.08**
Top dressing fertilizers (7+7+7); PC £0.77/kg; to seedbeds; by hand							
35 g/m²	2.49	0.17	5.94	–	2.86	100 m²	**8.80**
50 g/m²	3.55	0.17	5.94	–	4.08	100 m²	**10.03**
70 g/m²	4.97	0.17	5.94	–	5.72	100 m²	**11.66**

37 SOFT LANDSCAPING

Item	PC £	Labour hours	Labour £	Plant £	Material £	Unit	Total rate £
MAINTENANCE OF SEEDED/ TURFED AREAS – cont							
Watering turf; evenly; at a rate of 5 l/m^2							
using movable spray lines powering 3 nr sprinkler heads with a radius of 15 m and allowing for 60% overlap (irrigation machinery costs not included)	–	0.02	0.56	–	–	100 m^2	**0.56**
using sprinkler equipment and with sufficient water pressure to run 1 nr 15 m radius sprinkler	–	0.02	0.71	–	–	100 m^2	**0.71**
using hand-held watering equipment	–	0.25	8.91	–	–	100 m^2	**8.91**
PLANTING GENERAL							
Planting – General							
Preamble: Prices for all planting work are deemed to include carrying out planting in accordance with all good horticultural practice.							
MARKET PRICES OF MATERIALS							
Market prices of planting materials (Note: the rates shown generally reflect the manufacturer's recommended retail prices; trade and bulk discounts are often available on the prices shown)							
Imported topsoil; market prices; British Sugar Topsoil; sustainably sourced graded topsoil to British standards and with independent certification; rates shown allow for 20% settlement							
Landscape 20; BS3882:2007; sandy loam							
8 wheel delivery; 20 tonne loads (approximately 13.33 m^3)	–		–	–	61.31	m^3	**61.31**
articulated delivery	–		–	–	46.03	m^3	**46.03**
Topsoil prices; Average locally sourced prices							
Topsoil							
general purpose	–		–	–	29.59	m^3	**29.59**

37 SOFT LANDSCAPING

Item	PC £	Labour hours	Labour £	Plant £	Material £	Unit	Total rate £
Market prices of mulching materials							
Melcourt Industries Ltd; 25 m^3 loads (items labelled FSC are Forest Stewardship Council certified, items marked FT are certified fire tested)							
Ornamental Bark Mulch FT	–		–	–	68.66	m^3	**68.66**
Bark Nuggets® FT	–		–	–	66.28	m^3	**66.28**
Amenity Bark Mulch FSC FT	–		–	–	49.17	m^3	**49.17**
Contract Bark Mulch FSC	–		–	–	44.29	m^3	**44.29**
Spruce Ornamental FSC FT	–		–	–	50.52	m^3	**50.52**
Forest BioMulch®	–		–	–	46.35	m^3	**46.35**
Melcourt Industries Ltd; 70 m^3 loads							
Contract Bark Mulch FSC	–		–	–	30.87	m^3	**30.87**
Melcourt Industries Ltd; 80 m^3 loads							
Ornamental Bark Mulch FT	–		–	–	54.58	m^3	**54.58**
Bark Nuggets® FT	–		–	–	52.20	m^3	**52.20**
Amenity Bark Mulch FSC	–		–	–	35.20	m^3	**35.20**
Spruce Ornamental FSC FT	–		–	–	36.44	m^3	**36.44**
Forest BioMulch®	–		–	–	32.27	m^3	**32.27**
Mulch; Melcourt Industries Ltd; 25 m^3 loads							
Composted Fine Bark FSC	–		–	–	48.36	m^3	**48.36**
Spent Mushroom Compost	–		–	–	31.89	m^3	**31.89**
Topgrow	–		–	–	55.02	m^3	**55.02**
Mulch; Melcourt Industries Ltd; 50 m^3 loads							
Mulch; Melcourt Industries Ltd; 60 m^3 loads							
Spent Mushroom Compost	–		–	–	19.22	m^3	**19.22**
Mulch; Melcourt Industries Ltd; 65 m^3 loads							
Composted Fine Bark	–		–	–	35.36	m^3	**35.36**
Topgrow	–		–	–	43.64	m^3	**43.64**
Market prices of fertilizers							
Fertilizers; Germinal GB							
G1; 6+9+6	–		–	–	34.66	kg	**34.66**
G4; 11+5+5	–		–	–	36.88	kg	**36.88**
G6; 20+10+10	–		–	–	35.85	kg	**35.85**
G9; 12+2+9+Fe	–		–	–	38.45	kg	**38.45**
Fertilizers; Everris							
Enmag; controlled release fertilizer (8–9 months); 11+22+09; 70 g/m^2	–		–	–	21.61	100 m^2	**21.61**

37 SOFT LANDSCAPING

Item	PC £	Labour hours	Labour £	Plant £	Material £	Unit	Total rate £
MARKET PRICES OF MATERIALS – cont							
Market prices of fertilizers – cont							
Fertilizers; Enmag ; 11+21+9; controlled release fertilizer; costs for recommended application rates							
transplant	–		–	–	0.02	nr	**0.02**
whip	–		–	–	0.03	nr	**0.03**
feathered	–		–	–	0.06	nr	**0.06**
light standard	–		–	–	0.09	nr	**0.09**
standard	–		–	–	0.19	nr	**0.19**
selected standard	–		–	–	0.33	nr	**0.33**
heavy standard	–		–	–	0.38	nr	**0.38**
extra heavy standard	–		–	–	0.47	nr	**0.47**
16–18 cm girth	–		–	–	0.52	nr	**0.52**
18–20 cm girth	–		–	–	0.57	nr	**0.57**
20–25 cm girth	–		–	–	0.66	nr	**0.66**
25–30 cm girth	–		–	–	0.71	nr	**0.71**
30 −35 cm girth	–		–	–	0.76	nr	**0.76**
Fertilizers; Everris							
TPMC tree planting and mulching compost	–		–	–	3.79	bag	**3.79**
Fertilizers; Rigby Taylor Ltd; fertilizer application rates 35 g/m^2 unless otherwise shown							
liquid fertilizer; Mascot Microflow-C; 25+0+0 + trace elements; 200–1400 ml/100 m^2	–		–	–	8.05	100 m^2	**8.05**
liquid fertilizer; Mascot Microflow-CX; 4+3+16 + trace elements; 200–1400 ml/100 m^2	–		–	–	7.17	100 m^2	**7.17**
liquid fertilizer; Mascot Microflow-CX; 12+0+8 + trace elements; 200–1400 ml/100 m^2	–		–	–	6.43	100 m^2	**6.43**
liquid fertilizer; Mascot Microflow-CX; 17+2+5 + trace elements; 200–1400 ml/100 m^2	–		–	–	7.16	100 m^2	**7.16**
Wetting agents; Rigby Taylor Ltd							
wetting agent; Breaker Advance Granules; 20 kg	–		–	–	15.67	100 m^2	**15.67**

37 SOFT LANDSCAPING

Item	PC £	Labour hours	Labour £	Plant £	Material £	Unit	Total rate £
PLANTING PREPARATION AND CULTIVATION							
Site protection; temporary protective fencing							
Cleft chestnut rolled fencing; to 100 mm dia. chestnut posts; driving into firm ground at 3 m centres; pales at 50 mm centres							
900 mm high	19.95	0.11	3.80	–	30.51	m	**34.31**
1100 mm high	7.47	0.11	3.80	–	16.16	m	**19.96**
Extra over temporary protective fencing for removing and making good (no allowance for reuse of material)	–	0.07	2.37	0.29	–	m	**2.67**
Cultivation by tractor; Agripower Ltd; Specialist subcontract prepared turf and sports area preparation							
Ripping up subsoil; using approved subsoiling machine; minimum depth 250 mm below topsoil; at 1.20 m centres; in							
gravel or sandy clay	–		–	8.25	–	100 m²	**8.25**
soil compacted by machines	–		–	9.16	–	100 m²	**9.16**
clay	–		–	11.79	–	100 m²	**11.79**
chalk or other soft rock	–		–	13.75	–	100 m²	**13.75**
Extra for subsoiling at 1 m centres	–		–	2.36	–	100 m²	**2.36**
Breaking up existing ground; using tractor drawn tine cultivator or rotavator with stone burier							
Single pass							
100 mm deep	–		–	1.47	–	100 m²	**1.47**
150 mm deep	–		–	1.84	–	100 m²	**1.84**
200 mm deep	–		–	2.45	–	100 m²	**2.45**
Cultivating ploughed ground; using disc, drag or chain harrow							
4 passes	–		–	0.58	–	100 m²	**0.58**
Cultivation; pedestrian operated rotavator							
Breaking up existing ground; tine cultivator or rotavator							
100 mm deep	–	0.20	7.84	1.62	–	100 m²	**9.46**
150 mm deep	–	0.25	9.80	2.03	–	100 m²	**11.83**
200 mm deep	–	0.33	13.07	2.70	–	100 m²	**15.77**

37 SOFT LANDSCAPING

Item	PC £	Labour hours	Labour £	Plant £	Material £	Unit	Total rate £
PLANTING PREPARATION AND CULTIVATION – cont							
Cultivation – cont							
Breaking up existing ground; tine cultivator or rotavator in heavy clay or wet soils							
100 mm deep	–	0.40	15.69	3.24	–	100 m^2	**18.93**
150 mm deep	–	0.60	23.53	4.86	–	100 m^2	**28.39**
200 mm deep	–	0.75	29.41	6.08	–	100 m^2	**35.49**
Clearing stones; disposing off site							
by hand; stones not exceeding 50 mm in any direction; loading to skip 4.6 m^3	–	0.01	0.36	0.05	–	m^2	**0.42**
by mechanical stone rake; stones not exceeding 50 mm in any direction; loading to 15 m^3 truck by mechanical loader	–	–	0.07	0.17	–	m^2	**0.25**
Lightly cultivating; weeding; to fallow areas; disposing debris off site							
by hand	–	0.01	0.50	–	–	m^2	**0.69**
Cultivation by hand							
Dig over surface of area to be planted; turn soil and level for palnting to a depth of							
100 mm	–	0.04	1.43	–	–	m^2	**1.43**
200 mm	–	0.07	2.38	–	–	m^2	**2.38**
300 mm	–	0.13	4.46	–	–	m^2	**4.46**
500 mm	–	0.33	11.88	–	–	m^2	**11.88**
TOPSOIL FILLING							
Topsoil filling; Friedland Horticulture							
Market prices of topsoil; price shown includes a 20% settlement factor							
topsoil to BS3882	35.51		–	–	40.83	m^3	**40.83**
small quantities (less than 15 m^3)	54.58		–	–	62.77	m^3	**62.77**
British Sugar							
Horticultural Loam							
8 wheel delivery; 20 tonne loads (approximately 13.33 m^3)	79.54		–	–	91.47	m^3	**91.47**
articulated delivery	73.46		–	–	84.48	m^3	**84.48**
Sports and turf topsoil							
8 wheel delivery; 20 tonne loads (approximately 13.33 m^3)	68.12		–	–	78.34	m^3	**78.34**
articulated delivery	63.06		–	–	72.52	m^3	**72.52**

37 SOFT LANDSCAPING

Item	PC £	Labour hours	Labour £	Plant £	Material £	Unit	Total rate £
Specialist substrates and growing mediums; Boughton Loam Limited. Lightweight or horticulturally efficient growing mediums for green roof podiums (to G.R.O. guidelines), and horticultural environments. Prices for mechanical placement by lightweight machinery maximum 25 m from location of soils; Cranage priced separately. Delivery payload size shown							
Boughton Loam Ltd; Natural screened topsoil sourced from as dug soils, certified to NHBC BS3882;							
Mechanically placed to ground level planting areas or planters average 400 mm deep;							
Full loose loads.29 tonne articulated	49.60	0.13	3.88	11.43	49.60	m^3	**64.91**
Full loose loads.20 tonne rigid	52.32	0.13	3.88	11.43	52.32	m^3	**67.63**
Bulk bags; 29 tonne articulated	122.55	0.25	7.75	22.86	122.55	m^3	**153.16**
Bulk bags; 20 tonne rigid	134.20	0.25	7.75	22.86	134.20	m^3	**164.81**
By hand; to ground level planting areas or planters average 400 mm deep;							
Full loose loads.20 tonne rigid	52.32	0.40	37.20	–	52.32	m^3	**89.52**
Bulk bags; 20 tonne articulated	134.20	0.50	46.50	–	134.20	m^3	**180.70**
Extra over for cranage using main contractors tower crane; loose loaded material mechanically loaded to hopper; hopper loaded at soil stockpile location							
loose loaded material mechanically loaded to hopper.	–	0.50	31.00	11.92	–	m^3	**42.92**
single lift bulk bagged material; banksmen and slingers by main contractor (not included)	–	0.67	61.99	4.06	–	m^3	**66.05**
Topsoiling Operations							
Spreading and lightly consolidating approved topsoil (imported or from spoil heaps); in layers not exceeding 150 mm; travel distance from spoil heaps not exceeding 100 m; by machine (imported topsoil not included)							
minimum depth 100 mm	–	1.55	55.26	73.42	–	100 m^2	**128.68**
minimum depth 150 mm	–	2.33	83.19	110.49	–	100 m^2	**193.68**
minimum depth 300 mm	–	4.67	166.36	220.98	–	100 m^2	**387.34**
minimum depth 450 mm	–	6.99	249.19	331.11	–	100 m^2	**580.31**

37 SOFT LANDSCAPING

Item	PC £	Labour hours	Labour £	Plant £	Material £	Unit	Total rate £
TOPSOIL FILLING – cont							
Topsoiling Operations – cont							
Spreading and lightly consolidating approved topsoil (imported or from spoil heaps); in layers not exceeding 150 mm; travel distance from spoil heaps not exceeding 100 m; by hand (imported topsoil not included)							
minimum depth 100 mm	–	20.00	713.14	–	–	100 m²	**713.14**
minimum depth 150 mm	–	30.01	1069.71	–	–	100 m²	**1069.71**
minimum depth 300 mm	–	60.01	2139.43	–	–	100 m²	**2139.43**
minimum depth 450 mm	–	90.02	3209.14	–	–	100 m²	**3209.14**
Spreading topsoil; from dump not exceeding 100 m distance; by machine (topsoil not included)							
at 1 m³ per 13 m²; 75 mm thick	–	0.56	19.81	54.74	–	100 m²	**74.55**
at 1 m³ per 10 m²; 100 mm thick	–	0.89	31.69	55.59	–	100 m²	**87.27**
at 1 m³ per 6.50 m²; 150 mm thick	–	1.11	39.57	98.19	–	100 m²	**137.76**
at 1 m³ per 5 m²; 200 mm thick	–	1.48	52.76	130.52	–	100 m²	**183.28**
Spreading topsoil; from dump not exceeding 100 m distance; by hand (topsoil not included)							
at 1 m³ per 13 m²; 75 mm thick	–	15.00	534.86	–	–	100 m²	**534.86**
at 1 m³ per 10 m²; 100 mm thick	–	20.00	713.14	–	–	100 m²	**713.14**
at 1 m³ per 6.50 m²; 150 mm thick	–	30.01	1069.71	–	–	100 m²	**1069.71**
at 1 m³ per 5 m²; 200 mm thick	–	36.67	1307.43	–	–	100 m²	**1307.43**
Imported topsoil; tipped 100 m from area of application; by machine							
at 1 m³ per 13 m²; 75 mm thick	266.29	0.56	19.81	54.74	306.23	100 m²	**380.78**
at 1 m³ per 10 m²; 100 mm thick	355.05	0.89	31.69	55.59	408.31	100 m²	**495.58**
at 1 m³ per 6.50 m²; 150 mm thick	532.58	1.11	39.57	98.19	612.46	100 m²	**750.22**
at 1 m³ per 5 m²; 200 mm thick	710.10	1.48	52.76	130.52	816.62	100 m²	**999.90**
Imported topsoil; tipped 100 m from area of application; by hand							
at 1 m³ per 13 m²; 75 mm thick	266.29	15.00	534.86	–	306.23	100 m²	**841.09**
at 1 m³ per 10 m²; 100 mm thick	355.05	20.00	713.14	–	408.31	100 m²	**1121.45**
at 1 m³ per 6.50 m²; 150 mm thick	532.58	30.01	1069.71	–	612.46	100 m²	**1682.18**
at 1 m³ per 5 m²; 200 mm thick	710.10	36.67	1307.43	–	816.62	100 m²	**2124.04**
Extra over for spreading topsoil to slopes 15–30°; by machine or hand	–		–	–	–	10%	**–**
Extra over for spreading topsoil to slopes over 30°; by machine or hand	–		–	–	–	25%	**–**
Extra over for spreading topsoil from spoil heaps; travel exceeding 100 m; by machine							
100–150 m	–	0.01	0.44	0.10	–	m³	**0.54**
150–200 m	–	0.02	0.66	0.15	–	m³	**0.81**
200–300 m	–	0.03	0.99	0.22	–	m³	**1.21**

37 SOFT LANDSCAPING

Item	PC £	Labour hours	Labour £	Plant £	Material £	Unit	**Total rate £**
Extra over spreading topsoil for travel exceeding 100 m; by hand							
100 m	–	0.83	29.71	–	–	m³	**29.71**
200 m	–	1.67	59.42	–	–	m³	**59.42**
300 m	–	2.50	89.13	–	–	m³	**89.13**
Evenly grading; to general surfaces to bring to finished levels							
by machine (tractor mounted rotavator)	–		–	0.07	–	m²	**0.07**
by pedestrian operated rotavator	–	–	0.15	0.03	–	m²	**0.18**
by hand	–	0.01	0.36	–	–	m²	**0.36**
Extra over grading for slopes 15–30°; by machine or hand	–		–	–	–	10%	**–**
Extra over grading for slopes over 30°; by machine or hand	–		–	–	–	25%	**–**
Apply screened topdressing to grass surfaces; spread using Tru-Lute							
sand soil mixes 90/10 to 50/50	–	–	0.07	0.11	0.27	m²	**0.45**
Spread only existing cultivated soil to final levels using Tru-Lute							
cultivated soil	–	–	0.07	0.11	–	m²	**0.18**
SOIL IMPROVEMENTS							
Preparation of planting operations							
For the following topsoil improvement and planting operations add or subtract the following amounts for every £0.10 difference in the material cost price							
35 g/m²	–		–	–	0.40	100 m²	**0.40**
50 g/m²	–		–	–	0.58	100 m²	**0.58**
70 g/m²	–		–	–	0.81	100 m²	**0.81**
100 g/m²	–		–	–	1.15	100 m²	**1.15**
150 kg/ha	–		–	–	17.25	ha	**17.25**
200 kg/ha	–		–	–	23.00	ha	**23.00**
250 kg/ha	–		–	–	28.75	ha	**28.75**
300 kg/ha	–		–	–	34.50	ha	**34.50**
400 kg/ha	–		–	–	46.00	ha	**46.00**
500 kg/ha	–		–	–	57.50	ha	**57.50**
700 kg/ha	–		–	–	80.50	ha	**80.50**
1000 kg/ha	–		–	–	115.00	ha	**115.00**
1250 kg/ha	–		–	–	143.75	ha	**143.75**

37 SOFT LANDSCAPING

Item	PC £	Labour hours	Labour £	Plant £	Material £	Unit	Total rate £
SOIL IMPROVEMENTS – cont							
Preparation of planting operations – cont							
Fertilizers; in top 150 mm of topsoil; at 35 g/m^2							
Mascot Microfine; controlled release turf fertilizer; 8+0+6 + 2% Mg + 4% Fe	0.06	–	0.05	–	0.08	m^2	**0.12**
Enmag CRF 11–21–9+6MgO 8–9 Months Planting Fertilizer 25 kg	0.11	–	0.05	–	0.13	m^2	**0.18**
Mascot Outfield; turf fertilizer; 4 +10+10	–	–	0.05	–	0.05	m^2	**0.10**
Mascot Outfield; turf fertilizer; 9 +5+5	–	–	0.05	–	0.07	m^2	**0.11**
Fertilizers; in top 150 mm of topsoil at 70 g/m^2							
Mascot Microfine; controlled release turf fertilizer; 8+0+6 + 2% Mg + 4% Fe	0.12	–	0.05	–	0.15	m^2	**0.20**
Enmag; controlled release fertilizer (8–9 months); 11+22+09	0.23	–	0.05	–	0.26	m^2	**0.31**
Mascot Outfield; turf fertilizer; 4 +10+10	0.09	–	0.05	–	0.10	m^2	**0.15**
Mascot Outfield; turf fertilizer; 9 +5+5	0.10	–	0.05	–	0.12	m^2	**0.17**
'Topgrow'; Melcourt; peat free tree and shrub compost (25 m^3 loads); placing on beds by mechanical loader; spreading and rotavating into topsoil by tractor drawn machine							
50 mm thick	275.08		–	18.45	316.34	100 m^2	**334.80**
100 mm thick	577.67		–	27.08	664.32	100 m^2	**691.41**
150 mm thick	866.51		–	36.06	996.48	100 m^2	**1032.55**
200 mm thick	1155.34		–	44.87	1328.65	100 m^2	**1373.52**
'Topgrow'; Melcourt; peat free tree and shrub compost (65 m^3 loads); placing on beds by mechanical loader; spreading and rotavating into topsoil by tractor drawn machine							
50 mm thick	218.22		–	18.45	250.96	100 m^2	**269.41**
100 mm thick	458.27		–	27.08	527.01	100 m^2	**554.10**
150 mm thick	687.41		–	36.06	790.52	100 m^2	**826.58**
200 mm thick	916.54		–	44.87	1054.02	100 m^2	**1098.89**

37 SOFT LANDSCAPING

Item	PC £	Labour hours	Labour £	Plant £	Material £	Unit	Total rate £
Mushroom compost; Melcourt; (25 m³ loads); from not further than 25 m from location; cultivating into topsoil by pedestrian drawn machine							
50 mm thick	159.47	2.86	101.86	1.35	183.39	100 m²	**286.60**
100 mm thick	318.94	6.05	215.65	1.35	366.79	100 m²	**583.78**
150 mm thick	478.42	8.90	317.46	1.35	550.18	100 m²	**868.98**
200 mm thick	637.89	12.90	460.06	1.35	733.57	100 m²	**1194.98**
Mushroom compost; Melcourt (25 m³ loads); placing on beds by mechanical loader; spreading and rotavating into topsoil by tractor drawn cultivator							
50 mm thick	159.47		–	18.45	183.39	100 m²	**201.85**
100 mm thick	318.94		–	27.08	366.79	100 m²	**393.87**
150 mm thick	478.42		–	36.06	550.18	100 m²	**586.24**
200 mm thick	637.89		–	44.87	733.57	100 m²	**778.44**
Manure (60 m³ loads); from not further than 25 m from location; cultivating into topsoil by pedestrian drawn machine							
50 mm thick	124.55	2.86	101.86	1.35	143.23	100 m²	**246.44**
100 mm thick	261.54	6.05	215.65	1.35	300.78	100 m²	**517.77**
150 mm thick	392.32	8.90	317.46	1.35	451.16	100 m²	**769.97**
200 mm thick	523.09	12.90	460.06	1.35	601.55	100 m²	**1062.96**
Surface applications and soil additives; pre-planting; from not further than 25 m from location; by machine							
Medium bark soil conditioner; Melcourt Ltd; including turning in to cultivated ground; delivered in 25 m³ loads							
1 m³ per 40 m² = 25 mm thick	1.21		–	0.24	1.39	m²	**1.63**
1 m³ per 20 m² = 50 mm thick	2.42		–	0.35	2.78	m²	**3.13**
1 m³ per 13.33 m² = 75 mm thick	3.63		–	0.46	4.17	m²	**4.63**
1 m³ per 10 m² = 100 mm thick	4.84		–	0.58	5.56	m²	**6.14**
Works by hand; surface applications and soil additives; pre-planting; from not further than 25 m from location; by hand							
Mushroom compost soil conditioner; Melcourt Industries Ltd; including turning in to cultivated ground; delivered in 25 m³ loads							
1 m³ per 40 m² = 25 mm thick	0.80	0.02	0.79	–	0.92	m²	**1.71**
1 m³ per 20 m² = 50 mm thick	1.59	0.04	1.58	–	1.83	m²	**3.42**
1 m³ per 13.33 m² = 75 mm thick	2.39	0.07	2.37	–	2.75	m²	**5.13**
1 m³ per 10 m² = 100 mm thick	3.19	0.08	2.85	–	3.67	m²	**6.52**

37 SOFT LANDSCAPING

Item	PC £	Labour hours	Labour £	Plant £	Material £	Unit	Total rate £
SOIL IMPROVEMENTS – cont							
Works by hand – cont							
Medium bark soil conditioner; Melcourt Industries Ltd; including turning in to cultivated ground; delivered in 25 m^3 loads							
1 m^3 per 40 m^2 = 25 mm thick	1.21	0.02	0.79	–	1.39	m^2	**2.18**
1 m^3 per 20 m^2 = 50 mm thick	2.42	0.04	1.58	–	2.78	m^2	**4.36**
1 m^3 per 13.33 m^2 = 75 mm thick	3.63	0.07	2.37	–	4.17	m^2	**6.55**
1 m^3 per 10 m^2 = 100 mm thick	4.84	0.08	2.85	–	5.56	m^2	**8.41**
Manure; 20 m^3 loads; delivered not further than 25 m from location; cultivating into topsoil by pedestrian operated machine							
50 mm thick	135.38	2.86	101.86	1.35	155.68	100 m^2	**258.89**
100 mm thick	261.54	6.05	215.65	1.35	300.78	100 m^2	**517.77**
150 mm thick	392.32	8.90	317.46	1.35	451.16	100 m^2	**769.97**
200 mm thick	523.09	12.90	460.06	1.35	601.55	100 m^2	**1062.96**
Fertilizer (7+7+7); PC £0.77/kg; to beds; by hand							
35 g/m^2	2.49	0.17	5.94	–	2.86	100 m^2	**8.80**
50 g/m^2	3.55	0.17	5.94	–	4.08	100 m^2	**10.03**
70 g/m^2	4.97	0.17	5.94	–	5.72	100 m^2	**11.66**
Fertilizers; Enmag; PC £3.09/kg; controlled release fertilizer; to beds; by hand							
35 g/m^2	10.80	0.17	5.94	–	12.42	100 m^2	**18.37**
50 g/m^2	15.43	0.17	5.94	–	17.75	100 m^2	**23.69**
70 g/m^2	23.15	0.17	5.94	–	26.62	100 m^2	**32.56**
Note: For machine incorporation of fertilizers and soil conditioners see 'Cultivation'.							
TREE PLANTING PREPARATIONS							
Tree planting; pre-planting operations							
Excavating tree pits; depositing soil alongside pits; by machine							
600 × 600 × 600 mm deep	–	0.15	5.25	0.93	–	nr	**6.18**
900 × 900 × 600 mm deep	–	0.33	11.76	2.08	–	nr	**13.85**
1.00 × 1.00 × 600 mm deep	–	0.61	21.87	2.58	–	nr	**24.45**
1.25 × 1.25 × 600 mm deep	–	0.96	34.25	4.03	–	nr	**38.28**
1.00 × 1.00 × 1.00 m deep	–	1.02	36.46	4.30	–	nr	**40.76**
1.50 × 1.50 × 750 mm deep	–	1.73	61.53	7.25	–	nr	**68.78**
1.50 × 1.50 × 1.00 m deep	–	2.30	81.83	9.65	–	nr	**91.48**
1.75 × 1.75 × 1.00 m deep	–	3.13	111.66	13.17	–	nr	**124.83**
2.00 × 2.00 × 1.00 m deep	–	4.09	145.83	17.20	–	nr	**163.03**

37 SOFT LANDSCAPING

Item	PC £	Labour hours	Labour £	Plant £	Material £	Unit	**Total rate £**
Excavating tree pits; depositing soil alongside pits; by hand							
600 × 600 × 600 mm deep	–	0.44	15.69	–	–	nr	**15.69**
900 × 900 × 600 mm deep	–	1.00	35.65	–	–	nr	**35.65**
1.00 × 1.00x 600 mm deep	–	1.13	40.11	–	–	nr	**40.11**
1.25 × 1.25x 600 mm deep	–	1.93	68.80	–	–	nr	**68.80**
1.00 × 1.00 × 1.00 m deep	–	2.06	73.44	–	–	nr	**73.44**
1.50 × 1.50x 750 mm deep	–	3.47	123.71	–	–	nr	**123.71**
1.75 × 1.50x 750 mm deep	–	4.05	144.38	–	–	nr	**144.38**
1.50 × 1.50 × 1.00 m deep	–	4.63	165.06	–	–	nr	**165.06**
2.00 × 2.00x 750 mm deep	–	6.17	219.96	–	–	nr	**219.96**
2.00 × 2.00 × 1.00 m deep	–	8.23	293.42	–	–	nr	**293.42**
Breaking up subsoil in tree pits; to a depth of 200 mm	–	0.03	1.19	–	–	m^2	**1.19**
Spreading and lightly consolidating approved topsoil (imported or from spoil heaps); in layers not exceeding 150 mm; distance from spoil heaps not exceeding 100 m (imported topsoil not included); by machine							
minimum depth 100 mm	–	1.55	55.26	73.42	–	100 m^2	**128.68**
minimum depth 150 mm	–	2.33	83.19	110.49	–	100 m^2	**193.68**
minimum depth 300 mm	–	4.67	166.36	220.98	–	100 m^2	**387.34**
minimum depth 450 mm	–	6.99	249.19	331.11	–	100 m^2	**580.31**
Spreading and lightly consolidating approved topsoil (imported or from spoil heaps); in layers not exceeding 150 mm; distance from spoil heaps not exceeding 100 m (imported topsoil not included); by hand							
minimum depth 100 mm	–	20.00	713.14	–	–	100 m^2	**713.14**
minimum depth 150 mm	–	30.01	1069.71	–	–	100 m^2	**1069.71**
minimum depth 300 mm	–	60.01	2139.43	–	–	100 m^2	**2139.43**
minimum depth 450 mm	–	90.02	3209.14	–	–	100 m^2	**3209.14**
Extra for filling tree pits with imported topsoil; PC £29.59/m^3; calculations allow for 20% settlement							
depth 100 mm	–		–	–	4.90	m^2	**4.90**
depth 150 mm	–		–	–	7.35	m^2	**7.35**
depth 200 mm	–		–	–	9.80	m^2	**9.80**
depth 300 mm	–		–	–	14.70	m^2	**14.70**
depth 400 mm	–		–	–	19.60	m^2	**19.60**
depth 450 mm	–		–	–	22.05	m^2	**22.05**
depth 500 mm	–		–	–	24.50	m^2	**24.50**
depth 600 mm	–		–	–	29.40	m^2	**29.40**
depth 750 mm	–		–	–	30.62	m^2	**30.62**
depth 1.00 m	–		–	–	40.83	m^2	**40.83**

37 SOFT LANDSCAPING

Item	PC £	Labour hours	Labour £	Plant £	Material £	Unit	Total rate £
TREE PLANTING PREPARATIONS – cont							
Tree planting – cont							
Backfilling of tree pits with imported topsoil; compacting in layers							
depth 500 mm	–		–	6.86	20.42	m²	**27.27**
depth 600 mm	–		–	7.84	24.50	m²	**32.34**
depth 750 mm	–		–	9.14	30.62	m²	**39.77**
depth 1.00 m	–		–	13.71	40.83	m²	**54.54**
Add or deduct the following amounts for every £1.00 change in the material price of topsoil							
depth 100 mm	–		–	–	0.15	m²	**0.15**
depth 150 mm	–		–	–	0.22	m²	**0.22**
depth 200 mm	–		–	–	0.30	m²	**0.30**
depth 300 mm	–		–	–	0.45	m²	**0.45**
depth 400 mm	–		–	–	0.60	m²	**0.60**
depth 450 mm	–		–	–	0.67	m²	**0.67**
depth 500 mm	–		–	–	0.75	m²	**0.75**
depth 600 mm	–		–	–	0.90	m²	**0.90**
Structural soils (Amsterdam tree sand); Heicom; non compressive soil mixture for tree planting in areas to receive compressive surface treatments							
Backfilling and lightly compacting in layers; excavation, disposal, moving of material from delivery position and surface treatments not included; by machine							
individual treepits (29 t loads)	77.29	0.50	17.83	3.15	88.88	m³	**109.86**
individual treepits (20 t loads)	86.22	0.50	17.83	3.15	99.15	m³	**120.13**
in trenches (29 t loads)	77.29	0.42	14.86	2.63	88.88	m³	**106.36**
Backfilling and lightly compacting in layers; excavation, disposal, moving of material from delivery position and surface treatments not included; by hand							
individual treepits (29 t loads)	77.29	1.60	57.04	–	88.88	m³	**145.92**
individual treepits (20 t loads)	86.22	1.60	57.04	–	99.15	m³	**156.19**
in trenches	77.29	1.33	47.41	–	88.88	m³	**136.29**

37 SOFT LANDSCAPING

Item	PC £	Labour hours	Labour £	Plant £	Material £	Unit	**Total rate £**
TREE PLANTING							
Trees; planting labours only							
Bare root trees; including backfilling with previously excavated material (all other operations and materials not included)							
light standard; 6–8 cm girth	–	0.35	12.48	–	–	nr	**12.48**
standard; 8–10 cm girth	–	0.40	14.26	–	–	nr	**14.26**
selected standard; 10–12 cm girth	–	0.58	20.68	–	–	nr	**20.68**
heavy standard; 12–14 cm girth	–	0.83	29.71	–	–	nr	**29.71**
extra heavy standard; 14–16 cm girth	–	1.00	35.65	–	–	nr	**35.65**
Root balled trees; including backfilling with previously excavated material (all other operations and materials not included)							
standard; 8–10 cm girth	–	0.50	17.83	–	–	nr	**17.83**
selected standard; 10–12 cm girth	–	0.60	21.39	–	–	nr	**21.39**
heavy standard; 12–14 cm girth	–	0.80	28.52	–	–	nr	**20.52**
extra heavy standard;14–16 cm girth	–	1.50	53.48	–	–	nr	**53.48**
16–18 cm girth	–	1.30	46.20	43.46	–	nr	**89.66**
18–20 cm girth	–	1.60	57.04	53.65	–	nr	**110.69**
20–25 cm girth	–	4.50	160.43	167.82	–	nr	**328.25**
25–30 cm girth	–	6.00	213.90	221.50	–	nr	**435.40**
30–35 cm girth	–	11.00	392.15	394.24	–	nr	**786.39**
Treepits for urban planting							
Excavating tree pits; depositing soil alongside pits; by machine							
1.75 × 1.75 × 1.00 m deep	–	2.09	74.44	57.27	–	nr	**131.71**
2.00 × 2.00 × 1.00 m deep	–	2.73	97.22	74.60	–	nr	**171.81**
2.50 × 2.50 × 1.50 m deep	–	3.28	116.93	179.91	–	nr	**296.84**
The labour in this section is calculated on a 2 person team. The labour time below should be multiplied by 2 to calculate the cost as shown.							
Earthwork retention to tree pits							
bracing of deep treepits by plywood boards braced by timber stays							
Treepit 1.00 m × 1.00 m × 1.00 m	–	0.75	53.48	–	29.06	nr	**82.53**
Treepit 1.50 m × 1.50 m × 1.00 m	–	0.90	64.17	–	37.67	nr	**101.84**
Treepit 2.00 m × 2.00 m × 1.50 m	–	1.00	71.30	–	46.28	nr	**117.58**

37 SOFT LANDSCAPING

Item	PC £	Labour hours	Labour £	Plant £	Material £	Unit	Total rate £
TREE PLANTING – cont							
Earthwork retention to tree pits – cont							
Excavate tree break-out zone for tree roots to develop in restricted and compacted urban environments; depositing soil alongside pits by machine;							
Trenches 1.50 m wide × 1.50 m deep average	–	0.17	5.94	9.14	–	m³	**15.09**
Bases to treepits; levelling base breaking up base by hand and placing of granular drainage material 200 mm thick							
Treepit 1.00 m × 1.00 m	–	0.25	8.91	13.71	12.75	nr	**35.38**
Treepit 1.50 m × 1.50 m	–	0.45	16.04	24.68	28.69	nr	**69.42**
Treepit 2.00 m × 2.00 m	–	0.75	26.74	41.14	51.00	nr	**118.88**
Treepit 2.50 m × 2.50 m	–	0.75	26.74	41.14	79.70	nr	**147.57**
TREE PLANTING – NOTES							
Our supplier James Coles & Sons reports shortages of the stock represented and hence pricing pressures for 2021–2022. Please check with the supplier who has provided these prices at the time of compiling this year's publication. Quercus are in especially short supply due to processionary moth and some Quercus items have been removed from these pages for this year.							
Tree planting; containerized trees; nursery stock; James Coles & Sons (Nurseries) Ltd							
Acer platanoides 'Emerald Queen'; including backfillling with excavated material (other operations not included)							
standard; 8–10 cm girth	72.32	0.48	17.11	–	83.17	nr	**100.28**
selected standard; 10–12 cm girth	120.54	0.56	19.99	–	138.62	nr	**158.60**
heavy standard; 12–14 cm girth	136.61	0.76	27.25	–	157.10	nr	**184.35**
extra heavy standard; 14–16 cm girth	160.72	1.20	42.78	–	184.82	nr	**227.60**

37 SOFT LANDSCAPING

Item	PC £	Labour hours	Labour £	Plant £	Material £	Unit	Total rate £
Carpinus betulus; including backfillling with excavated material (other operations not included)							
standard; 8–10 cm girth	35.48	0.48	17.11	–	40.80	nr	**57.91**
selected standard; 10–12 cm girth	112.50	0.56	19.99	–	129.38	nr	**149.36**
heavy standard; 12–14 cm girth	136.61	0.76	27.25	–	157.10	nr	**184.35**
extra heavy standard; 14–16 cm girth	160.72	1.20	42.78	–	184.82	nr	**227.60**
Fagus sylvatica; including backfillling with excavated material (other operations not included)							
standard; 8–10 cm girth	72.32	0.48	17.11	–	83.17	nr	**100.28**
selected standard; 10–12 cm girth	128.57	0.56	19.99	–	147.86	nr	**167.85**
heavy standard; 12–14 cm girth	152.68	0.76	27.25	–	175.58	nr	**202.83**
extra heavy standard; 14–16 cm girth	160.72	1.20	42.78	–	184.82	nr	**227.60**
Prunus avium; including backfillling with excavated material (other operations not included)							
standard; 8–10 cm girth	72.32	0.40	14.26	–	83.17	nr	**97.43**
selected standard; 10–12 cm girth	96.43	0.56	19.99	–	110.89	nr	**130.88**
heavy standard; 12–14 cm girth	112.50	0.76	27.25	–	129.38	nr	**156.63**
extra heavy standard; 14–16 cm girth	136.61	1.20	42.78	–	157.10	nr	**199.88**
Quercus robur; including backfillling with excavated material (other operations not included)							
standard; 8–10 cm girth	77.38	0.48	17.11	–	88.99	nr	**106.10**
selected standard; 10–12 cm girth	111.78	0.56	19.99	–	128.54	nr	**148.53**
heavy standard; 12–14 cm girth	146.17	0.76	27.25	–	168.10	nr	**195.35**
extra heavy standard; 14–16 cm girth	171.97	1.20	42.78	–	197.77	nr	**240.55**
Betula utilis jaquemontii; multistemmed; including backfillling with excavated material (other operations not included)							
175/200 mm high	120.54	0.48	17.11	–	138.62	nr	**155.73**
200/250 mm high	200.90	0.56	19.99	–	231.03	nr	**251.02**
250/300 mm high	225.00	0.76	27.25	–	258.75	nr	**286.01**
300/350 mm high	304.21	1.20	42.78	–	349.85	nr	**392.63**
Tree planting; root balled trees, advanced nursery stock and semi-mature – General							
Preamble: The cost of planting semi-mature trees will depend on the size and species and on the access to the site for tree handling machines. Prices should be obtained for individual trees and planting.							

37 SOFT LANDSCAPING

Item	PC £	Labour hours	Labour £	Plant £	Material £	Unit	Total rate £
TREE PLANTING – cont							
Tree planting; bare root trees; nursery stock; James Coles & Sons (Nurseries) Ltd							
Acer platanoides; including backfillling with excavated material (other operations not included)							
light standard; 6–8 cm girth	22.67	0.35	12.48	–	26.07	nr	**38.54**
standard; 8–10 cm girth	29.57	0.40	14.26	–	34.00	nr	**48.26**
selected standard; 10–12 cm girth	39.42	0.58	20.68	–	45.33	nr	**66.01**
heavy standard; 12–14 cm girth	64.06	0.83	29.71	–	73.67	nr	**103.38**
extra heavy standard; 14–16 cm girth	78.84	1.00	35.65	–	90.67	nr	**126.32**
Carpinus betulus; including backfillling with excavated material (other operations not included)							
light standard; 6–8 cm girth	29.57	0.35	12.48	–	34.00	nr	**46.48**
standard; 8–10 cm girth	35.48	0.40	14.26	–	40.80	nr	**55.06**
selected standard; 10–12 cm girth	59.13	0.58	20.68	–	68.00	nr	**88.68**
heavy standard; 12–14 cm girth	108.41	0.83	29.72	–	124.67	nr	**154.39**
Fagus sylvatica; including backfillling with excavated material (other operations not included)							
light standard; 6–8 cm girth	32.02	0.35	12.48	–	36.83	nr	**49.31**
standard; 8–10 cm girth	39.42	0.40	14.26	–	45.33	nr	**59.59**
selected standard; 10–12 cm girth	68.99	0.58	20.68	–	79.34	nr	**100.01**
heavy standard; 12–14 cm girth	108.41	0.83	29.59	–	124.67	nr	**154.26**
extra heavy standard; 14–16 cm girth	133.75	1.00	35.65	–	153.81	nr	**189.46**
Prunus avium; including backfillling with excavated material (other operations not included)							
light standard; 6–8 cm girth	32.02	0.36	12.98	–	36.83	nr	**49.80**
standard; 8–10 cm girth	39.42	0.40	14.26	–	45.33	nr	**59.59**
selected standard; 10–12 cm girth	43.36	0.58	20.68	–	49.87	nr	**70.54**
heavy standard; 12–14 cm girth	59.13	0.83	29.71	–	68.00	nr	**97.71**
extra heavy standard; 14–16 cm girth	78.84	1.00	35.65	–	90.67	nr	**126.32**
Quercus robur; including backfillling with excavated material (other operations not included)							
light standard; 6–8 cm girth	30.36	0.35	12.48	–	34.91	nr	**47.39**
standard; 8–10 cm girth	47.63	0.40	14.26	–	54.77	nr	**69.03**
selected standard; 10–12 cm girth	65.05	0.58	20.68	–	74.80	nr	**95.48**
heavy standard; 12–14 cm girth	108.41	0.83	29.71	–	124.67	nr	**154.38**
Robinia pseudoacacia Frisia; including backfillling with excavated material (other operations not included)							
light standard; 6–8 cm girth	37.45	0.35	12.48	–	43.07	nr	**55.55**
standard; 8–10 cm girth	59.13	0.40	14.26	–	68.00	nr	**82.26**
selected standard; 10–12 cm girth	88.70	0.58	20.68	–	102.00	nr	**122.68**

37 SOFT LANDSCAPING

Item	PC £	Labour hours	Labour £	Plant £	Material £	Unit	**Total rate £**
Tree planting; root balled trees; nursery stock; James Coles & Sons (Nurseries) Ltd							
Acer platanoides; including backfillling with excavated material (other operations not included)							
standard; 8–10 cm girth	39.61	0.48	17.11	–	45.55	nr	**62.66**
selected standard; 10–12 cm girth	51.85	0.56	19.99	–	59.63	nr	**79.62**
heavy standard; 12–14 cm girth	61.60	0.76	27.25	–	90.72	nr	**117.97**
extra heavy standard; 14–16 cm girth	94.94	1.20	42.78	–	109.18	nr	**151.96**
Carpinus betulus; including backfillling with excavated material (other operations not included)							
standard; 8–10 cm girth	44.15	0.48	17.11	–	50.78	nr	**67.89**
selected standard; 10–12 cm girth	70.81	0.56	19.99	–	81.43	nr	**101.41**
heavy standard; 12–14 cm girth	80.55	0.76	27.25	–	112.51	nr	**139.76**
extra heavy standard; 14–16 cm girth	132.84	1.20	42.78	–	152.77	nr	**195.55**
Fagus sylvatica; including backfillling with excavated material (other operations not included)							
standard; 8–10 cm girth	49.46	0.48	17.11	–	56.88	nr	**73.99**
selected standard; 10–12 cm girth	82.94	0.56	19.99	–	95.38	nr	**115.36**
heavy standard; 12–14 cm girth	125.69	0.76	27.25	–	144.55	nr	**171.80**
extra heavy standard; 14–16 cm girth	155.68	1.20	42.78	–	179.03	nr	**221.81**
Prunus avium; including backfillling with excavated material (other operations not included)							
standard; 8–10 cm girth	35.63	0.40	14.26	–	40.98	nr	**55.24**
selected standard; 10–12 cm girth	57.31	0.56	19.99	–	65.91	nr	**85.89**
heavy standard; 12–14 cm girth	76.42	0.76	27.25	–	87.88	nr	**115.13**
extra heavy standard; 14–16 cm girth	19.13	1.20	42.78	–	112.66	nr	**155.44**
Quercus robur; including backfillling with excavated material (other operations not included)							
standard; 8–10 cm girth	65.14	0.48	17.11	–	74.91	nr	**92.03**
selected standard; 10–12 cm girth	76.49	0.56	19.99	–	87.97	nr	**107.95**
heavy standard; 12–14 cm girth	104.21	0.76	27.25	–	139.75	nr	**167.00**
extra heavy standard; 14–16 cm girth	177.21	1.20	42.78	–	203.79	nr	**246.57**
Robinia pseudoacacia; including backfillling with excavated material (other operations not included)							
standard; 8–10 cm girth	38.47	0.48	17.11	–	44.24	nr	**61.35**
selected standard; 10–12 cm girth	51.85	0.56	19.99	–	59.63	nr	**79.62**
heavy standard; 12–14 cm girth	66.33	0.76	27.25	–	96.16	nr	**123.41**
extra heavy standard; 14–16 cm girth	104.41	1.20	42.78	–	120.07	nr	**162.85**

37 SOFT LANDSCAPING

Item	PC £	Labour hours	Labour £	Plant £	Material £	Unit	Total rate £
TREE PLANTING – cont							
Tree planting; Airpot container grown trees; advanced nursery stock and semi-mature; Hillier Nurseries							
Acer platanoides 'Emerald Queen'; including backfilling with excavated material (other operations not included)							
16–18 cm girth	200.00	1.98	70.59	67.10	230.00	nr	**367.69**
18–20 cm girth	248.00	2.18	77.65	72.12	285.20	nr	**434.97**
20–25 cm girth	320.00	2.38	84.70	80.53	368.00	nr	**533.23**
25–30 cm girth	368.00	2.97	105.88	123.43	423.20	nr	**652.51**
30–35 cm girth	584.00	3.96	141.17	134.21	671.60	nr	**946.98**
Prunus avium 'Flora Plena'; including backfilling with excavated material (other operations not included)							
16–18 cm girth	194.00	1.98	70.59	67.10	223.10	nr	**360.79**
18–20 cm girth	241.00	1.60	57.04	72.12	277.15	nr	**406.31**
20–25 cm girth	310.00	2.38	84.70	80.53	356.50	nr	**521.73**
25–30 cm girth	357.00	2.97	105.88	123.43	410.55	nr	**639.86**
30–35 cm girth	566.00	3.96	141.17	134.21	650.90	nr	**926.28**
Quercus palustris 'Pin Oak'; including backfilling with excavated material (other operations not included)							
16–18 cm girth	294.00	1.98	70.59	67.10	338.10	nr	**475.79**
18–20 cm girth	365.00	1.60	57.04	72.12	419.75	nr	**548.91**
20–25 cm girth	470.00	2.38	84.70	80.53	540.50	nr	**705.73**
25–30 cm girth	541.00	2.97	105.88	123.43	622.15	nr	**851.46**
30–35 cm girth	900.00	3.96	141.17	134.21	1035.00	nr	**1310.38**
The labour in this section is calculated on a 3 person team. The labour time below should be multiplied by 3 to calculate the cost as shown.							
Tree planting; Multi-Stem trees; Hillier Nurseries; Airpot container grown trees; semi-mature; Betula Pendula; including backfilling with excavated material (other operations not included)							
300–400 cm	250.00	3.00	320.85	123.43	287.50	nr	**731.78**
400–500 cm	395.00	3.25	347.59	123.43	454.25	nr	**925.27**
500–600 cm	600.00	3.50	374.33	130.82	690.00	nr	**1195.15**
600–700 cm	1350.00	3.75	401.06	135.90	1552.50	nr	**2089.47**
700–800 cm	2000.00	4.00	427.80	138.44	2300.00	nr	**2866.24**
800–900 cm	2400.00	4.50	481.28	152.13	2760.00	nr	**3393.40**

37 SOFT LANDSCAPING

Item	PC £	Labour hours	Labour £	Plant £	Material £	Unit	Total rate £
Betula jacquemontii; including backfilling with excavated material (other operations not included)							
300–400 cm	565.00	3.00	320.85	123.43	649.75	nr	**1094.03**
400–500 cm	720.00	3.25	347.59	123.43	828.00	nr	**1299.02**
500–600 cm	1450.00	3.50	374.33	130.82	1667.50	nr	**2172.65**
600–700 cm	2400.00	4.00	427.80	138.44	2760.00	nr	**3326.24**
700–800 cm	2950.00	4.50	481.28	152.13	3392.50	nr	**4025.90**
Tree planting; Airpot container grown trees; semi-mature and mature trees; Hillier Nurseries; planting and back filling; planted by telehandler or by crane; delivery included; all other operations priced separately							
Semi-mature trees; indicative prices							
40–45 cm girth	1200.00	4.00	142.60	84.93	1380.00	nr	**1607.53**
45–50 cm girth	1700.00	4.00	142.60	84.93	1955.00	nr	**2182.53**
55–60 cm girth	2800.00	6.00	213.90	84.93	3220.00	nr	**3518.83**
60–70 cm girth	3600.00	7.00	249.55	114.72	4140.00	nr	**4504.27**
70–80 cm girth	4200.00	7.50	267.38	144.52	4830.00	nr	**5241.89**
80–90 cm girth	6000.00	8.00	285.20	169.85	6900.00	nr	**7355.05**
Tree planting; rootballed trees; advanced nursery stock and semi-mature; Lorenz von Ehren							
Acer platanoides 'Emerald Queen'; including backfilling with excavated material (other operations not included)							
16–18 cm girth	150.00	1.30	46.20	5.25	172.50	nr	**223.96**
18–20 cm girth	200.00	1.60	57.04	5.25	230.00	nr	**292.29**
20–25 cm girth	250.00	4.50	160.43	24.81	287.50	nr	**472.74**
25–30 cm girth	350.00	6.00	213.90	30.83	402.50	nr	**647.23**
30–35 cm girth	450.00	11.00	392.15	41.16	517.50	nr	**950.81**
Aesculus carnea 'Briotti'; including backfilling with excavated material (other operations not included)							
16–18 cm girth	250.00	1.30	46.20	5.25	287.50	nr	**338.96**
18–20 cm girth	350.00	1.60	57.04	5.25	402.50	nr	**464.79**
20–25 cm girth	450.00	4.50	160.43	24.81	517.50	nr	**702.74**
25–30 cm girth	550.00	6.00	213.90	30.83	632.50	nr	**877.23**
30–35 cm girth	700.00	11.00	392.15	41.16	805.00	nr	**1238.31**
Prunus avium 'Plena'; including backfilling with excavated material (other operations not included)							
16–18 cm girth	150.00	1.30	46.20	5.25	172.50	nr	**223.96**
18–20 cm girth	200.00	1.60	57.04	5.25	230.00	nr	**292.29**
20–25 cm girth	250.00	4.50	160.43	24.81	287.50	nr	**472.74**
25–30 cm girth	300.00	6.00	213.90	30.83	345.00	nr	**589.73**
30–35 cm girth	550.00	11.00	392.15	36.08	632.50	nr	**1060.73**

37 SOFT LANDSCAPING

Item	PC £	Labour hours	Labour £	Plant £	Material £	Unit	Total rate £
TREE PLANTING – cont							
Tree planting – cont							
Quercus palustris 'Pin Oak'; including backfilling with excavated material (other operations not included)							
16–18 cm girth	200.00	1.30	46.20	5.25	230.00	nr	**281.46**
18–20 cm girth	250.00	1.60	57.04	5.25	287.50	nr	**349.79**
20–25 cm girth	300.00	4.50	160.43	24.81	345.00	nr	**530.24**
25–30 cm girth	450.00	6.00	213.90	30.83	517.50	nr	**762.23**
30–35 cm girth	550.00	11.00	392.15	41.16	632.50	nr	**1065.81**
Tilia cordata 'Greenspire'; including backfilling with excavated material (other operations not included)							
16–18 cm girth	150.00	1.30	46.20	5.25	172.50	nr	**223.96**
18–20 cm girth	150.00	1.60	57.04	5.25	172.50	nr	**234.79**
20–25 cm girth	200.00	4.50	160.43	24.81	230.00	nr	**415.24**
25–30 cm girth; 5 × transplanted; 4.0–5.0 m tall	300.00	6.00	213.90	30.83	345.00	nr	**589.73**
30–35 cm girth; 5 × transplanted; 5.0–7.0 m tall	550.00	11.00	392.15	41.16	632.50	nr	**1065.81**
Betula pendula (3 stems); including backfilling with excavated material (other operations not included)							
3.0–3.5 m high	300.00	1.98	70.59	67.10	345.00	nr	**482.69**
3.5–4.0 m high	400.00	1.60	57.04	72.12	460.00	nr	**589.16**
4.0–4.5 m high	500.00	2.38	84.70	80.53	575.00	nr	**740.23**
4.5–5.0 m high	600.00	2.97	105.88	123.43	690.00	nr	**919.31**
5.0–6.0 m high	800.00	3.96	141.17	134.21	920.00	nr	**1195.38**
6.0–7.0 m high	1000.00	4.50	160.43	157.22	1150.00	nr	**1467.65**
Pinus sylvestris; including backfilling with excavated material (other operations not included)							
3.0–3.5 m high	950.00	1.98	70.59	67.10	1092.50	nr	**1230.19**
3.5–4.0 m high	1200.00	1.60	57.04	72.12	1380.00	nr	**1509.16**
4.0–4.5 m high	1500.00	2.38	84.70	80.53	1725.00	nr	**1890.23**
4.5–5.0 m high	2000.00	2.97	105.88	123.43	2300.00	nr	**2529.31**
5.0–6.0 m high	2500.00	3.96	141.17	134.21	2875.00	nr	**3150.38**
6.0–7.0 m high	3500.00	4.50	160.43	162.29	4025.00	nr	**4347.72**

37 SOFT LANDSCAPING

Item	PC £	Labour hours	Labour £	Plant £	Material £	Unit	Total rate £
Pleached trees; Carpinus betulus; specimen trees; Lorenz Von Ehren; supply and planting only; excavation, support and treepit additives not included							
Box shaped trees to provide 'floating hedge' or screen effect; planting distance 1 tree per m run; trunk and box alignment; wire root balls							
4 × transplanted; 20–25 cm; 1.00 m centres	450.00	5.00	178.25	24.81	517.50	m	**720.56**
5 × transplanted; 25–30 cm; 1.00 m centres	600.00	6.50	231.73	30.83	690.00	m	**952.55**
5 × transplanted; 30–35 cm; 1.20 m centres	1020.00	11.00	392.15	41.16	1173.00	m	**1606.31**
Pleached trees; Tilia europaea 'Pallida'; specimen trees; Lorenz Von Ehren; supply and planting only; excavation, support and treepit additives not included							
Box pleached trees to provide 'floating hedge' or screen effect; wire root balls							
4 × transplanted; 20–25 cm; 1.00 m centres	400.00	5.00	178.25	24.81	460.00	m	**663.06**
4 × transplanted; 25–30 cm; 1.00 m centres	500.00	5.00	178.25	24.81	575.00	m	**778.06**
5 × transplanted; 30–35 cm; 1.20 m centres	541.65	5.00	178.25	24.81	622.89	m	**825.96**
5 × transplanted; 35–40 cm; 1.50 m centres	566.70	5.00	178.25	24.81	651.70	m	**854.76**
6 × transplanted; 40–45 cm; 1.80 m centres	694.50	5.00	178.25	24.81	798.68	m	**1001.74**
6 × transplanted; 45–50 cm; 1.80 m centres	833.40	5.00	178.25	24.81	958.41	m	**1161.47**
6 × transplanted; 50–60 cm; 2.00 m centres	875.00	5.00	178.25	24.81	1006.25	m	**1209.31**

37 SOFT LANDSCAPING

Item	PC £	Labour hours	Labour £	Plant £	Material £	Unit	Total rate £
TREE PLANTING – cont							
Espalier pleached trees; specimen trees; Lorenz Von Ehren; supply and planting only; excavation, support and treepit additives not included							
Frame pleached trees; to provide floating screen effects; wire root balls							
Carpinus betulus; 20–25 cm; 1.00 m centres	550.00	5.00	178.25	24.81	632.50	m	**835.56**
Carpinus betulus; 25–30 cm; 1.00 m centres	750.00	5.00	178.25	24.81	862.50	m	**1065.56**
Tilia europaea ‘Pallida’; 20–25 cm; 1.00 m centres	500.00	5.00	178.25	24.81	575.00	m	**778.06**
Tilia europaea ‘Pallida’; 25–30 cm; 1.00 m centres	850.00	5.00	178.25	24.81	977.50	m	**1180.56**
Tilia europaea ‘Pallida’; 30–35 cm; 1.20 m centres	916.63	5.00	178.25	24.81	1054.12	m	**1257.19**
Umbrella shaped pleaches							
Umbrella or roof pleached trees to provide umbrella effect; wire root balls							
Platanus acerifolia; clear stem; 4 × transplanted; 20–25 cm	600.00	5.00	178.25	24.81	690.00	nr	**893.06**
Platanus acerifolia; specimen; 5 × transplanted; 30–35 cm	1000.00	5.00	178.25	24.81	1150.00	nr	**1353.06**
Platanus acerifolia; specimen; 5 × transplanted; 25–30 cm	800.00	5.00	178.25	24.81	920.00	nr	**1123.06**
Tree planting; containerized trees; Lorenz von Ehren; to the tree prices above add for Airpot containerization only							
Tree size							
20–25 cm	–		–	–	115.00	nr	**115.00**
25–30 cm	–		–	–	172.50	nr	**172.50**
30–35 cm	–		–	–	172.50	nr	**172.50**
35–40 cm	–		–	–	253.00	nr	**253.00**
40–45 cm	–		–	–	402.50	nr	**402.50**
45–50 cm	–		–	–	517.50	nr	**517.50**
50–60 cm	–		–	–	690.00	nr	**690.00**
60–70 cm	–		–	–	920.00	nr	**920.00**
70–80 cm	–		–	–	1150.00	nr	**1150.00**
80–90 cm	–		–	–	1380.00	nr	**1380.00**

37 SOFT LANDSCAPING

Item	PC £	Labour hours	Labour £	Plant £	Material £	Unit	Total rate £
Tree planting; rootballed trees; semi-mature and mature trees; Lorenz von Ehren; planting and back filling; planted by telehandler or by crane; delivery included; all other operations priced separately							
Semi-mature trees							
40–45 cm girth	1353.75	8.00	285.20	95.08	1556.81	nr	**1937.10**
45–50 cm girth	2707.50	8.00	285.20	122.51	3113.63	nr	**3521.33**
50–60 cm girth	3790.50	10.00	356.50	149.63	4359.08	nr	**4865.21**
60–70 cm girth	5415.00	15.00	534.75	363.09	6227.25	nr	**7125.09**
70–80 cm girth	8122.50	18.00	641.70	339.62	9340.88	nr	**10322.20**
80–90 cm girth	10288.50	18.00	641.70	339.62	11831.78	nr	**12813.10**
90–100 cm girth	12454.50	18.00	641.70	339.62	14322.68	nr	**15304.00**
TREE PLANTING ACCESSORIES							
Gt ArborRaft System; Greentech Ltd; Strong geocellular system replaces traditional subbase layer in urban tree planting enabling shallow tree planting in a nutrient rich uncompacted environment that better reflects natural environment encouraging tree establishment and growth							
70t Loading capacity; single layer installation to fully graded and levelled partially backfilled treepit							
ArborRaft 150 mm unit (708 mm × 354 mm x150 mm)	77.92	0.17	11.89	–	89.61	m²	**101.50**
ArborRaft 85 mm unit (708 mm × 354 mm x85 mm)	73.01	0.14	10.19	–	83.96	m²	**94.14**
tree planting soil with nutrients compaction resistance and water retention qualities; structure remains open for water, oxygen and nutrients to reach developing root system of the tree.							
ArborRaft treepit soil 29 t loads; by machine	45.32		–	17.65	52.12	m³	**69.77**
ArborRaft treepit soil 20 t loads; by machine	62.75		–	17.65	72.16	m³	**89.80**
ArborRaft treepit soil bagged; by machine	89.24		–	29.41	102.63	m³	**132.03**
ArborRaft treepit soil bagged; by hand	89.24	0.50	35.65	–	102.63	m³	**138.28**

37 SOFT LANDSCAPING

Item	PC £	Labour hours	Labour £	Plant £	Material £	Unit	Total rate £
TREE PLANTING ACCESSORIES – cont							
StrataCell; GreenBlue Urban; Tree planting accessories for Urban tree planting; Load bearing tree planting modules for load support in tree environments; Placed to excavated treepit; single layer; 500 mm × 500 mm × 250 mm deep; interlocking load bearing cells; backfilling with imported topsoil							
7.5 tonne load limit	192.47	0.25	8.91	3.38	236.89	m^2	**249.19**
40 tonne load limit	1042.42	0.25	8.91	3.38	1214.33	m^2	**1226.63**
double layer; 500 mm × 500 mm × 500 mm deep; interlocking load bearing cells; backfilling with imported topsoil							
7.5 tonne load limit	1539.73	0.50	17.83	6.77	1801.79	m^2	**1826.39**
40 tonne load limit	2084.83	0.50	17.83	6.77	2428.66	m^2	**2453.26**
Root director; GreenBlue Urban; preformed root barrier system with integral root deflecting ribs; Placed over tree rootball; under pavements; overall size;							
510 × 595 × 310 mm	64.82	0.25	17.83	–	74.54	nr	**92.37**
650 × 855 × 455 mm	106.93	0.33	23.53	–	122.97	nr	**146.50**
975 × 1370 × 545 mm	164.66	0.45	32.09	–	189.36	nr	**221.44**
1300 × 1805 × 500 mm	275.98	0.60	42.78	–	317.38	nr	**360.16**
Tree staking J Toms Ltd; extra over trees for tree stake(s); driving 500 mm into firm ground; trimming to approved height; including two tree ties to approved pattern							
one stake; 1.50 m long × 32 mm × 32 mm	2.04	0.20	7.13	–	2.34	nr	**9.47**
two stakes; 900 mm long × 25 mm × 25 mm	2.14	0.30	10.70	–	2.47	nr	**13.16**
two stakes; 1.50 m long × 32 mm × 32 mm	2.90	0.30	10.70	–	3.34	nr	**14.03**
three stakes; 1.50 m long × 32 mm × 32 mm	4.35	0.36	12.83	–	5.01	nr	**17.84**

37 SOFT LANDSCAPING

Item	PC £	Labour hours	Labour £	Plant £	Material £	Unit	Total rate £
Tree anchors							
Platipus Anchors Ltd; extra over trees for tree anchors							
RF1P rootball kit; for 75–220 mm girth ×2–4.5 m high; inclusive of Plati-Mat PM1	33.79	1.00	35.65	–	38.86	nr	**74.51**
RF2P; rootball kit; for 220–450 mm girth × 4.5–7.5 m high; inclusive of Plati-Mat PM2	57.45	1.33	47.41	–	66.07	nr	**113.49**
RF3P; rootball kit; for 450–750 mm girth × 7.5 to 12 m high; inclusive of Plati-Mat PM3	129.26	1.50	53.48	–	148.64	nr	**202.12**
installation tools; drive rod for RF1 kits	–		–	–	78.46	nr	**78.46**
installation tools; drive rod for RF2 kits	–		–	–	119.56	nr	**119.56**
Extra over trees for land drain to tree pits; 100 mm dia. perforated flexible agricultural drain; including excavating drain trench; laying pipe; backfilling	1.31	1.00	35.65	–	1.51	m	**37.16**
Platipus anchors; deadman anchoring							
Deadman kits; anchoring to concrete kerbs placed in base of tree pit							
trees 12–25 cm girth; 2.5–4.0 m high	30.27	0.75	26.74	–	65.34	nr	**92.08**
trees 25–45 cm girth; 4.0–7.5 m high	52.15	1.50	53.48	–	90.50	nr	**143.98**
trees 45–75 cm girth; 7.5–12.0 m high	114.04	2.00	71.30	–	161.68	nr	**232.98**
Tree planting; tree pit additives							
Melcourt Industries Ltd; Topgrow; incorporating into topsoil at 1 part Topgrow to 3 parts excavated topsoil; supplied in 75 l bags; pit size							
600 × 600 × 600 mm	6.16	0.02	0.71	–	7.08	nr	**7.80**
900 × 900 × 900 mm	20.79	0.06	2.14	–	23.91	nr	**26.05**
1.00 × 1.00 × 1.00	28.49	0.24	8.56	–	32.76	nr	**41.32**
1.25 × 1.25 × 1.25	55.70	0.40	14.26	–	64.05	nr	**78.31**
1.50 × 1.50 × 1.50	96.25	0.90	32.09	–	110.69	nr	**142.77**
Melcourt Industries Ltd; Topgrow; incorporating into topsoil at 1 part Topgrow to 3 parts excavated topsoil; supplied in 60 m³ loose loads; pit size							
600 × 600 × 600 mm	2.36	0.02	0.60	–	2.71	nr	**3.31**
900 × 900 × 900 mm	7.96	0.05	1.78	–	9.15	nr	**10.93**
1.00 × 1.00 × 1.00	10.91	0.20	7.13	–	12.55	nr	**19.68**

37 SOFT LANDSCAPING

Item	PC £	Labour hours	Labour £	Plant £	Material £	Unit	Total rate £
TREE PLANTING ACCESSORIES – cont							
Tree planting – cont							
Melcourt Industries Ltd – cont							
1.25 × 1.25 × 1.25	21.31	0.33	11.88	–	24.51	nr	**36.39**
1.50 × 1.50 × 1.50	36.83	0.75	26.74	–	42.35	nr	**69.09**
Tree planting; root barriers							
GreenBlue Urban; linear root deflection barriers; installed to trench measured separately							
Re-Root 2000; 2.0 mm thick × 2 m wide	24.42	0.05	1.78	–	28.08	m	**29.87**
Re-Root 2000; 1.0 mm thick × 1 m wide	8.33	0.05	1.78	–	9.58	m	**11.36**
Re-Root 600; 1.0 mm thick	12.76	0.05	1.78	–	14.67	m	**16.46**
GreenBlue Urban; irrigation systems; Root Rain tree pit irrigation systems							
Metro; small; 35 mm pipe dia. × 1.25 m long; for specimen shrubs and standard trees							
plastic	6.03	0.25	8.91	–	6.93	nr	**15.85**
plastic; with chain	11.29	0.25	8.91	–	12.98	nr	**21.90**
Metro; medium; 35 mm pipe dia. × 1.75 m long; for standard and selected standard trees							
plastic	7.11	0.29	10.19	–	8.18	nr	**18.36**
plastic; with chain	12.56	0.29	10.19	–	14.44	nr	**24.63**
Metro; large; 35 mm pipe dia. × 2.50 m long; for selected standards and extra heavy standards							
plastic	8.74	0.33	11.88	–	10.05	nr	**21.93**
plastic; with chain	14.17	0.33	11.88	–	16.30	nr	**28.18**
Urban; for large capacity general purpose irrigation to parkland and street verge planting							
RRUrb1; 3.0 m pipe	–	0.29	10.19	–	13.32	nr	**23.50**
RRUrb2; 5.0 m pipe	14.60	0.33	11.88	–	16.79	nr	**28.67**
RRUrb3; 8.0 m pipe	19.15	0.40	14.26	–	22.02	nr	**36.28**
Civic; heavy cast aluminium inlet; for heavily trafficked locations							
5.0 m pipe	53.46	0.33	11.88	–	61.48	nr	**73.36**
8.0 m pipe	60.05	0.40	14.26	–	69.06	nr	**83.32**
Tree irrigation kits; Platipus anchors; direct water delivery to the rootball area of trees							

37 SOFT LANDSCAPING

Item	PC £	Labour hours	Labour £	Plant £	Material £	Unit	**Total rate £**
Piddler tree irrigation system; permeable membrane delivering targeted irrigation to rootball surround							
root balls up to 550 mm dia.; Irrigation Kit PID0	7.03	0.25	8.91	–	8.08	nr	**17.00**
root balls up to 900 mm dia.; Irrigation Kit PID1	13.00	0.30	10.70	–	14.95	nr	**25.64**
root balls up to 1.55 m dia.; Irrigation Kit PID2	16.89	0.40	14.26	–	19.43	nr	**33.69**
root balls up to 2.40 m dia.; Irrigation Kit PID3	20.58	0.45	16.04	–	23.66	nr	**39.71**
root balls up to 3.10 m dia.; Irrigation Kit PID4	34.14	0.50	17.83	–	39.26	nr	**57.08**
GreenBlue Urban; Arboresin porous tree pit surfacing resin to galvanized tree pit frames							
Two part support frame system							
support frame 1.20 m × 1.20 m	230.86	1.20	85.56	–	1175.20	nr	**1260.76**
support frame 1.50 m × 1.50 m	360.72	1.40	99.82	–	1680.82	nr	**1780.64**
support frame 1.20 m diam	181.16	1.10	78.43	–	1218.37	nr	**1296.80**
TREE PLANTING OPERATIONS AFTER PLANTING							
Mulching of tree pits; Melcourt Industries Ltd; FSC (Forest Stewardship Council) and FT (fire tested) certified							
Spreading mulch; to individual trees; maximum distance 25 m (mulch not included)							
50 mm thick	–	0.05	1.73	–	–	m²	**1.73**
75 mm thick	–	0.07	2.60	–	–	m²	**2.60**
100 mm thick	–	0.10	3.47	–	–	m²	**3.47**
Mulch; Bark Nuggets®; to individual trees; delivered in 80 m³ loads; maximum distance 25 m							
50 mm thick	2.74	0.05	1.73	–	3.15	m²	**4.88**
75 mm thick	4.11	0.07	2.60	–	4.73	m²	**7.33**
100 mm thick	5.48	0.10	3.47	–	6.30	m²	**9.77**
Mulch; Bark Nuggets®; to individual trees; delivered in 25 m³ loads; maximum distance 25 m							
50 mm thick	3.48	0.05	1.73	–	4.00	m²	**5.73**
75 mm thick	5.22	0.05	1.78		6.00	m²	**7.79**
100 mm thick	6.96	0.07	2.38	–	8.00	m²	**10.38**

37 SOFT LANDSCAPING

Item	PC £	Labour hours	Labour £	Plant £	Material £	Unit	Total rate £
TREE PLANTING OPERATIONS AFTER PLANTING – cont							
Mulching of tree pits – cont							
Mulch; Amenity Bark Mulch; to individual trees; delivered in 80 m^3 loads; maximum distance 25 m							
50 mm thick	1.85	0.05	1.73	–	2.13	m^2	**3.86**
75 mm thick	2.77	0.07	2.60	–	3.19	m^2	**5.79**
100 mm thick	3.70	0.10	3.47	–	4.25	m^2	**7.72**
Mulch; Amenity Bark Mulch; to individual trees; delivered in 25 m^3 loads; maximum distance 25 m							
50 mm thick	2.58	0.05	1.73	–	2.97	m^2	**4.70**
75 mm thick	3.87	0.07	2.60	–	4.45	m^2	**7.05**
100 mm thick	5.16	0.07	2.38	–	5.94	m^2	**8.31**
The labour in this section is calculated on a 2 person team. The labour time below should be multiplied by 2 to calculate the cost as shown.							
Resin bound mulch or surface to tree pits; Dekorgrip Ltd							
resin bound system with a typical makeup of 10 mm aggregate to create voids to allow and promote the free flow of air and water around tree beds to surface of tree pit							
Dekortree permeable resin treepits 50 mm uV stabilized 10-25 m^2	–	0.75	51.89	–	–	m^2	**184.65**
Dekortree permeable resin treepits 50 mm uV stabilized over 25 m^2	–	0.75	51.96	–	–	m^2	**168.27**
Dekortree 50 mm; Standard resin 10- 25 m^2	–	0.75	52.03	–	–	m^2	**155.17**
Dekortree 50 mm uV Standard over 25 m^2	–	0.75	52.12	–	–	m^2	**139.94**
SHRUB PLANTING							
Shrub planting – General							
Preamble: For preparation of planting areas see 'Cultivation' at the beginning of the section on planting.							

37 SOFT LANDSCAPING

Item	PC £	Labour hours	Labour £	Plant £	Material £	Unit	Total rate £
Market prices of shrubs							
The prices for shrubs are so varied that it is not possible to provide lists in this publication. For general budgeting purposes however the following average prices may be used with deciduous shrubs being slightly less and evergreeens slighty more.							
1 litre shrub	–		–	–	1.99	nr	**1.99**
2 litre shrub	–		–	–	4.23	nr	**4.23**
3 litre shrub	–		–	–	6.23	nr	**6.23**
5 litre shrub	–		–	–	12.45	nr	**12.45**
10 litre specimen shrub	–		–	–	24.91	nr	**24.91**
25 litre specimen shrub	–		–	–	74.73	nr	**74.73**
Market prices of high impact specimen shrubs; Prices are varied and the prices below reflect the price ranges for plants as described; container sizes;							
25 litre container	47.65		–	–	54.80	nr	**54.80**
25 litre container; specialist feature plant specimen	67.15		–	–	77.22	nr	**77.22**
40 litre container	79.60		–	–	91.54	nr	**91.54**
40 litre container; specialist feature plant specimen	121.30		–	–	139.49	nr	**139.49**
80 litre container	107.22		–	–	123.30	nr	**123.30**
80 litre container; specialist feature plant specimen	243.68		–	–	280.23	nr	**280.23**
130 litre container; specialist feature plant specimen	351.98		–	–	404.77	nr	**404.77**
Shrub planting							
Setting out; selecting planting from holding area; loading to wheelbarrows; planting as plan or as directed; distance from holding area maximum 50 m; plants 2–3 litre containers							
single plants not grouped	–	0.04	1.43	–	–	nr	**1.43**
plants in groups of 3–5 nr	–	0.03	0.89	–	–	nr	**0.89**
plants in groups of 10–100 nr	–	0.02	0.59	–	–	nr	**0.59**
plants in groups of 100 nr minimum	–	0.01	0.41	–	–	nr	**0.41**
Forming planting holes; in cultivated ground (cultivating not included); by mechanical auger; trimming holes by hand; depositing excavated material alongside holes							
250 mm dia.	–	0.03	1.19	0.09	–	nr	**1.28**
250 × 250 mm	–	0.04	1.43	0.16	–	nr	**1.59**
300 × 300 mm	–	0.08	2.67	0.20	–	nr	**2.87**

37 SOFT LANDSCAPING

Item	PC £	Labour hours	Labour £	Plant £	Material £	Unit	Total rate £
SHRUB PLANTING – cont							
Shrub planting – cont							
Planting shrubs; hand excavation; forming planting holes; in cultivated ground (cultivating not included); depositing excavated material alongside holes; Placing plants previously set out alongside and backfilling (plants not included)							
100 × 100 × 100 mm deep; with mattock or hoe	–	0.01	0.40	–	–	nr	**0.40**
250 × 250 × 300 mm deep	–	0.01	0.47	–	–	nr	**0.47**
300 × 300 × 300 mm deep	–	0.02	0.60	–	–	nr	**0.60**
400 × 400 × 400 mm deep	–	0.08	2.97	–	–	nr	**2.97**
500 × 500 × 500 mm deep	–	0.17	5.94	–	–	nr	**5.94**
600 × 600 × 600 mm deep	–	0.25	8.91	–	–	nr	**8.91**
900 × 900 × 600 mm deep	–	0.50	17.83	–	–	nr	**17.83**
1.00 m × 1.00 m x 600 mm deep	–	1.00	35.65	–	–	nr	**35.65**
1.25 m × 1.25 m x 600 mm deep	–	1.33	47.53	–	–	nr	**47.53**
Hand excavation; forming planting holes; in uncultivated ground; depositing excavated material alongside holes							
100 × 100 × 100 mm deep; with mattock or hoe	–	0.01	0.47	–	–	nr	**0.47**
250 × 250 × 250 mm deep	–	0.03	0.89	–	–	nr	**0.89**
300 × 300 × 300 mm deep	–	0.06	2.23	–	–	nr	**2.23**
400 × 400 × 400 mm deep	–	0.25	8.91	–	–	nr	**8.91**
500 × 500 × 500 mm deep	–	0.33	11.60	–	–	nr	**11.60**
600 × 600 × 600 mm deep	–	0.55	19.61	–	–	nr	**19.61**
900 × 900 × 600 mm deep	–	1.25	44.56	–	–	nr	**44.56**
1.00 m × 1.00 m x 600 mm deep	–	1.54	54.81	–	–	nr	**54.81**
1.25 m × 1.25 m x 600 mm deep	–	2.41	86.01	–	–	nr	**86.01**
Bare root planting; to planting holes (forming holes not included); including backfilling with excavated material (bare root plants not included)							
bare root 1+1; 30–90 mm high	–	0.02	0.60	–	–	nr	**0.60**
bare root 1+2; 90–120 mm high	–	0.02	0.60	–	–	nr	**0.60**

37 SOFT LANDSCAPING

Item	PC £	Labour hours	Labour £	Plant £	Material £	Unit	Total rate £
Shrub planting where plants are set out at equal off set centres where the spacings are always the specified distance between the plant centres. This introduces more uniform but slightly higher distances than rectangular centres							
2 litre containerized plants; in cultivated ground (cultivating not included); PC £3.68/nr							
1.00 m centres; 1.43 plants/m^2	–	0.02	0.68	–	6.06	m^2	**6.73**
900 mm centres; 1.69 plants/m^2	–	0.02	0.80	–	7.16	m^2	**7.96**
750 mm centres; 2.6 plants/m^2	–	0.04	1.27	–	11.01	m^2	**12.28**
600 mm centres; 3.04 plants/m^2	–	0.04	1.44	–	12.87	m^2	**14.31**
500 mm centres; 4.5 plants/m^2	–	0.06	2.14	–	19.06	m^2	**21.19**
450 mm centres; 6.75 plants/m^2	–	0.09	3.20	–	28.58	m^2	**31.79**
400 mm centres; 7.50 plants/m^2	–	0.17	5.99	–	42.43	m^2	**48.42**
300 mm centres; 12.7 plants/m^2	–	0.17	6.02	–	54.07	m^2	**60.10**
Shrub planting; 3 litre containerized plants; in cultivated ground (cultivating not included); PC £5.00/nr							
1.00 m centres; 1.43 plants/m^2	–	0.02	0.68	–	8.91	m^2	**9.58**
900 mm centres; 1.69 plants/m^2	–	0.02	0.80	–	10.52	m^2	**11.33**
750 mm centres; 2.6 plants/m^2	–	0.04	1.27	–	16.19	m^2	**17.46**
600 mm centres; 3.04 plants/m^2	–	0.04	1.44	–	18.93	m^2	**20.37**
500 mm centres; 4.5 plants/m^2	–	0.06	2.14	–	28.02	m^2	**30.16**
450 mm centres; 6.75 plants/m^2	–	0.09	3.20	–	42.03	m^2	**45.24**
400 mm centres; 7.50 plants/m^2	–	0.17	5.99	–	62.40	m^2	**68.39**
300 mm centres; 12.7 plants/m^2	–	0.17	6.02	–	79.52	m^2	**85.54**
HEDGE PLANTING							
Hedges							
The labour in this section is calculated on a 2 person team. The labour time below should be multiplied by 2 to calculate the cost as shown.							
Excavating trench for hedges; depositing soil alongside trench; by machine							
300 mm deep × 300 mm wide	–	0.03	2.16	1.55	–	m	**3.71**
300 mm deep × 600 mm wide	–	0.05	3.24	2.35	–	m	**5.59**
300 mm deep × 450 mm wide	–	0.05	3.21	2.32	–	m	**5.53**
500 mm deep × 500 mm wide		0.06	4.47	3.23	–	m	**7.70**
500 mm deep × 700 mm wide	–	0.09	6.25	4.52	–	m	**10.77**
600 mm deep × 750 mm wide	–	0.11	8.04	5.81	–	m	**13.85**
750 mm deep × 900 mm wide	–	0.17	12.06	8.72	–	m	**20.79**

37 SOFT LANDSCAPING

Item	PC £	Labour hours	Labour £	Plant £	Material £	Unit	Total rate £
HEDGE PLANTING – cont							
The labour in this section is calculated on a 3 person team. The labour time below should be multiplied by 3 to calculate the cost as shown.							
Excavating trench for hedges; depositing soil alongside trench; by hand							
300 mm deep × 300 mm wide	–	0.08	8.91	–	–	m	**8.91**
300 mm deep × 450 mm wide	–	0.13	13.37	–	–	m	**13.37**
300 mm deep × 600 mm wide	–	0.17	17.83	–	–	m	**17.83**
500 mm deep × 500 mm wide	–	0.23	24.18	–	–	m	**24.18**
500 mm deep × 700 mm wide	–	0.32	33.91	–	–	m	**33.91**
600 mm deep × 750 mm wide	–	0.41	43.59	–	–	m	**43.59**
750 mm deep × 900 mm wide	–	0.61	65.27	–	–	m	**65.27**
Setting out; notching out; excavating trench; breaking up subsoil to minimum depth 300 mm							
minimum 400 mm deep	–	0.25	8.91	–	–	m	**8.91**
Disposal of excavated soil to stockpile							
Up to 25 m distant							
By hand	–	2.00	85.56	–	–	m^3	**85.56**
By machine	–		–	3.57	–	m^3	**3.57**
The labour in this section is calculated on a 2 person team. The labour time below should be multiplied by 2 to calculate the cost as shown.							
Instant Hedge planting; Readyhedge Ltd; continuous strips of instant hedges; delivered in 1.00 m lengths; planting and backfilling with excavated material of hedging plants to trenches (not included); including minor trimming and shaping after planting							
Beech (Fagus) hedges							
height 0.80–1.00 m × 200 mm deep	48.00	0.20	14.26	10.16	55.20	m	**79.62**
height 1.40–1.60 m × 400 mm deep	140.00	0.25	17.83	12.70	161.00	m	**191.53**
height 1.60 m × 400 mm deep	192.00	0.67	47.54	33.87	220.80	m	**302.21**
Purple Beech (Fagus) hedges							
height 0.60–0.70 m × 200 mm deep	73.64	0.20	14.26	10.16	84.69	m	**109.11**
height 1.50–1.60 m × 350 mm deep	192.00	0.33	23.76	16.93	220.80	m	**261.50**

37 SOFT LANDSCAPING

Item	PC £	Labour hours	Labour £	Plant £	Material £	Unit	Total rate £
Yew (Taxus) hedges							
700 mm high × 250 mm wide;	132.30	0.50	53.48	4.52	152.15	m	**210.14**
1.60–1.80 m high × 300 mm wide; 500 mm centres	151.20	0.57	61.11	4.52	173.88	m	**239.51**
Prunus lauruscerasus hedges							
height 1.40–1.60 m × 40 cm deep	155.00	0.25	17.83	12.70	178.25	m	**208.78**
Prunus lucitanica (Portuguese laurel) hedges							
height 70–90 cm × 20 cm deep	55.00	0.20	14.26	10.16	63.25	m	**87.67**
height 1.30–1.50 m × 35 cm deep	155.00	0.25	17.83	12.70	178.25	m	**208.78**
Carpinus (Hornbeam) hedges							
height 80–100 cm × 20 cm deep	186.00	0.20	14.26	10.16	213.90	m	**238.32**
height 1.60–1.80 m × 40 cm deep	105.00	0.25	17.83	12.70	120.75	m	**151.28**
Box (Buxus) hedges							
height 30–35 cm × 20 cm deep	50.00	0.10	7.13	5.08	57.50	m	**69.71**
height 40–45 cm × 25 cm deep	125.00	0.13	8.91	6.35	143.75	m	**159.01**
The labour in this section is calculated on a 3 person team. The labour time below should be multiplied by 3 to calculate the cost as shown.							
Instant Hedge planting; Readyhedge Ltd; individual pre-clipped hedge plants to form mature hedge; planting and backfilling of hedging plants to trenches excavated separately; including minor trimming and shaping after planting							
Feathered hedges; Beech or Hornbeam							
1.75 m high × 300 mm wide; 400 mm centres	52.50	0.20	21.39	2.26	60.38	m	**84.03**
1.75 m high × 300 mm wide; 500 mm centres	42.00	0.16	17.11	2.26	48.30	m	**67.67**
1.75 m high × 300 mm wide; 750 mm centres	29.82	0.11	12.22	2.26	34.29	m	**48.78**
1.75 m high × 300 mm wide; 900 mm centres	23.31	0.09	9.51	2.26	26.81	m	**38.58**
1.75/2.00 m high × 300 mm wide; 600 mm centres	35.00	0.33	35.65	21.30	40.25	m	**97.20**
1.75/2.00 m high × 300 mm wide; 900 mm centres	23.31	0.29	30.56	18.69	26.81	m	**76.05**
2.00/2.25 m high 300 wide; 600 mm centres	53.34	0.40	42.78	25.71	61.35	m	**129.83**
2.00/2.25 m high × 300 mm wide; 900 mm centres	35.20	0.33	35.65	21.30	40.48	m	**97.42**
2.00/2.25 m high × 300 mm wide; 1.20 m centres	26.67	0.25	26.74	16.73	30.67	m	**74.13**

37 SOFT LANDSCAPING

Item	PC £	Labour hours	Labour £	Plant £	Material £	Unit	Total rate £
HEDGE PLANTING – cont							
Hedge planting; Readyhedge Ltd; feathered hedge plants to form mature hedge; planting and backfilling of hedging plants to trenches excavated separately; including minor trimming and shaping after planting							
(Yew) Taxus hedging; root balled							
600/800 mm high × 300 mm wide; 250 mm centres	62.00	0.20	21.39	2.26	71.30	m	**94.95**
600/800 mm high × 300 mm wide; 500 mm centres	31.00	0.11	11.88	2.26	35.65	m	**49.79**
800 mm / 1.00 m high × 300 mm wide; 250 mm centres	80.00	0.20	21.39	2.26	92.00	m	**115.65**
800 mm / 1.00 m high × 300 mm wide; 500 mm centres	40.00	0.11	11.88	2.26	46.00	m	**60.14**
1.00/1.20 m high × 300/350 mm wide; 3 nr/m	86.58	0.17	17.83	2.26	99.57	m	**119.66**
1.00/1.20 m high × 300/350 mm wide; 2 nr/m	52.00	0.13	13.37	2.26	59.80	m	**75.43**
1.25/1.50 m high × 300/350 mm wide; 3 nr/m	116.55	0.17	17.83	2.26	134.03	m	**154.12**
1.25/1.50 m high × 300/350 mm wide; 2 nr/m	70.00	0.13	14.26	2.26	80.50	m	**97.02**
1.50/1.75 m high × 400 mm wide; 450 mm centres	144.30	0.33	35.65	15.97	165.95	m	**217.56**
1.50/1.75 m high × 400 mm wide; 600 mm centres	108.34	0.22	23.76	14.45	124.59	m	**162.80**
1.50/1.75 m high × 400 mm wide; 900 mm centres	72.15	0.20	21.39	13.23	82.97	m	**117.59**
1.75/2.00 m high × 500 mm wide; 600 mm centres	133.34	0.33	35.65	21.30	153.34	m	**210.28**
1.75/2.00 m high × 500 mm wide; 900 mm centres	88.80	0.29	30.56	18.69	102.12	m	**151.36**
1.75/2.00 m high × 500 mm wide; 1.20 m centres	66.66	0.25	26.74	16.73	76.66	m	**120.13**
2.00/2.25 m high × 600 mm wide; 600 mm centres	145.19	0.50	53.48	31.95	166.97	m	**252.39**
2.00/2.25 m high × 600 mm wide; 900 mm centres	96.74	0.40	42.78	17.48	111.25	m	**171.51**
2.00/2.25 m high × 600 mm wide; 1.20 m centres	72.33	0.25	26.74	12.16	83.18	m	**122.08**
(Yew) Taxus hedging; wire root balled; extra over for wire rootballing	5.25		–	–	6.04	ea	**6.04**

37 SOFT LANDSCAPING

Item	PC £	Labour hours	Labour £	Plant £	Material £	Unit	Total rate £
Buxus (Box) hedges; rootballed							
400/500 mm high; 200 mm rootball; 200 mm centres	47.00	0.09	3.17	–	54.05	m	**57.22**
400/500 mm high; 200 mm rootball; 300 mm centres	31.30	0.07	2.38	–	36.00	m	**38.38**
400/500 mm high; 200 mm rootball; 500 mm centres	18.80	0.05	1.78	–	21.62	m	**23.40**
500/600 mm high; 200 mm rootball; 200 mm centres	65.65	0.09	3.17	–	75.50	m	**78.67**
500/600 mm high; 200 mm rootball; 300 mm centres	43.72	0.07	2.38	–	50.28	m	**52.67**
500/600 mm high; 200 mm rootball; 500 mm centres	26.26	0.08	2.74	–	30.20	m	**32.94**
600/800 mm high; 250 mm rootball; 250 mm centres	73.52	0.09	3.24	–	84.55	m	**87.79**
600/800 mm high; 250 mm rootball; 400 mm centres	45.95	0.08	2.97	–	52.84	m	**55.81**
600/800 mm high; 250 mm rootball; 600 mm centres	21.88	0.08	2.74	–	25.17	m	**27.91**
800 mm / 1.00 m high; 300 mm rootball; 300 mm centres	87.41	0.10	3.57	–	100.52	m	**104.09**
800 mm / 1.00 m high; 300 mm rootball; 500 mm centres	52.50	0.11	3.96	–	60.38	m	**64.34**
800 mm / 1.00 m high; 300 mm rootball; 750 mm centres	34.91	0.13	4.46	–	40.15	m	**44.61**
1.00 m / 1.25 m high; 350 mm rootball; 400 mm centres	102.38	0.13	4.46	–	117.73	m	**122.19**
1.00 m / 1.25 m high; 350 mm rootball; 600 mm centres	68.26	0.11	3.96	–	78.50	m	**82.46**
1.00 m / 1.25 m high; 350 mm rootball; 900 mm centres	45.45	0.10	3.57	–	52.27	m	**55.84**
Native species and bare root hedge planting							
Hedge planting operations only (excavation of trenches priced separately) including backfill with excavated topsoil; PC £0.56/nr; 40–60 cm							
single row; 200 mm centres	2.82	0.06	2.23	–	3.24	m	**5.47**
single row; 300 mm centres	1.88	0.06	1.98	–	2.16	m	**4.14**
single row; 400 mm centres	1.41	0.04	1.49	–	1.62	m	**3.11**
single row; 500 mm centres	1.13	0.03	1.19	–	1.30	m	**2.48**
double row; 200 mm centres	5.63	0.17	5.94	–	6.48	m	**12.42**
double row; 300 mm centres	3.75	0.13	4.75	–	4.32	m	**9.07**
double row; 400 mm centres	2.82	0.08	2.97	–	3.24	m	**6.21**
double row; 500 mm centres	2.25	0.07	2.38	–	2.59	m	**4.97**
Extra over hedges for incorporating manure; at 1 m^3 per 30 m	0.83	0.03	0.89	–	0.95	m	**1.85**

37 SOFT LANDSCAPING

Item	PC £	Labour hours	Labour £	Plant £	Material £	Unit	Total rate £
TOPIARY							
Clipped topiary; Lorenz von Ehren; German field grown clipped and transplanted as detailed; planted to plant pit; including backfilling with excavated material and TPMC							
Buxus sempervirens (box); balls							
300 mm dia.; 3 × transplanted; container grown or rootballed	30.00	0.25	8.91	–	45.65	nr	**54.56**
500 mm dia.; 4 × transplanted; wire rootballed	80.00	1.50	53.48	6.03	176.86	nr	**236.37**
900 mm dia.; 5 × transplanted; wire rootballed	320.00	2.75	98.04	6.03	508.77	nr	**612.84**
1300 mm dia.; 6 × transplanted; wire rootballed	960.00	3.10	110.52	7.24	1242.59	nr	**1360.35**
Buxus sempervirens (box); pyramids							
500 mm high; 3 × transplanted; container grown or rootballed	50.00	1.50	53.48	6.03	142.36	nr	**201.87**
900 mm high; 4 × transplanted; wire rootballed	160.00	2.75	98.04	6.03	324.77	nr	**428.84**
1300 mm high; 5 × transplanted; wire rootballed	460.00	3.10	110.52	7.24	667.59	nr	**785.35**
Buxus sempervirens (box); truncated pyramids							
500 mm high; 3 × transplanted; container grown or rootballed	110.00	1.50	53.48	6.03	211.36	nr	**270.87**
900 mm high; 4 × transplanted; wire rootballed	380.00	2.75	98.04	6.03	577.77	nr	**681.84**
Buxus sempervirens (box); truncated cone							
1300 mm high; 5 × transplanted; wire rootballed	1130.00	3.10	110.52	7.24	1438.09	nr	**1555.85**
Buxus sempervirens (box); cubes							
500 mm square; 4 × transplanted; rootballed	140.00	1.50	53.48	6.03	245.86	nr	**305.37**
900 mm square; 5 × transplanted; wire rootballed	470.00	2.75	98.04	6.03	681.27	nr	**785.34**
Taxus baccata (yew); balls							
500 mm dia.; 4 × transplanted; rootballed	90.00	1.50	53.48	6.03	188.36	nr	**247.87**
900 mm dia.; 5 × transplanted; wire rootballed	250.00	2.75	98.04	6.03	428.27	nr	**532.34**
1300 mm dia.; 6 × transplanted; wire rootballed	750.00	3.10	110.52	7.24	1001.09	nr	**1118.85**

37 SOFT LANDSCAPING

Item	PC £	Labour hours	Labour £	Plant £	Material £	Unit	**Total rate £**
Taxus baccata (yew); cones							
800 mm high; 4 × transplanted; wire rootballed	80.00	1.50	53.48	6.03	176.86	nr	**236.37**
1500 mm high; 5 × transplanted; wire rootballed	180.00	2.00	71.30	6.03	311.98	nr	**389.32**
2500 mm high; 6 × transplanted; wire rootballed	580.00	3.10	110.52	7.24	805.59	nr	**923.35**
Taxus baccata (yew); cubes							
500 mm square; 4 × transplanted; rootballed	90.00	1.50	53.48	6.03	188.36	nr	**247.87**
900 mm square; 5 × transplanted; wire rootballed	350.00	2.75	98.04	6.03	543.27	nr	**647.34**
Taxus baccata (yew); pyramids							
900 mm high; 4 × transplanted; wire rootballed	150.00	1.50	53.48	6.03	257.36	nr	**316.87**
1500 mm high; 5 × transplanted; wire rootballed	300.00	3.10	110.52	7.24	483.59	nr	**601.35**
2500 mm high; 6 × transplanted; wire rootballed	840.00	4.00	142.60	12.07	1142.36	nr	**1297.02**
Carpinus betulus (common hornbeam); columns; round base							
800 mm wide × 2000 mm high; 4 × transplanted; wire rootballed	250.00	3.10	110.52	7.24	426.09	nr	**543.85**
800 mm wide × 2750 mm high; 4 × transplanted; wire rootballed	400.00	4.00	142.60	12.07	636.36	nr	**791.02**
Carpinus betulus (common hornbeam); cones							
3 m high; 5 × transplanted; wire rootballed	350.00	4.00	142.60	12.07	578.86	nr	**733.52**
4 m high; 5 × transplanted; wire rootballed	850.00	5.00	178.25	12.07	1187.47	nr	**1377.78**
Carpinus betulus 'Fastigiata'; pyramids							
4 m high; 6 × transplanted; wire rootballed	1170.00	5.00	178.25	12.07	1555.47	nr	**1745.78**
7 m high; 7 × transplanted; wire rootballed	3180.00	7.50	267.38	18.10	3917.28	nr	**4202.75**
LIVING WALLS							
Design to living walls; Scotscape Living Walls Ltd							
Fytotextile Living Wall Design							
10–30 m²	–		–	–	–	nr	**595.65**
30–70 m²	–		–	–	–	nr	**774.35**
70–100 m²	–		–	–	–	nr	**1012.61**
100–150 m²	–		–	–	–	nr	**1191.30**
over 150 m²	–		–	–	–	nr	**1786.95**

37 SOFT LANDSCAPING

Item	PC £	Labour hours	Labour £	Plant £	Material £	Unit	Total rate £
LIVING WALLS – cont							
Living wall; fytotextile and soil-based system; Scotscape Living Walls; self-contained, irrigated vertical planting system; Fabric modules, fully established upon installation; saturated weight 40 kg per m²; access costs excluded							
System units 250 × 500 × 100 mm overall; fixed to horizontal hanging rails on softwood batten and DPM; planting depth 150 mm at 30 degree angle to vertical surface; inclusive of automatic dripline irrigation system and guttering with running outlets to drain (not included); Area of installation;							
5–10 m²	–		–	–	–	m²	**536.09**
10–20 m²	–		–	–	–	m²	**488.43**
20–30 m²	–		–	–	–	m²	**476.52**
30–70 m²	–		–	–	–	m²	**476.52**
70–100 m²	–		–	–	–	m²	**470.56**
100–150 m²	–		–	–	–	m²	**464.61**
over 150 m²	–		–	–	–	m²	**464.61**
Fully automatic irrigation systems to the above; comprising breaktank, station controller, solenoid valves, plant feeders and dripline emitters							
10–30 m²	–		–	–	–	nr	**3276.08**
30–70 m²	–		–	–	–	nr	**4616.29**
70–100 m²	–		–	–	–	nr	**5896.94**
100–150 m²	–		–	–	–	nr	**7445.63**
over 150 m²	–		–	–	–	nr	**9232.58**
Rainwater Harvesting System	–		–	–	–	nr	**2680.43**
Preliminary costs for all living walls							
general preliminaries	–		–	–	–	nr	**416.96**
design costs; site visits, consultations and presentations	–		–	–	–	nr	**411.00**
Maintenance to living walls							
Annual maintenance to living walls; regular visits to maintain planting and systems; inclusive of feeding; pest control and calibration of irrigation systems							
10–30 m²	–		–	–	–	annum	**2978.25**
30–70 m²	–		–	–	–	annum	**5658.68**
70–100 m²	–		–	–	–	annum	**8934.75**
100–150 m²	–		–	–	–	annum	**10721.70**
over 150 m²	–		–	–	–	annum	**13104.30**

37 SOFT LANDSCAPING

Item	PC £	Labour hours	Labour £	Plant £	Material £	Unit	**Total rate £**
Green Screen; Mobilane Ltd							
Fully installed vertical green screen of Hedera hibernica; 65 plants per linear m; all on 5 mm galvanized steel weldmesh; attached to existing wall fence or hoarding							
screen 1.80 high	–		–	–	218.50	m	**218.50**
screen 2.20 m high	–		–	–	276.00	m	**276.00**
screen 4.00 m high	–		–	–	506.00	m	**506.00**
HERBACEOUS/GROUNDCOVER/ BULB/BEDDING PLANTING							
Groundcover plants							
Groundcover plants in massed areas; PC £1.73/nr; including forming planting holes in cultivated ground (cultivating not included); backfilling with excavated material; 1 litre containers							
average 4 plants per m^2; 500 mm centres	6.93	0.09	3.33	–	7.97	m^2	**11.30**
average 6 plants per m^2; 408 mm centres	10.40	0.14	4.99	–	11.96	m^2	**16.95**
average 8 plants per m^2; 354 mm centres	13.86	0.19	6.66	–	15.94	m^2	**22.60**
Groundcover plants in massed areas; PC £3.03/nr; including forming planting holes in cultivated ground (cultivating not included); backfilling with excavated material; 1 litre containers							
average 4 plants per m^2; 500 mm centres	12.13	0.09	3.33	–	13.95	m^2	**17.28**
average 6 plants per m^2; 408 mm centres	18.19	0.14	4.99	–	20.92	m^2	**25.92**
average 8 plants per m^2; 354 mm centres	24.26	0.19	6.66	–	27.90	m^2	**34.56**
Herbaceous planting							
Herbaceous plants; PC £1.73/nr; including forming planting holes in cultivated ground (cultivating not included); backfilling with excavated material; 1 litre containers							
average 4 plants per m^2; 500 mm centres	6.03	0.00	3.33		7.07	m^2	**11.30**
average 6 plants per m^2; 408 mm centres	10.40	0.14	4.99	–	11.96	m^2	**16.95**
average 8 plants per m^2; 354 mm centres	13.86	0.19	6.66	–	15.94	m^2	**22.60**

37 SOFT LANDSCAPING

Item	PC £	Labour hours	Labour £	Plant £	Material £	Unit	Total rate £
HERBACEOUS/GROUNDCOVER/ BULB/BEDDING PLANTING – cont							
Herbaceous planting – cont							
Note: For machine incorporation of fertilizers and soil conditioners see 'Cultivation'.							
Bulb planting							
Bulbs; including forming planting holes in cultivated area (cultivating not included); backfilling with excavated material							
small	–	0.83	29.71	–	–	100 nr	**29.71**
medium	22.00	0.83	29.71	–	25.30	100 nr	**55.01**
large	25.00	0.91	32.41	–	28.75	100 nr	**61.16**
Bulbs; in grassed area; using bulb planter; including backfilling with screened topsoil or peat and cut turf plug							
small	–	1.67	59.42	–	–	100 nr	**59.42**
medium	22.00	1.67	59.42	–	25.30	100 nr	**84.72**
large	25.00	2.00	71.30	–	28.75	100 nr	**100.05**
Aquatic planting							
Aquatic plants; Planting only in prepared growing medium in pool; plant size 2–3 litre containerized (plants not included)	–	0.04	1.43	–	–	nr	**1.43**
MARKET PRICES OF NATIVE SPECIES							
The following prices are typically some of the more popular species used in native plantations; prices vary between species; readers should check the catalogue of the supplier							
1+1; 1 year seedling transplanted and grown for a year							
30–40 cm	–		–	–	0.48	nr	**0.48**
40–60 cm	–		–	–	0.56	nr	**0.56**
1+2; 1 year seedling transplanted and grown for 2 years							
60–80 cm	–		–	–	0.81	nr	**0.81**

37 SOFT LANDSCAPING

Item	PC £	Labour hours	Labour £	Plant £	Material £	Unit	Total rate £
Whips							
80–100 cm	–		–	–	0.93	nr	**0.93**
100–125 cm	–		–	–	1.95	nr	**1.95**
Feathered trees							
125–150 cm	–		–	–	3.01	nr	**3.01**
150–180 cm	–		–	–	6.06	nr	**6.06**
180–200 cm	–		–	–	10.40	nr	**10.40**
200–250 cm	–		–	–	12.67	nr	**12.67**
Native species planting							
Forming planting holes; in cultivated ground (cultivating not included); by mechanical auger; trimming holes by hand; depositing excavated material alongside holes							
250 mm dia.	–	0.03	1.03	0.08	–	nr	**1.11**
250 × 250 mm	–	0.04	1.24	0.14	–	nr	**1.38**
300 × 300 mm	–	0.08	2.33	0.17	–	nr	**2.50**
Hand excavation; forming planting holes; in cultivated ground (cultivating not included); depositing excavated material alongside holes							
100 × 100 × 100 mm deep; with mattock or hoe	–	0.01	0.21	–	–	nr	**0.21**
250 × 250 × 300 mm deep	–	0.04	1.24	–	–	nr	**1.24**
300 × 300 × 300 mm deep	–	0.06	1.72	–	–	nr	**1.72**
TREE/SHRUB PROTECTION							
Tree planting; tree protection – General							
Preamble: Care must be taken to ensure that tree grids and guards are removed when trees grow beyond the specified dia. of guard.							
Tree planting; tree protection							
Expandable plastic tree guards; including 25 mm softwood stakes							
500 mm high	1.73	0.17	5.94	–	1.99	nr	**7.93**
1.00 m high	1.86	0.17	5.94	–	2.14	nr	**8.08**
English Woodlands; Weldmesh tree guards; nailing to tree stakes (tree stakes not included)							
1800 mm high × 300 mm dia.	30.32	0.25	8.91	–	34.87	nr	**43.79**
J. Toms Ltd; spiral rabbit guards; clear or brown							
610 × 38 mm	0.31	0.03	1.19	–	0.36	nr	**1.55**

37 SOFT LANDSCAPING

Item	PC £	Labour hours	Labour £	Plant £	Material £	Unit	Total rate £
TREE/SHRUB PROTECTION – cont							
Tree planting – cont							
English Woodlands; Plastic Mesh Tree Guards; black; supplied in 50 m rolls							
13 × 13 mm small mesh; roll width 60 cm	2.35	0.04	1.27	–	3.81	nr	**5.08**
13 × 13 mm small mesh; roll width 120 cm	4.91	0.06	2.10	–	6.75	nr	**8.85**
Tree guards of 3 nr 2.40 m × 100 mm stakes; driving 600 mm into firm ground; bracing with timber braces at top and bottom; including 3 strands barbed wire	0.93	1.00	35.65	–	1.07	nr	**36.72**
English Woodlands; strimmer guard in heavy duty black plastic; 225 mm high	5.94	0.07	2.38	–	6.83	nr	**9.21**
Standard Treeshelter inclusive of 25 mm stake; prices shown for quantities of 500 nr							
0.75 m high	1.84	0.05	1.78	–	3.33	nr	**5.11**
Tubex Ltd; Shrubshelter inclusive of 25 mm stake; prices shown for quantities of 500 nr							
0.6 m high	2.96	0.07	2.38	–	3.40	nr	**5.78**
Extra over trees for spraying with antidesiccant spray; Wiltpruf							
selected standards; standards; light standards	6.54	0.20	7.13	–	7.52	nr	**14.65**
standards; heavy standards	10.90	0.25	8.91	–	12.54	nr	**21.45**
FORESTRY PLANTING							
Forestry planting							
Deep ploughing rough ground to form planting ridges at							
2.00 m centres	–		–	64.94	–	100 m²	**64.94**
3.00 m centres	–		–	61.11	–	100 m²	**61.11**
4.00 m centres	–		–	41.56	–	100 m²	**41.56**
Notching plant forestry seedlings; T or L notch	56.32	0.75	26.74	–	64.76	100 nr	**91.50**
Turf planting forestry seedlings	56.32	2.00	71.30	–	64.76	100 nr	**136.06**
Cleaning and weeding around seedlings; once	–	0.50	17.83	–	–	100 nr	**17.83**
Treading in and firming ground around seedlings planted; at 2500 per ha after frost or other ground disturbance; once	–	0.33	11.88	–	–	100 nr	**11.88**
Beating up initial planting; once (including supply of replacement seedlings at 10% of original planting)	5.63	0.25	8.91	–	6.48	100 nr	**15.39**

37 SOFT LANDSCAPING

Item	PC £	Labour hours	Labour £	Plant £	Material £	Unit	Total rate £
OPERATIONS AFTER PLANTING							
Operations after planting							
Initial cutting back to shrubs and hedge plants; including disposal of all cuttings	–	1.00	35.65	–	–	100 m^2	**35.65**
Mulch; Melcourt Industries Ltd; Bark Nuggets®; to plant beds; delivered in 80 m^3 loads; maximum distance 25 m							
50 mm thick	2.74	0.04	1.58	–	3.15	m^2	**4.73**
75 mm thick	4.11	0.07	2.38	–	4.73	m^2	**7.11**
100 mm thick	5.48	0.09	3.17	–	6.30	m^2	**9.47**
Mulch; Melcourt Industries Ltd; Bark Nuggets®; to plant beds; delivered in 25 m^3 loads; maximum distance 25 m							
50 mm thick	3.09	0.03	1.04	–	3.55	m^2	**4.59**
75 mm thick	5.22	0.07	2.38	–	6.00	m^2	**8.38**
100 mm thick	6.96	0.09	3.17	–	8.00	m^2	**11.17**
Mulch; Melcourt Industries Ltd; Amenity Bark Mulch FSC; to plant beds; delivered in 80 m^3 loads; maximum distance 25 m							
50 mm thick	1.85	0.04	1.58	–	2.13	m^2	**3.71**
75 mm thick	2.77	0.07	2.38	–	3.19	m^2	**5.57**
100 mm thick	3.70	0.09	3.17	–	4.25	m^2	**7.42**
Mulch; Melcourt Industries Ltd; Amenity Bark Mulch FSC; to plant beds; delivered in 25 m^3 loads; maximum distance 25 m							
50 mm thick	2.58	0.04	1.58	–	2.97	m^2	**4.55**
75 mm thick	3.87	0.07	2.38	–	4.45	m^2	**6.83**
100 mm thick	5.16	0.09	3.17	–	5.94	m^2	**9.11**
Mulch mats; English Woodlands; lay mulch mat to planted area or plant station; mat secured with metal J pins 240 × 3 mm							
Mats to individual plants and plant stations							
Woven Polypropylene; 50 × 50 cm; square mat	0.74	0.05	1.78	–	2.26	nr	**4.04**
Woven Polypropylene; 1 × 1 m; square mat	2.28	0.07	2.38	–	4.03	nr	**6.41**
Hedging mats							
Woven Polypropylene; 1 × 100 m roll; hedge planting	0.64	0.01	0.45	–	1.44	m^2	**1.89**
General planting areas							
Plantex membrane; 1 × 14 m roll; weed control	2.01	0.01	0.36	–	3.02	m^2	**3.38**

37 SOFT LANDSCAPING

Item	PC £	Labour hours	Labour £	Plant £	Material £	Unit	Total rate £
OPERATIONS AFTER PLANTING – cont							
Fertilizers; application during aftercare period							
Fertilizers; in top 150 mm of topsoil at 35 g/m²							
Mascot Microfine; turf fertilizer; 8 +0+6 + 2% Mg + 4% Fe	6.23	0.12	4.37	–	7.17	100 m²	**11.54**
Enmag; controlled release fertilizer (8–9 months); 11+22+09	11.34	0.12	4.37	–	13.04	100 m²	**17.41**
Mascot Outfield; turf fertilizer; 8 +12+8	5.71	0.12	4.37	–	6.57	100 m²	**10.94**
Mascot Outfield; turf fertilizer; 9 +5+5	5.22	0.12	4.37	–	6.00	100 m²	**10.37**
Fertilizers; in top 150 mm of topsoil at 70 g/m²							
Mascot Microfine; turf fertilizer; 8 +0+6 + 2% Mg + 4% Fe	12.47	0.12	4.37	–	14.34	100 m²	**18.71**
Enmag; controlled release fertilizer (8–9 months); 11+22+09	22.69	0.12	4.37	–	26.09	100 m²	**30.46**
Mascot Outfield; turf fertilizer; 8 +12+8	11.43	0.12	4.37	–	13.14	100 m²	**17.51**
Mascot Outfield; turf fertilizer; 9 +5+5	10.43	0.12	4.37	–	12.00	100 m²	**16.37**
AFTERCARE AS PART OF A LANDSCAPE CONTRACT							
Maintenance operations (Note: the following rates apply to aftercare maintenance executed as part of a landscaping contract only.)							
Weeding and hand forking planted areas; including disposing weeds and debris on site; areas maintained weekly	–	–	0.14	–	–	m²	**0.14**
Weeding and hand forking planted areas; including disposing weeds and debris on site; areas maintained monthly	–	0.01	0.36	–	–	m²	**0.36**
Extra over weeding and hand forking planted areas for disposing excavated material off site; loaded to maintenance vehicle and transported to the contractors yard							
green waste for composting	–		–	–	–	m³	**71.48**
green waste to contractors yard deposited in skips	–		–	52.78	–	m³	**125.63**

37 SOFT LANDSCAPING

Item	PC £	Labour hours	Labour £	Plant £	Material £	Unit	Total rate £
Edging to plant beds; maintaing edge against turfed surround or boundary; edging by half moon tool							
planting beds							
maintaining edges	–	0.04	1.43	–	–	m	**1.43**
reforming edges; removing arisings	–	0.07	2.38	–	–	m	**2.38**
Treepits edges							
maintaining edges	–	0.10	3.57	–	–	nr	**3.57**
reforming edges; removing arisings	–	0.20	7.13	–	–	nr	**7.13**
Sweeping and cleaning pathways and surfaces							
Sweep paths or surfaces to high footfall public or commercial areas							
per visit – weekly visits	–	0.01	0.18	–	–	m²	**0.18**
sweeping snow	–	0.03	0.89	–	–	m²	**0.89**
gritting and clearing ice	–	0.03	0.89	–	0.07	m²	**0.96**
Top up mulch at the end of the 12 month maintenance period to the full thickness as shown below							
Mulch; Melcourt Industries Ltd; Bark Nuggets®; to plant beds; delivered in 80 m³ loads; maximum distance 25 m							
50 mm thick	0.55	0.03	1.02	–	0.63	m²	**1.65**
75 mm thick	0.82	0.01	0.18	–	0.95	m²	**1.12**
Mulch; Melcourt Industries Ltd; Bark Nuggets®; to plant beds; delivered in 25 m³ loads; maximum distance 25 m							
50 mm thick	0.70	0.03	1.02	–	0.80	m²	**1.82**
75 mm thick	1.04	0.03	1.19	–	1.20	m²	**2.39**
Mulch; Melcourt Industries Ltd; Amenity Bark Mulch FSC; to plant beds; delivered in 80 m³ loads; maximum distance 25 m							
50 mm thick	0.92	0.03	1.02	–	1.06	m²	**2.08**
75 mm thick	1.29	0.03	1.19	–	1.49	m²	**2.67**
Mulch; Melcourt Industries Ltd; Amenity Bark Mulch FSC; to plant beds; delivered in 25 m³ loads; maximum distance 25 m							
50 mm thick	1.29	0.03	0.89	–	1.48	m²	**2.38**
75 mm thick	1.68	0.03	1.02	–	1.93	m²	**2.95**

37 SOFT LANDSCAPING

Item	PC £	Labour hours	Labour £	Plant £	Material £	Unit	Total rate £
AFTERCARE AS PART OF A LANDSCAPE CONTRACT – cont							
Top up mulch at the end of the 12 month maintenance period to the full thickness as shown below – cont							
Fertilizers; at 35 g/m^2							
Mascot Microfine; turf fertilizer; 8 +0+6 + 2% Mg + 4% Fe	6.23	0.12	4.37	–	7.17	100 m^2	**11.54**
Enmag; controlled release fertilizer (8–9 months); 11+22+09	11.34	0.12	4.37	–	13.04	100 m^2	**17.41**
Mascot Outfield; turf fertilizer; 8+12+8	5.71	0.12	4.37	–	6.57	100 m^2	**10.94**
Mascot Outfield; turf fertilizer; 9 +5+5	5.22	0.12	4.37	–	6.00	100 m^2	**10.37**
Fertilizers; at 70 g/m^2							
Mascot Microfine; turf fertilizer; 8 +0+6 + 2% Mg + 4% Fe	12.47	0.12	4.37	–	14.34	100 m^2	**18.71**
Enmag; controlled release fertilizer (8–9 months); 11+22+09	22.69	0.12	4.37	–	26.09	100 m^2	**30.46**
Mascot Outfield; turf fertilizer; 8 +12+8	11.43	0.12	4.37	–	13.14	100 m^2	**17.51**
Mascot Outfield; turf fertilizer; 9 +5+5	10.43	0.12	4.37	–	12.00	100 m^2	**16.37**
Watering planting; evenly; at a rate of 5 l/m^2							
using hand-held watering equipment	–	0.25	8.91	–	–	100 m^2	**8.91**
using sprinkler equipment and with sufficient water pressure to run one 15 m radius sprinkler	–	0.14	4.96	–	–	100 m^2	**4.96**
using movable spray lines powering three sprinkler heads with a radius of 15 m and allowing for 60% overlap (irrigation machinery costs not included)	–	0.02	0.56	–	–	100 m^2	**0.56**

37 SOFT LANDSCAPING

Item	PC £	Labour hours	Labour £	Plant £	Material £	Unit	**Total rate £**
LONG TERM MAINTENANCE							
Preamble: Long-term landscape maintenance Maintenance on long-term contracts differs in cost from that of maintenanceas part of a landscape contract. In thissection the contract period is generally3–5 years. Staff are generally allocated to a single project only and therefore productivity is higher whilst overhead costs are lower. Labour costs in this section are lower than the costs used in other parts of the book. Machinery is assumed to be leased over a five year period and written off over the same period. The costs of maintenance and consumables for the various machinery types have been included in the information that follows. Finance costs for the machinery have not been allowed for. The rates shown below are for machines working in unconfined contiguous areas. Users should adjust the times and rates if working in smaller spaces or spaces with obstructions.							
MARKET PRICES OF LANDSCAPE CHEMICALS							
The following table provides the areas of spraying in each planting scenario for plants which require 1.00 m dia. weed free circles Area of weed free circles required in each planting density							
plants at 500 mm centres; 400 nr/100 m²	–		–	–	400.00	m²	**400.00**
plants at 600 m centres; 278 nr/100 m²	–		–	–	278.00	m²	**278.00**
plants at 750 mm centres; 178 nr/100 m²	–		–	–	178.00	m²	**178.00**
plants at 1.00 m centres; 100 nr/100 m²					100.00	m²	**100.00**
plants at 1.50 m centres; 4444 nr/ha	–		–	–	3490.00	m²/ha	**3490.00**
plants at 1.75 m centres; 3265 nr/ha	–		–	–	2564.00	m²/ha	**2564.00**
plants at 2.00 m centres; 2500 nr/ha	–		–	–	1963.00	m²/ha	**1963.00**

37 SOFT LANDSCAPING

Item	PC £	Labour hours	Labour £	Plant £	Material £	Unit	Total rate £
MARKET PRICES OF LANDSCAPE CHEMICALS – cont							
Market prices of landscape chemicals at suggested application rates							
Note: All chemicals are standard knapsack or backpack applied unless specifically stated as CDA or TDA.							
TOTAL HERBICIDES							
Herbicide applications; CDA; chemical application via low pressure specialized wands to landscape planting; application to maintain 1.00 m dia. clear circles (0.79 m²) around new planting							
Rigby Taylor CDA; Discman CDA Biograde enhanced movement glyphosate; application rate 7.5–10 litre/ha							
plants at 1.50 m centres; 4444 nr/ha	–		–	–	103.22	ha	**103.22**
plants at 1.75 m centres; 3265 nr/ha	–		–	–	71.05	ha	**71.05**
plants at 2.00 m centres; 2500 nr/ha	–		–	–	58.42	ha	**58.42**
mass spraying	–		–	–	197.36	ha	**197.36**
spot spraying;1 incident / 25 m²	–		–	–	7.89	ha	**7.89**
spot spraying;1 incident / 50 m²	–		–	–	0.39	ha	**0.39**
spot spraying;1 incident / 100 m²	–		–	–	0.20	ha	**0.20**
Total herbicides; CDA; low pressure sustainable applications							
Hilite; Nomix Enviro Ltd; glyphosate oil-emulsion; Total Droplet Control (TDC formerly CDA); low volume sustainable chemical applications; formulation controls annual and perennial grasses and non-woody broad-leaved weeds in amenity and industrial areas, parks, highways, paved areas and factory sites; also effective for clearing vegetation prior to planting of ornamental species, around the bases of trees, shrubs and roses, as well as in forestry areas							
10 l/ha	–		–	–	2.99	100 m²	**2.99**
7.5 l/ ha	–		–	–	2.24	100 m²	**2.24**

37 SOFT LANDSCAPING

Item	PC £	Labour hours	Labour £	Plant £	Material £	Unit	**Total rate £**
Gallup Biograde Amenity; Rigby Taylor Ltd; glyphosate 360 g/l formulation							
general use; woody weeds; ash, beech, bracken, bramble; 3 l/ha	–		–	–	0.36	100 m²	**0.36**
annual and perennial grasses; heather (peat soils); 4 l/ha	–		–	–	0.48	100 m²	**0.48**
pre-planting; general clearance; 5 l/ha	–		–	–	0.61	100 m²	**0.61**
heather; mineral soils; 6 l/ha	–		–	–	0.73	100 m²	**0.73**
rhododendron; 10 l/ha	–		–	–	1.21	100 m²	**1.21**
Gallup Hi-aktiv Amenity; Rigby Taylor Ltd; 490 g/l formulation							
general use; woody weeds; ash, beech, bracken, bramble; 2.2 l/ha	–		–	–	0.29	100 m²	**0.29**
annual and perennial grasses; heather (peat soils); 2.9 l/ha	–		–	–	0.38	100 m²	**0.38**
pre-planting; general clearance; 3.7 l/ha	–		–	–	0.48	100 m²	**0.48**
heather; mineral soils; 4.4 l/ha	–		–	–	0.58	100 m²	**0.58**
rhododendron; 7.3 l/ha	–		–	–	0.96	100 m²	**0.96**
Discman CDA Biograde; Rigby Taylor Ltd							
application rate; 7 l/ha; light vegetation and annual weeds	–		–	–	1.38	100 m²	**1.38**
application rate; 8.5 l/ha; forestry/ woodlands; before planting, to control broad-leaved and grass weeds	–		–	–	1.68	100 m²	**1.68**
application rate; 10 l/ha; established deep-rooted annuals, perennial grasses and broad-leaved weeds	–		–	–	1.97	100 m²	**1.97**
Roundup Pro Biactive 360; Everris							
application rate; 5 l/ha; land not intended for cropping	–		–	–	0.42	100 m²	**0.42**
application rate; 10 l/ha; forestry; weed control	–		–	–	0.84	100 m²	**0.84**
Roundup Pro Vantage 480 ; Everris; 480 g/litre glyphosate							
application rate 4 l/ha; natural surfaces not intended to bear vegetation; amenity vegetation control; permeable surfaces overlaying soil; hard surfaces	–		–	–	0.39	100 m²	**0.39**
application rate 8 l/ha; forest; forest nursery weed control	–		–	–	0.78	100 m²	**0.78**
application rate 16 mm/l; stump application	–		–	–	0.16	litre	**0.16**

37 SOFT LANDSCAPING

Item	PC £	Labour hours	Labour £	Plant £	Material £	Unit	Total rate £
MARKET PRICES OF LANDSCAPE CHEMICALS – cont							
Roundup Pro Biactive 450; Everris; enhanced movement glyphosate; application rate 4 litre/ha							
plants at 1.50 m centres; 4444 nr/ha	–		–	–	56.65	ha	**56.65**
plants at 1.75 m centres; 3265 nr/ha	–		–	–	41.67	ha	**41.67**
plants at 2.00 m centres; 2500 nr/ha	–		–	–	41.67	ha	**31.88**
mass spraying	–		–	–	41.67	ha	**43.08**
spot spraying	–		–	–	41.67	100 m²	**1.62**
Dual; Nomix Enviro Ltd; glyphosate/sulfosulfuron; TDC; low volume sustainable chemical applications; residual effect lasting up to 6 months; shrub beds, gravelled areas, tree bases, parks, industrial areas, hard areas and around obstacles							
application rate 9 l/ha	–		–	–	3.21	100 m²	**3.21**
Total herbicides with residual pre-emergent action							
Pistol; ALS; Bayer; barrier prevention for emergence control and glyphosate for control of emerged weeds at a single application per season; application rate 4.5 litre/ha; 40 g/l diflufenican and 250 g/l glyphosate							
plants at 1.50 m centres; 4444 nr/ha	–		–	–	45.97	ha	**45.97**
plants at 1.75 m centres; 3265 nr/ha	–		–	–	33.67	ha	**33.67**
plants at 2.00 m centres; 2500 nr/ha	–		–	–	33.67	ha	**32.94**
Residual herbicides							
Chikara; Rigby Taylor Ltd; Belchim Crop Protection; flazasulfuron; systemic pre-emergent and early post-emergent herbicide for the control of annual and perennial weeds on natural surfaces not intended to bear vegetation and permeable surfaces over-lying soil							
150 g/ha	–		–	–	2.26	100 m²	**2.26**

37 SOFT LANDSCAPING

Item	PC £	Labour hours	Labour £	Plant £	Material £	Unit	**Total rate £**
Contact herbicides							
Natural Weed and Moss Spray; Headland Amenity Ltd; acetic acid							
application rate; 100 l/ha	–		–	–	5.90	100 m^2	**5.90**
Selective herbicides CDA							
Hilite; Nomix Enviro; glyphosate oil-emulsion; TD; low volume sustainable chemical applications; formulation controls annual and perennial grasses and non-woody broad-leaved weeds in amenity and industrial areas, parks, highways, paved areas and factory sites; also effective for clearing vegetation prior to planting of ornamental species, around the bases of trees, shrubs and roses, as well as in forestry areas							
10 l/ha	–		–	–	2.99	100 m^2	**2.99**
7.5 l/ha	–		–	–	2.24	100 m^2	**2.24**
Selective herbicides; power or knapsack sprayer							
Junction; Rigby Taylor Ltd; selective post-emergence turf herbicide containing 6.25 g/litre florasulam plus 452 g/litre 2,4-D; specially designed for use by contractors for maintenance treatments of a wide range of broad-leaved weeds whilst giving excellent safety to turfgrass							
application rate; 1.2 l/ha	–		–	–	0.57	100 m^2	**0.57**
Greenor; Rigby Taylor Ltd; fluroxypyr, clopyralid and MCPA; systemic selective treatment of weeds in turfgrass							
application rate; 4 l/ha	–		–	–	1.94	100 m^2	**1.94**
Crossbar Herbicide; Rigby Taylor Ltd; fluroxypyr, 2,4-D and dicamba for control of weeds in amenity turf including fine turf							
application rate; 2 l/ha	–		–	–	0.95	100 m^2	**0.95**
Praxys; Everris; fluroxypyr-meptyl, clopyralid and florasulam; selective, systemic post-emergence herbicide for managed amenity turf, lawns and amenity grassland with high selectivity to established and young turf							
application rate; 9090–14,285 m^2/5 l	–		–	–	0.90	100 m^2	**0.90**

37 SOFT LANDSCAPING

Item	PC £	Labour hours	Labour £	Plant £	Material £	Unit	Total rate £
MARKET PRICES OF LANDSCAPE CHEMICALS – cont							
Selective herbicides – cont							
Cabadex; Headland Amenity Ltd; fluroxypyr and florasulam; newly sown and established turf (particularly speedwell, yarrow and yellow suckling clover)							
application rate 2 l/ha	–		–	–	1.13	100 m²	**1.13**
Blaster; Headland Amenity Ltd; triclopyr clopyralid; nettles, docks, thistles, brambles, woody weeds in amenity grassland							
application rate; 4 l/ha	–		–	–	1.45	100 m²	**1.45**
Aquatic herbicides; backpack or power sprayers							
Roundup Pro Biactive 360; Everris							
application rate; 6 l/ha; aquatic weed control	–		–	–	0.51	100 m²	**0.51**
Roundup Pro Biactive 450; Everris; 450 g/litre glyphosate							
application rate 4.8 l/ha; enclosed waters, open waters, land immediately adjacent to aquatic areas	–		–	–	0.47	100 m²	**0.47**
Aquatic herbicides; TDC							
Conqueror; Nomix Enviro; low volume sustainable chemical applications glyphosate with aquatic approval							
10 l/ha; established annuals, deep-rooted perennials and broad-leaved weeds	–		–	–	0.31	100 m²	**0.31**
7.5 l/ha; light vegetation and small annual weeds, aquatic situations	–		–	–	2.33	100 m²	**2.33**
12.5 l/ha; aquatic situations, emergent weeds e.g. reed, grasses and watercress	–		–	–	0.39	100 m²	**0.39**
15 l/ha; aquatic situations, floating weeds such as water lilies	–		–	–	0.47	100 m²	**0.47**

37 SOFT LANDSCAPING

Item	PC £	Labour hours	Labour £	Plant £	Material £	Unit	**Total rate £**
Fungicides							
Exterris; Rigby Taylor Ltd; broad spectrum fungicide displaying protective and curative activity against fusarium patch and red thread							
application rate; 700 g/ha	–		–	–	3.75	100 m²	**3.75**
Dedicate; Rigby Taylor Ltd; contact and systemic fungicide with curative and preventative activity against major turf diseases such as fusarium, red thread, dollar spot, anthracnose and rust							
application rate; 700 g/ha	–		–	–	7.20	100 m²	**7.20**
LANDSCAPE CHEMICAL APPLICATION							
Chemical applications labours only							
Controlled droplet application (CDA)							
plants at 1.50 m centres; 4444 nr/ha	–	9.26	536.71	–	–	ha	**536.71**
plants at 1.50 m centres; 44.44 nr/100 m²	–	0.93	53.67	–	–	100 m²	**53.67**
plants at 1.75 m centres; 3265 nr/ha	–	6.80	394.13	–	–	ha	**394.13**
plants at 1.75 m centres; 3265 nr/ha	–	0.68	39.41	–	–	100 m²	**39.41**
plants at 2.00 m centres; 2500 nr/ha	–	5.21	301.97	–	–	ha	**301.97**
plants at 2.00 m centres; 2500 nr/ha	–	0.52	30.20	–	–	100 m²	**30.20**
spot spraying to shrub beds	–	0.20	7.13	–	–	100 m²	**7.13**
spot spraying to hard surfaces	–	0.13	4.46	–	–	100 m²	**4.46**
mass spraying	–	2.50	144.90	–	–	ha	**144.90**
mass spraying	–	0.25	14.49	–	–	100 m²	**14.49**
Power spraying							
mass spraying to areas containing vegetation	–	2.00	71.30	219.20	–	ha	**290.50**
mass spraying to hard surfaces	–	–	0.15	0.45	–	100 m²	**0.60**
spraying to kerbs and channels along roadways 300 mm wide	–	1.25	44.56	137.00	–	1 km	**181.56**
General herbicides, in accordance with manufacturer's instructions; Knapsack spray application, see 'Market rates of landscape chemicals' for costs of specific applications							
knapsack sprayer; selective spraying around bases of plants	–	0.18	6.25	–	–	100 m²	**6.25**
knapsack sprayer; spot spraying		0.25	8.91			100 m²	**8.91**
knapsack sprayer; mass spraying	–	0.40	14.26	–	–	100 m²	**14.26**
granular distribution by hand or hand applicator	–	0.20	7.13	–	–	100 m²	**7.13**

37 SOFT LANDSCAPING

Item	PC £	Labour hours	Labour £	Plant £	Material £	Unit	Total rate £
LANDSCAPE CHEMICAL APPLICATION – cont							
Chemical applications labours only – cont							
Backpack spraying; keeping weed free at bases of plants							
plants at 1.50 m centres; 4444 nr/ha	–	13.89	805.06	–	–	ha	**805.06**
plants at 1.75 m centres; 3265 nr/ha	–	10.20	591.19	–	–	ha	**591.19**
plants at 2.00 m centres; 2500 nr/ha	–	7.82	452.96	–	–	ha	**452.96**
mass spraying	–	7.50	434.70	–	–	ha	**434.70**
Herbicide applications; standard backpack spray applicators; application to maintain 1.00 m dia. clear circles (0.79 m^2) around new planting							
Everris; Roundup Pro Biactive 360; glyphosate; enhanced movement glyphosate; application rate 5 litre/ha							
plants at 1.50 m centres; 4444 nr/ha	16.26	12.35	715.52	–	18.70	ha	**734.22**
plants at 1.75 m centres; 3265 nr/ha	11.89	9.07	525.70	–	13.68	ha	**539.38**
plants at 2.00 m centres; 2500 nr/ha	9.18	6.95	402.53	–	10.55	ha	**413.08**
mass spraying	–	–	0.04	–	0.01	m^2	**0.05**
GRASS CUTTING							
The labour in this section is calculated on a 2 person team. The labour time below should be multiplied by 2 to calculate the cost as shown.							
Grass cutting only; ride-on or tractor drawn equipment; works carried out on one site only such as large fields or amenity areas where labour and machinery is present for a full day; grass cutting only; strimming not included							
Using multiple-gang mower with cylindrical cutters; contiguous areas such as playing fields and the like larger than 3000 m^2							
3 gang; 2.13 m cutting width	–	0.01	0.30	0.15	–	100 m^2	**0.46**
5 gang; 3.40 m cutting width	–	–	0.27	0.19	–	100 m^2	**0.47**
7 gang; 4.65 m cutting width	–	–	0.17	0.17	–	100 m^2	**0.34**

37 SOFT LANDSCAPING

Item	PC £	Labour hours	Labour £	Plant £	Material £	Unit	Total rate £
Using multiple-gang mower with cylindrical cutters; non-contiguous areas such as verges and general turf areas							
3 gang	–	0.01	0.60	0.15	–	100 m²	**0.76**
5 gang	–	0.01	0.39	0.19	–	100 m²	**0.58**
Using multiple-rotary mower with vertical drive shaft and horizontally rotating bar or disc cutters; contiguous areas larger than 3000 m²							
cutting grass, overgrowth or the like using flail mower or reaper	–	0.01	0.58	0.37	–	100 m²	**0.95**
Grass cutting with strimming; ride-on or tractor drawn equipment; works carried out on one site only such as large fields or amenity areas where labour and machinery is present for a full day; includes accompanying strimmer operative							
Using multiple-gang mower with cylindrical cutters; contiguous areas such as playing fields and the like larger than 3000 m²							
3 gang; 2.13 m cutting width	–	0.01	0.60	0.18	–	100 m²	**0.78**
5 gang; 3.40 m cutting width	–	0.01	0.55	0.22	–	100 m²	**0.77**
7 gang; 4.65 m cutting width	–	0.01	0.34	0.22	–	100 m²	**0.56**
Using multiple-gang mower with cylindrical cutters; non-contiguous areas such as verges and general turf areas							
3 gang	–	0.02	1.21	0.26	–	100 m²	**1.47**
5 gang	–	0.01	0.78	0.23	–	100 m²	**1.00**
Using multiple-rotary mower with vertical drive shaft and horizontally rotating bar or disc cutters; contiguous areas larger than 3000 m²							
cutting grass, overgrowth or the like using flail mower or reaper	–	0.02	1.16	0.47	–	100 m²	**1.63**

37 SOFT LANDSCAPING

Item	PC £	Labour hours	Labour £	Plant £	Material £	Unit	Total rate £
GRASS CUTTING – cont							
Grass cutting only; ride-on or tractor drawn equipment; works carried out on multiple sites; machinery moved by trailer between sites; strimming not included							
Using multiple-gang mower with cylindrical cutters; contiguous areas such as playing fields and the like larger than 3000 m^2							
3 gang; 2.13 m cutting width	–	0.01	0.49	0.15	–	100 m^2	**0.65**
5 gang; 3.40 m cutting width	–	0.01	0.45	0.19	–	100 m^2	**0.64**
7 gang; 4.65 m cutting width	–	–	0.28	0.17	–	100 m^2	**0.45**
Using multiple-gang mower with cylindrical cutters; non-contiguous areas such as verges and general turf areas							
3 gang	–	0.02	0.98	0.15	–	100 m^2	**1.13**
5 gang	–	0.01	0.63	0.19	–	100 m^2	**0.82**
Using multiple-rotary mower with vertical drive shaft and horizontally rotating bar or disc cutters; contiguous areas larger than 3000 m^2							
cutting grass, overgrowth or the like using flail mower or reaper	–	0.02	0.94	0.37	–	100 m^2	**1.31**
Cutting grass, overgrowth or the like; using tractor-mounted side-arm flail mower; in areas inaccessible to alternative machine; on surface							
not exceeding 30° from horizontal	–	0.01	0.68	0.45	–	100 m^2	**1.13**
30–50° from horizontal	–	0.02	1.36	0.45	–	100 m^2	**1.81**
Grass cutting and strimming; ride-on or tractor drawn equipment; works carried out on multiple sites; machinery moved by trailer between sites; includes for accompanying strimmer operative							
Using multiple-gang mower with cylindrical cutters; contiguous areas such as playing fields and the like larger than 3000 m^2							
3 gang; 2.13 m cutting width	–	0.02	0.99	0.20	–	100 m^2	**1.18**
5 gang; 3.40 m cutting width	–	0.02	0.89	0.23	–	100 m^2	**1.12**
7 gang; 4.65 m cutting width	–	0.01	0.55	0.19	–	100 m^2	**0.74**

37 SOFT LANDSCAPING

Item	PC £	Labour hours	Labour £	Plant £	Material £	Unit	Total rate £
Using multiple-gang mower with cylindrical cutters; non-contiguous areas such as verges and general turf areas							
3 gang	–	0.03	1.96	0.24	–	100 m²	**2.20**
5 gang	–	0.02	1.26	0.30	–	100 m²	**1.57**
Using multiple-rotary mower with vertical drive shaft and horizontally rotating bar or disc cutters; contiguous areas larger than 3000 m²							
cutting grass, overgrowth or the like using flail mower or reaper	–	0.03	1.89	0.45	–	100 m²	**2.34**
Cutting grass, overgrowth or the like; using tractor-mounted side-arm flail mower; in areas inaccessible to alternative machine; on surface							
not exceeding 30° from horizontal	–	0.02	1.38	0.51	–	100 m²	**1.89**
30° to 50° from horizontal	–	0.05	2.71	0.57	–	100 m²	**3.28**
Grass cutting; pedestrian operated equipment; the following rates are for grass cutting only; there is no allowance for accompanying strimming operations							
Using cylinder lawn mower fitted with not less than five cutting blades, front and rear rollers; on surface not exceeding 30° from horizontal; arisings let fly; width of cut							
51 cm	–	0.03	1.77	0.21	–	100 m	**1.98**
61 cm	–	0.03	1.48	0.23	–	100 m	**1.71**
71 cm	–	0.02	1.27	0.25	–	100 m	**1.51**
91 cm	–	0.02	0.99	0.50	–	100 m	**1.49**
Using rotary self-propelled mower; width of cut							
45 cm	–	0.02	1.15	0.13	–	100 m²	**1.28**
81 cm	–	0.01	0.64	0.22	–	100 m²	**0.86**
120 cm	–	0.01	0.43	0.14	–	100 m²	**0.57**
Add for using grass box for collecting and depositing arisings							
removing and depositing arisings	–	0.03	1.45	–	–	100 m²	**1.45**

37 SOFT LANDSCAPING

Item	PC £	Labour hours	Labour £	Plant £	Material £	Unit	Total rate £
GRASS CUTTING – cont							
Grass cutting; pedestrian operated equipment; the following rates are for grass cutting with accompanying strimming operations							
Using cylinder lawn mower fitted with not less than five cutting blades, front and rear rollers; on surface not exceeding 30° from horizontal; arisings let fly; width of cut							
51 cm	–	0.06	3.53	0.37	–	100 m	**3.90**
61 cm	–	0.05	2.95	0.36	–	100 m	**3.31**
71 cm	–	0.04	2.53	0.35	–	100 m	**2.89**
91 cm	–	0.03	1.98	0.58	–	100 m	**2.56**
Using rotary self-propelled mower; width of cut							
45 cm	–	0.04	2.30	0.23	–	100 m^2	**2.52**
81 cm	–	0.02	1.27	0.27	–	100 m^2	**1.54**
91 cm	–	0.02	1.13	0.17	–	100 m^2	**1.30**
120 cm	–	0.01	0.86	0.18	–	100 m^2	**1.03**
Add for using grass box for collecting and depositing arisings							
removing and depositing arisings	–	0.03	1.45	–	–	100 m^2	**1.45**
Add for 30–50° from horizontal	–		–	–	–	33%	**–**
Add for slopes exceeding 50°	–		–	–	–	100%	**–**
Cutting grass or light woody undergrowth; using trimmer with nylon cord or metal disc cutter; on surface							
not exceeding 30° from horizontal	–	0.10	5.80	0.50	–	100 m^2	**6.29**
30–50° from horizontal	–	0.20	11.59	1.00	–	100 m^2	**12.59**
exceeding 50° from horizontal	–	0.25	14.49	1.25	–	100 m^2	**15.74**
Grass cutting; collecting arisings							
Extra over for tractor drawn and self-propelled machinery using attached grass boxes; depositing arisings							
22 cuts per year	–	0.03	1.45	–	–	100 m^2	**1.45**
18 cuts per year	–	0.04	2.17	–	–	100 m^2	**2.17**
12 cuts per year	–	0.05	2.90	–	–	100 m^2	**2.90**
4 cuts per year	–	0.13	7.25	–	–	100 m^2	**7.25**
Arisings collected by trailed sweepers							
22 cuts per year	–	0.01	0.49	0.32	–	100 m^2	**0.81**
18 cuts per year	–	0.01	0.64	0.32	–	100 m^2	**0.97**
12 cuts per year	–	0.02	0.97	0.32	–	100 m^2	**1.29**
4 cuts per year	–	0.03	1.94	0.32	–	100 m^2	**2.26**

37 SOFT LANDSCAPING

Item	PC £	Labour hours	Labour £	Plant £	Material £	Unit	Total rate £
Disposing arisings on site; 100 m distance maximum							
22 cuts per year	–	–	0.24	0.03	–	100 m²	**0.27**
18 cuts per year	–	–	0.29	0.04	–	100 m²	**0.33**
12 cuts per year	–	0.01	0.43	0.06	–	100 m²	**0.48**
4 cuts per year	–	0.02	1.26	0.18	–	100 m²	**1.45**
Disposal of arisings off site							
22 cuts per year	–	–	0.23	0.06	–	100 m²	**2.09**
18 cuts per year	–	0.01	0.28	0.07	–	100 m²	**2.75**
12 cuts per year	–	–	0.18	0.19	–	100 m²	**4.67**
4 cuts per year	–	0.01	0.56	1.13	–	100 m²	**8.90**
GROUNDCARE OPERATIONS							
The labour in this section is calculated on a 2 person team. The labour time below should be multiplied by 2 to calculate the cost as shown.							
Harrowing							
Harrowing grassed area with							
drag harrow	–	0.01	0.37	0.18	–	100 m²	**0.55**
chain or light flexible spiked harrow	–	0.01	0.49	0.18	–	100 m²	**0.67**
Scarifying							
Mechanical							
Sisis ARP4; including grass collection box; towed by tractor; area scarified annually	–	0.01	0.68	0.78	–	100 m²	**1.46**
Sisis ARP4; including grass collection box; towed by tractor; area scarified two years previously	–	0.01	0.81	0.95	–	100 m²	**1.76**
pedestrian operated self-powered equipment	–	0.04	2.03	1.25	–	100 m²	**3.27**
add for disposal of arisings	–	0.01	0.69	1.40	–	100 m²	**37.93**
By hand							
hand implement	–	0.25	14.49	–	–	100 m²	**14.49**
add for disposal of arisings	–	0.01	0.69	1.40	–	100 m²	**37.93**
Rolling							
Rolling grassed area; equipment towed by tractor; once over; using							
smooth roller	–	0.01	0.42	0.20	–	100 m²	**0.61**

37 SOFT LANDSCAPING

Item	PC £	Labour hours	Labour £	Plant £	Material £	Unit	Total rate £
GROUNDCARE OPERATIONS – cont							
Turf aeration							
By machine							
Vertidrain turf aeration equipment towed by tractor to effect a minimum penetration of 100–250 mm at 100 mm centres	–	0.02	1.13	3.04	–	100 m²	**4.17**
Ryan GA 30; self-propelled turf aerating equipment; to effect a minimum penetration of 100 mm at varying centres	–	0.02	1.13	1.30	–	100 m²	**2.43**
Groundsman; pedestrian operated; self-powered solid or slitting tine turf aerating equipment to effect a minimum penetration of 100 mm	–	0.07	4.06	1.18	–	100 m²	**5.25**
Cushman core harvester; self-propelled; for collection of arisings	–	0.01	0.56	0.91	–	100 m²	**1.47**
By hand							
hand fork; to effect a minimum penetration of 100 mm and spaced 150 mm apart	–	0.67	38.64	–	–	100 m²	**38.64**
hollow tine hand implement; to effect a minimum penetration of 100 mm and spaced 150 mm apart	–	1.00	57.96	–	–	100 m²	**57.96**
collection of arisings by hand	–	1.50	86.94	–	–	100 m²	**86.94**
Turf areas; surface treatments and top dressing;British sugar topsoil							
Apply screened topdressing to grass surfaces; spread using Tru-Lute							
Sand soil mixes 90/10 to 50/50	–	–	0.06	0.11	0.32	m²	**0.48**
Apply screened soil 3 mm Kettering loam to goal mouths and worn areas							
20 mm thick	–	0.01	0.29	–	1.26	m²	**1.55**
10 mm thick	1.64	0.01	0.29	–	1.88	m²	**2.17**
Leaf clearance; clearing grassed area of leaves and other extraneous debris							
Using equipment towed by tractor							
large grassed areas with perimeters of mature trees such as sports fields and amenity areas	–	0.01	0.37	0.09	–	100 m²	**0.45**
large grassed areas containing ornamental trees and shrub beds	–	–	0.12	1.22	–	100 m²	**1.34**

37 SOFT LANDSCAPING

Item	PC £	Labour hours	Labour £	Plant £	Material £	Unit	**Total rate £**
Using pedestrian operated mechanical equipment and blowers							
grassed areas with perimeters of mature trees such as sports fields and amenity areas	–	0.01	0.49	0.99	–	100 m²	**1.47**
grassed areas containing ornamental trees and shrub beds	–	0.05	2.90	0.42	–	100 m²	**3.32**
verges	–	0.03	1.94	0.06	–	100 m²	**2.00**
By hand							
grassed areas with perimeters of mature trees such as sports fields and amenity areas	–	0.05	2.90	0.44	–	100 m²	**3.33**
grassed areas containing ornamental trees and shrub beds	–	0.08	4.83	0.73	–	100 m²	**5.55**
verges	–	0.17	9.66	1.45	–	100 m²	**11.11**
Removal of arisings							
areas with perimeters of mature trees	–	–	0.19	0.12	–	100 m²	**3.19**
areas containing ornamental trees and shrub beds	–	0.01	0.56	0.47	–	100 m²	**8.22**
Litter clearance							
Collection and disposal of litter from grassed area	–	0.01	0.29	–	–	100 m²	**0.47**
Collection and disposal of litter from isolated grassed area not exceeding 1000 m²	–	0.02	1.15	–	–	100 m²	**1.34**
Tree guards, stakes and ties							
Adjusting existing tree tie	–	0.02	0.97	–	–	nr	**0.97**
Taking up single or double tree stake and ties; removing and disposing	–	0.03	0.89	–	–	nr	**0.89**
SHRUB BED MAINTENANCE							
The labour in this section is calculated on a 3 person team. The labour time below should be multiplied by 3 to calculate the cost as shown.							
Edge maintenance							
Maintain edges where lawn abuts pathway or hard surface using							
strimmer	–	–	0.16	0.01	–	m	**0.17**
shears	–	0.01	0.48	–	–	m	**0.48**
Maintain edges where lawn abuts plant bed using							
mechanical edging tool	–	–	0.20	0.04	–	m	**0.24**
shears	–	–	0.32	–	–	m	**0.32**
half moon edging tool	–	0.01	0.57	–	–	m	**0.57**

37 SOFT LANDSCAPING

Item	PC £	Labour hours	Labour £	Plant £	Material £	Unit	Total rate £
SHRUB BED MAINTENANCE – cont							
Pruning shrubs							
Trimming ground cover planting							
soft groundcover; vinca ivy and the like	–	0.33	28.69	–	–	100 m^2	**28.69**
woody groundcover; cotoneaster and the like	–	0.50	43.04	–	–	100 m^2	**43.04**
Pruning massed shrub border (measure ground area)							
shrub beds pruned annually	–	–	0.29	–	–	m^2	**0.29**
shrub beds pruned hard every 3 years	–	0.01	0.80	–	–	m^2	**0.80**
Cutting off dead heads							
bush or standard rose	–	0.02	1.43	–	–	nr	**1.43**
climbing rose	–	0.03	2.39	–	–	nr	**2.39**
Pruning roses							
bush or standard rose	–	0.02	1.43	–	–	nr	**1.43**
climbing rose or rambling rose; tying in as required	–	0.02	1.91	–	–	nr	**1.91**
Pruning ornamental shrub; height before pruning (increase these rates by 50% if pruning work has not been executed during the previous two years)							
not exceeding 1 m	–	0.01	1.15	–	–	nr	**1.15**
1 to 2 m	–	0.02	1.59	–	–	nr	**1.59**
exceeding 2 m	–	0.04	3.59	–	–	nr	**3.59**
Removing excess growth etc. from face of building etc.; height before pruning							
not exceeding 2 m	–	0.01	0.82	–	–	nr	**0.82**
2 to 4 m	–	0.02	1.43	–	–	nr	**1.43**
4 to 6 m	–	0.03	2.39	–	–	nr	**2.39**
6 to 8 m	–	0.04	3.59	–	–	nr	**3.59**
8 to 10 m	–	0.05	4.10	–	–	nr	**4.10**
Removing epicormic growth from base of shrub or trunk and base of tree (any height, any dia.); number of growths							
not exceeding 10 nr	–	0.02	1.43	–	–	nr	**1.43**
10 to 20 nr	–	0.02	1.91	–	–	nr	**1.91**
Shrub beds, borders and planters							
Lifting							
bulbs	–	0.17	14.35	–	–	100 nr	**14.35**
tubers or corms	–	0.13	11.48	–	–	100 nr	**11.48**
established herbaceous plants; hoeing and depositing for replanting	–	0.66	57.38	–	–	100 nr	**57.38**

37 SOFT LANDSCAPING

Item	PC £	Labour hours	Labour £	Plant £	Material £	Unit	**Total rate £**
Cutting down spent growth of herbaceous plant; clearing arisings							
unstaked	–	0.01	0.57	–	–	nr	**0.57**
staked; not exceeding 4 stakes per plant; removing stakes and putting into store	–	0.01	0.72	–	–	nr	**0.72**
Hand weeding; regular visits; costs per occasion							
newly planted areas; mulched beds;	–	0.03	2.39	–	–	100 m^2	**2.39**
newly planted areas; unmulched	–	0.04	3.83	–	–	100 m^2	**3.83**
established areas	–	0.01	0.96	–	–	100 m^2	**0.96**
Removing grasses from groundcover areas	–	0.02	1.91	–	–	100 m^2	**1.91**
Hand digging with fork between shrubs; not exceeding 150 mm deep; breaking down lumps; leaving surface with a medium tilth	–	0.44	38.25	–	–	100 m^2	**38.25**
Hand digging with fork or spade to an average depth of 230 mm; breaking down lumps; leaving surface with a medium tilth	–	0.66	57.38	–	–	100 m^2	**57.38**
Hand hoeing; not exceeding 50 mm deep; leaving surface with a medium tilth	–	0.13	11.48	–	–	100 m^2	**11.48**
Hand raking to remove stones etc.; breaking down lumps: leaving surface with a fine tilth prior to planting	–	0.22	19.13	–	–	100 m^2	**19.13**
Hand weeding; planter, window box; not exceeding 1.00 m^2							
ground level box	–	0.02	1.43	–	–	nr	**1.43**
box accessed by stepladder	–	0.03	2.39	–	–	nr	**2.39**
Spreading only compost, mulch or processed bark to a depth of 75 mm							
on shrub bed with existing mature planting	–	0.03	2.61	–	–	m^2	**2.61**
recently planted areas	–	0.02	1.91	–	–	m^2	**1.91**
groundcover and herbaceous areas	–	0.02	2.16	–	–	m^2	**2.16**
Clearing cultivated area of litter and other extraneous debris; using hand implement (exludes winter leaf clearance							
weekly maintenance; private areas	–	0.01	0.48	–	–	100 m^2	**0.48**
weekly maintenance; public areas	–	0.01	0.96	–	–	100 m^2	**0.96**
daily maintenance public areas	–	0.01	0.48	–	–	100 m^2	**0.48**

37 SOFT LANDSCAPING

Item	PC £	Labour hours	Labour £	Plant £	Material £	Unit	Total rate £
SHRUB BED MAINTENANCE – cont							
Shrub beds, borders and planters – cont							
Clearing cultivated area of winter leaf fall, extraneous debris; using hand implement; winter leaf; removal of cleared material to stockpile on site							
per occasion; occasional trees and deciduous shrubs within 100 m^2;	–	0.01	1.20	–	–	100 m^2	**1.20**
per occasion; dense tall trees and deciduous shrubs in close proximity	–	0.03	2.39	–	–	100 m^2	**2.39**
BEDDING MAINTENANCE							
The labour in this section is calculated on a 3 person team. The labour time below should be multiplied by 3 to calculate the cost as shown.							
Works to established bedding areas							
Lifting							
bedding plants; hoeing and depositing for disposal	–	0.90	78.03	–	–	100 m^2	**78.03**
Hand digging with fork; not exceeding 150 mm deep: breaking down lumps; leaving surface with a medium tilth	–	0.24	20.75	–	–	100 m^2	**20.75**
Hand weeding							
newly planted areas	–	0.64	55.34	–	–	100 m^2	**55.34**
established areas	–	0.16	13.83	–	–	100 m^2	**13.83**
Hand digging with fork or spade to an average depth of 230 mm; breaking down lumps; leaving surface with a medium tilth	–	0.16	13.83	–	–	100 m^2	**13.83**
Hand hoeing: not exceeding 50 mm deep; leaving surface with a medium tilth	–	0.13	11.07	–	–	100 m^2	**11.07**
Hand raking to remove stones etc.; breaking down lumps; leaving surface with a fine tilth prior to planting	–	0.21	18.44	–	–	100 m^2	**18.44**

37 SOFT LANDSCAPING

Item	PC £	Labour hours	Labour £	Plant £	Material £	Unit	**Total rate £**
Hand weeding; planter, window box; not exceeding 1.00 m^2							
ground level box	–	0.02	1.39	–	–	nr	**1.39**
box accessed by stepladder	–	0.03	2.30	–	–	nr	**2.30**
Spreading only; compost, mulch or processed bark to a depth of 75 mm							
on shrub bed with existing mature planting	–	0.03	2.52	–	–	m^2	**2.52**
recently planted areas	–	0.02	1.85	–	–	m^2	**1.85**
groundcover and herbaceous areas	–	0.02	2.07	–	–	m^2	**2.07**
Collecting bedding from nursery	–	0.95	83.00	19.51	–	100 m^2	**102.51**
Setting out							
mass planting single variety	–	0.04	3.46	–	–	m^2	**3.46**
pattern	–	0.11	9.22	–	–	m^2	**9.22**
Planting only							
massed bedding plants	–	0.06	5.54	–	–	m^2	**5.54**
Clearing cultivated area of leaves, litter and other extraneous debris; using hand implement							
weekly maintenance	–	0.04	3.46	–	–	100 m^2	**3.46**
daily maintenance	–	0.01	0.46	–	–	100 m^2	**0.46**
IRRIGATION AND WATERING							
Irrigation and watering							
Hand held hosepipe; flow rate 25 litres per minute; irrigation requirement							
10 litres/m^2	–	0.67	21.36	–	–	100 m^2	**21.36**
15 litres/m^2	–	1.00	31.88	–	–	100 m^2	**31.88**
20 litres/m^2	–	1.33	42.40	–	–	100 m^2	**42.40**
25 litres/m^2	–	1.67	53.24	–	–	100 m^2	**53.24**
Hand held hosepipe; flow rate 40 litres per minute; irrigation requirement							
10 litres/m^2	–	0.42	13.39	–	–	100 m^2	**13.39**
15 litres/m^2	–	0.63	20.08	–	–	100 m^2	**20.08**
20 litres/m^2	–	0.83	26.46	–	–	100 m^2	**26.46**
25 litres/m^2	–	1.04	33.22	–	–	100 m^2	**33.22**

37 SOFT LANDSCAPING

Item	PC £	Labour hours	Labour £	Plant £	Material £	Unit	Total rate £
HEDGE MAINTENANCE							
The labour in this section is calculated on a 2 person team. The labour time below should be multiplied by 2 to calculate the cost as shown.							
Hedge cutting; field hedges cut once or twice annually							
Trimming sides and top using hand tool or hand held mechanical tools							
not exceeding 2 m high	–	0.05	2.90	0.25	–	10 m²	**3.15**
2 to 4 m high	–	0.17	9.66	0.83	–	10 m²	**10.49**
Hedge cutting; ornamental							
Trimming sides and top using hand tool or hand held mechanical tools							
not exceeding 2 m high	–	0.06	3.62	0.31	–	10 m²	**3.93**
2 to 4 m high	–	0.25	14.49	1.25	–	10 m²	**15.74**
Hedge cutting; reducing width; hand tool or hand held mechanical tools							
Not exceeding 2 m high							
average depth of cut not exceeding 300 mm	–	0.01	0.73	0.06	–	m²	**0.79**
average depth of cut 300–600 mm	–	0.03	1.61	0.14	–	m²	**1.75**
average depth of cut 600–900 mm	–	0.03	1.81	0.16	–	m²	**1.97**
2–4 m high							
average depth of cut not exceeding 300 mm	–	0.02	1.16	0.10	–	m²	**1.26**
average depth of cut 300–600 mm	–	0.03	1.81	0.16	–	m²	**1.97**
average depth of cut 600–900 mm	–	0.05	2.90	0.25	–	m²	**3.15**
4–6 m high							
average depth of cut not exceeding 300 mm	–	0.05	2.90	0.25	–	m²	**3.15**
average depth of cut 300–600 mm	–	0.08	4.83	0.42	–	m²	**5.24**
average depth of cut 600–900 mm	–	0.25	14.49	1.25	–	m²	**15.74**
Hedge cutting; reducing width; tractor mounted hedge cutting equipment							
Not exceeding 2 m high							
average depth of cut not exceeding 300 mm	–	0.02	1.16	0.95	–	10 m²	**2.11**
average depth of cut 300–600 mm	–	0.03	1.45	1.18	–	10 m²	**2.63**
average depth of cut 600–900 mm	–	0.10	5.80	4.73	–	10 m²	**10.53**

37 SOFT LANDSCAPING

Item	PC £	Labour hours	Labour £	Plant £	Material £	Unit	Total rate £
2–4 m high							
average depth of cut not exceeding 300 mm	–	0.01	0.37	0.30	–	10 m^2	**0.66**
average depth of cut 300–600 mm	–	0.01	0.72	0.59	–	10 m^2	**1.32**
average depth of cut 600–900 mm	–	0.01	0.58	0.47	–	10 m^2	**1.05**
Hedge cutting; reducing height; hand tool or hand held mechanical tools							
Not exceeding 2 m high							
average depth of cut not exceeding 300 mm	–	0.03	1.94	0.17	–	10 m^2	**2.10**
average depth of cut 300–600 mm	–	0.07	3.87	0.33	–	10 m^2	**4.20**
average depth of cut 600–900 mm	–	0.20	11.59	1.00	–	10 m^2	**12.59**
2–4 m high							
average depth of cut not exceeding 300 mm	–	0.02	0.97	0.08	–	m^2	**1.05**
average depth of cut 300–600 mm	–	0.03	1.94	0.17	–	m^2	**2.10**
average depth of cut 600–900 mm	–	0.10	5.80	0.50	–	m^2	**6.29**
4–6 m high							
average depth of cut not exceeding 300 mm	–	0.03	1.94	0.17	–	m^2	**2.10**
average depth of cut 300–600 mm	–	0.06	3.62	0.31	–	m^2	**3.93**
average depth of cut 600–900 mm	–	0.13	7.25	0.62	–	m^2	**7.87**
Hedge cutting; removal and disposal of arisings							
Sweeping up and depositing arisings							
300 mm cut	–	0.03	1.45	–	–	10 m^2	**1.45**
600 mm cut	–	0.10	5.80	–	–	10 m^2	**5.80**
900 mm cut	–	0.20	11.59	–	–	10 m^2	**11.59**
Chipping arisings							
300 mm cut	–	0.01	0.58	0.47	–	10 m^2	**1.05**
600 mm cut	–	0.04	2.42	1.95	–	10 m^2	**4.36**
900 mm cut	–	0.10	5.80	4.67	–	10 m^2	**10.47**
Disposal of unchipped arisings							
300 mm cut	–	0.01	0.47	0.71	–	10 m^2	**5.99**
600 mm cut	–	0.02	0.94	1.42	–	10 m^2	**9.58**
900 mm cut	–	0.04	2.34	3.54	–	10 m^2	**23.96**
Disposal of chipped arisings							
300 mm cut	–	–	0.09	0.24	–	10 m^2	**7.49**
600 mm cut	–	0.01	0.47	0.24	–	10 m^2	**15.03**
900 mm cut	–	0.02	0.93	0.47	–	10 m^2	**37.20**

37 SOFT LANDSCAPING

Item	PC £	Labour hours	Labour £	Plant £	Material £	Unit	Total rate £
CYCLIC MAINTENANCE OPERATIONS							
Note: These items for landscape maintenance may be featured for maintenance in the section 'Aftercare As Part Of A Landscape Contract'							
Path or hard surface maintenance							
Cyclic maintenance to surfaces based on annual requirements for long term maintenance contractors							
Check or clear daily for litter or dog faeces	–	12.00	347.76	–	–	100 m²	**347.76**
Sweeping monthly	–	1.00	28.98	–	–	100 m²	**28.98**
Jet washing bi-annually	–	0.33	9.56	–	–	100 m²	**9.56**
Surface grit application; 6 occasions per annum	–	3.00	86.94	–	62.14	100 m²	**149.08**
Snow clearance; per occasion	–	1.25	36.23	–	–	100 m²	**36.23**
Childrens play area maintenance							
Visit daily to inspect for safety (Play area as part of larger maintained staffed area)	–	0.10	2.90	–	–	nr	**2.90**
Visit daily to inspect for safety (Play area as isolated play area)	–	0.50	14.49	–	–	nr	**14.49**
Monthly inspections of play equipment integrity and safety	–	1.00	28.98	–	–	nr	**28.98**
Annual repairs and maintenance to equipment in play areas; high density populations	–	4.00	115.92	–	–	nr	**115.92**
Annual repairs and maintenance to equipment in play areas; low density populations	–	1.00	28.98	–	–	nr	**28.98**
Street furniture maintenance – annual cost							
Clean benches weekly	–	2.60	92.69	–	–	nr	**92.69**
Sand timber bench bi-annually; remove splinters	–	0.50	17.83	–	–	nr	**17.83**
Maintain lighting bollard inclusive of cleaning and bulb replacement	–	0.50	17.83	–	4.93	nr	**22.76**
Maintain lighting standard inclusive of cleaning and bulb replacement	–	0.75	26.74	–	4.93	nr	**31.67**
Maintain clean exterior to litter and dog bins	–	3.00	86.94	–	–	nr	**86.94**

37 SOFT LANDSCAPING

Item	PC £	Labour hours	Labour £	Plant £	Material £	Unit	**Total rate £**
Litter maintenance							
Check and clear litter bins daily (high density) based on 240 days per annum	–	20.00	579.60	379.05	17.93	nr	**976.58**
Check and clear litter bins weekly	–	4.33	125.57	253.42	3.89	nr	**382.88**
Clear dog litter bins daily (in conjunction with bin clearance on site)	–	6.00	173.88	358.69	4.48	nr	**537.05**
Clear dog litter bins weekly (in conjunction with bin clearance on site)	–	1.00	28.98	77.72	17.93	nr	**124.63**
The labour in this section is calculated on a 2 person team. The labour time below should be multiplied by 2 to calculate the cost as shown.							
Sports maintenance							
Take football posts from store and erect	–	2.00	142.60	–	–	set	**142.60**
Remove football posts and remove to store	–	0.75	53.48	–	–	set	**53.48**
Mark football/rugby field initial mark	26.77	5.00	144.90	–	30.79	nr	**175.69**
Mark football/rugby field remark per occasion	6.56	1.00	28.98	–	7.55	nr	**36.53**
Graffiti maintenance per occasion based on term graffiti removal contract							
Remove graffiti from street furniture;	–		–	–	59.77	nr	**59.77**
Remove graffiti from masonry	–		–	–	59.77	nr	**59.77**
JAPANESE KNOTWEED							
Japanese Knotweed Management Plan; PBA Solutions Ltd							
Formulate Japanese Knotweed Management Plan (KMP) as recommended by the Environment Agency Code of Practice 2006 (EA CoP)	–		–	–	–	nr	**1472.46**

37 SOFT LANDSCAPING

Item	PC £	Labour hours	Labour £	Plant £	Material £	Unit	Total rate £
JAPANESE KNOTWEED – cont							
Treatment of Japanese Knotweed; PBA Solutions Ltd; removal and/or treatment in accordance with 'The Knotweed Code of Practice' published by the Environment Agency 2006 (EA CoP); Japanese knotweed must be treated by qualified practitioners							
Commercial chemical treatments based on areas less than 50 m^2 knotweed area; allows for the provision of contractual arrangements including method statements and spraying records; also allows for the provision of risk, environment and Coshh assessments; rates for 3–5 year programme, incl.... management plan and post treatment monitoring							
rate per visit	–		–	–	–	nr	**357.07**
rate for complete treatment programme lasting 3 years	–		–	–	–	nr	**3214.88**
Commercial chemical treatments based on areas between 50–300 m^2 knotweed area; allows for the provision of contractual arrangements including method statements and spraying records also allows for the provision of risk, environment and Coshh assessments; rates for 3–5 year programme, incl. management plan and post treatment monitoring							
rate per visit	–		–	–	–	nr	**442.96**
rate for complete treatment programme lasting 3 years	–		–	–	–	nr	**3987.91**
Commercial chemical treatments based on areas greater than 900 m^2 knotweed area; allows for the provision of contractual arrangements including method statements and spraying records also allows for the provision of risk, environment and Coshh assessments; rates for 3–5 year programme, incl.... management plan and post treatment monitoring							
rate per visit	–		–	–	–	m^2	**0.97**
Rate per year (3–5 year)	–		–	–	–	m^2	**8.78**

37 SOFT LANDSCAPING

Item	PC £	Labour hours	Labour £	Plant £	Material £	Unit	**Total rate £**
Extra over minimum rate above; for areas over 900 m^2							
rate per visit	–		–	–	–	m^2	**0.97**
rate for complete treatment programme lasting 3 years	–		–	–	–	m^2	**8.78**
Note: Taking waste off site will require pretreatment. Landfill operators are required by the Environment Agency to obtain evidence of pretreatment to show that the waste has been treated to either reduce its volume or hazardousness nature, or facilitate its handling, or enhance its recovery by the waste produce (site of origin) prior being taken to landfill. Generally pretreatment of Japanese Knotweed can include chemical treatment, cutting or burning. The cost to implement this operation would be similar to other minimum treatment costs above. To enable waste to go to landfill soil testing would need to be completed and consultation with a landfill operator or invasive weed specialist is recommended in the first instance.							
Japanese Knotweed contaminated waste removal; PBA Solutions Ltd (see note above)							
Removal off site to a licensed landfill site using 20 tonne gross vehicle weight lorries; price is for haulage and tipping only; the mileage stated below is the distance from point of collection to licensed landfill site; price excludes landfill tax and loading of material into lorries; rates will fluctuate between landfill operators and the figures given are for initial guidance only							
less than 20 miles	–		–	–	–	load	**981.64**
20–40 miles	–		–	–	–	load	**1043.00**
40–60 miles	–		–	–	–	load	**1411.11**

37 SOFT LANDSCAPING

Item	PC £	Labour hours	Labour £	Plant £	Material £	Unit	**Total rate £**
JAPANESE KNOTWEED – cont							
Root barriers; PBA Solutions Ltd; used in conjunction with on site control methods such as bund treatment, cell burial, vertical boundary protection and capping. The following barriers have been reliably used in connection with Japanese Knotweed control and as such are effective when correctly installed as part of a control programme which would usually include for chemical control.							
Linear root deflection barriers; installed to trench measured separately							
Flexiroot non permeable; reinforced LDPE coated polypropylene geomembrane	–		–	–	–	m	**5.52**
Cutex Bioroot-barrier X permeable root barrier; mechanically bonded geocomposite consisting of a copper-foil bonded between 2 layers of geotextile	–		–	–	–	m^2	**14.72**
Landfill tax rates; based on waste being within landfill operator thresholds for controlled waste and with Japanese Knotweed being the only form of contamination; the rates below are based on visual inspections of the load at the gate of the tip							
Rates from April 2018							
standard rate from April 2018 where knotweed is visible at more than 5% on inspection at landfill)	–		–	–	117.26	tonne	**117.26**
lower rate (where knotweed is visible at less than 5% on inspection at landfill)	–		–	–	3.74	tonne	**3.74**

37 SOFT LANDSCAPING

Item	PC £	Labour hours	Labour £	Plant £	Material £	Unit	Total rate £
ECOLOGICAL MITIGATION WORKS							
Ecological mitigation following Ecologist survey; Five Rivers Environmental Ltd; Operations shown below are for carrying out recommended operations on a site to receive treatments following or in accord with an ecological consultants survey and recommedations (not included); Items below all exlude any overnight accommodation or subsistence if the site area is remote; All arisings stacked or left on site as is good ecological practise							
Preliminaries							
Rams logistics and the like	–		–	–	–	nr	**824.24**
Badgers and Otters							
Badger artificial sett creation	–		–	–	–	nr	**15066.37**
Badger sett closure 9 entrances footprint 30 × 10 m	–		–	–	–	nr	**670.70**
Badger or Otter fencing 0–200 m	–		–	–	–	nr	**56.17**
Badger or Otter fencing 500–1000 m	–		–	–	–	nr	**51.35**
Amphibians							
Permanent Amphibian fencing to 200 m	–		–	–	–	m	**24.33**
Permanent Amphibian fencing 500–1000 m	–		–	–	–	m	**19.14**
Semi Permanent Amphibian fencing to 200 m	–		–	–	–	m	**17.37**
Semi Permanent Amphibian fencing 500–1000 m	–		–	–	–	m	**11.08**
Temporary Amphibian fencing to 200 m	–		–	–	–	m	**16.56**
Temporary Amphibian fencing 500–1000 m	–		–	–	–	m	**9.83**
Pitfall Trap installation to 200 traps	–		–	–	–	nr	**17.69**
Searches; Destructive searches for animals and reptiles							
Destructive search for mammals and wild life 1 ha	–		–	–	–	ha	**4091.04**
Destructive search for mammals and wild life 2 ha	–		–	–	–	ha	**3351.06**
Destructive search for mammals and wild life 5 ha +	–		–	–	–	ha	**2863.72**

37 SOFT LANDSCAPING

Item	PC £	Labour hours	Labour £	Plant £	Material £	Unit	Total rate £
ECOLOGICAL MITIGATION WORKS – cont							
Hibernacula construction							
Brick hibernacula	–		–	–	–	nr	**470.56**
Log pile hibernacula	–		–	–	–	nr	**351.43**
Pond construction							
Unlined pond 2.00 m^2 deep	–		–	–	–	450 m^2	**7392.02**
Pond lined with bentonite liner 2.00 m deep	–		–	–	–	450 m^2	**17601.46**
Watervoles; Watervole displacement 1000 m^2							
Watervole displacement including vegetation clearance and dewatering	–		–	–	–	nr	**10949.35**
Site preparation for Watervole displacement	–		–	–	–	nr	**10949.35**
Watervole displacement cutting operations	–		–	–	–	nr	**11257.79**
Watervole fencing							
Watervole fencing to 200 m^2	–		–	–	–	m^2	**45.40**
Watervole fencing to 201–500 m^2	–		–	–	–	m^2	**45.40**
Watervole fencing over 1500 m^2	–		–	–	–	m^2	**38.78**
Rabbits							
Destructive searches of rabbit burrows	–		–	–	–	ha	**5895.80**

39 ELECTRICAL SERVICES

Item	PC £	Labour hours	Labour £	Plant £	Material £	Unit	**Total rate £**
CLARIFICATION NOTES ON LABOUR COSTS IN THIS SECTION							
General groundworks team							
Generally a two man team is used in this section; The column Labour hours reports team hours. The column Labour £ reports the total cost of the team for the unit of work shown							
2 man expert team	–	1.00	71.30	–	–	hr	**71.30**
craftman	–	1.00	35.65	–	–	hr	**35.65**
subcontract electrician	–		–	–	–	hr	**65.52**
GENERAL							
Street area floodlighting – General							
Preamble: There are an enormous number of luminaires available which are designed for small scale urban and garden projects. The designs are continually changing and the landscape designer is advised to consult the manufacturer's latest catalogue. Most manufacturers supply light fittings suitable for column, bracket, bulkhead, wall or soffit mounting. Highway lamps and columns for trafficked roads are not included in this section as the design of highway lighting is a very specialized subject outside the scope of most landscape contracts. The IP reference number refers to the waterproof properties of the fitting; the higher the number the more waterproof the fitting. Most items can be fitted with time clocks or PIR controls.							
CABLES AND SWITCHING							
General groundworks team							
Generally a three man team is used in this section; The column Labour hours reports team hours. The column Labour £ reports the total cost of the team for the unit of work shown							
3 man team	–	1.00	106.95	–	–	hr	**106.95**

39 ELECTRICAL SERVICES

Item	PC £	Labour hours	Labour £	Plant £	Material £	Unit	Total rate £
CABLES AND SWITCHING – cont							
Trenching for electrical services; 3 tonne excavator (bucket volume 0.13 m³); arisings laid alongside							
straight clear runs in recently excavated or filled ground							
600 mm deep	–		–	2.45	–	m	**2.45**
800 mm deep	–		–	2.87	–	m	**2.87**
1.00 m deep	–		–	3.44	–	m	**3.44**
in brownfield sites or to multiple non- contiguous areas							
600 mm deep	–		–	4.29	–	m	**4.29**
800 mm deep	–		–	6.44	–	m	**6.44**
1.00 m deep	–		–	8.59	–	m	**8.59**
Extra over any types of excavating irrespective of depth for breaking out existing materials; heavy duty 110 volt breaker tool							
hard rock	–	5.00	178.25	23.41	–	m^3	**201.66**
concrete	–	3.00	106.95	14.05	–	m^3	**121.00**
reinforced concrete	–	4.00	142.60	24.71	–	m^3	**167.31**
brickwork, blockwork or stonework	–	1.50	53.48	7.02	–	m^3	**60.50**
By hand							
600 mm deep	–	0.24	8.57	–	–	m	**8.57**
800 mm deep	–	0.43	15.24	–	–	m	**15.24**
1.00 m deep	–	0.67	23.77	–	–	m	**23.77**
Backfilling of trenches; including laying of electrical marker tape; compacting lightly as work proceeds							
To trenches containing armoured cable							
by machine	0.23		–	3.53	0.26	m	**3.79**
by hand	0.23	0.20	7.13	–	0.26	m	**7.39**
To trenches containing ducted cable; including 150 mm sharp sand over the duct							
by machine	2.95		–	4.41	3.39	m	**7.80**
by hand	2.95	0.25	8.91	–	3.39	m	**12.30**
Cable to trenches (trenching operations not included); laying only cable							
Twin core steel wire armoured cable; 50 m drums							
1.5 mm core	1.11	0.02	1.43	–	1.28	m	**2.70**
2.5 mm core	1.52	0.03	2.37	–	1.75	m	**4.13**

39 ELECTRICAL SERVICES

Item	PC £	Labour hours	Labour £	Plant £	Material £	Unit	Total rate £
Twin core steel wire armoured cable; lengths less than 50 m runs							
1.5 mm core	1.24	0.02	1.43	–	1.43	m	**2.85**
2.5 mm core	1.64	0.03	2.37	–	1.89	m	**4.26**
Three core steel wire armoured cable; 50 m drums							
1.5 mm core	1.22	0.02	1.43	–	1.40	m	**2.83**
2.5 mm core	1.97	0.02	1.43	–	2.26	m	**3.69**
4.0 mm core	2.70	0.03	1.78	–	3.11	m	**4.89**
6.0 mm core	3.55	0.03	2.37	–	4.08	m	**6.46**
10.0 mm core	5.81	0.04	2.85	–	6.68	m	**9.53**
Three core steel wire armoured cable; lengths less than 50 m runs							
1.5 mm core	1.37	0.02	1.43	–	1.58	m	**3.01**
2.5 mm core	1.96	0.02	1.43	–	2.26	m	**3.68**
Ducts to trenches; twin wall flexible cable ducts with drawstrings; laid on 150 mm clean sharp sand							
Twin wall duct; laying to trenches							
63 mm × 50 m coils	1.03	0.02	0.71	–	1.19	m	**1.90**
110 mm × 50 m coils	1.94	0.02	0.71	–	2.23	m	**2.94**
Cable drawing through ducts							
straight runs up to 50 m lengths	–	1.00	35.65	–	–	nr	**35.65**
Terminations to armoured cables; cutting coiling and taping length of cable to receive connection to light fitting, transformer and the like; fixing to temporary post							
Armoured cable	–	0.33	11.87	–	–	m	**11.87**
Plain ducted cable	–	0.13	4.46	–	–	m	**4.46**
Junction boxes; fixing to cables to receive connections to light fittings or transformers; to IP68							
Underground jointing box							
2 way	37.52	0.33	15.10	–	43.15	m	**58.24**
3 way	75.98	0.57	25.91	–	87.38	m	**113.29**
Above-ground jointing box							
2 way	20.94	0.33	15.10	–	24.08	m	**39.18**
3 way	26.01	0.40	18.14	–	29.91	m	**48.05**
Electrical connections							
Connections to existing distribution boards							
per switching circuit located within 1 m of the distribution board; chasing of cables to walls not included	–		–	–	–	nr	**79.44**

39 ELECTRICAL SERVICES

Item	PC £	Labour hours	Labour £	Plant £	Material £	Unit	Total rate £
CABLES AND SWITCHING – cont							
Electrical connections – cont							
Connections to switches; external power cable circuits; connections to internal wall switches; drilling through walls and making good; chasing of cables to walls not included							
price for the first circuit	7.57	1.00	34.65	–	8.46	nr	**123.82**
additional circuits	7.57		–	–	8.44	nr	**27.76**
POP-UP AND BOLLARD POWER SUPPLY							
Note: All cables ducts and power supply and connections to the distribution boards are not included							
The labour in this section is calculated on a 2 person team. The labour time below should be multiplied by 2 to calculate the cost as shown.							
Power supply Bollards; Marshalls Bega; Permanent lockable connecting pillar with installation accessories. Strong construction for power supply in public or industrial facilities.							
3 nr safety socket outlets 16 A; 3 CEE-socket outlets 16 A; 2 CEE-socket outlets 16 A; Circuit breakers and connecting pole;							
M70 238 Connecting pillar	2958.05	2.00	124.00	–	2966.55	nr	**3308.61**
In-ground power units – Stainless steel ; Kent Stainless Ltd; recessed power supply unit with integrated recessed manhole cover in Stainless steel 316; inclusive of fixing and commissioning to existing electrical supply; excludes construction of housing pit, ducting, connections to distribution board and drainage to housing pit and paving to recess.							

39 ELECTRICAL SERVICES

Item	PC £	Labour hours	Labour £	Plant £	Material £	Unit	**Total rate £**
In-ground unit Type 2; FACTA and BSEN124 compliance; IP 67 with a combination of 3 or 4 sockets either 16 A or 32 A. or smaller IP67 electrical panel with 2 sockets plus a data panel with 1 to 4 RJ45 sockets.							
450 × 600 recessed	2176.83	5.00	310.00	–	2176.83	nr	**2920.03**
Power Bollard; Stainless steel; Kent Stainless Ltd; 16L stainless steel; 2 power sockets with locking access panel; inclusive of all excavations, disposal and bedding in 1:3:6 concrete surround. Excludes connections to ducting cables and connections to distribution board							
900 × 390 mm dia.	2924.10	3.00	186.00	–	2939.21	nr	**3214.56**
LOW VOLTAGE							
Low voltage lights; connecting to transformers (transformers shown separately); fixing as described							
Hunza Spike adjustable spotlight; 63.5 mm dia. × 75 mm long; c/w 20/35/50 w lamp							
stainless steel	224.73		–	–	300.67	nr	**323.41**
copper	139.73		–	–	202.93	nr	**225.64**
black	89.67		–	–	145.40	nr	**168.08**
Hunza Spike spotlight;; 63.5 mm dia. × 75 mm long; c/w 20/35/50 w lamp							
stainless steel	189.81		–	–	260.51	nr	**283.24**
copper	132.74		–	–	194.90	nr	**217.60**
black	76.85		–	–	130.66	nr	**153.33**
Hunza pole lights; single; fixing to concrete base 150 × 150 × 150 mm							
stainless steel	196.79	0.75	26.65	–	269.56	nr	**318.95**
copper	200.27	0.75	26.66	–	273.54	nr	**322.94**
black	126.93	0.75	26.63	–	189.24	nr	**238.58**
Hunza pole lights; double; fixing to concrete base 150 × 150 × 150 mm							
stainless steel	427.33	0.75	26.69	–	534.59	nr	**584.06**
copper	309.73	0.75	26.68	–	399.41	nr	**448.85**
black	427.33	0.75	26.60		531.59	nr	**584.06**

39 ELECTRICAL SERVICES

Item	PC £	Labour hours	Labour £	Plant £	Material £	Unit	Total rate £
LOW VOLTAGE – cont							
Low voltage lights – cont							
Recessed wall lights; eyelid steplights; fixing to brick or stone walls; inclusive of core drilling							
stainless steel	37.52	0.75	26.67	–	339.55	nr	**388.97**
copper	37.52	0.75	26.66	–	306.09	nr	**355.50**
black	37.52	0.75	26.66	–	306.09	nr	**355.50**
Hunza; recessed deck path or lawn lights; inclusive of core drilling excavation and all making good							
stainless steel light installed to deck	426.96	0.50	17.79	–	533.22	nr	**573.79**
stainless steel light installed to path	464.48	0.75	26.69	–	533.22	nr	**582.68**
stainless steel light installed to lawn	464.48	0.40	14.24	–	533.22	nr	**570.23**
stainless steel driveway light	464.48	0.50	17.79	–	533.22	nr	**573.79**
Exterior Lighting Transformers							
Outdoor transformers for 12 volt exterior lighting; connecting to junction box (not included); IP67							
Single light transformers; 100 × 68 × 72 mm; 50 vA							
fused	38.38		–	–	42.97	nr	**109.66**
Two light transformer; 130 × 85 × 85 mm; 100 vA							
fused	60.50		–	–	68.07	nr	**135.09**
Three light transformer; 130 × 85 × 85 mm; 150 vA							
fused	66.88		–	–	75.33	nr	**142.42**
Four light transformer; 130 × 85 × 100 mm; 200 vA							
fused	78.71		–	–	88.83	nr	**156.05**
MARKET PRICES OF LAMPS							
Market prices of lamps							
Lamps							
70 w HQIT-S	–		–	–	11.24	nr	**11.24**
70 w HQIT	–		–	–	28.20	nr	**28.20**
70 w SON	–		–	–	10.94	nr	**10.94**
70 w SONT	–		–	–	8.58	nr	**8.58**
100 w SONT	–		–	–	8.58	nr	**8.58**
150 w SONT	–		–	–	8.58	nr	**8.58**
28 w 2D	–		–	–	6.36	nr	**6.36**
100 w GLS\E27	–		–	–	3.18	nr	**3.18**

39 ELECTRICAL SERVICES

Item	PC £	Labour hours	Labour £	Plant £	Material £	Unit	**Total rate £**
RECESSED LIGHTING							
Bega Lighting; Recessed step or wall LED Lighting; AC Lighting installed to preformed recesses or cast in by landscape contractor ; connections by electrial specialist; (all supply cables ducts junctions not included)							
Recessed luminaires for walls and stairs; 65 mm deep; LED							
M 33 053 5.6 W 170 × 70 mm	190.45	0.50	17.67	–	259.91	nr	**345.49**
M 33 054 7.8 W 260 × 70 mm	210.15	0.50	17.68	–	282.52	nr	**368.14**
M 33 055 11.0 W 320 × 70 mm	231.17	0.50	17.69	–	306.66	nr	**392.33**
Recessed luminaires for walls and stairs; 90 mm deep; Remote controllable; LED							
M 33 058 14.5 W 330 x125 mm	305.67	0.50	17.69	–	391.72	nr	**494.39**
M 33 059 19.5 W 420 × 125 mm	341.39	0.50	17.70	–	432.72	nr	**535.45**
M 33 060 24.0 W 520 × 125 mm	374.33	0.50	17.71	–	470.57	nr	**573.36**

39 ELECTRICAL SERVICES

Item	PC £	Labour hours	Labour £	Plant £	Material £	Unit	Total rate £
STREET AND PRECINCT LIGHTING							
The labour in this section is calculated on a 3 person team. The labour time below should be multiplied by 3 to calculate the cost as shown.							
Pole lighting Bega; Bega lighting poles installed to precinct or parking area; Including all excavations, concrete footings and bases and connections to the underground cabling circuit; connections, underground cabling (not included); pre-treated with high-quality powder-coating.							
Aluminium luminaire poles with anchorage unit							
M 70 906 3.50 m high for 76 mm Ø × 135 mm luminaire fitting; 800 mm anchor root	800.52	1.50	160.01	25.71	983.54	nr	**1243.01**
M 70 743 4.50 m high for 76 mm Ø × 135 mm luminaire fitting; 800 mm anchor root	890.37	1.75	186.72	25.71	1086.88	nr	**1373.09**
M 70 749 6.00 m high for 89 mm Ø × 120 mm luminaire fitting; 1.00 m anchor root	1744.43	2.00	213.57	51.47	2105.09	nr	**2456.61**
Aluminium tube-bow poles with 800 mm anchorage unit; Surface lacquered; for traditional style luminaires							
M 70 990 4.30 m high120 Ø mm for 1 luminaire	1130.10	1.75	186.80	25.73	1362.52	nr	**1648.85**
M 70 991 4.30 m high120 Ø mm for 2 luminaires	1402.25	1.75	186.85	25.73	1675.33	nr	**1961.74**
M 70 993 4.30 m high120 Ø mm for 3 luminaires	1833.84	1.75	186.92	25.74	2171.60	nr	**2458.11**

39 ELECTRICAL SERVICES

Item	PC £	Labour hours	Labour £	Plant £	Material £	Unit	Total rate £
Pole lighting asymetrical; Bega; luminaires installed to Bega poles (not included); inclusive of all connections to supply via poles; Attack angle adjustable to 0° or 15°; dimmable							
Asymmetrical lighting; 12.0- 16.0 m centres							
M 99 515; LED 26 W; 4.0–6.0 m height	645.36	1.00	35.53	8.73	739.58	nr	**852.10**
M 99 519; LED 36.6 W; 6.0–8.0 m height	517.77	1.00	35.50	8.72	592.95	nr	**705.39**
Asymmetrical flat beam; maximum centres and light range shown							
M 99 446 LED 15.7 W; 8 × 18 m centres; 3.5–5.0 m height	706.90	1.00	35.54	8.73	810.31	nr	**922.86**
M 99 491 LED 18.8 W; 8 × 16 m centres; 4.0–6.0 m height	725.84	1.00	35.54	8.73	832.10	nr	**944.65**
M 99 499 LED 28.6w; 12 × 24 m max. centres; 5.0–7.0 m height	737.40	1.00	35.54	8.73	845.36	nr	**957.91**
Symmetrical; 4.00–6.00 m height							
M 84 401 LED 43.5 W; 11 m max. centres	1260.89	1.00	35.58	8.74	1447.25	nr	**1559.94**
M 84 402 LED 58.0 W; 22 m max. centres	1600.21	1.00	35.59	8.74	1837.36	nr	**1950.09**
Asymmetrical flat beam; 4.00–6.00 m high							
M 84 403 LED 38.5 W; 10 × 17 m max. centres	1380.55	1.00	35.59	8.74	1584.88	nr	**1697.59**
M 84 404 LED 58W; 14 × 27 m max. centres	1600.21	1.00	35.59	8.74	1837.36	nr	**1950.09**
Pole luminaires 7.00- 9.00 m high							
M 99 522; LED; 78.0 W; Single; 12 × 27 m max. centres	1241.18	1.25	44.46	17.47	1424.01	nr	**1571.36**
M 99 533 2 LED; 212.0 W; Double; 12 × 72 m max. centres	2432.78	2.50	88.91	21.84	2790.88	nr	**3072.46**
Traditional style luminaires; Asymmetrical flat beam; 4.0–6.0 m high; LED							
M 77 910; 19.0 W; 8 × 18 max. centres	1046.38	1.00	35.57	8.74	1200.62	nr	**1313.27**
M 77 911; 43.0 W; 12 × 22 m max. centres	1264.18	1.00	35.58	8.74	1451.02	nr	**1563.71**
Traditional style luminaires; Asymmetrical flat beam; 4.0–6.00 m high; Metal halide or Sodium vapour lamp; 50 W or 70 W							
M 77 994; 8 × 18 max. centres	821.41	1.00	35.55	8.73	942.00	nr	**1054.59**
M 77 996; 12 × 24 m max. centres	994.40	1.00	35.57	8.74	1140.87	nr	**1253.52**

39 ELECTRICAL SERVICES

Item	PC £	Labour hours	Labour £	Plant £	Material £	Unit	**Total rate £**
STREET AND PRECINCT LIGHTING – cont							
Pole lighting asymetrical – cont							
Extra over for lamps							
Metal halide; HCI -TT 70 W	45.97		–	–	52.87	nr	**52.87**
Metal Halide; CDO-TT Plus 50 W 828	29.04		–	–	33.39	nr	**33.39**
Sodium Vapour 50 W; SON-T PIA PLUS 50 W	10.28		–	–	11.82	nr	**11.82**
Sodium Vapour 70 W; SON-T PIA PLUS 70 W	20.08		–	–	23.09	nr	**23.09**

PART 4

Tables and Memoranda

This part contains the following sections:

CONVERSION TABLES

CONVERSION TABLES

Length	Unit	Conversion factors			
Millimetre	mm	1 in	= 25.4 mm	1 mm	= 0.0394 in
Centimetre	cm	1 in	= 2.54 cm	1 cm	= 0.3937 in
Metre	m	1 ft	= 0.3048 m	1 m	= 3.2808 ft
		1 yd	= 0.9144 m		= 1.0936 yd
Kilometre	km	1 mile	= 1.6093 km	1 km	= 0.6214 mile

Note:	1 cm	= 10 mm	1 ft	= 12 in
	1 m	= 1 000 mm	1 yd	= 3 ft
	1 km	= 1 000 m	1 mile	= 1 760 yd

Area	Unit	Conversion factors			
Square Millimetre	mm^2	1 in^2	= 645.2 mm^2	1 mm^2	= 0.0016 in^2
Square Centimetre	cm^2	1 in^2	= 6.4516 cm^2	1 cm^2	= 1.1550 in^2
Square Metre	m^2	1 ft^2	= 0.0929 m^2	1 m^2	= 10.764 ft^2
		1 yd^2	= 0.8361 m^2	1 m^2	= 1.1960 yd^2
Square Kilometre	km^2	1 $mile^2$	= 2.590 km^2	1 km^2	= 0.3861 $mile^2$

Note:	1 cm^2	= 100 mm^2	1 ft^2	= 144 in^2
	1 m^2	= 10 000 cm^2	1 yd^2	= 9 ft^2
	1 km^2	= 100 hectares	1 acre	= 4 840 yd^2
			1 $mile^2$	= 640 acres

Volume	Unit	Conversion factors			
Cubic Centimetre	cm^3	1 cm^3	= 0.0610 in^3	1 in^3	= 16.387 cm^3
Cubic Decimetre	dm^3	1 dm^3	= 0.0353 ft^3	1 ft^3	= 28.329 dm^3
Cubic Metre	m^3	1 m^3	= 35.3147 ft^3	1 ft^3	= 0.0283 m^3
		1 m^3	= 1.3080 yd^3	1 yd^3	= 0.7646 m^3
Litre	l	1 l	= 1.76 pint	1 pint	= 0.5683 l
			= 2.113 US pt		= 0.4733 US l

Note:	1 dm^3	= 1 000 cm^3	1 ft^3	= 1 728 in^3	1 pint	= 20 fl oz
	1 m^3	= 1 000 dm^3	1 yd^3	= 27 ft^3	1 gal	= 8 pints
	1 l	= 1 dm^3				

Neither the Centimetre nor Decimetre are SI units, and as such their use, particularly that of the Decimetre, is not widespread outside educational circles.

Mass	Unit	Conversion factors			
Milligram	mg	1 mg	= 0.0154 grain	1 grain	= 64.935 mg
Gram	g	1 g	= 0.0353 oz	1 oz	= 28.35 g
Kilogram	kg	1 kg	= 2.2046 lb	1 lb	= 0.4536 kg
Tonne	t	1 t	= 0.9842 ton	1 ton	= 1.016 t

Note:	1 g	= 1000 mg	1 oz	= 437.5 grains	1 cwt	= 112 lb
	1 kg	= 1000 g	1 lb	= 16 oz	1 ton	= 20 cwt
	1 t	= 1000 kg	1 stone	= 14 lb		

Force	Unit	Conversion factors			
Newton	N	1 lbf	= 4.448 N	1 kgf	= 9.807 N
Kilonewton	kN	1 lbf	= 0.004448 kN	1 ton f	= 9.964 kN
Meganewton	MN	100 tonf	= 0.9964 MN		

CONVERSION TABLES

Pressure and stress	**Unit**	**Conversion factors**	
Kilonewton per square metre	kN/m^2	1 lbf/in^2	= 6.895 kN/m^2
		1 bar	= 100 kN/m^2
Meganewton per square metre	MN/m^2	1 $tonf/ft^2$	= 107.3 kN/m^2 = 0.1073 MN/m^2
		1 kgf/cm^2	= 98.07 kN/m^2
		1 lbf/ft^2	= 0.04788 kN/m^2

Coefficient of consolidation (Cv) or swelling	**Unit**	**Conversion factors**	
Square metre per year	m^2/year	1 cm^2/s	= 3 154 m^2/year
		1 ft^2/year	= 0.0929 m^2/year

Coefficient of permeability	**Unit**	**Conversion factors**	
Metre per second	m/s	1 cm/s	= 0.01 m/s
Metre per year	m/year	1 ft/year	= 0.3048 m/year
			= 0.9651 × $(10)^8$m/s

Temperature	**Unit**	**Conversion factors**	
Degree Celsius	°C	°C = 5/9 × (°F − 32)	°F = (9 × °C)/ 5 + 32

CONVERSION TABLES

SPEED CONVERSION

km/h	m/min	mph	fpm
1	16.7	0.6	54.7
2	33.3	1.2	109.4
3	50.0	1.9	164.0
4	66.7	2.5	218.7
5	83.3	3.1	273.4
6	100.0	3.7	328.1
7	116.7	4.3	382.8
8	133.3	5.0	437.4
9	150.0	5.6	492.1
10	166.7	6.2	546.8
11	183.3	6.8	601.5
12	200.0	7.5	656.2
13	216.7	8.1	710.8
14	233.3	8.7	765.5
15	250.0	9.3	820.2
16	266.7	9.9	874.9
17	283.3	10.6	929.6
18	300.0	11.2	984.3
19	316.7	11.8	1038.9
20	333.3	12.4	1093.6
21	350.0	13.0	1148.3
22	366.7	13.7	1203.0
23	383.3	14.3	1257.7
24	400.0	14.9	1312.3
25	416.7	15.5	1367.0
26	433.3	16.2	1421.7
27	450.0	16.8	1476.4
28	466.7	17.4	1531.1
29	483.3	18.0	1585.7
30	500.0	18.6	1640.4
31	516.7	19.3	1695.1
32	533.3	19.9	1749.8
33	550.0	20.5	1804.5
34	566.7	21.1	1859.1
35	583.3	21.7	1913.8
36	600.0	22.4	1968.5
37	616.7	23.0	2023.2
38	633.3	23.6	2077.9
39	650.0	24.2	2132.5
40	666.7	24.9	2187.2
41	683.3	25.5	2241.9
42	700.0	26.1	2296.6
43	716.7	26.7	2351.3
44	733.3	27.3	2405.9
45	750.0	28.0	2460.6

CONVERSION TABLES

km/h	m/min	mph	fpm
46	766.7	28.6	2515.3
47	783.3	29.2	2570.0
48	800.0	29.8	2624.7
49	816.7	30.4	2679.4
50	833.3	31.1	2734.0

GEOMETRY

GEOMETRY

Two dimensional figures

Figure	Diagram of figure	Surface area	Perimeter
Square	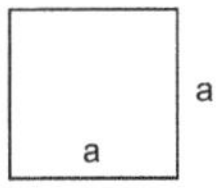	a^2	4a
Rectangle	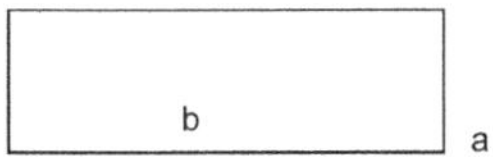	ab	$2(a + b)$
Triangle	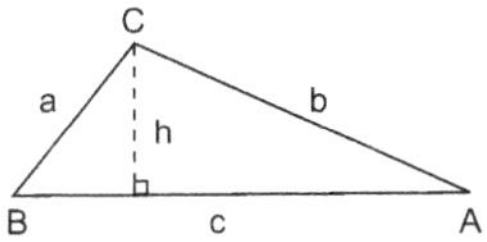	$\frac{1}{2}ch$	$a + b + c$
Circle	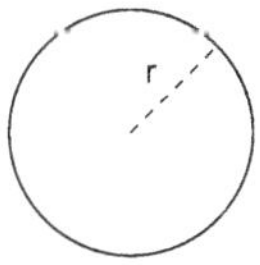	πr^2 $\frac{1}{4}\pi d^2$ where $2r = d$	$2\pi r$ πd
Parallelogram	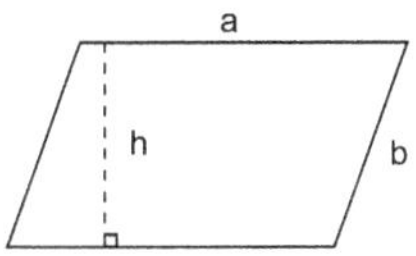	ah	$2(a + b)$
Trapezium	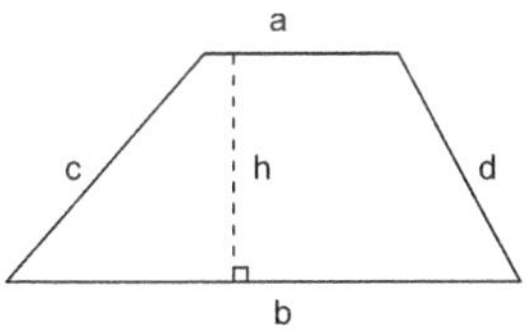	$\frac{1}{2}h(a + b)$	$a + b + c + d$
Ellipse	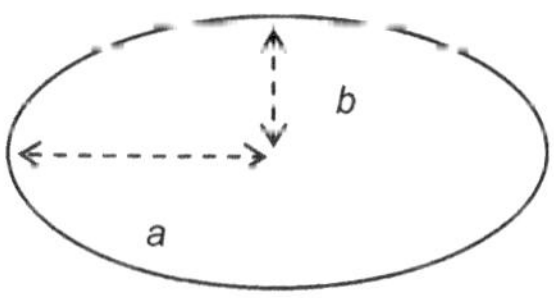	Approximately πab	$\pi(a + b)$
Hexagon	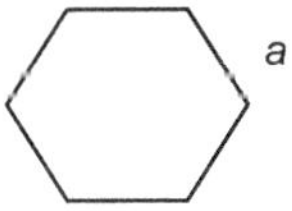	$2.6 \times a^2$	

GEOMETRY

Figure	Diagram of figure	Surface area	Perimeter
Octagon	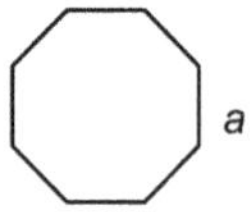	$4.83 \times a^2$	6a
Sector of a circle	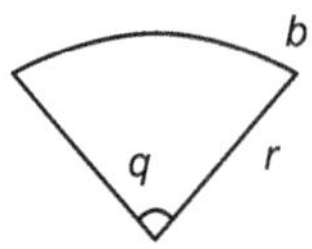	$\frac{1}{2}rb$ or $\frac{q}{360}\pi r^2$ note $b = \text{angle } \frac{q}{360} \times \pi 2r$	
Segment of a circle	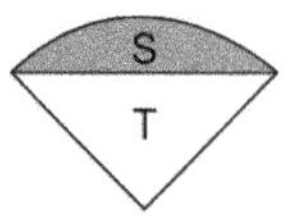	S – T where S = area of sector, T = area of triangle	
Bellmouth	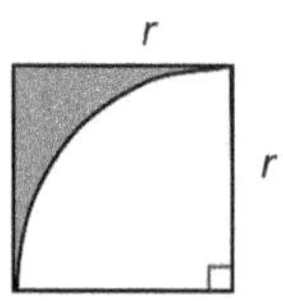	$\frac{3}{14} \times r^2$	

GEOMETRY

Three dimensional figures

Figure	Diagram of figure	Surface area	Volume
Cube	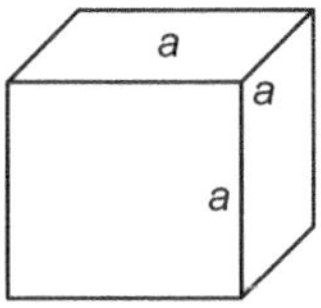	$6a^2$	a^3
Cuboid/ rectangular block	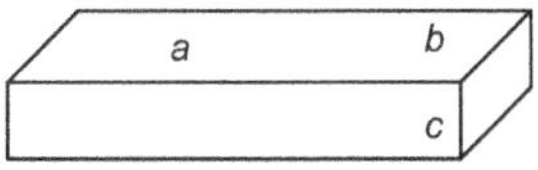	$2(ab + ac + bc)$	abc
Prism/ triangular block	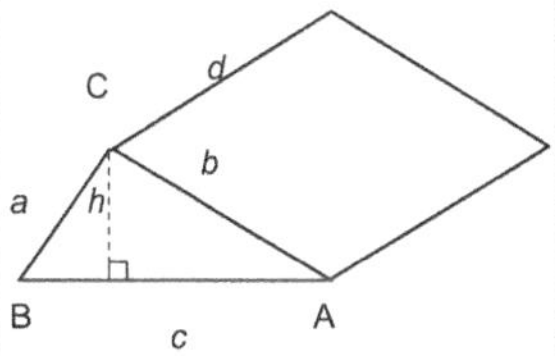	$bd + hc + dc + ad$	$\frac{1}{2}hcd$
Cylinder	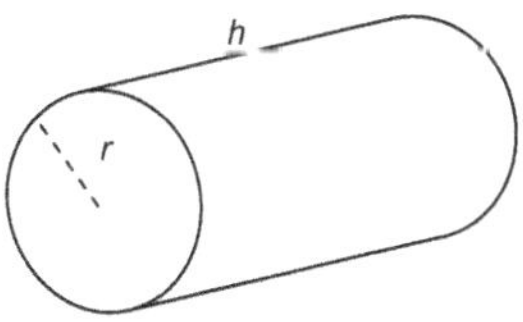	$2\pi r^2 + 2\pi h$	$\pi r^2 h$
Sphere	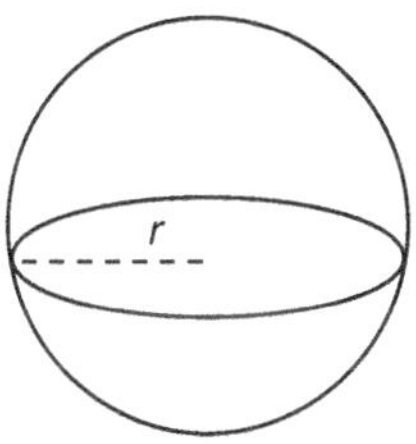	$4\pi r^2$	$\frac{4}{3}\pi r^3$
Segment of sphere	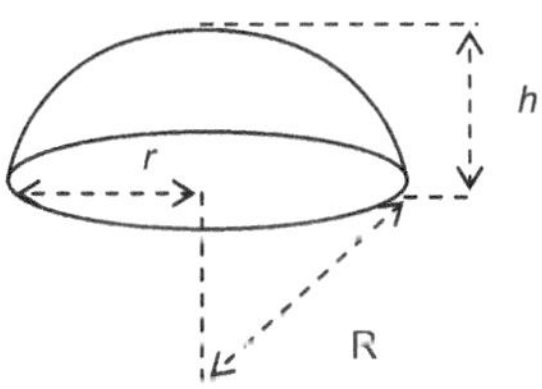	$2\pi Rh$	$\frac{1}{6}\pi h(3r^2 + h^2)$ $\frac{1}{3}\pi h^2(3R - H)$
Pyramid	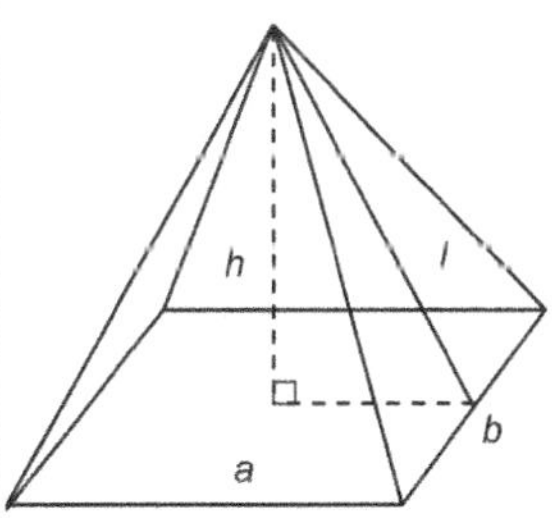	$(a + b)l + ab$	$\frac{1}{3}abh$

GEOMETRY

Figure	Diagram of figure	Surface area	Volume
Frustum of a pyramid	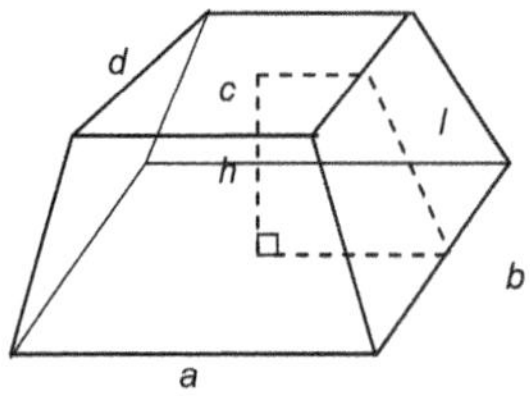	$l(a+b+c+d)+\sqrt{(ab+cd)}$ [rectangular figure only]	$\frac{h}{3}(ab+cd+\sqrt{abcd})$
Cone	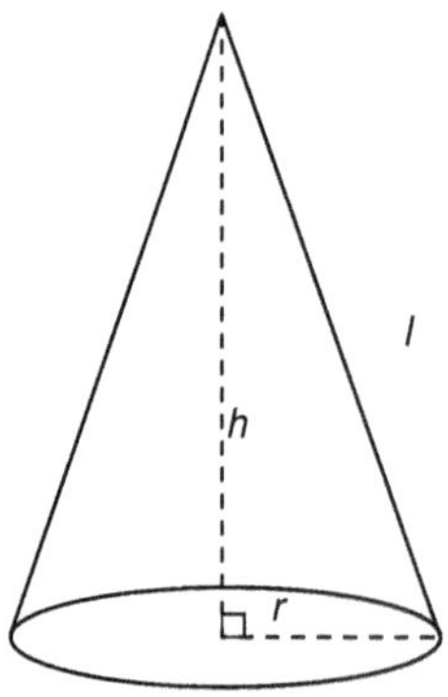	πrl (excluding base) $\pi rl+\pi r^2$ (including base)	$\frac{1}{3}\pi r^2h$ $\frac{1}{12}\pi d^2h$
Frustum of a cone	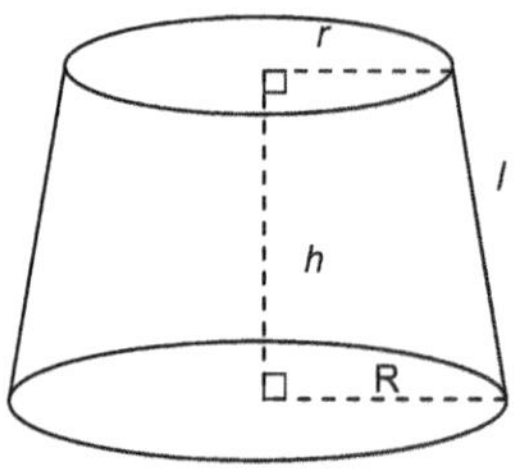	$\pi r^2+\pi R^2+\pi l(R+r)$	$\frac{1}{3}\pi(R^2+Rr+r^2)$

FORMULAE

Formulae

Formula	Description
Pythagoras Theorem	$A^2 = B^2 + C^2$ where A is the hypotenuse of a right-angled triangle and B and C are the two adjacent sides
Simpsons Rule	The Area is divided into an even number of strips of equal width, and therefore has an odd number of ordinates at the division points area = $\frac{S(A+2B+4C)}{3}$ where S = common interval (strip width) A = sum of first and last ordinates B = sum of remaining odd ordinates C = sum of the even ordinates The Volume can be calculated by the same formula, but by substituting the area of each coordinate rather than its length
Trapezoidal Rule	A given trench is divided into two equal sections, giving three ordinates, the first, the middle and the last volume = $\frac{S \times (A+B+2C)}{2}$ where S = width of the strips A = area of the first section B = area of the last section C = area of the rest of the sections
Prismoidal Rule	A given trench is divided into two equal sections, giving three ordinates, the first, the middle and the last volume = $\frac{L \times (A+4B+C)}{6}$ where L = total length of trench A = area of the first section B = area of the middle section C = area of the last section

TYPICAL THERMAL CONDUCTIVITY OF BUILDING MATERIALS

(Always check manufacturer's details – variation will occur depending on product and nature of materials)

	Thermal conductivity (W/mK)		Thermal conductivity (W/mK)
Acoustic plasterboard	0.25	Oriented strand board	0.13
Aerated concrete slab (500 kg/m^3)	0.16	Outer leaf brick	0.77
Aluminium	237	Plasterboard	0.22
Asphalt (1700 kg/m^3)	0.5	Plaster dense (1300 kg/m^3)	0.5
Bitumen-impregnated fibreboard	0.05	Plaster lightweight (600 kg/m^3)	0.16
Blocks (standard grade 600 kg/m^3)	0.15	Plywood (950 kg/m^3)	0.16
Blocks (solar grade 460 kg/m^3)	0.11	Prefabricated timber wall panels (check manufacturer)	0.12
Brickwork (outer leaf 1700 kg/m^3)	0.84	Screed (1200 kg/m^3)	0.41
Brickwork (inner leaf 1700 kg/m^3)	0.62	Stone chippings (1800 kg/m^3)	0.96
Dense aggregate concrete block 1800 kg/m^3 (exposed)	1.21	Tile hanging (1900 kg/m^3)	0.84
Dense aggregate concrete block 1800 kg/m^3 (protected)	1.13	Timber (650 kg/m^3)	0.14
Calcium silicate board (600 kg/m^3)	0.17	Timber flooring (650 kg/m^3)	0.14
Concrete general	1.28	Timber rafters	0.13
Concrete (heavyweight 2300 kg/m^3)	1.63	Timber roof or floor joists	0.13
Concrete (dense 2100 kg/m^3 typical floor)	1.4	Roof tile (1900 kg/m^3)	0.84
Concrete (dense 2000 kg/m^3 typical floor)	1.13	Timber blocks (650 kg/m^3)	0.14
Concrete (medium 1400 kg/m^3)	0.51	Cellular glass	0.045
Concrete (lightweight 1200 kg/m^3)	0.38	Expanded polystyrene	0.034
Concrete (lightweight 600 kg/m^3)	0.19	Expanded polystyrene slab (25 kg/m^3)	0.035
Concrete slab (aerated 500 kg/m^3)	0.16	Extruded polystyrene	0.035
Copper	390	Glass mineral wool	0.04
External render sand/cement finish	1	Mineral quilt (12 kg/m^3)	0.04
External render (1300 kg/m^3)	0.5	Mineral wool slab (25 kg/m^3)	0.035
Felt – Bitumen layers (1700 kg/m^3)	0.5	Phenolic foam	0.022
Fibreboard (300 kg/m^3)	0.06	Polyisocyanurate	0.025
Glass	0.93	Polyurethane	0.025
Marble	3	Rigid polyurethane	0.025
Metal tray used in wriggly tin concrete floors (7800 kg/m^3)	50	Rock mineral wool	0.038
Mortar (1750 kg/m^3)	0.8		

EARTHWORK

Weights of Typical Materials Handled by Excavators

The weight of the material is that of the state in its natural bed and includes moisture. Adjustments should be made to allow for loose or compacted states

Material	Mass (kg/m³)	Mass (lb/cu yd)
Ashes, dry	610	1028
Ashes, wet	810	1365
Basalt, broken	1954	3293
Basalt, solid	2933	4943
Bauxite, crushed	1281	2159
Borax, fine	849	1431
Caliche	1440	2427
Cement, clinker	1415	2385
Chalk, fine	1221	2058
Chalk, solid	2406	4055
Cinders, coal, ash	641	1080
Cinders, furnace	913	1538
Clay, compacted	1746	2942
Clay, dry	1073	1808
Clay, wet	1602	2700
Coal, anthracite, solid	1506	2538
Coal, bituminous	1351	2277
Coke	610	1028
Dolomite, lumpy	1522	2565
Dolomite, solid	2886	4864
Earth, dense	2002	3374
Earth, dry, loam	1249	2105
Earth, Fullers, raw	673	1134
Earth, moist	1442	2430
Earth, wet	1602	2700
Felsite	2495	4205
Fieldspar, solid	2613	4404
Fluorite	3093	5213
Gabbro	3093	5213
Gneiss	2696	4544
Granite	2690	4534
Gravel, dry ¼ to 2 inch	1682	2835
Gravel, dry, loose	1522	2565
Gravel, wet ¼ to 2 inch	2002	3374
Gypsum, broken	1450	2444
Gypsum, solid	2787	4697
Hardcore (consolidated)	1928	3249
Lignite, dry	801	1350
Limestone, broken	1554	2619
Limestone, solid	2596	4375
Magnesite, magnesium ore	2993	5044
Marble	2679	4515
Marl, wet	2216	3735
Mica, broken	1602	2700
Mica, solid	2883	4859
Peat, dry	400	674
Peat, moist	700	1179

EARTHWORK

Material	Mass (kg/m^3)	Mass (lb/cu yd)
Peat, wet	1121	1889
Potash	1281	2159
Pumice, stone	640	1078
Quarry waste	1438	2423
Quartz sand	1201	2024
Quartz, solid	2584	4355
Rhyolite	2400	4045
Sand and gravel, dry	1650	2781
Sand and gravel, wet	2020	3404
Sand, dry	1602	2700
Sand, wet	1831	3086
Sandstone, solid	2412	4065
Shale, solid	2637	4444
Slag, broken	2114	3563
Slag, furnace granulated	961	1619
Slate, broken	1370	2309
Slate, solid	2667	4495
Snow, compacted	481	810
Snow, freshly fallen	160	269
Taconite	2803	4724
Trachyte	2400	4045
Trap rock, solid	2791	4704
Turf	400	674
Water	1000	1685

Transport Capacities

Type of vehicle	Capacity of vehicle	
	Payload	Heaped capacity
Wheelbarrow	150	0.10
1 tonne dumper	1250	1.00
2.5 tonne dumper	4000	2.50
Articulated dump truck (Volvo A20 6 × 4)	18500	11.00
Articulated dump truck (Volvo A35 6 × 6)	32000	19.00
Large capacity rear dumper (Euclid R35)	35000	22.00
Large capacity rear dumper (Euclid R85)	85000	50.00

EARTHWORK

Machine Volumes for Excavating and Filling

Machine type	Cycles per minute	Volume per minute (m^3)
1.5 tonne excavator	1	0.04
	2	0.08
	3	0.12
3 tonne excavator	1	0.13
	2	0.26
	3	0.39
5 tonne excavator	1	0.28
	2	0.56
	3	0.84
7 tonne excavator	1	0.28
	2	0.56
	3	0.84
21 tonne excavator	1	1.21
	2	2.42
	3	3.63
Backhoe loader JCB3CX excavator	1	0.28
Rear bucket capacity 0.28 m^3	2	0.56
	3	0.84
Backhoe loader JCB3CX loading	1	1.00
Front bucket capacity 1.00 m^3	2	2.00

Machine Volumes for Excavating and Filling

Machine type	Loads per hour	Volume per hour (m^3)
1 tonne high tip skip loader	5	2.43
Volume 0.485 m^3	7	3.40
	10	4.85
3 tonne dumper	4	7.60
Max volume 2.40 m^3	5	9.50
Available volume 1.9 m^3	7	13.30
	10	19.00
6 tonne dumper	4	15.08
Max volume 3.40 m^3	5	18.85
Available volume 3.77 m^3	7	26.39
	10	37.70

EARTHWORK

Bulkage of Soils (after excavation)

Type of soil	Approximate bulking of 1 m^3 after excavation
Vegetable soil and loam	25–30%
Soft clay	30–40%
Stiff clay	10–20%
Gravel	20–25%
Sand	40–50%
Chalk	40–50%
Rock, weathered	30–40%
Rock, unweathered	50–60%

Shrinkage of Materials (on being deposited)

Type of soil	Approximate bulking of 1 m^3 after excavation
Clay	10%
Gravel	8%
Gravel and sand	9%
Loam and light sandy soils	12%
Loose vegetable soils	15%

Voids in Material Used as Subbases or Beddings

Material	m^3 of voids/m^3
Alluvium	0.37
River grit	0.29
Quarry sand	0.24
Shingle	0.37
Gravel	0.39
Broken stone	0.45
Broken bricks	0.42

Angles of Repose

Type of soil		Degrees
Clay	– dry	30
	– damp, well drained	45
	– wet	15–20
Earth	– dry	30
	– damp	45
Gravel	– moist	48
Sand	– dry or moist	35
	– wet	25
Loam		40

EARTHWORK

Slopes and Angles

Ratio of base to height	Angle in degrees
5:1	11
4:1	14
3:1	18
2:1	27
1½:1	34
1:1	45
1:1½	56
1:2	63
1:3	72
1:4	76
1:5	79

Grades (in Degrees and Percents)

Degrees	Percent	Degrees	Percent
1	1.8	24	44.5
2	3.5	25	46.6
3	5.2	26	48.8
4	7.0	27	51.0
5	8.8	28	53.2
6	10.5	29	55.4
7	12.3	30	57.7
8	14.0	31	60.0
9	15.8	32	62.5
10	17.6	33	64.9
11	19.4	34	67.4
12	21.3	35	70.0
13	23.1	36	72.7
14	24.9	37	75.4
15	26.8	38	78.1
16	28.7	39	81.0
17	30.6	40	83.9
18	32.5	41	86.9
19	34.4	42	90.0
20	36.4	43	93.3
21	38.4	44	96.6
22	40.4	45	100.0
23	42.4		

EARTHWORK

Bearing Powers

Ground conditions		Bearing power		
		kg/m^2	lb/in^2	Metric t/m^2
Rock,	broken	483	70	50
	solid	2415	350	240
Clay,	dry or hard	380	55	40
	medium dry	190	27	20
	soft or wet	100	14	10
Gravel,	cemented	760	110	80
Sand,	compacted	380	55	40
	clean dry	190	27	20
Swamp and alluvial soils		48	7	5

Earthwork Support

Maximum depth of excavation in various soils without the use of earthwork support

Ground conditions	Feet (ft)	Metres (m)
Compact soil	12	3.66
Drained loam	6	1.83
Dry sand	1	0.3
Gravelly earth	2	0.61
Ordinary earth	3	0.91
Stiff clay	10	3.05

It is important to note that the above table should only be used as a guide. Each case must be taken on its merits and, as the limited distances given above are approached, careful watch must be kept for the slightest signs of caving in

CONCRETE WORK

Weights of Concrete and Concrete Elements

Type of material		kg/m³	lb/cu ft
Ordinary concrete (dense aggregates)			
Non-reinforced plain or mass concrete			
Nominal weight		2305	144
Aggregate	– limestone	2162 to 2407	135 to 150
	– gravel	2244 to 2407	140 to 150
	– broken brick	2000 (av)	125 (av)
	– other crushed stone	2326 to 2489	145 to 155
Reinforced concrete			
Nominal weight		2407	150
Reinforcement	– 1%	2305 to 2468	144 to 154
	– 2%	2356 to 2519	147 to 157
	– 4%	2448 to 2703	153 to 163
Special concretes			
Heavy concrete			
Aggregates	– barytes, magnetite	3210 (min)	200 (min)
	– steel shot, punchings	5280	330
Lean mixes			
Dry-lean (gravel aggregate)		2244	140
Soil-cement (normal mix)		1601	100

CONCRETE WORK

Type of material		kg/m² per mm thick	lb/sq ft per inch thick
Ordinary concrete (dense aggregates)			
Solid slabs (floors, walls etc.)			
Thickness:	75 mm or 3 in	184	37.5
	100 mm or 4 in	245	50
	150 mm or 6 in	378	75
	250 mm or 10 in	612	125
	300 mm or 12 in	734	150
Ribbed slabs			
Thickness:	125 mm or 5 in	204	42
	150 mm or 6 in	219	45
	225 mm or 9 in	281	57
	300 mm or 12 in	342	70
Special concretes			
Finishes etc.			
	Rendering, screed etc. Granolithic, terrazzo	1928 to 2401	10 to 12.5
	Glass-block (hollow) concrete	1734 (approx)	9 (approx)
Prestressed concrete		Weights as for reinforced concrete (upper limits)	
Air-entrained concrete		Weights as for plain or reinforced concrete	

CONCRETE WORK

Average Weight of Aggregates

Materials	Voids %	Weight kg/m³
Sand	39	1660
Gravel 10–20 mm	45	1440
Gravel 35–75 mm	42	1555
Crushed stone	50	1330
Crushed granite (over 15 mm)	50	1345
(n.e. 15 mm)	47	1440
'All-in' ballast	32	1800–2000

Material	kg/m³	lb/cu yd
Vermiculite (aggregate)	64–80	108–135
All-in aggregate	1999	125

Applications and Mix Design

Site mixed concrete

Recommended mix	Class of work suitable for	Cement (kg)	Sand (kg)	Coarse aggregate (kg)	Nr 25 kg bags cement per m³ of combined aggregate
1:3:6	Roughest type of mass concrete such as footings, road haunching over 300 mm thick	208	905	1509	8.30
1:2.5:5	Mass concrete of better class than 1:3:6 such as bases for machinery, walls below ground etc.	249	881	1474	10.00
1:2:4	Most ordinary uses of concrete, such as mass walls above ground, road slabs etc. and general reinforced concrete work	304	889	1431	12.20
1:1.5:3	Watertight floors, pavements and walls, tanks, pits, steps, paths, surface of 2 course roads, reinforced concrete where extra strength is required	371	801	1336	14.90
1:1:2	Works of thin section such as fence posts and small precast work	511	720	1206	20.40

CONCRETE WORK

Ready mixed concrete

Application	Designated concrete	Standardized prescribed concrete	Recommended consistence (nominal slump class)
Foundations			
Mass concrete fill or blinding	GEN 1	ST2	S3
Strip footings	GEN 1	ST2	S3
Mass concrete foundations			
Single storey buildings	GEN 1	ST2	S3
Double storey buildings	GEN 3	ST4	S3
Trench fill foundations			
Single storey buildings	GEN 1	ST2	S4
Double storey buildings	GEN 3	ST4	S4
General applications			
Kerb bedding and haunching	GEN 0	ST1	S1
Drainage works – immediate support	GEN 1	ST2	S1
Other drainage works	GEN 1	ST2	S3
Oversite below suspended slabs	GEN 1	ST2	S3
Floors			
Garage and house floors with no embedded steel	GEN 3	ST4	S2
Wearing surface: Light foot and trolley traffic	RC30	ST4	S2
Wearing surface: General industrial	RC40	N/A	S2
Wearing surface: Heavy industrial	RC50	N/A	S2
Paving			
House drives, domestic parking and external parking	PAV 1	N/A	S2
Heavy-duty external paving	PAV 2	N/A	S2

CONCRETE WORK

Prescribed Mixes for Ordinary Structural Concrete

Weights of cement and total dry aggregates in kg to produce approximately one cubic metre of fully compacted concrete together with the percentages by weight of fine aggregate in total dry aggregates

Conc. grade	Nominal max size of aggregate (mm)	40		20		14		10	
	Workability	**Med.**	**High**	**Med.**	**High**	**Med.**	**High**	**Med.**	**High**
	Limits to slump that may be expected (mm)	**50–100**	**100–150**	**25–75**	**75–125**	**10–50**	**50–100**	**10–25**	**25–50**
7	Cement (kg)	180	200	210	230	–	–	–	–
	Total aggregate (kg)	1950	1850	1900	1800	–	–	–	–
	Fine aggregate (%)	30–45	30–45	35–50	35–50	–	–	–	–
10	Cement (kg)	210	230	240	260	–	–	–	–
	Total aggregate (kg)	1900	1850	1850	1800	–	–	–	–
	Fine aggregate (%)	30–45	30–45	35–50	35–50	–	–	–	–
15	Cement (kg)	250	270	280	310	–	–	–	–
	Total aggregate (kg)	1850	1800	1800	1750	–	–	–	–
	Fine aggregate (%)	30–45	30–45	35–50	35–50	–	–	–	–
20	Cement (kg)	300	320	320	350	340	380	360	410
	Total aggregate (kg)	1850	1750	1800	1750	1750	1700	1750	1650
	Sand								
	Zone 1 (%)	35	40	40	45	45	50	50	55
	Zone 2 (%)	30	35	35	40	40	45	45	50
	Zone 3 (%)	30	30	30	35	35	40	40	45
25	Cement (kg)	340	360	360	390	380	420	400	450
	Total aggregate (kg)	1800	1750	1750	1700	1700	1650	1700	1600
	Sand								
	Zone 1 (%)	35	40	40	45	45	50	50	55
	Zone 2 (%)	30	35	35	40	40	45	45	50
	Zone 3 (%)	30	30	30	35	35	40	40	45
30	Cement (kg)	370	390	400	430	430	470	460	510
	Total aggregate (kg)	1750	1700	1700	1650	1700	1600	1650	1550
	Sand								
	Zone 1 (%)	35	40	40	45	45	50	50	55
	Zone 2 (%)	30	35	35	40	40	45	45	50
	Zone 3 (%)	30	30	30	35	35	40	40	45

REINFORCEMENT

Weights of Bar Reinforcement

Nominal sizes (mm)	Cross-sectional area (mm^2)	Mass (kg/m)	Length of bar (m/tonne)
6	28.27	0.222	4505
8	50.27	0.395	2534
10	78.54	0.617	1622
12	113.10	0.888	1126
16	201.06	1.578	634
20	314.16	2.466	405
25	490.87	3.853	260
32	804.25	6.313	158
40	1265.64	9.865	101
50	1963.50	15.413	65

Weights of Bars (at specific spacings)

Weights of metric bars in kilogrammes per square metre

Size	Spacing of bars in millimetres									
(mm)	75	100	125	150	175	200	225	250	275	300
6	2.96	2.220	1.776	1.480	1.27	1.110	0.99	0.89	0.81	0.74
8	5.26	3.95	3.16	2.63	2.26	1.97	1.75	1.58	1.44	1.32
10	8.22	6.17	4.93	4.11	3.52	3.08	2.74	2.47	2.24	2.06
12	11.84	8.88	7.10	5.92	5.07	4.44	3.95	3.55	3.23	2.96
16	21.04	15.78	12.63	10.52	9.02	7.89	7.02	6.31	5.74	5.26
20	32.88	24.66	19.73	16.44	14.09	12.33	10.96	9.87	8.97	8.22
25	51.38	38.53	30.83	25.69	22.02	19.27	17.13	15.41	14.01	12.84
32	84.18	63.13	50.51	42.09	36.08	31.57	28.06	25.25	22.96	21.04
40	131.53	98.65	78.92	65.76	56.37	49.32	43.84	39.46	35.87	32.88
50	205.51	154.13	123.31	102.76	88.08	77.07	68.50	61.65	56.05	51.38

Basic weight of steelwork taken as 7850 kg/m^3
Basic weight of bar reinforcement per metre run = 0.00785 kg/mm^2
The value of π has been taken as 3.141592654

REINFORCEMENT

Fabric Reinforcement

Preferred range of designated fabric types and stock sheet sizes

Fabric reference	Longitudinal wires			Cross wires			
	Nominal wire size (mm)	Pitch (mm)	Area (mm^2/m)	Nominal wire size (mm)	Pitch (mm)	Area (mm^2/m)	Mass (kg/m^2)
Square mesh							
A393	10	200	393	10	200	393	6.16
A252	8	200	252	8	200	252	3.95
A193	7	200	193	7	200	193	3.02
A142	6	200	142	6	200	142	2.22
A98	5	200	98	5	200	98	1.54
Structural mesh							
B1131	12	100	1131	8	200	252	10.90
B785	10	100	785	8	200	252	8.14
B503	8	100	503	8	200	252	5.93
B385	7	100	385	7	200	193	4.53
B283	6	100	283	7	200	193	3.73
B196	5	100	196	7	200	193	3.05
Long mesh							
C785	10	100	785	6	400	70.8	6.72
C636	9	100	636	6	400	70.8	5.55
C503	8	100	503	5	400	49.0	4.34
C385	7	100	385	5	400	49.0	3.41
C283	6	100	283	5	400	49.0	2.61
Wrapping mesh							
D98	5	200	98	5	200	98	1.54
D49	2.5	100	49	2.5	100	49	0.77

Stock sheet size 4.8 m × 2.4 m, Area 11.52 m^2

Average weight kg/m^3 of steelwork reinforcement in concrete for various building elements

Substructure	kg/m^3 concrete	Substructure	kg/m^3 concrete
Pile caps	110–150	Plate slab	150–220
Tie beams	130–170	Cant slab	145–210
Ground beams	230–330	Ribbed floors	130–200
Bases	125–180	Topping to block floor	30–40
Footings	100–150	Columns	210–310
Retaining walls	150–210	Beams	250–350
Raft	60–70	Stairs	130–170
Slabs – one way	120–200	Walls – normal	40–100
Slabs – two way	110–220	Walls – wind	70–125

Note: For exposed elements add the following %:
Walls 50%, Beams 100%, Columns 15%

FORMWORK

Formwork Stripping Times – Normal Curing Periods

Conditions under which concrete is maturing	Minimum periods of protection for different types of cement					
	Number of days (where the average surface temperature of the concrete exceeds 10°C during the whole period)			Equivalent maturity (degree hours) calculated as the age of the concrete in hours multiplied by the number of degrees Celsius by which the average surface temperature of the concrete exceeds 10°C		
	Other	SRPC	OPC or RHPC	Other	SRPC	OPC or RHPC
1. Hot weather or drying winds	7	4	3	3500	2000	1500
2. Conditions not covered by 1	4	3	2	2000	1500	1000

KEY
OPC – Ordinary Portland Cement
RHPC – Rapid-hardening Portland Cement
SRPC – Sulphate-resisting Portland Cement

Minimum Period before Striking Formwork

	Minimum period before striking		
	Surface temperature of concrete		
	16°C	17°C	t°C (0–25)
Vertical formwork to columns, walls and large beams	12 hours	18 hours	300 hours t+10
Soffit formwork to slabs	4 days	6 days	100 days t+10
Props to slabs	10 days	15 days	250 days t+10
Soffit formwork to beams	9 days	14 days	230 days t+10
Props to beams	14 days	21 days	360 days t+10

MASONRY

Number of Bricks Required for Various Types of Work per m^2 of Walling

Description	Brick size	
	215 × 102.5 × 50 mm	215 × 102.5 × 65 mm
Half brick thick		
Stretcher bond	74	59
English bond	108	86
English garden wall bond	90	72
Flemish bond	96	79
Flemish garden wall bond	83	66
One brick thick and cavity wall of two half brick skins		
Stretcher bond	148	119

Quantities of Bricks and Mortar Required per m^2 of Walling

	Unit	No of bricks required	Mortar required (cubic metres)		
Standard bricks			No frogs	Single frogs	Double frogs
Brick size 215 × 102.5 × 50 mm					
half brick wall (103 mm)	m^2	72	0.022	0.027	0.032
2 × half brick cavity wall (270 mm)	m^2	144	0.044	0.054	0.064
one brick wall (215 mm)	m^2	144	0.052	0.064	0.076
one and a half brick wall (322 mm)	m^2	216	0.073	0.091	0.108
Mass brickwork	m^3	576	0.347	0.413	0.480
Brick size 215 × 102.5 × 65 mm					
half brick wall (103 mm)	m^2	58	0.019	0.022	0.026
2 × half brick cavity wall (270 mm)	m^2	116	0.038	0.045	0.055
one brick wall (215 mm)	m^2	116	0.046	0.055	0.064
one and a half brick wall (322 mm)	m^2	174	0.063	0.074	0.088
Mass brickwork	m^3	464	0.307	0.360	0.413
Metric modular bricks			Perforated		
Brick size 200 × 100 × 75 mm					
90 mm thick	m^2	67	0.016	0.019	
190 mm thick	m^2	133	0.042	0.048	
290 mm thick	m^2	200	0.068	0.078	
Brick size 200 × 100 × 100 mm					
90 mm thick	m^2	50	0.013	0.016	
190 mm thick	m^2	100	0.036	0.041	
290 mm thick	m^2	150	0.059	0.067	
Brick size 300 × 100 × 75 mm					
90 mm thick	m^2	33	–	0.015	
Brick size 300 × 100 × 100 mm					
90 mm thick	m^2	44	0.015	0.018	

Note: Assuming 10 mm thick joints

MASONRY

Mortar Required per m² Blockwork (9.88 blocks/m²)

Wall thickness	75	90	100	125	140	190	215
Mortar m^3/m^2	0.005	0.006	0.007	0.008	0.009	0.013	0.014

Mortar Group	Cement: lime: sand	Masonry cement: sand	Cement: sand with plasticizer
1	1:0–0.25:3		
2	1:0.5:4–4.5	1:2.5-3.5	1:3–4
3	1:1:5–6	1:4–5	1:5–6
4	1:2:8–9	1:5.5–6.5	1:7–8
5	1:3:10–12	1:6.5–7	1:8

Group 1: strong inflexible mortar
Group 5: weak but flexible

All mixes within a group are of approximately similar strength
Frost resistance increases with the use of plasticizers
Cement: lime: sand mixes give the strongest bond and greatest resistance to rain penetration
Masonry cement equals ordinary Portland cement plus a fine neutral mineral filler and an air entraining agent

Calcium Silicate Bricks

Type	Strength	Location
Class 2 crushing strength	14.0 N/mm²	not suitable for walls
Class 3	20.5 N/mm²	walls above dpc
Class 4	27.5 N/mm²	cappings and copings
Class 5	34.5 N/mm²	retaining walls
Class 6	41.5 N/mm²	walls below ground
Class 7	48.5 N/mm²	walls below ground

The Class 7 calcium silicate bricks are therefore equal in strength to Class B bricks
Calcium silicate bricks are not suitable for DPCs

Durability of Bricks	
FL	Frost resistant with low salt content
FN	Frost resistant with normal salt content
ML	Moderately frost resistant with low salt content
MN	Moderately frost resistant with normal salt content

MASONRY

Brickwork Dimensions

No. of horizontal bricks	Dimensions (mm)	No. of vertical courses	Height of vertical courses (mm)
½	112.5	1	75
1	225.0	2	150
1½	337.5	3	225
2	450.0	4	300
2½	562.5	5	375
3	675.0	6	450
3½	787.5	7	525
4	900.0	8	600
4½	1012.5	9	675
5	1125.0	10	750
5½	1237.5	11	825
6	1350.0	12	900
6½	1462.5	13	975
7	1575.0	14	1050
7½	1687.5	15	1125
8	1800.0	16	1200
8½	1912.5	17	1275
9	2025.0	18	1350
9½	2137.5	19	1425
10	2250.0	20	1500
20	4500.0	24	1575
40	9000.0	28	2100
50	11250.0	32	2400
60	13500.0	36	2700
75	16875.0	40	3000

TIMBER

Weights of Timber

Material	kg/m^3	lb/cu ft
General	806 (avg)	50 (avg)
Douglas fir	479	30
Yellow pine, spruce	479	30
Pitch pine	673	42
Larch, elm	561	35
Oak (English)	724 to 959	45 to 60
Teak	643 to 877	40 to 55
Jarrah	959	60
Greenheart	1040 to 1204	65 to 75
Quebracho	1285	80
Material	**kg/m^2 per mm thickness**	**lb/sq ft per inch thickness**
Wooden boarding and blocks		
Softwood	0.48	2.5
Hardwood	0.76	4
Hardboard	1.06	5.5
Chipboard	0.76	4
Plywood	0.62	3.25
Blockboard	0.48	2.5
Fibreboard	0.29	1.5
Wood-wool	0.58	3
Plasterboard	0.96	5
Weather boarding	0.35	1.8

TIMBER

Conversion Tables (for timber only)

Inches	Millimetres	Feet	Metres
1	25	1	0.300
2	50	2	0.600
3	75	3	0.900
4	100	4	1.200
5	125	5	1.500
6	150	6	1.800
7	175	7	2.100
8	200	8	2.400
9	225	9	2.700
10	250	10	3.000
11	275	11	3.300
12	300	12	3.600
13	325	13	3.900
14	350	14	4.200
15	375	15	4.500
16	400	16	4.800
17	425	17	5.100
18	450	18	5.400
19	475	19	5.700
20	500	20	6.000
21	525	21	6.300
22	550	22	6.600
23	575	23	6.900
24	600	24	7.200

Planed Softwood

The finished end section size of planed timber is usually 3/16" less than the original size from which it is produced. This however varies slightly depending upon availability of material and origin of the species used.

Standards (timber) to cubic metres and cubic metres to standards (timber)

Cubic metres	Cubic metres standards	Standards
4.672	1	0.214
9.344	2	0.428
14.017	3	0.642
18.689	4	0.856
23.361	5	1.070
28.033	6	1.284
32.706	7	1.498
37.378	8	1.712
42.050	9	1.926
46.722	10	2.140
93.445	20	4.281
140.167	30	6.421
186.890	40	8.561
233.612	50	10.702
280.335	60	12.842
327.057	70	14.982
373.779	80	17.122

TIMBER

1 cu metre = 35.3148 cu ft = 0.21403 std

1 cu ft = 0.028317 cu metres

1 std = 4.67227 cu metres

Basic sizes of sawn softwood available (cross-sectional areas)

Thickness (mm)	Width (mm)								
	75	100	125	150	175	200	225	250	300
16	X	X	X	X					
19	X	X	X	X					
22	X	X	X	X					
25	X	X	X	X	X	X	X	X	X
32	X	X	X	X	X	X	X	X	X
36	X	X	X	X					
38	X	X	X	X	X	X	X		
44	X	X	X	X	X	X	X	X	X
47*	X	X	X	X	X	X	X	X	X
50	X	X	X	X	X	X	X	X	X
63	X	X	X	X	X	X	X		
75	X	X	X	X	X	X	X	X	
100		X		X		X		X	X
150				X		X			X
200						X			
250								X	
300									X

* This range of widths for 47 mm thickness will usually be found to be available in construction quality only

Note: The smaller sizes below 100 mm thick and 250 mm width are normally but not exclusively of European origin. Sizes beyond this are usually of North and South American origin

Basic lengths of sawn softwood available (metres)

1.80	2.10	3.00	4.20	5.10	6.00	7.20
	2.40	3.30	4.50	5.40	6.30	
	2.70	3.60	4.80	5.70	6.60	
		3.90			6.90	

Note: Lengths of 6.00 m and over will generally only be available from North American species and may have to be recut from larger sizes

TIMBER

Reductions from basic size to finished size by planning of two opposed faces

Purpose	Reductions from basic sizes for timber			
	15–35 mm	36–100 mm	101–150 mm	over 150 mm
a) Constructional timber	3 mm	3 mm	5 mm	6 mm
b) Matching interlocking boards	4 mm	4 mm	6 mm	6 mm
c) Wood trim not specified in BS 584	5 mm	7 mm	7 mm	9 mm
d) Joinery and cabinet work	7 mm	9 mm	11 mm	13 mm

Note: The reduction of width or depth is overall the extreme size and is exclusive of any reduction of the face by the machining of a tongue or lap joints

Maximum Spans for Various Roof Trusses

Maximum permissible spans for rafters for Fink trussed rafters

Basic size (mm)	Actual size (mm)	Pitch (degrees)								
		15 (m)	17.5 (m)	20 (m)	22.5 (m)	25 (m)	27.5 (m)	30 (m)	32.5 (m)	35 (m)
38 × 75	35 × 72	6.03	6.16	6.29	6.41	6.51	6.60	6.70	6.80	6.90
38 × 100	35 × 97	7.48	7.67	7.83	7.97	8.10	8.22	8.34	8.47	8.61
38 × 125	35 × 120	8.80	9.00	9.20	9.37	9.54	9.68	9.82	9.98	10.16
44 × 75	41 × 72	6.45	6.59	6.71	6.83	6.93	7.03	7.14	7.24	7.35
44 × 100	41 × 97	8.05	8.23	8.40	8.55	8.68	8.81	8.93	9.09	9.22
44 × 125	41 × 120	9.38	9.60	9.81	9.99	10.15	10.31	10.45	10.64	10.81
50 × 75	47 × 72	6.87	7.01	7.13	7.25	7.35	7.45	7.53	7.67	7.78
50 × 100	47 × 97	8.62	8.80	8.97	9.12	9.25	9.38	9.50	9.66	9.80
50 × 125	47 × 120	10.01	10.24	10.44	10.62	10.77	10.94	11.00	11.00	11.00

TIMBER

Sizes of Internal and External Doorsets

Description	Internal size (mm)	Permissible deviation	External size (mm)	Permissible deviation
Coordinating dimension: height of door leaf height sets	2100		2100	
Coordinating dimension: height of ceiling height set	2300 2350 2400 2700 3000		2300 2350 2400 2700 3000	
Coordinating dimension: width of all doorsets S = Single leaf set D = Double leaf set	600 S 700 S 800 S&D 900 S&D 1000 S&D 1200 D 1500 D 1800 D 2100 D		900 S 1000 S 1200 D 1800 D 2100 D	
Work size: height of door leaf height set	2090	± 2.0	2095	± 2.0
Work size: height of ceiling height set	2285 2335 2385 2685 2985	± 2.0	2295 2345 2395 2695 2995	± 2.0
Work size: width of all doorsets S = Single leaf set D = Double leaf set	590 S 690 S 790 S&D 890 S&D 990 S&D 1190 D 1490 D 1790 D 2090 D	± 2.0	895 S 995 S 1195 D 1495 D 1795 D 2095 D	± 2.0
Width of door leaf in single leaf sets F = Flush leaf P = Panel leaf	526 F 626 F 726 F&P 826 F&P 926 F&P	± 1.5	806 F&P 906 F&P	± 1.5
Width of door leaf in double leaf sets F = Flush leaf P = Panel leaf	362 F 412 F 426 F 562 F&P 712 F&P 826 F&P 1012 F&P	± 1.5	552 F&P 702 F&P 852 F&P 1002 F&P	± 1.5
Door leaf height for all doorsets	2040	± 1.5	1994	± 1.5

ROOFING

Total Roof Loadings for Various Types of Tiles/Slates

		Roof load (slope) kg/m^2		
		Slate/Tile	Roofing underlay and battens^2	Total dead load kg/m
Asbestos cement slate (600 × 300)		21.50	3.14	24.64
Clay tile	interlocking	67.00	5.50	72.50
	plain	43.50	2.87	46.37
Concrete tile	interlocking	47.20	2.69	49.89
	plain	78.20	5.50	83.70
Natural slate (18" × 10")		35.40	3.40	38.80
		Roof load (plan) kg/m^2		
Asbestos cement slate (600 × 300)		28.45	76.50	104.95
Clay tile	interlocking	53.54	76.50	130.04
	plain	83.71	76.50	60.21
Concrete tile	interlocking	57.60	76.50	134.10
	plain	96.64	76.50	173.14

ROOFING

Tiling Data

Product		Lap (mm)	Gauge of battens	No. slates per m²	Battens (m/m²)	Weight as laid (kg/m²)
CEMENT SLATES						
Eternit slates	600 × 300 mm	100	250	13.4	4.00	19.50
(Duracem)		90	255	13.1	3.92	19.20
		80	260	12.9	3.85	19.00
		70	265	12.7	3.77	18.60
	600 × 350 mm	100	250	11.5	4.00	19.50
		90	255	11.2	3.92	19.20
	500 × 250 mm	100	200	20.0	5.00	20.00
		90	205	19.5	4.88	19.50
		80	210	19.1	4.76	19.00
		70	215	18.6	4.65	18.60
	400 × 200 mm	90	155	32.3	6.45	20.80
		80	160	31.3	6.25	20.20
		70	165	30.3	6.06	19.60
CONCRETE TILES/SLATES						
Redland Roofing						
Stonewold slate	430 × 380 mm	75	355	8.2	2.82	51.20
Double Roman tile	418 × 330 mm	75	355	8.2	2.91	45.50
Grovebury pantile	418 × 332 mm	75	343	9.7	2.91	47.90
Norfolk pantile	381 × 227 mm	75	306	16.3	3.26	44.01
		100	281	17.8	3.56	48.06
Renown interlocking tile	418 × 330 mm	75	343	9.7	2.91	46.40
'49' tile	381 × 227 mm	75	306	16.3	3.26	44.80
		100	281	17.8	3.56	48.95
Plain, vertical tiling	265 × 165 mm	35	115	52.7	8.70	62.20
Marley Roofing						
Bold roll tile	420 × 330 mm	75	344	9.7	2.90	47.00
		100	–	10.5	3.20	51.00
Modern roof tile	420 × 330 mm	75	338	10.2	3.00	54.00
		100	–	11.0	3.20	58.00
Ludlow major	420 × 330 mm	75	338	10.2	3.00	45.00
		100	–	11.0	3.20	49.00
Ludlow plus	387 × 229 mm	75	305	16.1	3.30	47.00
		100	–	17.5	3.60	51.00
Mendip tile	420 × 330 mm	75	338	10.2	3.00	47.00
		100	–	11.0	3.20	51.00
Wessex	413 × 330 mm	75	338	10.2	3.00	54.00
		100	–	11.0	3.20	58.00
Plain tile	267 × 165 mm	65	100	60.0	10.00	76.00
		75	95	64.0	10.50	81.00
		85	90	68.0	11.30	86.00
Plain vertical tiles (feature)	267 × 165 mm	35	110	53.0	8.70	67.00
		34	115	56.0	9.10	71.00

ROOFING

Slate Nails, Quantity per Kilogram

Length	Type			
	Plain wire	Galvanized wire	Copper nail	Zinc nail
28.5 mm	325	305	325	415
34.4 mm	286	256	254	292
50.8 mm	242	224	194	200

Metal Sheet Coverings

Thicknesses and weights of sheet metal coverings								
Lead to BS 1178								
BS Code No	3	4	5	6	7	8		
Colour code	Green	Blue	Red	Black	White	Orange		
Thickness (mm)	1.25	1.80	2.24	2.50	3.15	3.55		
Density kg/m^2	14.18	20.41	25.40	30.05	35.72	40.26		
Copper to BS 2870								
Thickness (mm)		0.60	0.70					
Bay width								
Roll (mm)		500	650					
Seam (mm)		525	600					
Standard width to form bay	600	750						
Normal length of sheet	1.80	1.80						
Zinc to BS 849								
Zinc Gauge (Nr)	9	10	11	12	13	14	15	16
Thickness (mm)	0.43	0.48	0.56	0.64	0.71	0.79	0.91	1.04
Density (kg/m^2)	3.1	3.2	3.8	4.3	4.8	5.3	6.2	7.0
Aluminium to BS 4868								
Thickness (mm)	0.5	0.6	0.7	0.8	0.9	1.0	1.2	
Density (kg/m^2)	12.8	15.4	17.9	20.5	23.0	25.6	30.7	

ROOFING

Type of felt	Nominal mass per unit area (kg/10 m)	Nominal mass per unit area of fibre base (g/m^2)	Nominal length of roll (m)
Class 1			
1B fine granule	14	220	10 or 20
surfaced bitumen	18	330	10 or 20
	25	470	10
1E mineral surfaced bitumen	38	470	10
1F reinforced bitumen	15	160 (fibre) 110 (hessian)	15
1F reinforced bitumen, aluminium faced	13	160 (fibre) 110 (hessian)	15
Class 2			
2B fine granule surfaced bitumen asbestos	18	500	10 or 20
2E mineral surfaced bitumen asbestos	38	600	10
Class 3			
3B fine granule surfaced bitumen glass fibre	18	60	20
3E mineral surfaced bitumen glass fibre	28	60	10
3E venting base layer bitumen glass fibre	32	60*	10
3H venting base layer bitumen glass fibre	17	60*	20

* Excluding effect of perforations

GLAZING

GLAZING

Nominal thickness (mm)	Tolerance on thickness (mm)	Approximate weight (kg/m²)	Normal maximum size (mm)
Float and polished plate glass			
3	+ 0.2	7.50	2140 × 1220
4	+ 0.2	10.00	2760 × 1220
5	+ 0.2	12.50	3180 × 2100
6	+ 0.2	15.00	4600 × 3180
10	+ 0.3	25.00)	6000 × 3300
12	+ 0.3	30.00)	
15	+ 0.5	37.50	3050 × 3000
19	+ 1.0	47.50)	3000 × 2900
25	+ 1.0	63.50)	
Clear sheet glass			
2 *	+ 0.2	5.00	1920 × 1220
3	+ 0.3	7.50	2130 × 1320
4	+ 0.3	10.00	2760 × 1220
5 *	+ 0.3	12.50)	2130 × 2400
6 *	+ 0.3	15.00)	
Cast glass			
3	+ 0.4 − 0.2	6.00)	2140 × 1280
4	+ 0.5	7.50)	
5	+ 0.5	9.50	2140 × 1320
6	+ 0.5	11.50)	3700 × 1280
10	+ 0.8	21.50)	
Wired glass (Cast wired glass)			
6	+ 0.3 − 0.7	–))	3700 × 1840
7	+ 0.7	–)	
(Polished wire glass)			
6	+ 1.0	–	330 × 1830

* The 5 mm and 6 mm thickness are known as *thick drawn sheet*. Although 2 mm sheet glass is available it is not recommended for general glazing purposes

METAL

Weights of Metals

Material		kg/m^3	lb/cu ft
Metals, steel construction, etc.			
Iron			
	– cast	7207	450
	– wrought	7687	480
	– ore – general	2407	150
	– (crushed) Swedish	3682	230
Steel		7854	490
Copper			
	– cast	8731	545
	– wrought	8945	558
Brass		8497	530
Bronze		8945	558
Aluminium		2774	173
Lead		11322	707
Zinc (rolled)		7140	446
		g/mm^2 per metre	**lb/sq ft per foot**
Steel bars		7.85	3.4
Structural steelwork		Net weight of member @ 7854 kg/m^3	
	riveted	+ 10% for cleats, rivets, bolts, etc.	
	welded	+ 1.25% to 2.5% for welds, etc.	
Rolled sections			
	beams	+ 2.5%	
	stanchions	+ 5% (extra for caps and bases)	
Plate			
	web girders	+ 10% for rivets or welds, stiffeners, etc.	
		kg/m	**lb/ft**
Steel stairs: industrial type			
	1 m or 3 ft wide	84	56
Steel tubes			
	50 mm or 2 in bore	5 to 6	3 to 4
Gas piping			
	20 mm or ¾ in	2	1¼

METAL

Universal Beams BS 4: Part 1: 2005

Designation	Mass (kg/m)	Depth of section (mm)	Width of section (mm)	Thickness		Surface area (m²/m)
				Web (mm)	Flange (mm)	
1016 × 305 × 487	487.0	1036.1	308.5	30.0	54.1	3.20
1016 × 305 × 438	438.0	1025.9	305.4	26.9	49.0	3.17
1016 × 305 × 393	393.0	1016.0	303.0	24.4	43.9	3.15
1016 × 305 × 349	349.0	1008.1	302.0	21.1	40.0	3.13
1016 × 305 × 314	314.0	1000.0	300.0	19.1	35.9	3.11
1016 × 305 × 272	272.0	990.1	300.0	16.5	31.0	3.10
1016 × 305 × 249	249.0	980.2	300.0	16.5	26.0	3.08
1016 × 305 × 222	222.0	970.3	300.0	16.0	21.1	3.06
914 × 419 × 388	388.0	921.0	420.5	21.4	36.6	3.44
914 × 419 × 343	343.3	911.8	418.5	19.4	32.0	3.42
914 × 305 × 289	289.1	926.6	307.7	19.5	32.0	3.01
914 × 305 × 253	253.4	918.4	305.5	17.3	27.9	2.99
914 × 305 × 224	224.2	910.4	304.1	15.9	23.9	2.97
914 × 305 × 201	200.9	903.0	303.3	15.1	20.2	2.96
838 × 292 × 226	226.5	850.9	293.8	16.1	26.8	2.81
838 × 292 × 194	193.8	840.7	292.4	14.7	21.7	2.79
838 × 292 × 176	175.9	834.9	291.7	14.0	18.8	2.78
762 × 267 × 197	196.8	769.8	268.0	15.6	25.4	2.55
762 × 267 × 173	173.0	762.2	266.7	14.3	21.6	2.53
762 × 267 × 147	146.9	754.0	265.2	12.8	17.5	2.51
762 × 267 × 134	133.9	750.0	264.4	12.0	15.5	2.51
686 × 254 × 170	170.2	692.9	255.8	14.5	23.7	2.35
686 × 254 × 152	152.4	687.5	254.5	13.2	21.0	2.34
686 × 254 × 140	140.1	383.5	253.7	12.4	19.0	2.33
686 × 254 × 125	125.2	677.9	253.0	11.7	16.2	2.32
610 × 305 × 238	238.1	635.8	311.4	18.4	31.4	2.45
610 × 305 × 179	179.0	620.2	307.1	14.1	23.6	2.41
610 × 305 × 149	149.1	612.4	304.8	11.8	19.7	2.39
610 × 229 × 140	139.9	617.2	230.2	13.1	22.1	2.11
610 × 229 × 125	125.1	612.2	229.0	11.9	19.6	2.09
610 × 229 × 113	113.0	607.6	228.2	11.1	17.3	2.08
610 × 229 × 101	101.2	602.6	227.6	10.5	14.8	2.07
533 × 210 × 122	122.0	544.5	211.9	12.7	21.3	1.89
533 × 210 × 109	109.0	539.5	210.8	11.6	18.8	1.88
533 × 210 × 101	101.0	536.7	210.0	10.8	17.4	1.87
533 × 210 × 92	92.1	533.1	209.3	10.1	15.6	1.86
533 × 210 × 82	82.2	528.3	208.8	9.6	13.2	1.85
457 × 191 × 98	98.3	467.2	192.8	11.4	19.6	1.67
457 × 191 × 89	89.3	463.4	191.9	10.5	17.7	1.66
457 × 191 × 82	82.0	460.0	191.3	9.9	16.0	1.65
457 × 191 × 74	74.3	457.0	190.4	9.0	14.5	1.64
457 × 191 × 67	67.1	453.4	189.9	8.5	12.7	1.63
457 × 152 × 82	82.1	465.8	155.3	10.5	18.9	1.51
457 × 152 × 74	74.2	462.0	154.4	9.6	17.0	1.50
457 × 152 × 67	67.2	458.0	153.8	9.0	15.0	1.50
457 × 152 × 60	59.8	454.6	152.9	8.1	13.3	1.50
457 × 152 × 52	52.3	449.8	152.4	7.6	10.9	1.48
406 × 178 × 74	74.2	412.8	179.5	9.5	16.0	1.51
406 × 178 × 67	67.1	409.4	178.8	8.8	14.3	1.50
406 × 178 × 60	60.1	406.4	177.9	7.9	12.8	1.49

METAL

Designation	Mass (kg/m)	Depth of section (mm)	Width of section (mm)	Thickness		Surface area (m²/m)
				Web (mm)	Flange (mm)	
406 × 178 × 50	54.1	402.6	177.7	7.7	10.9	1.48
406 × 140 × 46	46.0	403.2	142.2	6.8	11.2	1.34
406 × 140 × 39	39.0	398.0	141.8	6.4	8.6	1.33
356 × 171 × 67	67.1	363.4	173.2	9.1	15.7	1.38
356 × 171 × 57	57.0	358.0	172.2	8.1	13.0	1.37
356 × 171 × 51	51.0	355.0	171.5	7.4	11.5	1.36
356 × 171 × 45	45.0	351.4	171.1	7.0	9.7	1.36
356 × 127 × 39	39.1	353.4	126.0	6.6	10.7	1.18
356 × 127 × 33	33.1	349.0	125.4	6.0	8.5	1.17
305 × 165 × 54	54.0	310.4	166.9	7.9	13.7	1.26
305 × 165 × 46	46.1	306.6	165.7	6.7	11.8	1.25
305 × 165 × 40	40.3	303.4	165.0	6.0	10.2	1.24
305 × 127 × 48	48.1	311.0	125.3	9.0	14.0	1.09
305 × 127 × 42	41.9	307.2	124.3	8.0	12.1	1.08
305 × 127 × 37	37.0	304.4	123.3	7.1	10.7	1.07
305 × 102 × 33	32.8	312.7	102.4	6.6	10.8	1.01
305 × 102 × 28	28.2	308.7	101.8	6.0	8.8	1.00
305 × 102 × 25	24.8	305.1	101.6	5.8	7.0	0.992
254 × 146 × 43	43.0	259.6	147.3	7.2	12.7	1.08
254 × 146 × 37	37.0	256.0	146.4	6.3	10.9	1.07
254 × 146 × 31	31.1	251.4	146.1	6.0	8.6	1.06
254 × 102 × 28	28.3	260.4	102.2	6.3	10.0	0.904
254 × 102 × 25	25.2	257.2	101.9	6.0	8.4	0.897
254 × 102 × 22	22.0	254.0	101.6	5.7	6.8	0.890
203 × 133 × 30	30.0	206.8	133.9	6.4	9.6	0.923
203 × 133 × 25	25.1	203.2	133.2	5.7	7.8	0.915
203 × 102 × 23	23.1	203.2	101.8	5.4	9.3	0.790
178 × 102 × 19	19.0	177.8	101.2	4.8	7.9	0.738
152 × 89 × 16	16.0	152.4	88.7	4.5	7.7	0.638
127 × 76 × 13	13.0	127.0	76.0	4.0	7.6	0.537

METAL

Universal Columns BS 4: Part 1: 2005

Designation	Mass (kg/m)	Depth of section (mm)	Width of section (mm)	Thickness		Surface area (m²/m)
				Web (mm)	Flange (mm)	
356 × 406 × 634	633.9	474.7	424.0	47.6	77.0	2.52
356 × 406 × 551	551.0	455.6	418.5	42.1	67.5	2.47
356 × 406 × 467	467.0	436.6	412.2	35.8	58.0	2.42
356 × 406 × 393	393.0	419.0	407.0	30.6	49.2	2.38
356 × 406 × 340	339.9	406.4	403.0	26.6	42.9	2.35
356 × 406 × 287	287.1	393.6	399.0	22.6	36.5	2.31
356 × 406 × 235	235.1	381.0	384.8	18.4	30.2	2.28
356 × 368 × 202	201.9	374.6	374.7	16.5	27.0	2.19
356 × 368 × 177	177.0	368.2	372.6	14.4	23.8	2.17
356 × 368 × 153	152.9	362.0	370.5	12.3	20.7	2.16
356 × 368 × 129	129.0	355.6	368.6	10.4	17.5	2.14
305 × 305 × 283	282.9	365.3	322.2	26.8	44.1	1.94
305 × 305 × 240	240.0	352.5	318.4	23.0	37.7	1.91
305 × 305 × 198	198.1	339.9	314.5	19.1	31.4	1.87
305 × 305 × 158	158.1	327.1	311.2	15.8	25.0	1.84
305 × 305 × 137	136.9	320.5	309.2	13.8	21.7	1.82
305 × 305 × 118	117.9	314.5	307.4	12.0	18.7	1.81
305 × 305 × 97	96.9	307.9	305.3	9.9	15.4	1.79
254 × 254 × 167	167.1	289.1	265.2	19.2	31.7	1.58
254 × 254 × 132	132.0	276.3	261.3	15.3	25.3	1.55
254 × 254 × 107	107.1	266.7	258.8	12.8	20.5	1.52
254 × 254 × 89	88.9	260.3	256.3	10.3	17.3	1.50
254 × 254 × 73	73.1	254.1	254.6	8.6	14.2	1.49
203 × 203 × 86	86.1	222.2	209.1	12.7	20.5	1.24
203 × 203 × 71	71.0	215.8	206.4	10.0	17.3	1.22
203 × 203 × 60	60.0	209.6	205.8	9.4	14.2	1.21
203 × 203 × 52	52.0	206.2	204.3	7.9	12.5	1.20
203 × 203 × 46	46.1	203.2	203.6	7.2	11.0	1.19
152 × 152 × 37	37.0	161.8	154.4	8.0	11.5	0.912
152 × 152 × 30	30.0	157.6	152.9	6.5	9.4	0.901
152 × 152 × 23	23.0	152.4	152.2	5.8	6.8	0.889

METAL

Joists BS 4: Part 1: 2005 (retained for reference, Corus have ceased manufacture in UK)

Designation	Mass (kg/m)	Depth of section (mm)	Width of section (mm)	Thickness		Surface area (m^2/m)
				Web (mm)	Flange (mm)	
254 × 203 × 82	82.0	254.0	203.2	10.2	19.9	1.210
203 × 152 × 52	52.3	203.2	152.4	8.9	16.5	0.932
152 × 127 × 37	37.3	152.4	127.0	10.4	13.2	0.737
127 × 114 × 29	29.3	127.0	114.3	10.2	11.5	0.646
127 × 114 × 27	26.9	127.0	114.3	7.4	11.4	0.650
102 × 102 × 23	23.0	101.6	101.6	9.5	10.3	0.549
102 × 44 × 7	7.5	101.6	44.5	4.3	6.1	0.350
89 × 89 × 19	19.5	88.9	88.9	9.5	9.9	0.476
76 × 76 × 13	12.8	76.2	76.2	5.1	8.4	0.411

Parallel Flange Channels

Designation	Mass (kg/m)	Depth of section (mm)	Width of section (mm)	Thickness		Surface area (m^2/m)
				Web (mm)	Flange (mm)	
430 × 100 × 64	64.4	430	100	11.0	19.0	1.23
380 × 100 × 54	54.0	380	100	9.5	17.5	1.13
300 × 100 × 46	45.5	300	100	9.0	16.5	0.969
300 × 90 × 41	41.4	300	90	9.0	15.5	0.932
260 × 90 × 35	34.8	260	90	8.0	14.0	0.854
260 × 75 × 28	27.6	260	75	7.0	12.0	0.79
230 × 90 × 32	32.2	230	90	7.5	14.0	0.795
230 × 75 × 26	25.7	230	75	6.5	12.5	0.737
200 × 90 × 30	29.7	200	90	7.0	14.0	0.736
200 × 75 × 23	23.4	200	75	6.0	12.5	0.678
180 × 90 × 26	26.1	180	90	6.5	12.5	0.697
180 × 75 × 20	20.3	180	75	6.0	10.5	0.638
150 × 90 × 24	23.9	150	90	6.5	12.0	0.637
150 × 75 × 18	17.9	150	75	5.5	10.0	0.579
125 × 65 × 15	14.8	125	65	5.5	9.5	0.489
100 × 50 × 10	10.2	100	50	5.0	8.5	0.382

METAL

Equal Angles BS EN 10056-1

Designation	Mass (kg/m)	Surface area (m^2/m)
200 × 200 × 24	71.1	0.790
200 × 200 × 20	59.9	0.790
200 × 200 × 18	54.2	0.790
200 × 200 × 16	48.5	0.790
150 × 150 × 18	40.1	0.59
150 × 150 × 15	33.8	0.59
150 × 150 × 12	27.3	0.59
150 × 150 × 10	23.0	0.59
120 × 120 × 15	26.6	0.47
120 × 120 × 12	21.6	0.47
120 × 120 × 10	18.2	0.47
120 × 120 × 8	14.7	0.47
100 × 100 × 15	21.9	0.39
100 × 100 × 12	17.8	0.39
100 × 100 × 10	15.0	0.39
100 × 100 × 8	12.2	0.39
90 × 90 × 12	15.9	0.35
90 × 90 × 10	13.4	0.35
90 × 90 × 8	10.9	0.35
90 × 90 × 7	9.61	0.35
90 × 90 × 6	8.30	0.35

Unequal Angles BS EN 10056-1

Designation	Mass (kg/m)	Surface area (m^2/m)
200 × 150 × 18	47.1	0.69
200 × 150 × 15	39.6	0.69
200 × 150 × 12	32.0	0.69
200 × 100 × 15	33.7	0.59
200 × 100 × 12	27.3	0.59
200 × 100 × 10	23.0	0.59
150 × 90 × 15	26.6	0.47
150 × 90 × 12	21.6	0.47
150 × 90 × 10	18.2	0.47
150 × 75 × 15	24.8	0.44
150 × 75 × 12	20.2	0.44
150 × 75 × 10	17.0	0.44
125 × 75 × 12	17.8	0.40
125 × 75 × 10	15.0	0.40
125 × 75 × 8	12.2	0.40
100 × 75 × 12	15.4	0.34
100 × 75 × 10	13.0	0.34
100 × 75 × 8	10.6	0.34
100 × 65 × 10	12.3	0.32
100 × 65 × 8	9.94	0.32
100 × 65 × 7	8.77	0.32

METAL

Structural Tees Split from Universal Beams BS 4: Part 1: 2005

Designation	Mass (kg/m)	Surface area (m^2/m)
305 × 305 × 90	89.5	1.22
305 × 305 × 75	74.6	1.22
254 × 343 × 63	62.6	1.19
229 × 305 × 70	69.9	1.07
229 × 305 × 63	62.5	1.07
229 × 305 × 57	56.5	1.07
229 × 305 × 51	50.6	1.07
210 × 267 × 61	61.0	0.95
210 × 267 × 55	54.5	0.95
210 × 267 × 51	50.5	0.95
210 × 267 × 46	46.1	0.95
210 × 267 × 41	41.1	0.95
191 × 229 × 49	49.2	0.84
191 × 229 × 45	44.6	0.84
191 × 229 × 41	41.0	0.84
191 × 229 × 37	37.1	0.84
191 × 229 × 34	33.6	0.84
152 × 229 × 41	41.0	0.76
152 × 229 × 37	37.1	0.76
152 × 229 × 34	33.6	0.76
152 × 229 × 30	29.9	0.76
152 × 229 × 26	26.2	0.76

Universal Bearing Piles BS 4: Part 1: 2005

Designation	Mass (kg/m)	Depth of Section (mm)	Width of section (mm)	Thickness	
				Web (mm)	Flange (mm)
356 × 368 × 174	173.9	361.4	378.5	20.3	20.4
356 × 368 × 152	152.0	356.4	376.0	17.8	17.9
356 × 368 × 133	133.0	352.0	373.8	15.6	15.7
356 × 368 × 109	108.9	346.4	371.0	12.8	12.9
305 × 305 × 223	222.9	337.9	325.7	30.3	30.4
305 × 305 × 186	186.0	328.3	320.9	25.5	25.6
305 × 305 × 149	149.1	318.5	316.0	20.6	20.7
305 × 305 × 126	126.1	312.3	312.9	17.5	17.6
305 × 305 × 110	110.0	307.9	310.7	15.3	15.4
305 × 305 × 95	94.9	303.7	308.7	13.3	13.3
305 × 305 × 88	88.0	301.7	307.8	12.4	12.3
305 × 305 × 79	78.9	299.3	306.4	11.0	11.1
254 × 254 × 85	85.1	254.3	260.4	14.4	14.3
254 × 254 × 71	71.0	249.7	258.0	12.0	12.0
254 × 254 × 63	63.0	247.1	256.6	10.6	10.7
203 × 203 × 54	53.9	204.0	207.7	11.3	11.4
203 × 203 × 45	44.9	200.2	205.9	9.5	9.5

METAL

Hot Formed Square Hollow Sections EN 10210 S275J2H & S355J2H

Size (mm)	Wall thickness (mm)	Mass (kg/m)	Superficial area (m²/m)
40 × 40	2.5	2.89	0.154
	3.0	3.41	0.152
	3.2	3.61	0.152
	3.6	4.01	0.151
	4.0	4.39	0.150
	5.0	5.28	0.147
50 × 50	2.5	3.68	0.194
	3.0	4.35	0.192
	3.2	4.62	0.192
	3.6	5.14	0.191
	4.0	5.64	0.190
	5.0	6.85	0.187
	6.0	7.99	0.185
	6.3	8.31	0.184
60 × 60	3.0	5.29	0.232
	3.2	5.62	0.232
	3.6	6.27	0.231
	4.0	6.90	0.230
	5.0	8.42	0.227
	6.0	9.87	0.225
	6.3	10.30	0.224
	8.0	12.50	0.219
70 × 70	3.0	6.24	0.272
	3.2	6.63	0.272
	3.6	7.40	0.271
	4.0	8.15	0.270
	5.0	9.99	0.267
	6.0	11.80	0.265
	6.3	12.30	0.264
	8.0	15.00	0.259
80 × 80	3.2	7.63	0.312
	3.6	8.53	0.311
	4.0	9.41	0.310
	5.0	11.60	0.307
	6.0	13.60	0.305
	6.3	14.20	0.304
	8.0	17.50	0.299
90 × 90	3.6	9.66	0.351
	4.0	10.70	0.350
	5.0	13.10	0.347
	6.0	15.50	0.345
	6.3	16.20	0.344
	8.0	20.10	0.339
100 × 100	3.6	10.80	0.391
	4.0	11.90	0.390
	5.0	14.70	0.387
	6.0	17.40	0.385
	6.3	18.20	0.384
	8.0	22.60	0.379
	10.0	27.40	0.374
120 × 120	4.0	14.40	0.470
	5.0	17.80	0.467
	6.0	21.20	0.465

METAL

Size (mm)	Wall thickness (mm)	Mass (kg/m)	Superficial area (m²/m)
	6.3	22.20	0.464
	8.0	27.60	0.459
	10.0	33.70	0.454
	12.0	39.50	0.449
	12.5	40.90	0.448
140 × 140	5.0	21.00	0.547
	6.0	24.90	0.545
	6.3	26.10	0.544
	8.0	32.60	0.539
	10.0	40.00	0.534
	12.0	47.00	0.529
	12.5	48.70	0.528
150 × 150	5.0	22.60	0.587
	6.0	26.80	0.585
	6.3	28.10	0.584
	8.0	35.10	0.579
	10.0	43.10	0.574
	12.0	50.80	0.569
	12.5	52.70	0.568
Hot formed from seamless hollow	16.0	65.2	0.559
160 × 160	5.0	24.10	0.627
	6.0	28.70	0.625
	6.3	30.10	0.624
	8.0	37.60	0.619
	10.0	46.30	0.614
	12.0	54.60	0.609
	12.5	56.60	0.608
	16.0	70.20	0.599
180 × 180	5.0	27.30	0.707
	6.0	32.50	0.705
	6.3	34.00	0.704
	8.0	42.70	0.699
	10.0	52.50	0.694
	12.0	62.10	0.689
	12.5	64.40	0.688
	16.0	80.20	0.679
200 × 200	5.0	30.40	0.787
	6.0	36.20	0.785
	6.3	38.00	0.784
	8.0	47.70	0.779
	10.0	58.80	0.774
	12.0	69.60	0.769
	12.5	72.30	0.768
	16.0	90.30	0.759
250 × 250	5.0	38.30	0.987
	6.0	45.70	0.985
	6.3	47.90	0.984
	8.0	60.30	0.979
	10.0	74.50	0.974
	12.0	88.50	0.969
	12.5	91.90	0.968
	16.0	115.00	0.959

METAL

Size (mm)	Wall thickness (mm)	Mass (kg/m)	Superficial area (m²/m)
300 × 300	6.0	55.10	1.18
	6.3	57.80	1.18
	8.0	72.80	1.18
	10.0	90.20	1.17
	12.0	107.00	1.17
	12.5	112.00	1.17
	16.0	141.00	1.16
350 × 350	8.0	85.40	1.38
	10.0	106.00	1.37
	12.0	126.00	1.37
	12.5	131.00	1.37
	16.0	166.00	1.36
400 × 400	8.0	97.90	1.58
	10.0	122.00	1.57
	12.0	145.00	1.57
	12.5	151.00	1.57
	16.0	191.00	1.56
(Grade S355J2H only)	20.00*	235.00	1.55

Note: * SAW process

METAL

Hot Formed Square Hollow Sections JUMBO RHS: JIS G3136

Size (mm)	Wall thickness (mm)	Mass (kg/m)	Superficial area (m^2/m)
350 × 350	19.0	190.00	1.33
	22.0	217.00	1.32
	25.0	242.00	1.31
400 × 400	22.0	251.00	1.52
	25.0	282.00	1.51
450 × 450	12.0	162.00	1.76
	16.0	213.00	1.75
	19.0	250.00	1.73
	22.0	286.00	1.72
	25.0	321.00	1.71
	28.0 *	355.00	1.70
	32.0 *	399.00	1.69
500 × 500	12.0	181.00	1.96
	16.0	238.00	1.95
	19.0	280.00	1.93
	22.0	320.00	1.92
	25.0	360.00	1.91
	28.0 *	399.00	1.90
	32.0 *	450.00	1.89
	36.0 *	498.00	1.88
550 × 550	16.0	263.00	2.15
	19.0	309.00	2.13
	22.0	355.00	2.12
	25.0	399.00	2.11
	28.0 *	443.00	2.10
	32.0 *	500.00	2.09
	36.0 *	555.00	2.08
	40.0 *	608.00	2.06
600 × 600	25.0 *	439.00	2.31
	28.0 *	487.00	2.30
	32.0 *	550.00	2.29
	36.0 *	611.00	2.28
	40.0 *	671.00	2.26
700 × 700	25.0 *	517.00	2.71
	28.0 *	575.00	2.70
	32.0 *	651.00	2.69
	36.0 *	724.00	2.68
	40.0 *	797.00	2.68

Note: * SAW process

METAL

Hot Formed Rectangular Hollow Sections: EN10210 S275J2h & S355J2H

Size (mm)	Wall thickness (mm)	Mass (kg/m)	Superficial area (m²/m)
50 × 30	2.5	2.89	0.154
	3.0	3.41	0.152
	3.2	3.61	0.152
	3.6	4.01	0.151
	4.0	4.39	0.150
	5.0	5.28	0.147
60 × 40	2.5	3.68	0.194
	3.0	4.35	0.192
	3.2	4.62	0.192
	3.6	5.14	0.191
	4.0	5.64	0.190
	5.0	6.85	0.187
	6.0	7.99	0.185
	6.3	8.31	0.184
80 × 40	3.0	5.29	0.232
	3.2	5.62	0.232
	3.6	6.27	0.231
	4.0	6.90	0.230
	5.0	8.42	0.227
	6.0	9.87	0.225
	6.3	10.30	0.224
	8.0	12.50	0.219
76.2 × 50.8	3.0	5.62	0.246
	3.2	5.97	0.246
	3.6	6.66	0.245
	4.0	7.34	0.244
	5.0	8.97	0.241
	6.0	10.50	0.239
	6.3	11.00	0.238
	8.0	13.40	0.233
90 × 50	3.0	6.24	0.272
	3.2	6.63	0.272
	3.6	7.40	0.271
	4.0	8.15	0.270
	5.0	9.99	0.267
	6.0	11.80	0.265
	6.3	12.30	0.264
	8.0	15.00	0.259
100 × 50	3.0	6.71	0.292
	3.2	7.13	0.292
	3.6	7.96	0.291
	4.0	8.78	0.290
	5.0	10.80	0.287
	6.0	12.70	0.285
	6.3	13.30	0.284
	8.0	16.30	0.279

METAL

Size (mm)	Wall thickness (mm)	Mass (kg/m)	Superficial area (m²/m)
100 × 60	3.0	7.18	0.312
	3.2	7.63	0.312
	3.6	8.53	0.311
	4.0	9.41	0.310
	5.0	11.60	0.307
	6.0	13.60	0.305
	6.3	14.20	0.304
	8.0	17.50	0.299
120 × 60	3.6	9.70	0.351
	4.0	10.70	0.350
	5.0	13.10	0.347
	6.0	15.50	0.345
	6.3	16.20	0.344
	8.0	20.10	0.339
120 × 80	3.6	10.80	0.391
	4.0	11.90	0.390
	5.0	14.70	0.387
	6.0	17.40	0.385
	6.3	18.20	0.384
	8.0	22.60	0.379
	10.0	27.40	0.374
150 × 100	4.0	15.10	0.490
	5.0	18.60	0.487
	6.0	22.10	0.485
	6.3	23.10	0.484
	8.0	28.90	0.479
	10.0	35.30	0.474
	12.0	41.40	0.469
	12.5	42.80	0.468
160 × 80	4.0	14.40	0.470
	5.0	17.80	0.467
	6.0	21.20	0.465
	6.3	22.20	0.464
	8.0	27.60	0.459
	10.0	33.70	0.454
	12.0	39.50	0.449
	12.5	40.90	0.448
200 × 100	5.0	22.60	0.587
	6.0	26.80	0.585
	6.3	28.10	0.584
	8.0	35.10	0.579
	10.0	43.10	0.574
	12.0	50.80	0.569
	12.5	52.70	0.568
	16.0	65.20	0.559
250 × 150	5.0	30.40	0.787
	6.0	36.20	0.785
	6.3	38.00	0.784
	8.0	47.70	0.779
	10.0	58.80	0.774
	12.0	69.60	0.769
	12.5	72.30	0.768
	16.0	90.30	0.759

METAL

Size (mm)	Wall thickness (mm)	Mass (kg/m)	Superficial area (m^2/m)
300 × 200	5.0	38.30	0.987
	6.0	45.70	0.985
	6.3	47.90	0.984
	8.0	60.30	0.979
	10.0	74.50	0.974
	12.0	88.50	0.969
	12.5	91.90	0.968
	16.0	115.00	0.959
400 × 200	6.0	55.10	1.18
	6.3	57.80	1.18
	8.0	72.80	1.18
	10.0	90.20	1.17
	12.0	107.00	1.17
	12.5	112.00	1.17
	16.0	141.00	1.16
450 × 250	8.0	85.40	1.38
	10.0	106.00	1.37
	12.0	126.00	1.37
	12.5	131.00	1.37
	16.0	166.00	1.36
500 × 300	8.0	98.00	1.58
	10.0	122.00	1.57
	12.0	145.00	1.57
	12.5	151.00	1.57
	16.0	191.00	1.56
	20.0	235.00	1.55

METAL

Hot Formed Circular Hollow Sections EN 10210 S275J2H & S355J2H

Outside diameter (mm)	Wall thickness (mm)	Mass (kg/m)	Superficial area (m^2/m)
21.3	3.2	1.43	0.067
26.9	3.2	1.87	0.085
33.7	3.0	2.27	0.106
	3.2	2.41	0.106
	3.6	2.67	0.106
	4.0	2.93	0.106
42.4	3.0	2.91	0.133
	3.2	3.09	0.133
	3.6	3.44	0.133
	4.0	3.79	0.133
48.3	2.5	2.82	0.152
	3.0	3.35	0.152
	3.2	3.56	0.152
	3.6	3.97	0.152
	4.0	4.37	0.152
	5.0	5.34	0.152
60.3	2.5	3.56	0.189
	3.0	4.24	0.189
	3.2	4.51	0.189
	3.6	5.03	0.189
	4.0	5.55	0.189
	5.0	6.82	0.189
76.1	2.5	4.54	0.239
	3.0	5.41	0.239
	3.2	5.75	0.239
	3.6	6.44	0.239
	4.0	7.11	0.239
	5.0	8.77	0.239
	6.0	10.40	0.239
	6.3	10.80	0.239
88.9	2.5	5.33	0.279
	3.0	6.36	0.279
	3.2	6.76	0.27
	3.6	7.57	0.279
	4.0	8.38	0.279
	5.0	10.30	0.279
	6.0	12.30	0.279
	6.3	12.80	0.279
114.3	3.0	8.23	0.359
	3.2	8.77	0.359
	3.6	9.83	0.359
	4.0	10.09	0.359
	5.0	13.50	0.359
	6.0	16.00	0.359
	6.3	16.80	0.359

METAL

Outside diameter (mm)	Wall thickness (mm)	Mass (kg/m)	Superficial area (m^2/m)
139.7	3.2	10.80	0.439
	3.6	12.10	0.439
	4.0	13.40	0.439
	5.0	16.60	0.439
	6.0	19.80	0.439
	6.3	20.70	0.439
	8.0	26.00	0.439
	10.0	32.00	0.439
168.3	3.2	13.00	0.529
	3.6	14.60	0.529
	4.0	16.20	0.529
	5.0	20.10	0.529
	6.0	24.00	0.529
	6.3	25.20	0.529
	8.0	31.60	0.529
	10.0	39.00	0.529
	12.0	46.30	0.529
	12.5	48.00	0.529
193.7	5.0	23.30	0.609
	6.0	27.80	0.609
	6.3	29.10	0.609
	8.0	36.60	0.609
	10.0	45.30	0.609
	12.0	53.80	0.609
	12.5	55.90	0.609
219.1	5.0	26.40	0.688
	6.0	31.50	0.688
	6.3	33.10	0.688
	8.0	41.60	0.688
	10.0	51.60	0.688
	12.0	61.30	0.688
	12.5	63.70	0.688
	16.0	80.10	0.688
244.5	5.0	29.50	0.768
	6.0	35.30	0.768
	6.3	37.00	0.768
	8.0	46.70	0.768
	10.0	57.80	0.768
	12.0	68.80	0.768
	12.5	71.50	0.768
	16.0	90.20	0.768
273.0	5.0	33.00	0.858
	6.0	39.50	0.858
	6.3	41.40	0.858
	8.0	52.30	0.858
	10.0	64.90	0.858
	12.0	77.20	0.858
	12.5	80.30	0.858
	16.0	101.00	0.858
323.9	5.0	39.30	1.02
	6.0	47.00	1.02
	6.3	49.30	1.02
	8.0	62.30	1.02
	10.0	77.40	1.02

METAL

Outside diameter (mm)	Wall thickness (mm)	Mass (kg/m)	Superficial area (m^2/m)
	12.0	92.30	1.02
	12.5	96.00	1.02
	16.0	121.00	1.02
355.6	6.3	54.30	1.12
	8.0	68.60	1.12
	10.0	85.30	1.12
	12.0	102.00	1.12
	12.5	106.00	1.12
	16.0	134.00	1.12
406.4	6.3	62.20	1.28
	8.0	79.60	1.28
	10.0	97.80	1.28
	12.0	117.00	1.28
	12.5	121.00	1.28
	16.0	154.00	1.28
457.0	6.3	70.00	1.44
	8.0	88.60	1.44
	10.0	110.00	1.44
	12.0	132.00	1.44
	12.5	137.00	1.44
	16.0	174.00	1.44
508.0	6.3	77.90	1.60
	8.0	98.60	1.60
	10.0	123.00	1.60
	12.0	147.00	1.60
	12.5	153.00	1.60
	16.0	194.00	1.60

METAL

Spacing of Holes in Angles

Nominal leg length (mm)	Spacing of holes						Maximum diameter of bolt or rivet		
	A	B	C	D	E	F	A	B and C	D, E and F
200		75	75	55	55	55		30	20
150		55	55					20	
125		45	60					20	
120									
100	55						24		
90	50						24		
80	45						20		
75	45						20		
70	40						20		
65	35						20		
60	35						16		
50	28						12		
45	25								
40	23								
30	20								
25	15								

KERBS, PAVING, ETC.

KERBS/EDGINGS/CHANNELS

Precast Concrete Kerbs to BS 7263

Straight kerb units: length from 450 to 915 mm

150 mm high × 125 mm thick		
bullnosed	type BN	
half battered	type HB3	
255 mm high × 125 mm thick		
45° splayed	type SP	
half battered	type HB2	
305 mm high × 150 mm thick		
half battered	type HB1	
Quadrant kerb units		
150 mm high × 305 and 455 mm radius to match	type BN	type QBN
150 mm high × 305 and 455 mm radius to match	type HB2, HB3	type QHB
150 mm high × 305 and 455 mm radius to match	type SP	type QSP
255 mm high × 305 and 455 mm radius to match	type BN	type QBN
255 mm high × 305 and 455 mm radius to match	type HB2, HB3	type QHB
225 mm high × 305 and 455 mm radius to match	type SP	type QSP
Angle kerb units		
305 × 305 × 225 mm high × 125 mm thick		
bullnosed external angle	type XA	
splayed external angle to match type SP	type XA	
bullnosed internal angle	type IA	
splayed internal angle to match type SP	type IA	
Channels		
255 mm wide × 125 mm high flat	type CS1	
150 mm wide × 125 mm high flat type	CS2	
255 mm wide × 125 mm high dished	type CD	

KERBS, PAVING, ETC.

Transition kerb units			
from kerb type SP to HB	left handed	type TL	
	right handed	type TR	
from kerb type BN to HB	left handed	type DL1	
	right handed	type DR1	
from kerb type BN to SP	left handed	type DL2	
	right handed	type DR2	

Number of kerbs required per quarter circle (780 mm kerb lengths)

Radius (m)	Number in quarter circle
12	24
10	20
8	16
6	12
5	10
4	8
3	6
2	4
1	2

Precast Concrete Edgings

Round top type ER	Flat top type EF	Bullnosed top type EBN
150 × 50 mm	150 × 50 mm	150 × 50 mm
200 × 50 mm	200 × 50 mm	200 × 50 mm
250 × 50 mm	250 × 50 mm	250 × 50 mm

KERBS, PAVING, ETC.

BASES

Cement Bound Material for Bases and Subbases

CBM1:	very carefully graded aggregate from 37.5–75 mm, with a 7-day strength of 4.5 N/mm^2
CBM2:	same range of aggregate as CBM1 but with more tolerance in each size of aggregate with a 7-day strength of 7.0 N/mm^2
CBM3:	crushed natural aggregate or blast furnace slag, graded from 37.5–150 mm for 40 mm aggregate, and from 20–75 mm for 20 mm aggregate, with a 7-day strength of 10 N/mm^2
CBM4:	crushed natural aggregate or blast furnace slag, graded from 37.5–150 mm for 40 mm aggregate, and from 20–75 mm for 20 mm aggregate, with a 7-day strength of 15 N/mm^2

INTERLOCKING BRICK/BLOCK ROADS/PAVINGS

Sizes of Precast Concrete Paving Blocks

Type R blocks
200 × 100 × 60 mm
200 × 100 × 65 mm
200 × 100 × 80 mm
200 × 100 × 100 mm

Type S
Any shape within a 295 mm space

Sizes of clay brick pavers
200 × 100 × 50 mm
200 × 100 × 65 mm
210 × 105 × 50 mm
210 × 105 × 65 mm
215 × 102.5 × 50 mm
215 × 102.5 × 65 mm

Type PA: 3 kN
Footpaths and pedestrian areas, private driveways, car parks, light vehicle traffic and over-run

Type PB: 7 kN
Residential roads, lorry parks, factory yards, docks, petrol station forecourts, hardstandings, bus stations

KERBS, PAVING, ETC.

PAVING AND SURFACING

Weights and Sizes of Paving and Surfacing

Description of item	Size	Quantity per tonne
Paving 50 mm thick	900 × 600 mm	15
Paving 50 mm thick	750 × 600 mm	18
Paving 50 mm thick	600 × 600 mm	23
Paving 50 mm thick	450 × 600 mm	30
Paving 38 mm thick	600 × 600 mm	30
Path edging	914 × 50 × 150 mm	60
Kerb (including radius and tapers)	125 × 254 × 914 mm	15
Kerb (including radius and tapers)	125 × 150 × 914 mm	25
Square channel	125 × 254 × 914 mm	15
Dished channel	125 × 254 × 914 mm	15
Quadrants	300 × 300 × 254 mm	19
Quadrants	450 × 450 × 254 mm	12
Quadrants	300 × 300 × 150 mm	30
Internal angles	300 × 300 × 254 mm	30
Fluted pavement channel	255 × 75 × 914 mm	25
Corner stones	300 × 300 mm	80
Corner stones	360 × 360 mm	60
Cable covers	914 × 175 mm	55
Gulley kerbs	220 × 220 × 150 mm	60
Gulley kerbs	220 × 200 × 75 mm	120

KERBS, PAVING, ETC.

Weights and Sizes of Paving and Surfacing

Material	kg/m^3	lb/cu yd
Tarmacadam	2306	3891
Macadam (waterbound)	2563	4325
Vermiculite (aggregate)	64–80	108–135
Terracotta	2114	3568
Cork – compressed	388	24
	kg/m^2	lb/sq ft
Clay floor tiles, 12.7 mm	27.3	5.6
Pavement lights	122	25
Damp-proof course	5	1
	kg/m^2 per mm thickness	lb/sq ft per inch thickness
Paving slabs (stone)	2.3	12
Granite setts	2.88	15
Asphalt	2.30	12
Rubber flooring	1.68	9
Polyvinyl chloride	1.94 (avg)	10 (avg)

Coverage (m^2) Per Cubic Metre of Materials Used as Subbases or Capping Layers

Consolidated thickness laid in (mm)	Square metre coverage		
	Gravel	Sand	Hardcore
50	15.80	16.50	–
75	10.50	11.00	–
100	7.92	8.20	7.42
125	6.34	6.60	5.90
150	5.28	5.50	4.95
175	–	–	4.23
200	–	–	3.71
225	–	–	3.30
300	–	–	2.47

KERBS, PAVING, ETC.

Approximate Rate of Spreads

Average thickness of course (mm)	Description	Approximate rate of spread			
		Open Textured		Dense, Medium & Fine Textured	
		(kg/m²)	(m²/t)	(kg/m²)	(m²/t)
35	14 mm open textured or dense wearing course	60–75	13–17	70–85	12–14
40	20 mm open textured or dense base course	70–85	12–14	80–100	10–12
45	20 mm open textured or dense base course	80–100	10–12	95–100	9–10
50	20 mm open textured or dense, or 28 mm dense base course	85–110	9–12	110–120	8–9
60	28 mm dense base course, 40 mm open textured of dense base course or 40 mm single course as base course		8–10	130–150	7–8
65	28 mm dense base course, 40 mm open textured or dense base course or 40 mm single course	100–135	7–10	140–160	6–7
75	40 mm single course, 40 mm open textured or dense base course, 40 mm dense roadbase	120–150	7–8	165–185	5–6
100	40 mm dense base course or roadbase	–	–	220–240	4–4.5

KERBS, PAVING, ETC.

Surface Dressing Roads: Coverage (m^2) per Tonne of Material

Size in mm	Sand	Granite chips	Gravel	Limestone chips
Sand	168	–	–	–
3	–	148	152	165
6	–	130	133	144
9	–	111	114	123
13	–	85	87	95
19	–	68	71	78

Sizes of Flags

Reference	Nominal size (mm)	Thickness (mm)
A	600 × 450	50 and 63
B	600 × 600	50 and 63
C	600 × 750	50 and 63
D	600 × 900	50 and 63
E	450 × 450	50 and 70 chamfered top surface
F	400 × 400	50 and 65 chamfered top surface
G	300 × 300	50 and 60 chamfered top surface

Sizes of Natural Stone Setts

Width (mm)		Length (mm)		Depth (mm)
100	×	100	×	100
75	×	150 to 250	×	125
75	×	150 to 250	×	150
100	×	150 to 250	×	100
100	×	150 to 250	×	150

SEEDING/TURFING AND PLANTING

Topsoil Quality

Topsoil grade	Properties
Premium	Natural topsoil, high fertility, loamy texture, good soil structure, suitable for intensive cultivation.
General purpose	Natural or manufactured topsoil of lesser quality than Premium, suitable for agriculture or amenity landscape, may need fertilizer or soil structure improvement.
Economy	Selected subsoil, natural mineral deposit such as river silt or greensand. The grade comprises two subgrades; 'Low clay' and 'High clay' which is more liable to compaction in handling. This grade is suitable for low-production agricultural land and amenity woodland or conservation planting areas.

Forms of Trees

Standards:	Shall be clear with substantially straight stems. Grafted and budded trees shall have no more than a slight bend at the union. Standards shall be designated as Half, Extra light, Light, Standard, Selected standard, Heavy, and Extra heavy.
Sizes of Standards	
Heavy standard	12–14 cm girth × 3.50 to 5.00 m high
Extra Heavy standard	14–16 cm girth × 4.25 to 5.00 m high
Extra Heavy standard	16–18 cm girth × 4.25 to 6.00 m high
Extra Heavy standard	18–20 cm girth × 5.00 to 6.00 m high
Semi-mature trees:	Between 6.0 m and 12.0 m tall with a girth of 20 to 75 cm at 1.0 m above ground.
Feathered trees:	Shall have a defined upright central leader, with stem furnished with evenly spread and balanced lateral shoots down to or near the ground.
Whips:	Shall be without significant feather growth as determined by visual inspection.
Multi-stemmed trees:	Shall have two or more main stems at, near, above or below ground.

Seedlings grown from seed and not transplanted shall be specified when ordered for sale as:

1+0	one year old seedling
2+0	two year old seedling
1+1	one year seed bed, one year transplanted = two year old seedling
1+2	one year seed bed, two years transplanted = three year old seedling
2+1	two years seed bed, one year transplanted = three year old seedling
1u1	two years seed bed, undercut after 1 year = two year old seedling
2u2	four years seed bed, undercut after 2 years = four year old seedling

SEEDING/TURFING AND PLANTING

Cuttings

The age of cuttings (plants grown from shoots, stems, or roots of the mother plant) shall be specified when ordered for sale. The height of transplants and undercut seedlings/cuttings (which have been transplanted or undercut at least once) shall be stated in centimetres. The number of growing seasons before and after transplanting or undercutting shall be stated.

0 + 1	one year cutting
0 + 2	two year cutting
0 + 1 + 1	one year cutting bed, one year transplanted = two year old seedling
0 + 1 + 2	one year cutting bed, two years transplanted = three year old seedling

Grass Cutting Capacities in m² per hour

Speed mph	Width of cut in metres												
	0.5	0.7	1.0	1.2	1.5	1.7	2.0	2.0	2.1	2.5	2.8	3.0	3.4
1.0	724	1127	1529	1931	2334	2736	3138	3219	3380	4023	4506	4828	5472
1.5	1086	1690	2293	2897	3500	4104	4707	4828	5069	6035	6759	7242	8208
2.0	1448	2253	3058	3862	4667	5472	6276	6437	6759	8047	9012	9656	10944
2.5	1811	2816	3822	4828	5834	6840	7846	8047	8449	10058	11265	12070	13679
3.0	2173	3380	4587	5794	7001	8208	9415	9656	10139	12070	13518	14484	16415
3.5	2535	3943	5351	6759	8167	9576	10984	11265	11829	14082	15772	16898	19151
4.0	2897	4506	6115	7725	9334	10944	12553	12875	13518	16093	18025	19312	21887
4.5	3259	5069	6880	8690	10501	12311	14122	14484	15208	18105	20278	21726	24623
5.0	3621	5633	7644	9656	11668	13679	15691	16093	16898	20117	22531	24140	27359
5.5	3983	6196	8409	10622	12834	15047	17260	17703	18588	22128	24784	26554	30095
6.0	4345	6759	9173	11587	14001	16415	18829	19312	20278	24140	27037	28968	32831
6.5	4707	7322	9938	12553	15168	17783	20398	20921	21967	26152	29290	31382	35566
7.0	5069	7886	10702	13518	16335	19151	21967	22531	23657	28163	31543	33796	38302

Number of Plants per m² at the following offset spacings

(All plants equidistant horizontally and vertically)

Distance mm	Nr of Plants
100	114.43
200	28.22
250	18.17
300	12.35
400	6.72
500	4.29
600	3.04
700	2.16
750	1.88
900	1.26
1000	1.05
1200	0.68
1500	0.42
2000	0.23

SEEDING/TURFING AND PLANTING

Grass Clippings Wet: Based on 3.5 m^3/tonne

Annual kg/100 m^2	Average 20 cuts kg/100 m^2	m^2/tonne	m^2/m^3
32.0	1.6	61162.1	214067.3

Nr of cuts	22	20	18	16	12	4
kg/cut	1.45	1.60	1.78	2.00	2.67	8.00
Area capacity of 3 tonne vehicle per load						
m^2	206250	187500	168750	150000	112500	37500
Load m^3	**100 m^2 units/m^3 of vehicle space**					
1	196.4	178.6	160.7	142.9	107.1	35.7
2	392.9	357.1	321.4	285.7	214.3	71.4
3	589.3	535.7	482.1	428.6	321.4	107.1
4	785.7	714.3	642.9	571.4	428.6	142.9
5	982.1	892.9	803.6	714.3	535.7	178.6

Transportation of Trees

To unload large trees a machine with the necessary lifting strength is required. The weight of the trees must therefore be known in advance. The following table gives a rough overview. The additional columns with root ball dimensions and the number of plants per trailer provide additional information, for example about preparing planting holes and calculating unloading times.

Girth in cm	Rootball diameter in cm	Ball height in cm	Weight in kg	Numbers of trees per trailer
16–18	50–60	40	150	100–120
18–20	60–70	40–50	200	80–100
20–25	60–70	40–50	270	50–70
25–30	80	50–60	350	50
30–35	90–100	60–70	500	12–18
35–40	100–110	60–70	650	10–15
40–45	110–120	60–70	850	8–12
45–50	110–120	60–70	1100	5–7
50–60	130–140	60–70	1600	1–3
60–70	150–160	60–70	2500	1
70–80	180–200	70	4000	1
80–90	200–220	70–80	5500	1
90–100	230–250	80–90	7500	1
100–120	250–270	80–90	9500	1

Data supplied by Lorenz von Ehren GmbH
The information in the table is approximate; deviations depend on soil type, genus and weather

FENCING AND GATES

Types of Preservative

Creosote (tar oil) can be 'factory' applied	by pressure to BS 144: pts 1&2 by immersion to BS 144: pt 1 by hot and cold open tank to BS 144: pts 1&2
Copper/chromium/arsenic (CCA)	by full cell process to BS 4072 pts 1&2
Organic solvent (OS)	by double vacuum (vacvac) to BS 5707 pts 1&3 by immersion to BS 5057 pts 1&3
Pentachlorophenol (PCP)	by heavy oil double vacuum to BS 5705 pts 2&3
Boron diffusion process (treated with disodium octaborate to BWPA Manual 1986)	

Note: Boron is used on green timber at source and the timber is supplied dry

Cleft Chestnut Pale Fences

Pales	Pale spacing	Wire lines	
900 mm	75 mm	2	temporary protection
1050 mm	75 or 100 mm	2	light protective fences
1200 mm	75 mm	3	perimeter fences
1350 mm	75 mm	3	perimeter fences
1500 mm	50 mm	3	narrow perimeter fences
1800 mm	50 mm	3	light security fences

Close-Boarded Fences

Close-boarded fences 1.05 to 1.8 m high
Type BCR (recessed) or BCM (morticed) with concrete posts 140 × 115 mm tapered and Type BW with timber posts

Palisade Fences

Wooden palisade fences
Type WPC with concrete posts 140 × 115 mm tapered and Type WPW with timber posts

For both types of fence:
Height of fence 1050 mm: two rails
Height of fence 1200 mm: two rails
Height of fence 1500 mm: three rails
Height of fence 1650 mm: three rails
Height of fence 1800 mm: three rails

FENCING AND GATES

Post and Rail Fences

Wooden post and rail fences Type MPR 11/3 morticed rails and Type SPR 11/3 nailed rails Height to top of rail 1100 mm Rails: three rails 87 mm, 38 mm
Type MPR 11/4 morticed rails and Type SPR 11/4 nailed rails Height to top of rail 1100 mm Rails: four rails 87 mm, 38 mm
Type MPR 13/4 morticed rails and Type SPR 13/4 nailed rails Height to top of rail 1300 mm Rail spacing 250 mm, 250 mm, and 225 mm from top Rails: four rails 87 mm, 38 mm

Steel Posts

Rolled steel angle iron posts for chain link fencing

Posts	Fence height	Strut	Straining post
1500 × 40 × 40 × 5 mm	900 mm	1500 × 40 × 40 × 5 mm	1500 × 50 × 50 × 6 mm
1800 × 40 × 40 × 5 mm	1200 mm	1800 × 40 × 40 × 5 mm	1800 × 50 × 50 × 6 mm
2000 × 45 × 45 × 5 mm	1400 mm	2000 × 45 × 45 × 5 mm	2000 × 60 × 60 × 6 mm
2600 × 45 × 45 × 5 mm	1800 mm	2600 × 45 × 45 × 5 mm	2600 × 60 × 60 × 6 mm
3000 × 50 × 50 × 6 mm with arms	1800 mm	2600 × 45 × 45 × 5 mm	3000 × 60 × 60 × 6 mm

Concrete Posts

Concrete posts for chain link fencing

Posts and straining posts	Fence height	Strut
1570 mm 100 × 100 mm	900 mm	1500 mm × 75 × 75 mm
1870 mm 125 × 125 mm	1200 mm	1830 mm × 100 × 75 mm
2070 mm 125 × 125 mm	1400 mm	1980 mm × 100 × 75 mm
2620 mm 125 × 125 mm	1800 mm	2590 mm × 100 × 85 mm
3040 mm 125 × 125 mm	1800 mm	2590 mm × 100 × 85 mm (with arms)

FENCING AND GATES

Rolled Steel Angle Posts

Rolled steel angle posts for rectangular wire mesh (field) fencing

Posts	Fence height	Strut	Straining post
1200 × 40 × 40 × 5 mm	600 mm	1200 × 75 × 75 mm	1350 × 100 × 100 mm
1400 × 40 × 40 × 5 mm	800 mm	1400 × 75 × 75 mm	1550 × 100 × 100 mm
1500 × 40 × 40 × 5 mm	900 mm	1500 × 75 × 75 mm	1650 × 100 × 100 mm
1600 × 40 × 40 × 5 mm	1000 mm	1600 × 75 × 75 mm	1750 × 100 × 100 mm
1750 × 40 × 40 × 5 mm	1150 mm	1750 × 75 × 100 mm	1900 × 125 × 125 mm

Concrete Posts

Concrete posts for rectangular wire mesh (field) fencing

Posts	Fence height	Strut	Straining post
1270 × 100 × 100 mm	600 mm	1200 × 75 × 75 mm	1420 × 100 × 100 mm
1470 × 100 × 100 mm	800 mm	1350 × 75 × 75 mm	1620 × 100 × 100 mm
1570 × 100 × 100 mm	900 mm	1500 × 75 × 75 mm	1720 × 100 × 100 mm
1670 × 100 × 100 mm	600 mm	1650 × 75 × 75 mm	1820 × 100 × 100 mm
1820 × 125 × 125 mm	1150 mm	1830 × 75 × 100 mm	1970 × 125 × 125 mm

Cleft Chestnut Pale Fences

Timber Posts

Timber posts for wire mesh and hexagonal wire netting fences

Round timber for general fences

Posts	Fence height	Strut	Straining post
1300 × 65 mm dia.	600 mm	1200 × 80 mm dia.	1450 × 100 mm dia.
1500 × 65 mm dia.	800 mm	1400 × 80 mm dia.	1650 × 100 mm dia.
1600 × 65 mm dia.	900 mm	1500 × 80 mm dia.	1750 × 100 mm dia.
1700 × 65 mm dia.	1050 mm	1600 × 80 mm dia.	1850 × 100 mm dia.
1800 × 65 mm dia.	1150 mm	1750 × 80 mm dia.	2000 × 120 mm dia.

Squared timber for general fences

Posts	Fence height	Strut	Straining post
1300 × 75 × 75 mm	600 mm	1200 × 75 × 75 mm	1450 × 100 × 100 mm
1500 × 75 × 75 mm	800 mm	1400 × 75 × 75 mm	1650 × 100 × 100 mm
1600 × 75 × 75 mm	900 mm	1500 × 75 × 75 mm	1750 × 100 × 100 mm
1700 × 75 × 75 mm	1050 mm	1600 × 75 × 75 mm	1850 × 100 × 100 mm
1800 × 75 × 75 mm	1150 mm	1750 × 75 × 75 mm	2000 × 125 × 100 mm

FENCING AND GATES

Steel Fences to BS 1722: Part 9: 1992

	Fence height	Top/bottom rails and flat posts	Vertical bars
Light	1000 mm	40 × 10 mm 450 mm in ground	12 mm dia. at 115 mm cs
	1200 mm	40 × 10 mm 550 mm in ground	12 mm dia. at 115 mm cs
	1400 mm	40 × 10 mm 550 mm in ground	12 mm dia. at 115 mm cs
Light	1000 mm	40 × 10 mm 450 mm in ground	16 mm dia. at 120 mm cs
	1200 mm	40 × 10 mm 550 mm in ground	16 mm dia. at 120 mm cs
	1400 mm	40 × 10 mm 550 mm in ground	16 mm dia. at 120 mm cs
Medium	1200 mm	50 × 10 mm 550 mm in ground	20 mm dia. at 125 mm cs
	1400 mm	50 × 10 mm 550 mm in ground	20 mm dia. at 125 mm cs
	1600 mm	50 × 10 mm 600 mm in ground	22 mm dia. at 145 mm cs
	1800 mm	50 × 10 mm 600 mm in ground	22 mm dia. at 145 mm cs
Heavy	1600 mm	50 × 10 mm 600 mm in ground	22 mm dia. at 145 mm cs
	1800 mm	50 × 10 mm 600 mm in ground	22 mm dia. at 145 mm cs
	2000 mm	50 × 10 mm 600 mm in ground	22 mm dia. at 145 mm cs
	2200 mm	50 × 10 mm 600 mm in ground	22 mm dia. at 145 mm cs

Notes: Mild steel fences: round or square verticals; flat standards and horizontals. Tops of vertical bars may be bow-top, blunt, or pointed. Round or square bar railings

Timber Field Gates to BS 3470: 1975

Gates made to this standard are designed to open one way only
All timber gates are 1100 mm high
Width over stiles 2400, 2700, 3000, 3300, 3600, and 4200 mm
Gates over 4200 mm should be made in two leaves

Steel Field Gates to BS 3470: 1975

All steel gates are 1100 mm high	
Heavy duty:	width over stiles 2400, 3000, 3600 and 4500 mm
Light duty:	width over stiles 2400, 3000, and 3600 mm

FENCING AND GATES

Domestic Front Entrance Gates to BS 4092: Part 1: 1966

Metal gates:	Single gates are 900 mm high minimum, 900 mm, 1000 mm and 1100 mm wide

Domestic Front Entrance Gates to BS 4092: Part 2: 1966

Wooden gates:	All rails shall be tenoned into the stiles Single gates are 840 mm high minimum, 801 mm and 1020 mm wide Double gates are 840 mm high minimum, 2130, 2340 and 2640 mm wide

Timber Bridle Gates to BS 5709:1979 (Horse or Hunting Gates)

Gates open one way only	
Minimum width between posts	1525 mm
Minimum height	1100 mm

Timber Kissing Gates to BS 5709:1979

Minimum width	700 mm
Minimum height	1000 mm
Minimum distance between shutting posts	600 mm
Minimum clearance at mid-point	600 mm

Metal Kissing Gates to BS 5709:1979

Sizes are the same as those for timber kissing gates
Maximum gaps between rails 120 mm

Categories of Pedestrian Guard Rail to BS 3049:1976

Class A for normal use
Class B where vandalism is expected
Class C where crowd pressure is likely

DRAINAGE

Width Required for Trenches for Various Diameters of Pipes

Pipe diameter (mm)	Trench n.e. 1.50 m deep	Trench over 1.50 m deep
n.e. 100 mm	450 mm	600 mm
100–150 mm	500 mm	650 mm
150–225 mm	600 mm	750 mm
225–300 mm	650 mm	800 mm
300–400 mm	750 mm	900 mm
400–450 mm	900 mm	1050 mm
450–600 mm	1100 mm	1300 mm

Weights and Dimensions – Vitrified Clay Pipes

Product	Nominal diameter	Effective length	BS 65 limits of tolerance		Crushing strength	Weight	
			min	max			
	(mm)	(mm)	(mm)	(mm)	(kN/m)	(kg/pipe)	(kg/m)
Supersleve	100	1600	96	105	35.00	14.71	9.19
	150	1750	146	158	35.00	29.24	16.71
Hepsleve	225	1850	221	236	28.00	84.03	45.42
	300	2500	295	313	34.00	193.05	77.22
	150	1500	146	158	22.00	37.04	24.69
Hepseal	225	1750	221	236	28.00	85.47	48.84
	300	2500	295	313	34.00	204.08	81.63
	400	2500	394	414	44.00	357.14	142.86
	450	2500	444	464	44.00	454.55	181.63
	500	2500	494	514	48.00	555.56	222.22
	600	2500	591	615	57.00	796.23	307.69
	700	3000	689	719	67.00	1111.11	370.45
	800	3000	788	822	72.00	1351.35	450.45
Hepline	100	1600	95	107	22.00	14.71	9.19
	150	1750	145	160	22.00	29.24	16.71
	225	1850	219	239	28.00	84.03	45.42
	300	1850	292	317	34.00	142.86	77.22
Hepduct (conduit)	90	1500	–	–	28.00	12.05	8.03
	100	1600	–	–	28.00	14.71	9.19
	125	1750	–	–	28.00	20.73	11.84
	150	1750	–	–	28.00	29.24	16.71
	225	1850	–	–	28.00	84.03	45.42
	300	1850	–	–	34.00	142.86	77.22

DRAINAGE

Weights and Dimensions – Vitrified Clay Pipes

Nominal internal diameter (mm)	Nominal wall thickness (mm)	Approximate weight (kg/m)
150	25	45
225	29	71
300	32	122
375	35	162
450	38	191
600	48	317
750	54	454
900	60	616
1200	76	912
1500	89	1458
1800	102	1884
2100	127	2619

Wall thickness, weights and pipe lengths vary, depending on type of pipe required

The particulars shown above represent a selection of available diameters and are applicable to strength class 1 pipes with flexible rubber ring joints

Tubes with Ogee joints are also available

DRAINAGE

Weights and Dimensions – PVC-u Pipes

	Nominal size	Mean outside diameter (mm)		Wall thickness	Weight
		min	max	(mm)	(kg/m)
Standard pipes	82.4	82.4	82.7	3.2	1.2
	110.0	110.0	110.4	3.2	1.6
	160.0	160.0	160.6	4.1	3.0
	200.0	200.0	200.6	4.9	4.6
	250.0	250.0	250.7	6.1	7.2
Perforated pipes heavy grade	As above	As above	As above	As above	As above
thin wall	82.4	82.4	82.7	1.7	–
	110.0	110.0	110.4	2.2	–
	160.0	160.0	160.6	3.2	–

Width of Trenches Required for Various Diameters of Pipes

Pipe diameter (mm)	Trench n.e. 1.5 m deep (mm)	Trench over 1.5 m deep (mm)
n.e. 100	450	600
100–150	500	650
150–225	600	750
225–300	650	800
300–400	750	900
400–450	900	1050
450–600	1100	1300

DRAINAGE

DRAINAGE BELOW GROUND AND LAND DRAINAGE

Flow of Water Which Can Be Carried by Various Sizes of Pipe

Clay or Concrete Pipes

	Gradient of pipeline							
	1:10	1:20	1:30	1:40	1:50	1:60	1:80	1:100
Pipe size	**Flow in litres per second**							
DN 100 15.0	8.5	6.8	5.8	5.2	4.7	4.0	3.5	
DN 150 28.0	19.0	16.0	14.0	12.0	11.0	9.1	8.0	
DN 225 140.0	95.0	76.0	66.0	58.0	53.0	46.0	40.0	

Plastic Pipes

	Gradient of pipeline							
	1:10	1:20	1:30	1:40	1:50	1:60	1:80	1:100
Pipe size	**Flow in litres per second**							
82.4 mm i/dia.	12.0	8.5	6.8	5.8	5.2	4.7	4.0	3.5
110 mm i/dia.	28.0	19.0	16.0	14.0	12.0	11.0	9.1	8.0
160 mm i/dia.	76.0	53.0	43.0	37.0	33.0	29.0	25.0	22.0
200 mm i/dia.	140.0	95.0	76.0	66.0	58.0	53.0	46.0	40.0

Vitrified (Perforated) Clay Pipes and Fittings to BS En 295-5 1994

Length not specified		
75 mm bore	**250 mm bore**	**600 mm bore**
100	300	700
125	350	800
150	400	1000
200	450	1200
225	500	

Precast Concrete Pipes: Prestressed Non-pressure Pipes and Fittings: Flexible Joints to BS 5911: Pt. 103: 1994

Rationalized metric nominal sizes: 450, 500	
Length:	500–1000 by 100 increments 1000–2200 by 200 increments 2200–2800 by 300 increments
Angles: length:	450–600 angles 45, 22.5,11.25° 600 or more angles 22.5, 11.25°

DRAINAGE

Precast Concrete Pipes: Unreinforced and Circular Manholes and Soakaways to BS 5911: Pt. 200: 1994

Nominal sizes:	
Shafts:	675, 900 mm
Chambers:	900, 1050, 1200, 1350, 1500, 1800, 2100, 2400, 2700, 3000 mm
Large chambers:	To have either tapered reducing rings or a flat reducing slab in order to accept the standard cover
Ring depths:	1. 300–1200 mm by 300 mm increments except for bottom slab and rings below cover slab, these are by 150 mm increments 2. 250–1000 mm by 250 mm increments except for bottom slab and rings below cover slab, these are by 125 mm increments
Access hole:	750 × 750 mm for DN 1050 chamber 1200 × 675 mm for DN 1350 chamber

Calculation of Soakaway Depth

The following formula determines the depth of concrete ring soakaway that would be required for draining given amounts of water.

$$\mathbf{h} = \frac{\mathbf{4ar}}{\mathbf{3\pi D^2}}$$

h = depth of the chamber below the invert pipe
a = the area to be drained
r = the hourly rate of rainfall (50 mm per hour)
π = pi
D = internal diameter of the soakaway

This table shows the depth of chambers in each ring size which would be required to contain the volume of water specified. These allow a recommended storage capacity of ⅓ (one third of the hourly rainfall figure).

Table Showing Required Depth of Concrete Ring Chambers in Metres

Area m²	**50**	**100**	**150**	**200**	**300**	**400**	**500**
Ring size							
0.9	1.31	2.62	3.93	5.24	7.86	10.48	13.10
1.1	0.96	1.92	2.89	3.85	5.77	7.70	9.62
1.2	0.74	1.47	2.21	2.95	4.42	5.89	7.37
1.4	0.58	1.16	1.75	2.33	3.49	4.66	5.82
1.5	0.47	0.94	1.41	1.89	2.83	3.77	4.72
1.8	0.33	0.65	0.98	1.31	1.96	2.62	3.27
2.1	0.24	0.48	0.72	0.90	1.44	1.92	2.41
2.4	0.18	0.37	0.55	0.74	1.11	1.47	1.84
2.7	0.15	0.29	0.44	0.58	0.87	1.18	1.40
3.0	0.12	0.24	0.35	0.47	0.71	0.94	1.18

DRAINAGE

Precast Concrete Inspection Chambers and Gullies to BS 5911: Part 230: 1994

Nominal sizes:	375 diameter, 750, 900 mm deep 450 diameter, 750, 900, 1050, 1200 mm deep
Depths:	from the top for trapped or untrapped units: centre of outlet 300 mm invert (bottom) of the outlet pipe 400 mm
Depth of water seal for trapped gullies:	85 mm, rodding eye int. dia. 100 mm
Cover slab:	65 mm min

Bedding Flexible Pipes: PVC-u Or Ductile Iron

Type 1 =	100 mm fill below pipe, 300 mm above pipe: single size material
Type 2 =	100 mm fill below pipe, 300 mm above pipe: single size or graded material
Type 3 =	100 mm fill below pipe, 75 mm above pipe with concrete protective slab over
Type 4 =	100 mm fill below pipe, fill laid level with top of pipe
Type 5 =	200 mm fill below pipe, fill laid level with top of pipe
Concrete =	25 mm sand blinding to bottom of trench, pipe supported on chocks, 100 mm concrete under the pipe, 150 mm concrete over the pipe

DRAINAGE

Bedding Rigid Pipes: Clay or Concrete
(for vitrified clay pipes the manufacturer should be consulted)

Class D:	Pipe laid on natural ground with cut-outs for joints, soil screened to remove stones over 40 mm and returned over pipe to 150 m min depth. Suitable for firm ground with trenches trimmed by hand.
Class N:	Pipe laid on 50 mm granular material of graded aggregate to Table 4 of BS 882, or 10 mm aggregate to Table 6 of BS 882, or as dug light soil (not clay) screened to remove stones over 10 mm. Suitable for machine dug trenches.
Class B:	As Class N, but with granular bedding extending half way up the pipe diameter.
Class F:	Pipe laid on 100 mm granular fill to BS 882 below pipe, minimum 150 mm granular fill above pipe: single size material. Suitable for machine dug trenches.
Class A:	Concrete 100 mm thick under the pipe extending half way up the pipe, backfilled with the appropriate class of fill. Used where there is only a very shallow fall to the drain. Class A bedding allows the pipes to be laid to an exact gradient.
Concrete surround:	25 mm sand blinding to bottom of trench, pipe supported on chocks, 100 mm concrete under the pipe, 150 mm concrete over the pipe. It is preferable to bed pipes under slabs or wall in granular material.

PIPED SUPPLY SYSTEMS

Identification of Service Tubes From Utility to Dwellings

Utility	Colour	Size	Depth
British Telecom	grey	54 mm od	450 mm
Electricity	black	38 mm od	450 mm
Gas	yellow	42 mm od rigid 60 mm od convoluted	450 mm
Water	may be blue	(normally untubed)	750 mm

ELECTRICAL SUPPLY/POWER/LIGHTING SYSTEMS

Electrical Insulation Class En 60.598 BS 4533

Class 1:	luminaires comply with class 1 (I) earthed electrical requirements
Class 2:	luminaires comply with class 2 (II) double insulated electrical requirements
Class 3:	luminaires comply with class 3 (III) electrical requirements

Protection to Light Fittings

BS EN 60529:1992 Classification for degrees of protection provided by enclosures.
(IP Code – International or ingress Protection)

1st characteristic: against ingress of solid foreign objects		
The figure	2	indicates that fingers cannot enter
	3	that a 2.5 mm diameter probe cannot enter
	4	that a 1.0 mm diameter probe cannot enter
	5	the fitting is dust proof (no dust around live parts)
	6	the fitting is dust tight (no dust entry)
2nd characteristic: ingress of water with harmful effects		
The figure	0	indicates unprotected
	1	vertically dripping water cannot enter
	2	water dripping 15° (tilt) cannot enter
	3	spraying water cannot enter
	4	splashing water cannot enter
	5	jetting water cannot enter
	6	powerful jetting water cannot enter
	7	proof against temporary immersion
	8	proof against continuous immersion
Optional additional codes:		A–D protects against access to hazardous parts
	H	high voltage apparatus
	M	fitting was in motion during water test
	S	fitting was static during water test
	W	protects against weather
Marking code arrangement:		(example) IPX5S = IP (International or Ingress Protection) X (denotes omission of first characteristic) 5 = jetting S = static during water test

RAIL TRACKS

	kg/m of track	lb/ft of track
Standard gauge		
Bull-head rails, chairs, transverse timber (softwood) sleepers etc.	245	165
Main lines		
Flat-bottom rails, transverse prestressed concrete sleepers, etc.	418	280
Add for electric third rail	51	35
Add for crushed stone ballast	2600	1750
	kg/m²	**lb/sq ft**
Overall average weight – rails connections, sleepers, ballast, etc.	733	150
	kg/m of track	**lb/ft of track**
Bridge rails, longitudinal timber sleepers, etc.	112	75

RAIL TRACKS

Heavy Rails

British Standard Section No.	Rail height (mm)	Foot width (mm)	Head width (mm)	Min web thickness (mm)	Section weight (kg/m)
		Flat Bottom Rails			
60 A	114.30	109.54	57.15	11.11	30.62
70 A	123.82	111.12	60.32	12.30	34.81
75 A	128.59	114.30	61.91	12.70	37.45
80 A	133.35	117.47	63.50	13.10	39.76
90 A	142.88	127.00	66.67	13.89	45.10
95 A	147.64	130.17	69.85	14.68	47.31
100 A	152.40	133.35	69.85	15.08	50.18
110 A	158.75	139.70	69.85	15.87	54.52
113 A	158.75	139.70	69.85	20.00	56.22
50 'O'	100.01	100.01	52.39	10.32	24.82
80 'O'	127.00	127.00	63.50	13.89	39.74
60R	114.30	109.54	57.15	11.11	29.85
75R	128.59	122.24	61.91	13.10	37.09
80R	133.35	127.00	63.50	13.49	39.72
90R	142.88	136.53	66.67	13.89	44.58
95R	147.64	141.29	68.26	14.29	47.21
100R	152.40	146.05	69.85	14.29	49.60
95N	147.64	139.70	69.85	13.89	47.27
		Bull Head Rails			
95R BH	145.26	69.85	69.85	19.05	47.07

Light Rails

British Standard Section No.	Rail height (mm)	Foot width (mm)	Head width (mm)	Min web thickness (mm)	Section weight (kg/m)
		Flat Bottom Rails			
20M	65.09	55.56	30.96	6.75	9.88
30M	75.41	69.85	38.10	9.13	14.79
35M	80.96	76.20	42.86	9.13	17.39
35R	85.73	82.55	44.45	8.33	17.40
40	88.11	80.57	45.64	12.3	19.89
		Bridge Rails			
13	48.00	92	36.00	18.0	13.31
16	54.00	108	44.50	16.0	16.06
20	55.50	127	50.00	20.5	19.86
28	67.00	152	50.00	31.0	28.62
35	76.00	160	58.00	34.5	35.38
50	76.00	165	58.50	–	50.18
		Crane Rails			
A65	75.00	175.00	65.00	38.0	43.10
A75	85.00	200.00	75.00	45.0	56.20
A100	95.00	200.00	100.00	60.0	74.30
A120	105.00	220.00	120.00	72.0	100.00
175CR	152.40	152.40	107.95	38.1	86.92

RAIL TRACKS

Fish Plates

British Standard Section No.	Overall plate length		Hole diameter	Finished weight per pair	
	4 Hole (mm)	6 Hole (mm)	(mm)	4 Hole (kg/pair)	6 Hole (kg/pair)
For British Standard Heavy Rails: Flat Bottom Rails					
60 A	406.40	609.60	20.64	9.87	14.76
70 A	406.40	609.60	22.22	11.15	16.65
75 A	406.40	–	23.81	11.82	17.73
80 A	406.40	609.60	23.81	13.15	19.72
90 A	457.20	685.80	25.40	17.49	26.23
100 A	508.00	–	pear	25.02	–
110 A (shallow)	507.00	–	27.00	30.11	54.64
113 A (heavy)	507.00	–	27.00	30.11	54.64
50 'O' (shallow)	406.40	–	–	6.68	10.14
80 'O' (shallow)	495.30	–	23.81	14.72	22.69
60R (shallow)	406.40	609.60	20.64	8.76	13.13
60R (angled)	406.40	609.60	20.64	11.27	16.90
75R (shallow)	406.40	–	23.81	10.94	16.42
75R (angled)	406.40	–	23.81	13.67	–
80R (shallow)	406.40	609.60	23.81	11.03	17.89
80R (angled)	406.40	609.60	23.81	14.90	22.33
For British Standard Heavy Rails: Bull Head Rails					
95R BH (shallow)	–	457.20	27.00	14.59	14.61
For British Standard Light Rails: Flat Bottom Rails					
30M	355.6	–	–	–	2.72
35M	355.6	–	–	–	2.83
40	355.6	–	–	3.76	–

FRACTIONS, DECIMALS AND MILLIMETRE EQUIVALENTS

Fractions	Decimals	(mm)		Fractions	Decimals	(mm)
1/64	0.015625	0.396875		33/64	0.515625	13.096875
1/32	0.03125	0.79375		17/32	0.53125	13.49375
3/64	0.046875	1.190625		35/64	0.546875	13.890625
1/16	0.0625	1.5875		9/16	0.5625	14.2875
5/64	0.078125	1.984375		37/64	0.578125	14.684375
3/32	0.09375	2.38125		19/32	0.59375	15.08125
7/64	0.109375	2.778125		39/64	0.609375	15.478125
1/8	0.125	3.175		5/8	0.625	15.875
9/64	0.140625	3.571875		41/64	0.640625	16.271875
5/32	0.15625	3.96875		21/32	0.65625	16.66875
11/64	0.171875	4.365625		43/64	0.671875	17.065625
3/16	0.1875	4.7625		11/16	0.6875	17.4625
13/64	0.203125	5.159375		45/64	0.703125	17.859375
7/32	0.21875	5.55625		23/32	0.71875	18.25625
15/64	0.234375	5.953125		47/64	0.734375	18.653125
1/4	0.25	6.35		3/4	0.75	19.05
17/64	0.265625	6.746875		49/64	0.765625	19.446875
9/32	0.28125	7.14375		25/32	0.78125	19.84375
19/64	0.296875	7.540625		51/64	0.796875	20.240625
5/16	0.3125	7.9375		13/16	0.8125	20.6375
21/64	0.328125	8.334375		53/64	0.828125	21.034375
11/32	0.34375	8.73125		27/32	0.84375	21.43125
23/64	0.359375	9.128125		55/64	0.859375	21.828125
3/8	0.375	9.525		7/8	0.875	22.225
25/64	0.390625	9.921875		57/64	0.890625	22.621875
13/32	0.40625	10.31875		29/32	0.90625	23.01875
27/64	0.421875	10.71563		59/64	0.921875	23.415625
7/16	0.4375	11.1125		15/16	0.9375	23.8125
29/64	0.453125	11.50938		61/64	0.953125	24.209375
15/32	0.46875	11.90625		31/32	0.96875	24.60625
31/64	0.484375	12.30313		63/64	0.984375	25.003125
1/2	0.5	12.7		1.0	1	25.4

IMPERIAL STANDARD WIRE GAUGE (SWG)

IMPERIAL STANDARD WIRE GAUGE (SWG)

SWG	Diameter		SWG	Diameter	
No.	(inches)	(mm)	No.	(inches)	(mm)
7/0	0.5	12.7	23	0.024	0.61
6/0	0.464	11.79	24	0.022	0.559
5/0	0.432	10.97	25	0.02	0.508
4/0	0.4	10.16	26	0.018	0.457
3/0	0.372	9.45	27	0.0164	0.417
2/0	0.348	8.84	28	0.0148	0.376
1/0	0.324	8.23	29	0.0136	0.345
1	0.3	7.62	30	0.0124	0.315
2	0.276	7.01	31	0.0116	0.295
3	0.252	6.4	32	0.0108	0.274
4	0.232	5.89	33	0.01	0.254
5	0.212	5.38	34	0.009	0.234
6	0.192	4.88	35	0.008	0.213
7	0.176	4.47	36	0.008	0.193
8	0.16	4.06	37	0.007	0.173
9	0.144	3.66	38	0.006	0.152
10	0.128	3.25	39	0.005	0.132
11	0.116	2.95	40	0.005	0.122
12	0.104	2.64	41	0.004	0.112
13	0.092	2.34	42	0.004	0.102
14	0.08	2.03	43	0.004	0.091
15	0.072	1.83	44	0.003	0.081
16	0.064	1.63	45	0.003	0.071
17	0.056	1.42	46	0.002	0.061
18	0.048	1.22	47	0.002	0.051
19	0.04	1.016	48	0.002	0.041
20	0.036	0.914	49	0.001	0.031
21	0.032	0.813	50	0.001	0.025
22	0.028	0.711			

PIPES, WATER, STORAGE, INSULATION

WATER PRESSURE DUE TO HEIGHT

Imperial

Head (Feet)	Pressure (lb/in^2)		Head (Feet)	Pressure (lb/in^2)
1	0.43		70	30.35
5	2.17		75	32.51
10	4.34		80	34.68
15	6.5		85	36.85
20	8.67		90	39.02
25	10.84		95	41.18
30	13.01		100	43.35
35	15.17		105	45.52
40	17.34		110	47.69
45	19.51		120	52.02
50	21.68		130	56.36
55	23.84		140	60.69
60	26.01		150	65.03
65	28.18			

Metric

Head (m)	Pressure (bar)		Head (m)	Pressure (bar)
0.5	0.049		18.0	1.766
1.0	0.098		19.0	1.864
1.5	0.147		20.0	1.962
2.0	0.196		21.0	2.06
3.0	0.294		22.0	2.158
4.0	0.392		23.0	2.256
5.0	0.491		24.0	2.354
6.0	0.589		25.0	2.453
7.0	0.687		26.0	2.551
8.0	0.785		27.0	2.649
9.0	0.883		28.0	2.747
10.0	0.981		29.0	2.845
11.0	1.079		30.0	2.943
12.0	1.177		32.5	3.188
13.0	1.275		35.0	3.434
14.0	1.373		37.5	3.679
15.0	1.472		40.0	3.924
16.0	1.57		42.5	4.169
17.0	1.668		45.0	4.415

1 bar	=	14.5038 lbf/in^2
1 lbf/in^2	=	0.06895 bar
1 metre	=	3.2808 ft or 39.3701 in
1 foot	=	0.3048 metres
1 in wg	=	2.5 mbar (249.1 N/m^2)

PIPES, WATER, STORAGE, INSULATION

Dimensions and Weights of Copper Pipes to BSEN 1057, BSEN 12499, BSEN 14251

Outside diameter (mm)	Internal diameter (mm)	Weight per metre (kg)	Internal diameter (mm)	Weight per metre (kg)	Internal siameter (mm)	Weight per metre (kg)
	Formerly Table X		Formerly Table Y		Formerly Table Z	
6	4.80	0.0911	4.40	0.1170	5.00	0.0774
8	6.80	0.1246	6.40	0.1617	7.00	0.1054
10	8.80	0.1580	8.40	0.2064	9.00	0.1334
12	10.80	0.1914	10.40	0.2511	11.00	0.1612
15	13.60	0.2796	13.00	0.3923	14.00	0.2031
18	16.40	0.3852	16.00	0.4760	16.80	0.2918
22	20.22	0.5308	19.62	0.6974	20.82	0.3589
28	26.22	0.6814	25.62	0.8985	26.82	0.4594
35	32.63	1.1334	32.03	1.4085	33.63	0.6701
42	39.63	1.3675	39.03	1.6996	40.43	0.9216
54	51.63	1.7691	50.03	2.9052	52.23	1.3343
76.1	73.22	3.1287	72.22	4.1437	73.82	2.5131
108	105.12	4.4666	103.12	7.3745	105.72	3.5834
133	130.38	5.5151	–	–	130.38	5.5151
159	155.38	8.7795	–	–	156.38	6.6056

Dimensions of Stainless Steel Pipes to BS 4127

Outside siameter (mm)	Maximum outside siameter (mm)	Minimum outside diameter (mm)	Wall thickness (mm)	Working pressure (bar)
6	6.045	5.940	0.6	330
8	8.045	7.940	0.6	260
10	10.045	9.940	0.6	210
12	12.045	11.940	0.6	170
15	15.045	14.940	0.6	140
18	18.045	17.940	0.7	135
22	22.055	21.950	0.7	110
28	28.055	27.950	0.8	121
35	35.070	34.965	1.0	100
42	42.070	41.965	1.1	91
54	54.090	53.940	1.2	77

PIPES, WATER, STORAGE, INSULATION

Dimensions of Steel Pipes to BS 1387

Nominal Size	Approx. Outside Diameter	Outside diameter				Thickness		
		Light		Medium & Heavy		Light	Medium	Heavy
		Max	Min	Max	Min			
(mm)	(mm)	(mm)	(mm)	(mm)	(mm)	(mm)	(mm)	(mm)
6	10.20	10.10	9.70	10.40	9.80	1.80	2.00	2.65
8	13.50	13.60	13.20	13.90	13.30	1.80	2.35	2.90
10	17.20	17.10	16.70	17.40	16.80	1.80	2.35	2.90
15	21.30	21.40	21.00	21.70	21.10	2.00	2.65	3.25
20	26.90	26.90	26.40	27.20	26.60	2.35	2.65	3.25
25	33.70	33.80	33.20	34.20	33.40	2.65	3.25	4.05
32	42.40	42.50	41.90	42.90	42.10	2.65	3.25	4.05
40	48.30	48.40	47.80	48.80	48.00	2.90	3.25	4.05
50	60.30	60.20	59.60	60.80	59.80	2.90	3.65	4.50
65	76.10	76.00	75.20	76.60	75.40	3.25	3.65	4.50
80	88.90	88.70	87.90	89.50	88.10	3.25	4.05	4.85
100	114.30	113.90	113.00	114.90	113.30	3.65	4.50	5.40
125	139.70	–	–	140.60	138.70	–	4.85	5.40
150	165.1*	–	–	166.10	164.10	–	4.85	5.40

* 165.1 mm (6.5in) outside diameter is not generally recommended except where screwing to BS 21 is necessary
All dimensions are in accordance with ISO R65 except approximate outside diameters which are in accordance with ISO R64
Light quality is equivalent to ISO R65 Light Series II

Approximate Metres Per Tonne of Tubes to BS 1387

Nom. size	BLACK						GALVANIZED					
	Plain/screwed ends			Screwed & socketed			Plain/screwed ends			Screwed & socketed		
	L	M	H	L	M	H	L	M	H	L	M	H
(mm)	(m)	(m)	(m)	(m)	(m)	(m)	(m)	(m)	(m)	(m)	(m)	(m)
6	2765	2461	2030	2743	2443	2018	2604	2333	1948	2584	2317	1937
8	1936	1538	1300	1920	1527	1292	1826	1467	1254	1811	1458	1247
10	1483	1173	979	1471	1165	974	1400	1120	944	1386	1113	939
15	1050	817	688	1040	811	684	996	785	665	987	779	661
20	712	634	529	704	628	525	679	609	512	673	603	508
25	498	410	336	494	407	334	478	396	327	474	394	325
32	388	319	260	384	316	259	373	308	254	369	305	252
40	307	277	226	303	273	223	296	268	220	292	264	217
50	244	196	162	239	194	160	235	191	158	231	188	157
65	172	153	127	169	151	125	167	149	124	163	146	122
80	147	118	99	143	116	98	142	115	97	139	113	96
100	101	82	69	98	81	68	98	81	68	95	79	67
125	–	62	56	–	60	55	–	60	55	–	59	54
150	–	52	47	–	50	46	–	51	46	–	49	45

The figures for 'plain or screwed ends' apply also to tubes to BS 1775 of equivalent size and thickness
Key:
L – Light
M – Medium
H – Heavy

PIPES, WATER, STORAGE, INSULATION

Flange Dimension Chart to BS 4504 & BS 10

Normal Pressure Rating (PN 6) 6 Bar

Nom. size	Flange outside dia.	Table 6/2 Forged Welding Neck	Table 6/3 Plate Slip on	Table 6/4 Forged Bossed Screwed	Table 6/5 Forged Bossed Slip on	Table 6/8 Plate Blank	Raised face		Nr. bolt hole	Size of bolt
							Dia.	T'ness		
15	80	12	12	12	12	12	40	2	4	M10 × 40
20	90	14	14	14	14	14	50	2	4	M10 × 45
25	100	14	14	14	14	14	60	2	4	M10 × 45
32	120	14	16	14	14	14	70	2	4	M12 × 45
40	130	14	16	14	14	14	80	3	4	M12 × 45
50	140	14	16	14	14	14	90	3	4	M12 × 45
65	160	14	16	14	14	14	110	3	4	M12 × 45
80	190	16	18	16	16	16	128	3	4	M16 × 55
100	210	16	18	16	16	16	148	3	4	M16 × 55
125	240	18	20	18	18	18	178	3	8	M16 × 60
150	265	18	20	18	18	18	202	3	8	M16 × 60
200	320	20	22	–	20	20	258	3	8	M16 × 60
250	375	22	24	–	22	22	312	3	12	M16 × 65
300	440	22	24	–	22	22	365	4	12	M20 × 70

Normal Pressure Rating (PN 16) 16 Bar

Nom. size	Flange outside dia.	Table 6/2 Forged Welding Neck	Table 6/3 Plate Slip on	Table 6/4 Forged Bossed Screwed	Table 6/5 Forged Bossed Slip on	Table 6/8 Plate Blank	Raised face		Nr. bolt hole	Size of bolt
							Dia.	T'ness		
15	95	14	14	14	14	14	45	2	4	M12 × 45
20	105	16	16	16	16	16	58	2	4	M12 × 50
25	115	16	16	16	16	16	68	2	4	M12 × 50
32	140	16	16	16	16	16	78	2	4	M16 × 55
40	150	16	16	16	16	16	88	3	4	M16 × 55
50	165	18	18	18	18	18	102	3	4	M16 × 60
65	185	18	18	18	18	18	122	3	4	M16 × 60
80	200	20	20	20	20	20	138	3	8	M16 × 60
100	220	20	20	20	20	20	158	3	8	M16 × 65
125	250	22	22	22	22	22	188	3	8	M16 × 70
150	285	22	22	22	22	22	212	3	8	M20 × 70
200	340	24	24	–	24	24	268	3	12	M20 × 75
250	405	26	26	–	26	26	320	3	12	M24 × 90
300	460	28	28	–	28	28	378	4	12	M24 × 90

PIPES, WATER, STORAGE, INSULATION

Minimum Distances Between Supports/Fixings

Material	BS Nominal pipe size		Pipes – Vertical	Pipes – Horizontal on to low gradients
	(inch)	(mm)	Support distance in metres	Support distance in metres
Copper	0.50	15.00	1.90	1.30
	0.75	22.00	2.50	1.90
	1.00	28.00	2.50	1.90
	1.25	35.00	2.80	2.50
	1.50	42.00	2.80	2.50
	2.00	54.00	3.90	2.50
	2.50	67.00	3.90	2.80
	3.00	76.10	3.90	2.80
	4.00	108.00	3.90	2.80
	5.00	133.00	3.90	2.80
	6.00	159.00	3.90	2.80
muPVC	1.25	32.00	1.20	0.50
	1.50	40.00	1.20	0.50
	2.00	50.00	1.20	0.60
Polypropylene	1.25	32.00	1.20	0.50
	1.50	40.00	1.20	0.50
uPVC	–	82.40	1.20	0.50
	–	110.00	1.80	0.90
	–	160.00	1.80	1.20
Steel	0.50	15.00	2.40	1.80
	0.75	20.00	3.00	2.40
	1.00	25.00	3.00	2.40
	1.25	32.00	3.00	2.40
	1.50	40.00	3.70	2.40
	2.00	50.00	3.70	2.40
	2.50	65.00	4.60	3.00
	3.00	80.40	4.60	3.00
	4.00	100.00	4.60	3.00
	5.00	125.00	5.50	3.70
	6.00	150.00	5.50	4.50
	8.00	200.00	8.50	6.00
	10.00	250.00	9.00	6.50
	12.00	300.00	10.00	7.00
	16.00	400.00	10.00	8.25

PIPES, WATER, STORAGE, INSULATION

Litres of Water Storage Required Per Person Per Building Type

Type of building	Storage (litres)
Houses and flats (up to 4 bedrooms)	120/bedroom
Houses and flats (more than 4 bedrooms)	100/bedroom
Hostels	90/bed
Hotels	200/bed
Nurses homes and medical quarters	120/bed
Offices with canteen	45/person
Offices without canteen	40/person
Restaurants	7/meal
Boarding schools	90/person
Day schools – Primary	15/person
Day schools – Secondary	20/person

Recommended Air Conditioning Design Loads

Building type	Design loading
Computer rooms	500 W/m² of floor area
Restaurants	150 W/m² of floor area
Banks (main area)	100 W/m² of floor area
Supermarkets	25 W/m² of floor area
Large office block (exterior zone)	100 W/m² of floor area
Large office block (interior zone)	80 W/m² of floor area
Small office block (interior zone)	80 W/m² of floor area

PIPES, WATER, STORAGE, INSULATION

Capacity and Dimensions of Galvanized Mild Steel Cisterns – BS 417

Capacity (litres)	BS type (SCM)	Dimensions		
		Length (mm)	Width (mm)	Depth (mm)
18	45	457	305	305
36	70	610	305	371
54	90	610	406	371
68	110	610	432	432
86	135	610	457	482
114	180	686	508	508
159	230	736	559	559
191	270	762	584	610
227	320	914	610	584
264	360	914	660	610
327	450/1	1220	610	610
336	450/2	965	686	686
423	570	965	762	787
491	680	1090	864	736
709	910	1070	889	889

Capacity of Cold Water Polypropylene Storage Cisterns – BS 4213

Capacity (litres)	BS type (PC)	Maximum height (mm)
18	4	310
36	8	380
68	15	430
91	20	510
114	25	530
182	40	610
227	50	660
273	60	660
318	70	660
455	100	760

PIPES, WATER, STORAGE, INSULATION

Minimum Insulation Thickness to Protect Against Freezing for Domestic Cold Water Systems (8 Hour Evaluation Period)

Pipe size (mm)	Insulation thickness (mm)					
	Condition 1			Condition 2		
	λ = 0.020	λ = 0.030	λ = 0.040	λ = 0.020	λ = 0.030	λ = 0.040
Copper pipes						
15	11	20	34	12	23	41
22	6	9	13	6	10	15
28	4	6	9	4	7	10
35	3	5	7	4	5	7
42	3	4	5	8	4	6
54	2	3	4	2	3	4
76	2	2	3	2	2	3
Steel pipes						
15	9	15	24	10	18	29
20	6	9	13	6	10	15
25	4	7	9	5	7	10
32	3	5	6	3	5	7
40	3	4	5	3	4	6
50	2	3	4	2	3	4
65	2	2	3	2	3	3

Condition 1: water temperature 7°C; ambient temperature –6°C; evaluation period 8 h; permitted ice formation 50%; normal installation, i.e. inside the building and inside the envelope of the structural insulation
Condition 2: water temperature 2°C; ambient temperature –6°C; evaluation period 8 h; permitted ice formation 50%; extreme installation, i.e. inside the building but outside the envelope of the structural insulation
λ = thermal conductivity [W/(mK)]

Insulation Thickness for Chilled And Cold Water Supplies to Prevent Condensation

On a Low Emissivity Outer Surface (0.05, i.e. Bright Reinforced Aluminium Foil) with an Ambient Temperature of +25°C and a Relative Humidity of 80%

Steel pipe size (mm)	*t* = +10			*t* = +5			*t* = 0		
	Insulation thickness (mm)			Insulation thickness (mm)			Insulation thickness (mm)		
	λ = 0.030	λ = 0.040	λ = 0.050	λ = 0.030	λ = 0.040	λ = 0.050	λ = 0.030	λ = 0.040	λ = 0.050
15	16	20	25	22	28	34	28	36	43
25	18	24	29	25	32	39	32	41	50
50	22	28	34	30	39	47	38	49	60
100	26	34	41	30	47	57	46	60	73
150	29	38	46	40	52	64	51	67	82
250	33	43	53	48	60	74	59	77	94
Flat surfaces	39	52	65	56	75	93	73	97	122

t = temperature of contents (°C)
λ = thermal conductivity at mean temperature of insulation [W/(mK)]

PIPES, WATER, STORAGE, INSULATION

Insulation Thickness for Non-domestic Heating Installations to Control Heat Loss

Steel pipe size (mm)	*t* = 75			*t* = 100			*t* = 150		
	Insulation thickness (mm)			Insulation thickness (mm)			Insulation thickness (mm)		
	λ = 0.030	λ = 0.040	λ = 0.050	λ = 0.030	λ = 0.040	λ = 0.050	λ = 0.030	λ = 0.040	λ = 0.050
10	18	32	55	20	36	62	23	44	77
15	19	34	56	21	38	64	26	47	80
20	21	36	57	23	40	65	28	50	83
25	23	38	58	26	43	68	31	53	85
32	24	39	59	28	45	69	33	55	87
40	25	40	60	29	47	70	35	57	88
50	27	42	61	31	49	72	37	59	90
65	29	43	62	33	51	74	40	63	92
80	30	44	62	35	52	75	42	65	94
100	31	46	63	37	54	76	45	68	96
150	33	48	64	40	57	77	50	73	100
200	35	49	65	42	59	79	53	76	103
250	36	50	66	43	61	80	55	78	105

t = hot face temperature (°C)
λ = thermal conductivity at mean temperature of insulation [W/(mK)]

Index

Ebook Single-User Licence Agreement

We welcome you as a user of this Spon Price Book ebook and hope that you find it a useful and valuable tool. Please read this document carefully. **This is a legal agreement** between you (hereinafter referred to as the "Licensee") and Taylor and Francis Books Ltd. (the "Publisher"), which defines the terms under which you may use the Product. **By accessing and retrieving the access code on the label inside the front cover of this book you agree to these terms and conditions outlined herein. If you do not agree to these terms you must return the Product to your supplier intact, with the seal on the label unbroken and with the access code not accessed**.

1. Definition of the Product

The product which is the subject of this Agreement, (the "Product") consists of online and offline access to the VitalSource ebook edition of *Spon's External Works and Landscape Price Book 2023.*

2. Commencement and licence

2.1 This Agreement commences upon the breaking open of the document containing the access code by the Licensee (the "Commencement Date").

2.2 This is a licence agreement (the "Agreement") for the use of the Product by the Licensee, and not an agreement for sale.

2.3 The Publisher licenses the Licensee on a non-exclusive and non-transferable basis to use the Product on condition that the Licensee complies with this Agreement. The Licensee acknowledges that it is only permitted to use the Product in accordance with this Agreement.

3. Multiple use

Use of the Product is not provided or allowed for more than one user or for a wide area network or consortium.

4. Installation and Use

4.1 The Licensee may provide access to the Product for individual study in the following manner: The Licensee may install the Product on a secure local area network on a single site for use by one user.

4.2 The Licensee shall be responsible for installing the Product and for the effectiveness of such installation.

4.3 Text from the Product may be incorporated in a coursepack. Such use is only permissible with the express permission of the Publisher in writing and requires the payment of the appropriate fee as specified by the Publisher and signature of a separate licence agreement.

4.4 The Product is a free addition to the book and the Publisher is under no obligation to provide any technical support.

5. Permitted Activities

5.1 The Licensee shall be entitled to use the Product for its own internal purposes;

5.2 The Licensee acknowledges that its rights to use the Product are strictly set out in this Agreement, and all other uses (whether expressly mentioned in Clause 6 below or not) are prohibited.

6. Prohibited Activities

The following are prohibited without the express permission of the Publisher:

6.1 The commercial exploitation of any part of the Product.

6.2 The rental, loan, (free or for money or money's worth) or hire purchase of this product, save with the express consent of the Publisher.

6.3 Any activity which raises the reasonable prospect of impeding the Publisher's ability or opportunities to market the Product.

6.4 Any networking, physical or electronic distribution or dissemination of the product save as expressly permitted by this Agreement.

6.5 Any reverse engineering, decompilation, disassembly or other alteration of the Product save in accordance with applicable national laws.

6.6 The right to create any derivative product or service from the Product save as expressly provided for in this Agreement.

6.7 Any alteration, amendment, modification or deletion from the Product, whether for the purposes of error correction or otherwise.

7. **General Responsibilities of the License**

7.1 The Licensee will take all reasonable steps to ensure that the Product is used in accordance with the terms and conditions of this Agreement.

7.2 The Licensee acknowledges that damages may not be a sufficient remedy for the Publisher in the event of breach of this Agreement by the Licensee, and that an injunction may be appropriate.

7.3 The Licensee undertakes to keep the Product safe and to use its best endeavours to ensure that the product does not fall into the hands of third parties, whether as a result of theft or otherwise.

7.4 Where information of a confidential nature relating to the product of the business affairs of the Publisher comes into the possession of the Licensee pursuant to this Agreement (or otherwise), the Licensee agrees to use such information solely for the purposes of this Agreement, and under no circumstances to disclose any element of the information to any third party save strictly as permitted under this Agreement. For the avoidance of doubt, the Licensee's obligations under this sub-clause 7.4 shall survive the termination of this Agreement.

8. **Warrant and Liability**

8.1 The Publisher warrants that it has the authority to enter into this agreement and that it has secured all rights and permissions necessary to enable the Licensee to use the Product in accordance with this Agreement.

8.2 The Publisher warrants that the Product as supplied on the Commencement Date shall be free of defects in materials and workmanship, and undertakes to replace any defective Product within 28 days of notice of such defect being received provided such notice is received within 30 days of such supply. As an alternative to replacement, the Publisher agrees fully to refund the Licensee in such circumstances, if the Licensee so requests, provided that the Licensee returns this copy of *Spon's Mechanical & Electrical Services Price Book 2023* to the Publisher. The provisions of this sub-clause 8.2 do not apply where the defect results from an accident or from misuse of the product by the Licensee.

8.3 Sub-clause 8.2 sets out the sole and exclusive remedy of the Licensee in relation to defects in the Product.

8.4 The Publisher and the Licensee acknowledge that the Publisher supplies the Product on an "as is" basis. The Publisher gives no warranties:

8.4.1 that the Product satisfies the individual requirements of the Licensee; or

8.4.2 that the Product is otherwise fit for the Licensee's purpose; or

8.4.3 that the Product is compatible with the Licensee's hardware equipment and software operating environment.

8.5 The Publisher hereby disclaims all warranties and conditions, express or implied, which are not stated above.

8.6 Nothing in this Clause 8 limits the Publisher's liability to the Licensee in the event of death or personal injury resulting from the Publisher's negligence.

8.7 The Publisher hereby excludes liability for loss of revenue, reputation, business, profits, or for indirect or consequential losses, irrespective of whether the Publisher was advised by the Licensee of the potential of such losses.

8.8 The Licensee acknowledges the merit of independently verifying the price book data prior to taking any decisions of material significance (commercial or otherwise) based on such data. It is agreed that the Publisher shall not be liable for any losses which result from the Licensee placing reliance on the data under any circumstances.

8.9 Subject to sub-clause 8.6 above, the Publisher's liability under this Agreement shall be limited to the purchase price.

9. **Intellectual Property Rights**

9.1 Nothing in this Agreement affects the ownership of copyright or other intellectual property rights in the Product.

9.2 The Licensee agrees to display the Publishers' copyright notice in the manner described in the Product.

9.3 The Licensee hereby agrees to abide by copyright and similar notice requirements required by the Publisher, details of which are as follows:

"© 2023 Taylor & Francis. All rights reserved. All materials in *Spon's External Works and Landscape Price Book 2023* are copyright protected. All rights reserved. No such materials may be used, displayed, modified, adapted, distributed, transmitted, transferred, published or otherwise reproduced in any form or by any means now or hereafter developed other than strictly in accordance with the terms of the licence agreement enclosed with *Spon's External Works and Landscape Services Price Book 2023.* However, text and images may be printed and copied for research and private study within the preset program limitations. Please note the copyright notice above, and that any text or images printed or copied must credit the source."

9.4 This Product contains material proprietary to and copyedited by the Publisher and others. Except for the licence granted herein, all rights, title and interest in the Product, in all languages, formats and media throughout the world, including copyrights therein, are and remain the property of the Publisher or other copyright holders identified in the Product.

10. Non-assignment

This Agreement and the licence contained within it may not be assigned to any other person or entity without the written consent of the Publisher.

11. Termination and Consequences of Termination.

11.1 The Publisher shall have the right to terminate this Agreement if:

11.1.1 the Licensee is in material breach of this Agreement and fails to remedy such breach (where capable of remedy) within 14 days of a written notice from the Publisher requiring it to do so; or

11.1.2 the Licensee becomes insolvent, becomes subject to receivership, liquidation or similar external administration; or

11.1.3 the Licensee ceases to operate in business.

11.2 The Licensee shall have the right to terminate this Agreement for any reason upon two month's written notice. The Licensee shall not be entitled to any refund for payments made under this Agreement prior to termination under this sub-clause 11.2.

11.3 Termination by either of the parties is without prejudice to any other rights or remedies under the general law to which they may be entitled, or which survive such termination (including rights of the Publisher under sub-clause 7.4 above).

11.4 Upon termination of this Agreement, or expiry of its terms, the Licensee must destroy all copies and any back up copies of the product or part thereof.

12. General

12.1 *Compliance with export provisions*

The Publisher hereby agrees to comply fully with all relevant export laws and regulations of the United Kingdom to ensure that the Product is not exported, directly or indirectly, in violation of English law.

12.2 *Force majeure*

The parties accept no responsibility for breaches of this Agreement occurring as a result of circumstances beyond their control.

12.3 *No waiver*

Any failure or delay by either party to exercise or enforce any right conferred by this Agreement shall not be deemed to be a waiver of such right.

12.4 *Entire agreement*

This Agreement represents the entire agreement between the Publisher and the Licensee concerning the Product. The terms of this Agreement supersede all prior purchase orders, written terms and conditions, written or verbal representations, advertising or statements relating in any way to the Product.

12.5 *Severability*

If any provision of this Agreement is found to be invalid or unenforceable by a court of law of competent jurisdiction, such a finding shall not affect the other provisions of this Agreement and all provisions of this Agreement unaffected by such a finding shall remain in full force and effect.

12.6 *Variations*

This agreement may only be varied in writing by means of variation signed in writing by both parties.

12.7 *Notices*

All notices to be delivered to: Spon's Price Books, Taylor & Francis Books Ltd., 4 Park Square, Milton Park, Abingdon, Oxfordshire, OX14 4RN, UK.

12.8 *Governing law*

This Agreement is governed by English law and the parties hereby agree that any dispute arising under this Agreement shall be subject to the jurisdiction of the English courts.

If you have any queries about the terms of this licence, please contact:

Spon's Price Books
Taylor & Francis Books Ltd.
4 Park Square, Milton Park, Abingdon, Oxfordshire, OX14 4RN
Tel: +44 (0) 20 7017 6000
www.sponpress.com

Spon Press
an imprint of Taylor & Francis

Ebook options

Your print copy comes with a free ebook on the VitalSource® Bookshelf platform. Further copies of the ebook are available from https://www.routledge.com/
Pick your price book and select the ebook under the 'Select Format' option.

To buy Spon's ebooks for five or more users of in your organisation please contact:

Spon's Price Books
eBooks & Online Sales
Taylor & Francis Books Ltd.
4 Park Square, Milton Park, Oxfordshire, OX14 4RN
Tel: (020) 7017 6000
onlinesales@informa.com

Free Updates

with three easy steps...

1. Register today on www.pricebooks.co.uk/updates

2. We'll alert you by email when new updates are posted on our website

3. Then go to www.pricebooks.co.uk/updates
   and download the update.

All four Spon Price Books – *Architects' and Builders'*, *Civil Engineering and Highway Works*, *External Works and Landscape* and *Mechanical and Electrical Services* – are supported by an updating service. Two or three updates are loaded on our website during the year, typically in November, February and May. Each gives details of changes in prices of materials, wage rates and other significant items, with regional price level adjustments for Northern Ireland, Scotland and Wales and regions of England. The updates terminate with the publication of the next annual edition.

As a purchaser of a Spon Price Book you are entitled to this updating service for this 2023 edition – free of charge. Simply register via the website www.pricebooks.co.uk/updates and we will send you an email when each update becomes available.

If you haven't got internet access or if you've some questions about the updates please write to us at Spon Price Book Updates, Spon Press Marketing Department, 4 Park Square, Milton Park, Abingdon, Oxfordshire, OX14 4RN.

Find out more about Spon books
Visit www.pricebooks.co.uk for more details.

Ebook Set up and Use Instructions

Use of the ebook is subject to the single user licence at the back of this book.

Electronic access to your price book is now provided as an ebook on the VitalSource® Bookshelf platform. You can access it online or offline on your PC/Mac, smartphone or tablet. You can browse and search the content across all the books you've got, make notes and highlights and share these notes with other users.

Setting up

1. Create a VitalSource Bookshelf account at https://online.vitalsource.com/user/new (or log into your existing account if you already have one).

2. Retrieve the code by scratching off the security-protected label inside the front cover of this book. Log in to Bookshelf and click the **Redeem Codes link in the Tools** menu at the top right of the screen. Enter the code in the **Redeem code** box and press **Redeem**. Once the code has been redeemed your Spon's Price Book will download and appear in your library. N.B. the code in the scratch-off panel can only be used once.

When you have created a Bookshelf account and redeemed the code you will be able to access the ebook online or offline on your smartphone, tablet or PC/Mac. Your notes and highlights will automatically stay in sync no matter where you make them.

Use ONLINE

1. Log in to your Bookshelf account at https://online.vitalsource.com.
2. Double-click on the title in your **library** to open the ebook.

Use OFFLINE

Download BookShelf to your PC, Mac, iOS device, Android device or Kindle Fire, and log in to your Bookshelf account to access your ebook, as follows:

On your PC/Mac
Go to https://support.vitalsource.com/hc/en-us/categories/360001056774-Bookshelf#360002383594 and follow the instructions to download the free VitalSource Bookshelf app to your PC or Mac. Double-click the VitalSource Bookshelf icon that appears on your desktop and log into your Bookshelf account. Select **All Titles** from the menu on the left – you should see your price book on your Bookshelf. If your Price Book does not appear, select **Update Booklist** from the **Account** menu. Double-click the price book to open it.

On your iPhone/iPod Touch/iPad
Download the free Bookshelf App available via the iTunes App Store. Open the Bookshelf app and log into your Bookshelf account. Select **All Titles** – you should see your price book on your Bookshelf. Select the price book to open it. You can find more information at https://support.vitalsource.com/hc/en-us/articles/360014105113-iOS

On your Android™ smartphone or tablet
Download the free Bookshelf App available via Google Play. Open the Bookshelf app and log into your Bookshelf account. You should see your price book on your Bookshelf. Select the price book to open it. You can find more information at https://support.vitalsource.com/hc/en-us/articles/360014007294-Android

On your Kindle Fire
Download the free Bookshelf App from Amazon. Open the Bookshelf app and log into your Bookshelf account. Select All Titles - you should see your price book on your Bookshelf. Select the price book to open it. You can find more information at https://support.vitalsource.com/hc/en-us/articles/360014107953-Kindle-Fire

Support

If you have any questions about downloading Bookshelf, creating your account, or accessing and using your ebook edition, please visit http://support.vitalsource.com/

For questions or comments on content, please contact us on sponsonline@tandf.co.uk